Springer Series in

# MATERIALS SCIENCE 62

Springer
*Berlin*
*Heidelberg*
*New York*
*Hong Kong*
*London*
*Milan*
*Paris*
*Tokyo*

Physics and Astronomy ONLINE LIBRARY

springeronline.com

Springer Series in
# MATERIALS SCIENCE

*Editors:* R. Hull  R. M. Osgood, Jr.  J. Parisi  H. Warlimont

The Springer Series in Materials Science covers the complete spectrum of materials physics, including fundamental principles, physical properties, materials theory and design. Recognizing the increasing importance of materials science in future device technologies, the book titles in this series reflect the state-of-the-art in understanding and controlling the structure and properties of all important classes of materials.

61 **Fatigue in Ferroelectric Ceramics and Related Issues**
By D.C. Lupascu

62 **Epitaxy**
Physical Principles and Technical Implementation
By M.A. Herman, W. Richter, and H. Sitter

63 **Fundamentals of Ion Irradiation of Polymers**
By D. Fink

64 **Morphology Control of Materials and Nanoparticles**
Advanced Materials Processing and Characterization
Editors: Y. Waseda and A. Muramatsu

65 **Transport Processes in Ion Irradiated Polymers**
By D. Fink

66 **Multiphased Ceramic Materials**
Processing and Potential
Editors: W.-H. Tuan and J.-K. Guo

67 **Nondestructive Materials Characterization**
With Applications to Aerospace Materials
Editors: N.G.H. Meyendorf, P.B. Nagy, and S.I. Rokhlin

68 **Diffraction Analysis of the Microstructure of Materials**
Editors: E.J. Mittemeijer and P. Scardi

69 **Chemical–Mechanical Planarization of Semiconductor Materials**
Editor: M.R. Oliver

70 **Isotope Effect Applications in Solids**
By G.V. Plekhanov

71 **Dissipative Phenomena in Condensed Matter**
Some Applications
By S. Dattagupta and S. Puri

72 **Predictive Simulation of Semiconductor Processing**
Status and Challenges
Editors: J. Dabrowski and E.R. Weber

Series homepage – springer.de

Volumes 10–60 are listed at the end of the book.

Marian A. Herman
Wolfgang Richter
Helmut Sitter

# Epitaxy

## Physical Principles and Technical Implementation

With 305 Figures

Springer

Professor Dr. Marian A. Herman
Wissenschaftliches Zentrum der Polnischen Akademie der Wissenschaften
Boerhaavengasse 25, 1030 Wien, Austria E-mail: herman.viennapan@ycn.com

Professor Dr. Wolfgang Richter
Institut für Festkörperphysik, Technische Universität Berlin
Hardenbergstr. 36, 10623 Berlin, Germany E-mail: wolfgang.richter@tu-berlin.de
and
INFM, Dipartimento di Fisica, Università di Roma Tor Vergata
Via della Ricerca Scientifica 1, 00133 Roma, Italy E-mail: wolfgang.richter@roma2infm.it

Professor Dr. Helmut Sitter
Johannes-Kepler-Universität, Altenbergerstr. 69
4040 Linz, Austria E-mail: Helmut.Sitter@jku.at

*Series Editors:*



ISSN 0933-033X

ISBN 978-3-642-08737-0

Library of Congress Cataloging-in-Publication Data. Herman, Marian A. Epitaxy: physical principles and technical implementation/M.A. Herman, W. Richter, H. Sitter. p. cm. – (Springer series in materials science, ISSN 0933-033X; 62). Includes bibliographical references and index. ISBN 3-540-67821-2 (acid-free paper). 1. Epitaxy. 2. Thin films, Multilayered–Technique. I. Richter, Wolfgang, 1940– . II. Sitter, Helmut, 1951– . III. Title. IV. Series. V. Springer series in materials science; v. 62. QD921.H48 2004 621.3815'2–dc22 2003059146



Springer-Verlag is a part of Springer Science+Business Media

springeronline.com



Cover concept: eStudio Calamar Steinen
Cover production: *design & production* GmbH, Heidelberg

To Olivia

# Preface

Epitaxy, namely, the growth process of a solid film on a crystalline substrate in which the atoms of the growing film mimic the arrangement of the atoms of the substrate, is one of the most important issues in thin film technology. Especially important is the case of heteroepitaxy, i.e., the epitaxial growth of a solid film differing from the substrate crystal with respect to its chemical structure. The importance of epitaxy concerns both fundamental research on thin film growth processes and the application of these procedures to grow high quality crystal layers from different materials for the realization of technically important functions. This concerns also the development of a series of epitaxial growth techniques applied in different branches of solid state electronics, optoelectronics and photonics in manufacturing processes of discrete as well as integrated devices.

Coincident with this development, original research papers and reviews devoted to problems concerning epitaxial growth techniques have rapidly grown in number and, in addition, they have become very diversified. At present several hundred original papers on this subject appear in the literature each year. Also a large number of textbooks and scientific monographs have been published in recent years on different aspects of epitaxy, which is further evidence of the importance of this crystallization phenomenon. The importance of epitaxy is also evident in view of the currently observed intensive development of crystalline nanostructures consisting of quantum wells, superlattices, quantum wires and quantum dots grown by epitaxy. These structures exhibit a lot of extraordinary properties, known previously only through theoretical predictions of the quantum mechanics of solids. In contrast to the large number of publications appearing in the literature each year, there is a lack of comprehensive monographs comprising the whole variety of problems related to physical foundations and to technical implementation of the phenomenon of epitaxial crystallization.

Taking this into consideration and having learned from our pedagogical experience gained by university teaching practice, we decided to write an advanced textbook intended for undergraduate students, doctoral students, research scientists and practicing engineers interested in material science, solid state physics, crystal growth, thin film technology and solid state electronics, including microelectronics, optoelectronics and sensorics. The idea was to provide these readers with a comprehensive, up-dated text of a reasonably limited volume, from which they can learn the basic models and modifications of epitaxy, together with the relevant experimental and technological framework of it.

This textbook, as part of the *Springer Series on Materials Processing*, covers both experimental and theoretical aspects of the subject. Its contents can roughly be divided into four parts. The first part serves as a general introduction giving background information on epitaxy. It begins with a discussion, indicating the specificity and different modes of the epitaxial crystallization process. This is followed by a detailed presentation of the characteristic features of homo- and heteroepitaxial growth processes as well as of epitaxially grown films (epilayers) and film structures. This part of the book is concluded with a comprehensive survey of application areas of epitaxy, with special emphasis put on epitaxially grown multilayer structures, including the nanoscale low dimensional structures.

The second part consists of a review of the most frequently used epitaxial growth techniques and their technical implementations. Solid phase epitaxy (SPE), liquid phase epitaxy (LPE), vapor phase epitaxy (VPE) especially the metalorganic vapor phase epitaxy (MOVPE) variant, and molecular beam epitaxy (MBE) are presented and discussed in detail.

The third part introduces the basic ideas related to physical understanding of the phenomenon of epitaxial crystallization. It begins with an overview of the experimental techniques most frequently used for *in situ* analysis of the epitaxial growth processes. Optical and mass spectrometric techniques as well as diffraction-based techniques are presented and discussed there in detail. Then physical models of epitaxy in the thermodynamic and atomistic approaches are reviewed,with emphasis put on the current status of understanding of substrate surface structures and of their influence on epitaxial growth processes.

The fourth part of the book, concluding the main body of it, presents a detailed discussion of heteroepitaxial growth processes. This part introduces the reader into peculiarities of the growth phenomena of nearly-lattice-matched heterostructures and of highly strained heterostructures. It presents also the principles of artificial epitaxy (graphoepitaxy). Considerations concerning the materials-related peculiarities of heteroepitaxy conclude this part of the book.

At no stage have we made an attempt to trace the history of the described phenomena and technological processes, or to refer the reader to all existing literature sources. What is done is to focus the reader's attention on various types of observations related to epitaxial growth phenomena and to provide their possible explanation in terms of the basic ideas.

In consideration of the continuing stream of new information concerning different epitaxial growth techniques, which not only provides additional data but also often modifies the interpretation of a particular concept or technological solution, we have attempted to be sufficiently fundamental in our treatment of the subject that the conclusions drawn here will not be superseded. We hope therefore that future publications occurring in the field will be easily understood by readers referring to the principles presented here. It is worthwhile to emphasize that we did not intend to write this book in the form of a critical review of the different subjects of epitaxy. However, our evaluation of what is important for understanding epitaxy and recognition of its current status is introduced by selection of the relevant topics which are presented and discussed in detail.

In the course of writing we have experienced numerous useful discussions and help from many colleagues and coworkers. The cooperation with Springer Verlag was very pleasant and the patience of C. Ascheron with the authors is noteworthy to mention. In technically finishing the book we would like especially to thank Karsten Fleischer for his extraordinary help in mastering the LaTeX-machinery.

Berlin, Linz, Roma, Wien
October, 2003

*Marian A. Herman*
*Wolfgang Richter*
*Helmut Sitter*

# Contents

## Part IV. Physics of Epitaxy

# Part I

# Basic Concepts

# 1. Introduction

In the 19th century mineralogists noticed that two different naturally occurring crystal species sometimes grew together with some definite and unique orientation relationship, as revealed by their external forms [1.1]. These observations led to attempts to reproduce the effect artificially, during crystal growth from solution, and the first recorded successful attempt was reported in 1836 by Frankenheim [1.2], when the now well-known case of parallel oriented growth, or "parallel overgrowth" [1.3], of sodium nitrate on calcite was observed for the first time. Based on the reviews on natural overgrowth phenomena [1.4] and on structural data from X–ray diffraction studies, Royer established in 1928 the conditions for oriented overgrowth, defining the term epitaxy ("arrangement on"), too [1.5]. He formulated the following rule of epitaxy: "epitaxy occurs only when it involves the parallelism of two lattice planes that have networks of identical or quasi-identical form and of closely similar spacings". Experimental data gained later indicated that epitaxy occurs if the lattice misfit, defined as $100\,(a_\mathrm{f}-a_\mathrm{s})/a_\mathrm{s}$, where $a_\mathrm{s}$ and $a_\mathrm{f}$ are the corresponding network spacings (lattice constants) in the substrate and film, respectively, is not larger than 15 %. This geometrical approach to the understanding of epitaxy, introduced by Royer, has remained prominent to the present day [1.6–8].

## 1.1 Epitaxial Crystallization Process

Epitaxy, in common with all forms of crystal growth, is in fact a well-controlled phase transition which leads to a single crystalline solid [1.9]. Consequently, formation of an epitaxially grown deposit constitutes the creation of a new phase [1.10]. This is accomplished through a nucleation and growth relationship between two crystalline phases, which makes it possible for a crystalline phase "e" (epilayer) to grow in a structure-dependent manner onto a crystalline phase "s" (substrate) of given structure. In general, an interfacial region which is chemically and structurally inhomogeneous is then developed. In principle, in the completed epitaxial growth reaction, there exists a two-phase system consisting of two adjacent heterochemical (different chemical species) or isochemical (the same chemical species) epitaxial partners, i.e., the epilayer "e" and the substrate "s" [1.11]. When the two-phase

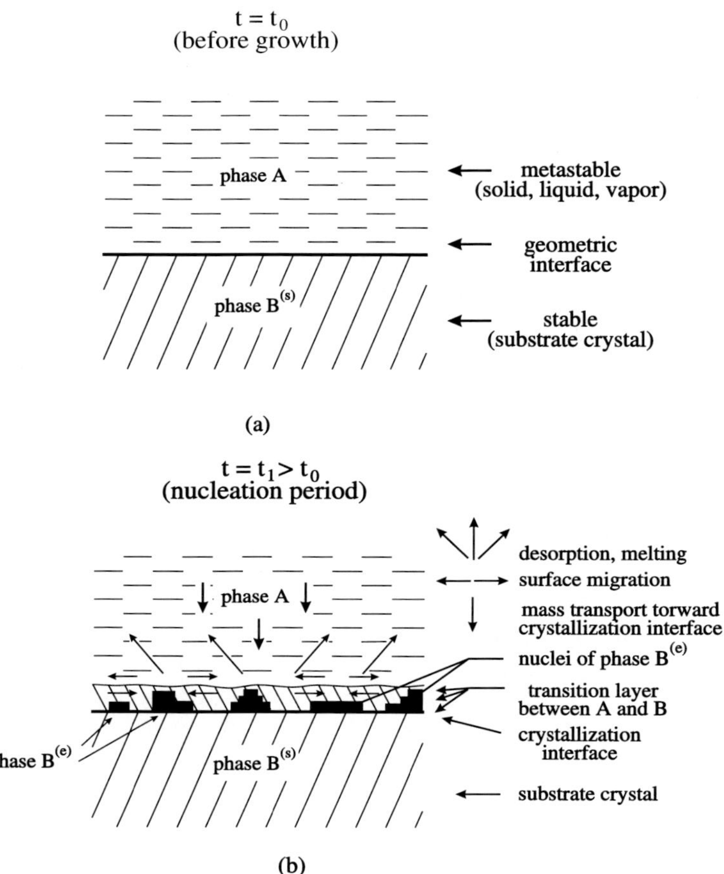


**Fig. 1.1.** Schematic illustration of the geometrical configuration of the epitaxial growth system in different time periods of crystallization; (**a**) before the growth started, (**b**) in the nucleation period related to growth of the first monolayer, (**c**) at early stage of epitaxy, when a thin epilayer has already been grown, and (**d**) in the time period when regular epitaxial growth proceeds, and a fairly thick epilayer has already been grown. (Figure continued next page)

system is isochemical in composition, then the epitaxial growth process is called homoepitaxy. On the other hand, in the case of a heterochemical system, the epitaxial growth process is called heteroepitaxy.

The phenomenon of epitaxial crystallization is based on a few key processes, which lead to "parallel oriented" growth of a single crystalline layer on a crystallographically oriented single crystal surface. First, and most general, is the phase transition between the metastable phase (gas, liquid, or solid) and the epilayer which has to be grown. This is usually related to mass transport of the constituent species from the bulk of the metastable phase toward

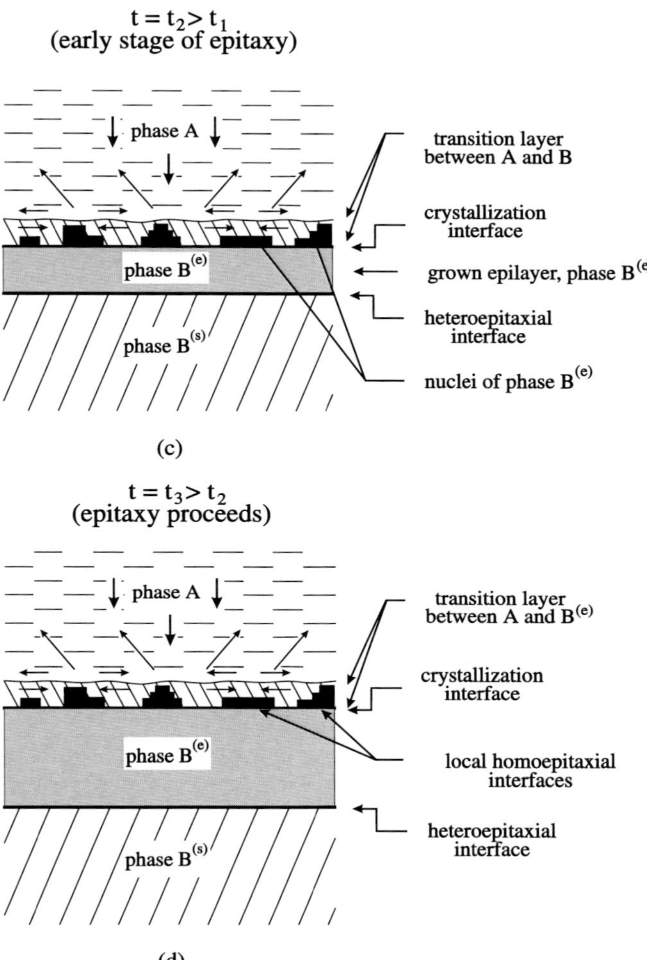


**Fig. 1.1.** (continued) Note that at the stage (**d**) the nuclei grow by homoepitaxial mode, while the heteroepitaxial interface has largely lost its influence on the growth process occurring on the upper surface of the epilayer. The key processes of epitaxial growth, i.e., perpendicular mass transport from the bulk of the phase A toward the crystallization interface, lateral mass transport by surface migration related to epitaxial ordering, desorption/melting from the crystallization area (the transition layer) toward the bulk of the metastable phase A, are indicated; however, the dimensions of the areas shown in the figure are not to scale

the growth front, which we will call hereafter the crystallization interface. In order to keep the crystallization process running, the driving force of crystallization, i.e., the local supersaturation of the metastable phase, should be ensured in the area of this interface. The growth process in epitaxy is by definition related to atomic ordering, which leads to creation of the first atomic or molecular monolayer of the growing film. Atomic ordering is a surface kinetic process, which is strongly dependent on the structure and the chemical activity of the substrate surface. A schematic illustration of the geometrical configuration of the epitaxial growth system, in different time periods, which indicates the key processes related to epitaxy, is shown in Fig. 1.1.

The epilayer grown on the substrate crystal surface may be formed from amorphous solid deposits, from a liquid phase, i.e., a solution or a melt, from the vapor or gas (consisting of neutral or ionized particles) and from atomic or molecular beams (in a high- or ultra high vacuum environment (UHV)). With respect to the crystallization phase involved in the growth of the "e" phase, the following classification names are used, at present, for the epitaxial crystallization processes: solid phase epitaxy (SPE), liquid phase epitaxy (LPE), vapor phase epitaxy (VPE), with its special modification called metalorganic chemical vapor phase Epitaxy (MOVPE) or organo–metallic VPE (OMVPE), and molecular beam epitaxy (MBE).

## 1.2 Growth Modes in Epitaxy

The growth process of thin epitaxial films is essentially the same as that of bulk crystals [1.12], except for the influence of the substrate at the initial stages. This influence comes from the misfit and thermal stress, from the defects appearing at the crystal–film interface and from the chemical interactions between the film and the substrate including segregation of the substrate elements towards the film surface [1.13]. Five possible modes of crystal growth may be distinguished in epitaxy. These are: the Volmer–Weber mode (VW-mode), the Frank–van der Merwe mode (FM-mode), the Stranski–Krastanov mode (SK-mode), the columnar growth mode (CG-mode), and the step flow mode (SF-mode). The mode by which the epitaxial film grows depends upon the lattice misfit between substrate and film, the supersaturation (the flux) of the crystallizing phase, the growth temperature and the adhesion energy.

The five most frequently occurring modes are illustrated schematically in Fig. 1.2 [1.14–17]. In the VW-mode, or island growth mode, small clusters are nucleated directly on the substrate surface and then grow into islands of the condensed phase. This happens when the atoms, or molecules, of the deposit are more strongly bound to each other than to the substrate. This mode is displayed by many systems of metals growing on insulators, including many metals on alkali halides, graphite and other layer compounds such as mica.

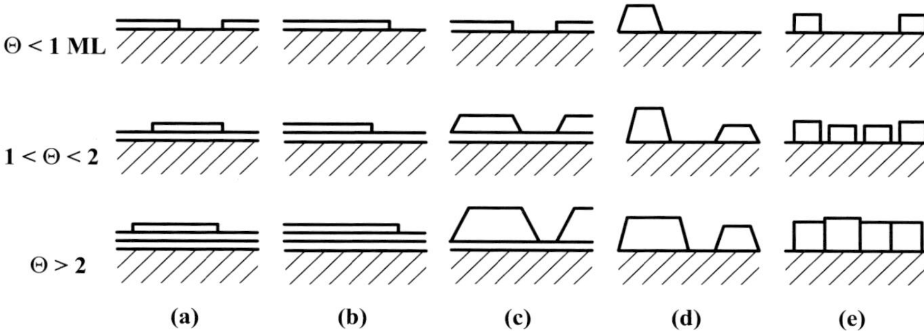


**Fig. 1.2.** Schematic representation of the five crystal growth modes most frequently occurring on flat surfaces of substrate crystals. (**a**) Layer-by-layer or Frank–van der Merwe (FM-mode); (**b**) step flow (SF-mode); (**c**) layer plus island or Stranski–Krastanov (SK-mode); (**d**) island or Volmer–Weber (VW-mode); (**e**) columnar growth mode (CG-mode). $\Theta$ represents the coverage in monolayers

The FM-mode displays the opposite characteristics. Because the atoms are more strongly bound to the substrate than to each other, the first atoms to condense form a complete monolayer on the surface, which becomes covered with a somewhat less tightly bound second layer. Provided the decrease in binding strength is monotonic toward the value for a bulk crystal of the deposit, the layer growth mode is obtained. This growth mode is observed in the case of adsorbed gases, such as several rare gases on graphite and on several metals, in some metal–metal systems, and in semiconductor growth on semiconductors.

The SK-mode, or layer plus island growth mode, is an "intermediate" case. After forming the first monolayer, or a few monolayers, subsequent layer growth is unfavorable and islands are formed on top of this intermediate layer. There are many possible reasons for this mode to occur and almost any factor which disturbs the monotonic decrease in binding energy characteristic for layer-by-layer growth may be the cause [1.15]. It occurs especially in cases when the interface energy is high (allowing for initial layer-by-layer growth) and the strain energy of the film is also high (making reduction of the strain energy by islanding favorable). In the InAs/GaAs material system the SK-mode leads, for example, to formation of dot arrays (Fig. 1.3) on the substrate surface [GaAs(100)], as an energetically metastable system with a preferred island size [1.18].

The fifth growth mode, the CG-mode, shows some similarities with the SK and VW modes, however, it is fundamentally different. In SK-mode as well as in VW-mode, when the film thickens, the condensed phase islands characteristic to these modes tend to merge and to cover the whole substrate surface. Although the grown film may exhibit variations in its thickness and presence of structural defects at the interfaces where adjacent islands merge, it forms a connected structure, in which the density of the film is a contin-

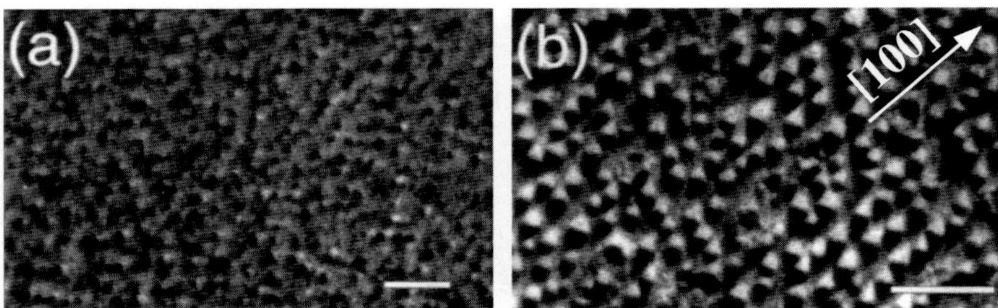


**Fig. 1.3.** Plan-view TEM images of InAs quantum dots in a GaAs matrix. Average thickness of InAs deposited is: (**a**) $d_{\mathrm{av}} = 0.6\,\mathrm{nm}$ and (**b**) $d_{\mathrm{av}} = 1.2\,\mathrm{nm}$. Note the preferential alignment of dots in rows parallel to <100> in (**b**). The markers represent 100 nm (taken from [1.18])

uous function of the position. In contrast, the film grown by the CG-mode usually thickens without the merger of columns. As a result, columns usually remain separated throughout the growth process of the film, and the films grown in this way are easily fractured [1.19]. The CG-mode occurs where low atomic mobility over the substrate surface leads to the formation of highly defective atomic columns of the deposited material on this surface [1.8]. In special growth conditions, however, the film grown by the CG-mode on a highly lattice mismatched substrate crystal may consist of an array of whisker-like nanocrystals (typically about 60 nm in diameter) of high crystal quality, as indicated by X–ray diffraction data [1.20], isolated from each other. Figure 1.4 shows a SEM micrograph of a GaN layer exhibiting columnar structure in the form of whisker-like nanocrystals, grown directly on Si(111) surface.

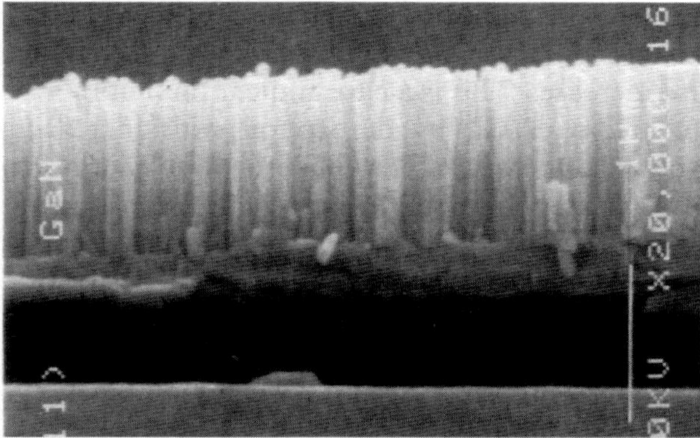


**Fig. 1.4.** SEM micrograph of a GaN layer grown directly on Si(111) substrate, exhibiting a columnar structure in the form of whisker-like nanocrystals (taken from [1.20])

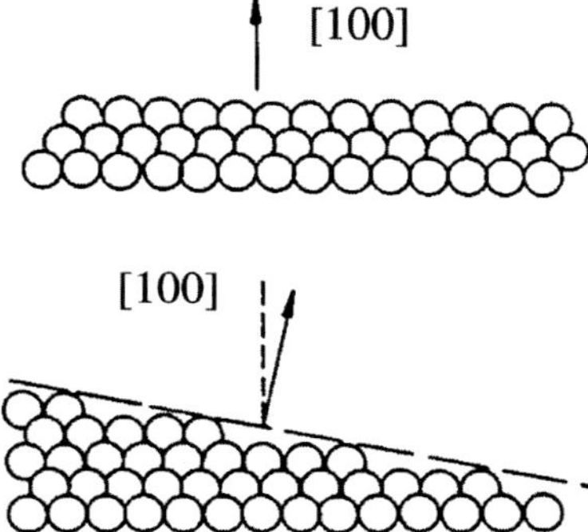


**Fig. 1.5.** Schematic illustration of how surface steps occur (lower panel) by slightly misorienting the (100) surface (upper panel) of a substrate crystal (taken from [1.16])

Beside the described growth modes, in many cases of high quality epitaxy the second mode, the so-called step-flow growth mode (SF-mode), is observed [1.21–24]. When the substrate wafer is cut slightly misoriented from a low-index plane in a specific direction, its surface breaks up into monoatomic steps with precisely oriented low-index terraces and edges (Fig. 1.5). The terrace surfaces, free of steps, have an average width of $l$ (Fig. 1.7) depending on crystallographic misorientation. They are created by the vicinal atomic planes of the crystal lattice of the substrate wafer. Two-dimensional (2D) nucleation may occur on the vicinal planes (the terraces) when the substrate temperature is sufficiently low, or the flux of the constituent elements of the growing film is high enough to prohibit fast surface migration of the species adsorbed on the terraces. In this case the film may grow on the terraces in the FM or SK mode (Fig. 1.7, right). However, when the substrate temperature is high enough or the flux is sufficiently low, then the adatoms can be so mobile, in comparison with their encounter probability, that they become incorporated directly into the step edges. In this case growth of the epitaxial film occurs by the advancement of steps along the terraces (Fig. 1.7, left).

The transition between FM and SF growth may be easily controlled by monitoring intensities of either reflection high energy electron diffraction (RHEED) [1.16], reflectance anisotropy spectroscopy (RAS) [1.25] or grazing incidence X-ray scattering (GIXS) [1.26]. In the FM mode intensity oscillations occur which are caused by the periodic morphological change (see Sect. 7.3.2 and 10.2.2) between (2D) nucleation and completion of a monolayer. As a direct consequence the RHEED intensity oscillates and similar oscillations occur in the RAS signal which samples the morphologically induced changes in the anisotropic dielectric function. Harbison et al. were the first to observe the analogy between RHEED oscillations and RAS oscillations during MBE growth of AlAs on a AlAs(001) surface [1.27], as shown in Fig. 1.6. The RAS oscillation period observed is identical to that of RHEED and thus proves that RAS is also sampling monolayer growth. There is, however, a phase shift between both oscillations. This is not surprising since

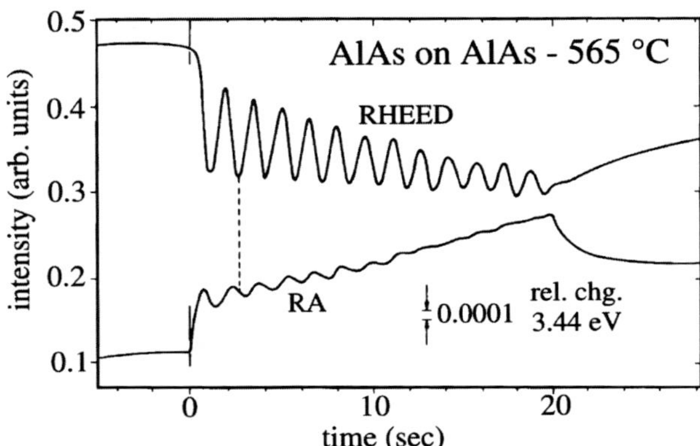


**Fig. 1.6.** Averages of nine RHEED and RAS traces upon initiation of AlAs growth on an As-stabilized (2×4) AlAs surface. The time needed to grow one monolayer is 1.5 s (taken from [1.27])

RHEED samples the variation in surface roughness during island formation, while RAS measures the corresponding variation in the anisotropic dielectric response. The latter exhibits dispersion, and consequently the phase shift is experimentally observed to change with wavelength, too [1.25].

However, under conditions where adatoms are predominantly incorporated into existing steps (SF-mode) the surface morphology does not change (left panel) and the RHEED/RAS/GIXS intensity from the growing surface is approximately constant. This relation between the surface morphology and the RHEED/RAS/GIXS intensity behavior is shown schematically in Fig. 1.7.

While in a UHV environment, all three techniques may be equally well applied, in non-UHV environments (VPE) only the RAS or GIXS technique are able to monitor the morphological changes of the growing epilayer. Thereby, the RAS technique being the technically most simple method.

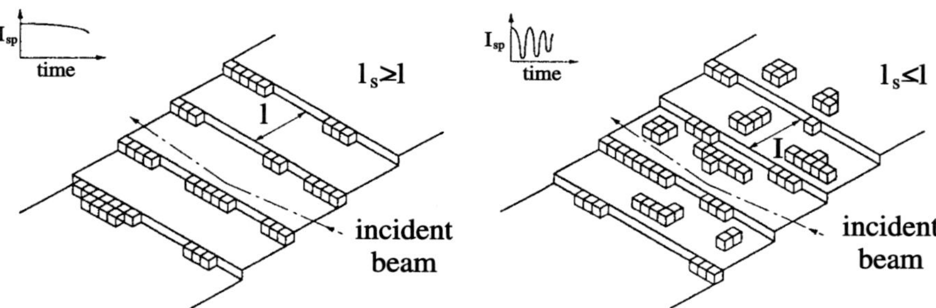


**Fig. 1.7.** Schematic illustration of the two layer-by-layer growth modes. Left panel: step flow (diffusion length $l_s$ > terrace width $l$); right panel: island nucleation on the terraces ($l_s < l$) (taken from [1.21])

# 2. Homo- and Heteroepitaxial Crystallization Phenomena

It is experimentally well documented that thin film deposits are formed by a "nucleation and growth" mechanism [2.1]. Due to their growth, nuclei, i.e., small embryonic clusters of atoms or molecules, agglomerate to form "islands". As growth proceeds, agglomeration increases, chains of islands are later formed and join up to produce a continuous deposit which, however, still contains channels and holes. These holes eventually fill to give a continuous and complete film and further growth leads to smoothing of the surface irregularities present in the deposit.

It has also evolved from thin film growth experiments that recrystallization by a grain boundary migration process occurs during growth [2.1]. Grain boundaries are formed in the necks between coalescing islands of differing orientations and are constrained to remain in an island until the coalescence neck is eliminated.

In principle, epitaxy involves similar nucleation and nucleus-growth processes; however, in this case these processes occur on a given crystalline substrate surface. In consequence, understanding and controlling the process of layer formation by epitaxy implies in each case a knowledge of the nucleation and nucleus-growth laws [2.2].

## 2.1 Nucleation and Epitaxy

Nucleation is the spontaneous formation of small embryonic clusters (nuclei), with some critical size determined by the equilibrium between their vapor pressure and the environmental pressure (Fig. 2.1). The nuclei form in the metastable supersaturated or undercooled medium (phase $A$). Their appearance is a prerequisite for a macroscopic phase transformation to take place [2.2–6]. This means that nucleation is the precursor of crystal growth and of the overall crystallization processes. Due to their increased surface/volume ratio such clusters, termed critical nuclei, have more energy than the bulk phases of the same mass. Hence, they have a chance to survive and to produce macroscopic entities of the new stable phase ($B$) only within the supersaturated/undercooled homogeneous medium. It should be noticed that the equilibrium between the critical nucleus and its environment is unstable provided that the volume of the whole system is not very small. Figure 2.1 represents

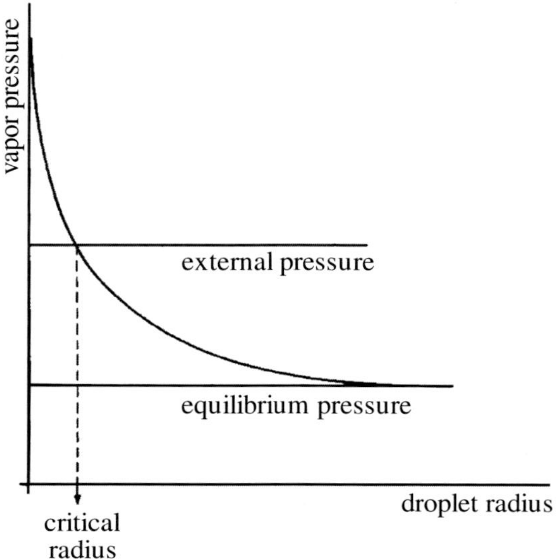


**Fig. 2.1.** Vapor pressure of small droplets in relation to their size (taken from [2.2])

the relationship between the vapor pressure of liquid droplets and their size [2.5]. If by any chance the radius $r_{cr}$ of the droplet, defining its critical size, increases slightly, the droplet will continue to grow until a macroscopic two phase equilibrium is attained. In the reverse case, the droplet will disappear. In other words, minor fluctuations of the critical radius release an irreversible process destroying the initial state of the system. This indicates that the energetics of phase formation in an initially homogeneous medium is governed by the energetics of the critical nucleus. The described process occurring in a homogeneous medium is called homogeneous nucleation. Despite the fact that homogeneous nucleation is a comparatively rare occurrence, its basic principles form the necessary background for the understanding of thin film formation, in general, and epitaxy in particular.

In the case of heterogeneous nucleation, the particle clusters forming the critical nuclei of phase $A$ do not occur in the homogeneous medium $A$ but within the matrix of adsorbed particles of medium $A$ on the substrate crystal, i.e., on the phase $B$. Consequently, the heterogeneous nucleation should be treated as a temporal sequence of two processes, namely: (i) formation of adsorbed material of the phase $A$ on the surface of substrate $B$, which is in sufficiently long-lasting contact with the surrounding phase $A$, and (ii) formation of a critical nucleus from the adsorbed particles and its subsequent growth to a supercritical size [2.6]. These two processes, i.e., nucleation and growth, can be juxtaposed in time and the amount of the condensed phase at any time will be naturally determined by some sort of convolution.

The relevance of nucleation processes to epitaxy, with respect to the production of high quality epitaxial layers (epilayers) concerns, first of all, the

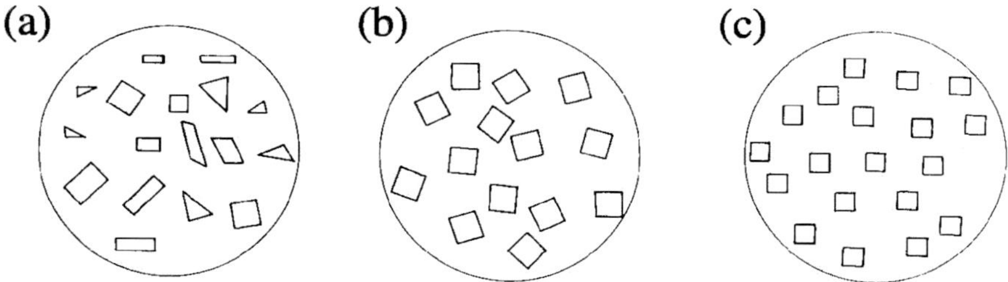


**Fig. 2.2.** Schematic illustration of the three intergrowth relations between "$e$" and "$s$": (**a**) deposit is fully non-oriented, (**b**) texture orientation where the deposit plane is parallel to the substrate surface, (**c**) deposit exhibits texture and azimuthal orientation, i.e., an epitaxial relation, to the substrate (taken from [2.7])

island growth (VW-mode and CG-mode) and layer growth (FM-mode and SK-mode) cases. In contrast to this the step-flow growth mode (SF-mode) does not require nucleation.

Epitaxial nucleation is a special case of heterogeneous nucleation in which different kinds of intergrowth relations between the deposit phase $A$ and the substrate phase $B$ may occur. The orientation behavior of the growing layer "$e$" on a substrate single crystal "$s$" may not only be influenced by nucleation of the deposited phase $A$, but also by the subsequent growth of the phase $A$ on the crystalline surface of the phase $B$ [2.1]. In fact, it is possible to divide epitaxial layers into two groups, for the first of which the orientation is controlled by initial nuclei, and in the second of which the subsequent growth is the controlling parameter [2.2]. In Fig. 2.2 three possible, basically different kinds of intergrowth relations between "$e$" and "$s$" are schematically shown. According to Fig. 2.2a the deposit crystallites are completely non-oriented with respect to the substrate crystal surface. According to Fig. 2.2b an orientation of texture may exist, that is, the deposit crystallites all grow with the same stable low index lattice plane on the substrate surface. They are, however, not oriented with respect to each other. Finally, in Fig. 2.2c deposit crystallites and the crystallites of the substrate show both the textural and azimuthal orientation towards one another. It has to be emphasized that only these latter orientation relationships are designated as epitaxial. This holds, in principle, despite the fact that in many cases of so called hard heteroepitaxy [2.8] (this means epitaxial growth of a solid film differing strongly from the substrate crystal in, at least, one of the following parameters: lattice constant, crystal structure, and chemical structure) neither the crystallographic orientation of the epilayer can easily be predicted, nor can simple solutions be given a priori to problems like, e.g., how one can get a certain surface orientation of the epilayer by choosing suitable growth conditions if different surface orientations may grow [2.9].

Let us now discuss briefly the two groups of epitaxial growth modes where nucleation plays a considerable role. In the island growth case, there has been a strong move in recent years to consider epitaxy as a postnucleation phenomenon [2.10]. Indeed, since critical nuclei are quite stable, can reorient themselves and move over the substrate, one can conclude that it is largely the migration, rotation and coalescence of small stable crystallites that eventually produce an epitaxial layer. This type of movement has been seen by many authors and has been made into a quantitative study by Masson, Metois, Kern [2.11] and co-workers. Metois et al. [2.12] have shown that in the case of Au-on-KCl(100) growth at low temperature (20°C) the (100) gold orientation results, on annealing at $T \geq 200°C$ from the coalescence of (111) oriented islands that migrate into each other and coalesce. There is thus no doubt that cluster (nuclei) rotation and migration processes are important in establishing epitaxial orientation. All the experiments performed by Masson et al. [2.11] are in the form of a low-temperature deposition followed by annealing. If deposition were carried out at higher temperature, as would normally be the case in the production of epitaxial layers, these processes must be going on during deposition. However, migration-coalescence processes are usually not necessary to establish epitaxy in a high-temperature deposition. Clusters are often seen to be in epitaxial orientation at all stages of the deposition, and cluster rotation and some migration without significant coalescence is all that is required [2.10].

In the layer-growth regime, the indications are that epitaxy results in almost all circumstances provided the substrate surface is clean enough. In this regime the substrate has a very strong influence on the form of the thin layer produced, and the growing layer has little option but to choose the best (i.e., necessarily epitaxial) orientation in which to grow. The example of this is a low energy electron diffraction (LEED) examination of the growth of xenon crystals on iridium [2.10]. The xenon crystals grow in (111) orientation on both the $1\times1$ and the $1\times5$ Ir(100) surface structures but in quite different azimuthal orientations in the two cases. The slowest growing face of xenon will also be (111), which is parallel to the substrate. Another case which is similar to the Xe-on-Ir growth is the growth of cadmium crystals on tungsten. The only difference is that the slowest growing face (0001) is not necessarily parallel to the tungsten surface. Thus, these epitaxial layers will not be smooth as in the case of Xe-on-Ir growth.

In the case of epitaxial growth on a substrate surface which is contaminated by precovering adsorbates, the impurities, even in very small concentration (e.g., less than 0.01 monolayer), can destroy smooth layer-by-layer growth (the FM-mode). The impurities could affect the nucleation kinetics, or the subsequent growth by reducing the binding energy at kink sites, or conceivably by favoring twin or stacking fault formation. This case has been exhaustively analyzed by Paunov [2.13], who has applied a simple lattice model to consider the influence of impurity adsorption on the epitaxial layer growth mode. The most important conclusions resulting from this work may be formulated as follows:

(i) the growth on a preadsorbed foreign substrate (on the adsorbed layer) takes place practically as on a clean substrate, because the exchange processes between incoming atoms and adsorbate particles keep the adsorbate predominantly on the top of the growing surface rather than in the bulk of the layer (compare this with the role of surfactants on the growth mode of lattice mismatched epilayers, which is discussed in Sect. 14.1.4);

(ii) the adsorbate changes drastically the chemical potential of the underlying layer rendering it close to the chemical potential of an infinitely large three-dimensional crystal;

(iii) the driving force of the surface segregation process is the free energy difference between the bulk and the "on the surface" configuration of the system.

The relation between nucleation processes and occurrence of epitaxy is still a subject of investigation. This concerns both the theoretical as well as experimental work. However, in the island growth case, experimental data strongly suggest that epitaxy is a postnucleation phenomenon involving rotation, migration, and rearrangement of stable critical nuclei. In the layer growth case, it seems probable that one can destroy epitaxy by influencing the growth process at a later stage with impurities or defects incorporated into the growing layer. The initial layers are, however, more or less forced to be related epitaxially to the substrate, provided that surface diffusion is sufficiently rapid that one is not dealing with the growth of amorphous layers.

## 2.2 Defects in Epitaxial Layers

Epitaxially grown layers usually contain many crystalline defects, a fact which results, first of all, from constraints imposed on the deposit by the substrate. Their physical properties and their densities, however, determine to a large extent the physical properties of the real epitaxial layer. It is therefore essential to know their properties, the way they originate and how they can be controlled or utilized perhaps in a useful manner in the epitaxial growth procedure.

The defects considered here are the usual crystalline imperfections such as point defects, dislocations, stacking faults and twins. In ordered alloys and in compound epilayers superdislocations and antiphase domain boundaries are also found [2.12]. The definite types of defects occurring in the epilayer, especially in the interface area between the epilayer and the substrate, are to some extent influenced by the growth mode by which the epilayer has been crystallized. However, the dominant source of imperfections in epilayers is the misfit between the layer and the substrate crystal lattices. If the misfit is small, the defects arise mainly as a result of coherence loss between "$e$" and "$s$"; however, when the misfit is large, the defects arise predominantly from the lack of exact lattice registry between heterogeneously nucleated epilayer particles. In intermediate situations both factors are expected to be important. Therefore, unless special heat treatments are employed following the growth process, it is unrealistic to expect to obtain defect-free epilayers. In order to be more precise, let us define the most important defects occurring in epilayers. In doing this we follow the reviews of Stowell [2.14] and Sharan et al. [2.15], as well as the book by Watts [2.16].

### 2.2.1 Point Defects

If the regular array of atoms of a crystal (epilayer) is interrupted by an imperfection that can be inscribed in a small sphere, the imperfection is called a point defect. Intrinsic point defects involve atoms of the host crystal only. Examples are vacancies, which are missing host atoms, and self-interstitials, consisting of squeezed-in additional host atoms. Extrinsic point defects involve atoms chemically different from the host crystal, such as unintentionally introduced impurities or intentionally introduced atoms used for electrical doping [2.16]. Much is known about the structure of point defects in crystals, and also in epilayers, largely through spectroscopic studies. However, extended studies are still undertaken for relating the occurrence of defects to the growth parameters chosen for crystallization, what is especially true when growth of epitaxial structures has to be performed [2.17].

One of the simplest point defects are impurity atoms in an otherwise perfect crystal lattice. The important question which arises in the case of impurity point defects is how many of the impurities "$I$" can be incorporated in a perfect crystal "$C$". Answering this question is often a serious thermodynamic problem, to which the solution may be found through consideration of the relevant phase diagrams (i.e. the relation of temperature versus composition of the crystal) describing the thermodynamic properties of the $I$-$C$ system, in which I is soluble in $C$ to some extent, but $C$ is not soluble in $I$ at all. The incorporation mechanism and the atomic structure of a point defect (which can be created, in general, by other imperfections of the crystal lattice) can be definitely established only by analysis of spectra (optical and electrical) characteristic of this defect. The knowledge of the behavior of point defects in epilayers is extremely important in cases when structures consisting of differently doped layers, like p-n junctions or n-i-p-i superlattices, have to be grown epitaxially.

### 2.2.2 Dislocations

A dislocation is a line defect in a crystal and is characterized by a vector $\boldsymbol{l}$ which defines (locally) the direction of the dislocation line, and by a vector $\boldsymbol{b}$, the Burgers vector, which defines the atomic displacement needed to generate the dislocation from a perfect lattice (Fig. 2.3). The vector $\boldsymbol{l}$ may

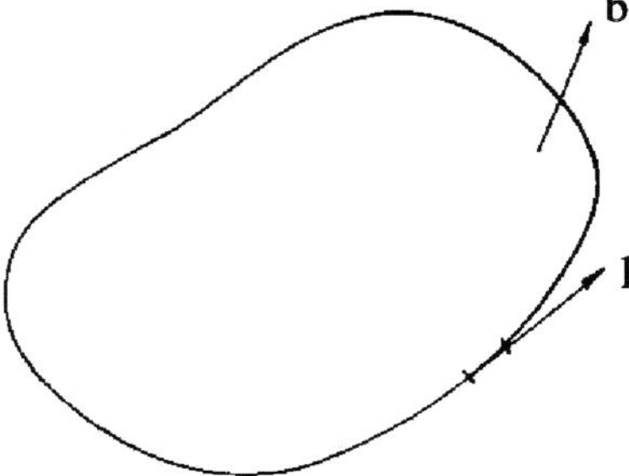


**Fig. 2.3.** Dislocation loop showing line vector $\boldsymbol{l}$ and Burgers vector $\boldsymbol{b}$ (taken from [2.14])

vary along the dislocation line, but $\boldsymbol{b}$ is invariant. Perfect dislocations have Burgers vectors that are translation vectors of the lattice. Imperfect or partial dislocations can exist, having Burgers vectors that are not translation vectors of the lattice; they are always associated with other imperfections such as stacking faults or antiphase boundaries. An important feature of a dislocation is its elastic strain field, and because of this dislocations are observable in an electron microscope. Atomic planes are significantly distorted in the vicinity of a dislocation, causing electrons to be diffracted more or less strongly than from the surrounding, more perfect crystal.

In the case of epitaxy one is concerned with dislocations in very thin crystals. In this case the method of imaging dislocations by the Moiré fringe technique (Fig. 2.4) is very useful [2.18]. The possibilities for dislocations to occur in epilayers during the growth have been discussed by Pashley in [2.19]. These are:

(i) the extension of substrate dislocations,
(ii) the accommodation of translational and rotational displacements between agglomerating islands that are close to epitaxial orientation,
(iii) the formation of dislocation loops by the aggregation of point defects, and

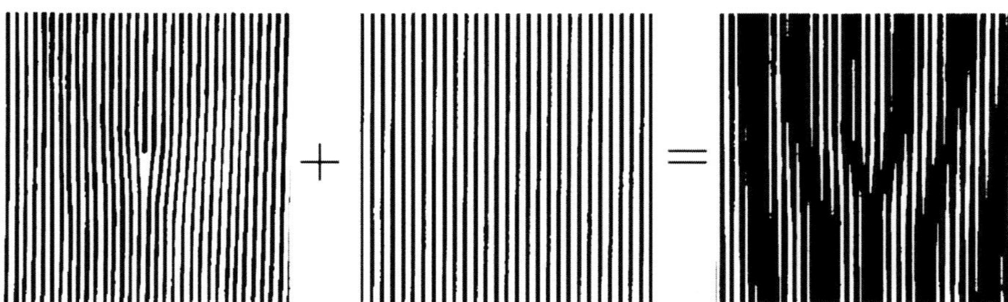

**Fig. 2.4.** Optical analog showing the appearance of a dislocation in a Moiré image (taken from [2.14])

(iv) plastic deformation of the epilayer, both during the growth and subsequent cooling and removal from the substrate.

### 2.2.3 Stacking Faults

A stacking fault is a planar defect across which the crystal has been displaced by a vector that is not a lattice translation vector. If a single stacking fault terminates within a perfect crystal, it must be bounded by a dislocation loop, the Burgers vector of which is also not a translation vector of the lattice; such dislocations are called "imperfect" or "partial". For example, in the fcc (face-centered cubic) lattice, stacking faults occur on {111} planes and are bounded by dislocations having Burgers vectors of the type 1/6 <112> or 1/3 <111>. In the former case, the fault is produced by shear on a {111} plane, whereas in the latter case removal or insertion of a layer of atoms is required so that the displacement is normal to the faulting plane. This is illustrated in Fig. 2.5, in which the conventional "abc" terminology representing the stacking sequence of atom planes in the fcc lattice is used. A fault made by the removal of a partial atomic layer is called "intrinsic" (Fig. 2.5a), and the insertion of an additional layer of atoms yields an "extrinsic" stacking fault (Fig. 2.5b). From the view point of the stacking sequence, an extrinsic fault in the fcc lattice is equivalent to two intrinsic faults on adjacent {111} planes. Stacking faults can be created by the dissociation of perfect dislocations into two partial dislocations bounding a stacking fault; in this case, the sum of the Burgers vectors of the partials equals that of the perfect dislocation.

Stacking faults are made visible in transmission electron microscope images by virtue of the phase shift, which occurs across the fault; this gives rise to a series of interference fringes parallel to the crystal surface. In Moiré images, stacking faults produce a displacement of the fringes across the fault. This is illustrated in Fig. 2.6.

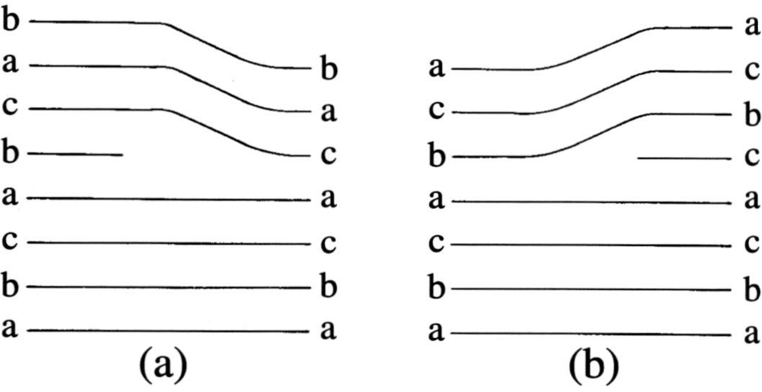


**Fig. 2.5.** Stacking sequence of (111) planes giving rise to: (**a**) intrinsic, and (**b**) extrinsic stacking faults (taken from [2.14])

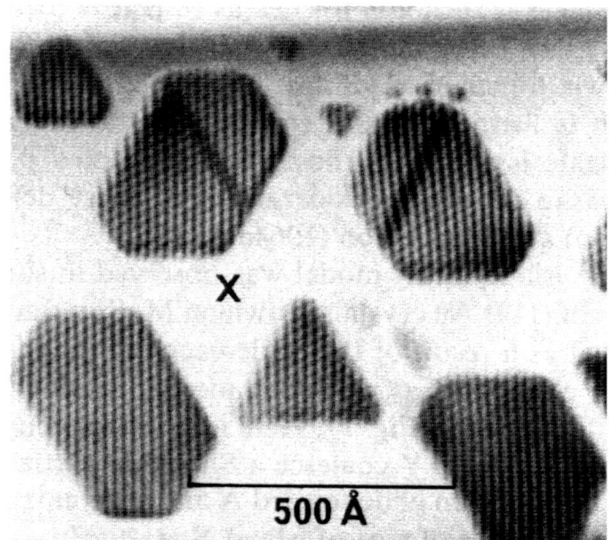


**Fig. 2.6.** Moiré images containing stacking faults. Note that in island X the fringes normal to the stacking fault are not displaced (taken from [2.14])

Electron microscope observations of epitaxial metal films revealed that stacking faults were often observed in very large numbers. Also observations of Si layers grown on Si demonstrated that stacking faults were the predominant type of defect. Several possibilities have been suggested for the origin of these faults, most of which coincide with those that have already been discussed in relation to dislocations. The most likely explanations are the following two:

(a) the accommodation of misfit between coalescing islands and

(b) the aggregation of point defects to form loops or tetrahedra of stacking faults.

### 2.2.4 Twins

In perfect epitaxial layers all crystal unit cell sides are strictly parallel to the corresponding unit cell sides of the substrate. However, frequently also symmetrical overgrowth on single crystalline substrate surfaces are observed, in which occasionally only one of the cell sides (on the face or edge) is in common with the relevant substrate cell side, though the one individual may be oppositely directed relative to the other. Such symmetric overgrowths, which are not accidental but statistically determined, are called twins . The twinned individuals may be transformed into one another by operation on symmetry elements of the twin, i.e., twin planes and/or twin axes [2.18].

A twin is characterized by a reflection of atomic positions across the twinning plane. For example, in fcc crystals and those with diamond or sphalerite

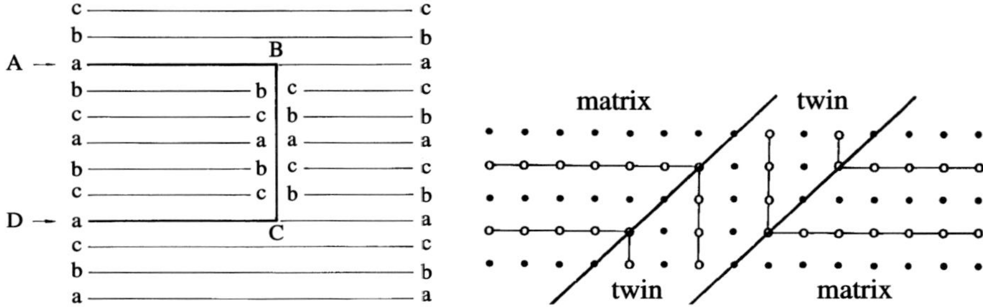


**Fig. 2.7.** (**a**) Stacking of (111) planes in a fcc crystal showing an embedded twin, (**b**) schematic diagram of an order twin for A (●) and B (○) atoms in a compound AB (taken from [2.14])

structures (called also zincblende structures), the twinning plane is {111}, and one may use the "abc" stacking notation to represent the relative atom positions in matrix and twin. The stacking sequence of a (111) twin embedded in a matrix is shown in Fig. 2.7a. Here the (111) planes *AB* and *DC* are called "coherent" twin planes; nearest neighbor atom distances are unchanged across these planes. Planes such as *BC* are called "incoherent" twin planes in view of the fact that the two lattices do not fit exactly at them, with the consequence that there is considerable atomic misfit. If a twin platelet terminates within a crystal, or in a fairly thick epilayer, it must do so at an incoherent boundary.

As already mentioned, in fcc crystals and those having the diamond or sphalerite structures, twins always occur on (111) planes. In (001)-face crystals, however, four sets of orthogonal twins are usually seen in epitaxial deposits. These can often be mistaken for thin stacking faults when viewed in an electron microscope because they give rise to fringes in the electron image that are similar to (but not identical with) those produced by stacking faults. Differentiation between stacking faults and very thin twins demands that detailed contrast experiments be carried out [2.14]. However, thick twins can readily be distinguished by reference to the electron diffraction pattern, which will exhibit extra reflections if twins are present.

There is evidence that, for both dislocations and stacking faults, the translational and rotational misfits between coalescing islands play an important part in the generation of these defects; it appears that in many cases twins originate as a consequence of these misfits. Twins might form to accommodate translational misfit between coalescing islands, and, although twins have been observed by *in situ* electron microscope methods to form when islands coalesce, the details of the way displacement misfits are manifested in microtwins are still a subject of investigation. This concerns especially the cases when alloys and compounds are grown epitaxially. Figure 2.7b shows schematically a twin in a layered structure of the CuAuI crystal, where the boundary between neighboring layers is a twin boundary. Another example

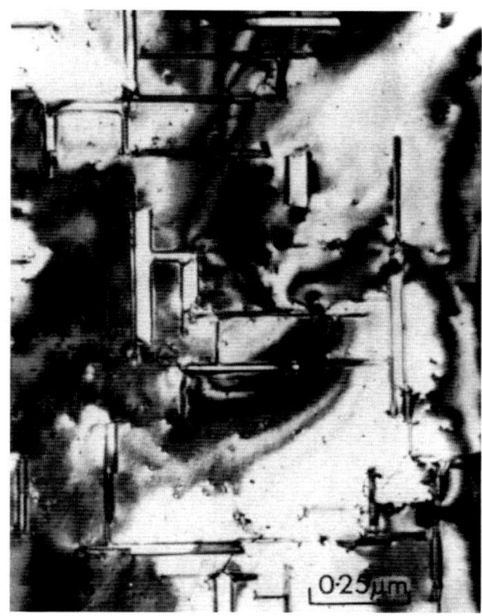


**Fig. 2.8.** Electron micrographs of microtwins in an epitaxially grown Au(100) layer (taken from [2.14])

which is shown in Fig. 2.8 for an Au(100) epilayer concerns twins occurring in metallic epitaxial layers.

Let us now discuss some of the defects peculiar to alloys and compound structures, namely, the antiphase domain boundaries (APB) and so called "superdislocations". It is obvious that normal dislocations, stacking faults and twins may also occur in these epitaxially grown material systems.

### 2.2.5 Antiphase Domain Boundaries

Consider first what is meant by an antiphase boundary. This is illustrated for a two-dimensional simple cubic compound of $A$ and $B$ atoms in Fig. 2.9. Within the region outlined by the dashed square, $A$ and $B$ atoms have been interchanged so that across the dashed boundary $A$ atoms face $A$ atoms. This interface is an APB and encloses an antiphase domain.

Additional factors evolve from consideration of APB defects in non-centrosymmetrical ordered crystals, such as compounds of group III–V (e.g., GaAs or InP) and II–VI (e.g., CdTe or ZnSe) elements, which can solidify with the sphalerite structure; this gives an ordered superlattice in which interpenetrating fcc lattices of $A$ and $B$ atoms are displaced by $0.25\,a_0$ <111> ($a_0$ means here the lattice constant of the compound $AB$). The $(1\bar{1}0)$ section of the ordered sphalerite lattice is shown in Fig. 2.10. Two distinct surfaces can exist that contain either all $A$ or all $B$ atoms. Holt [2.20] examined theoretically the defect structures in this lattice and showed that two different types of APB are possible. These are illustrated in Fig. 2.11. An APB

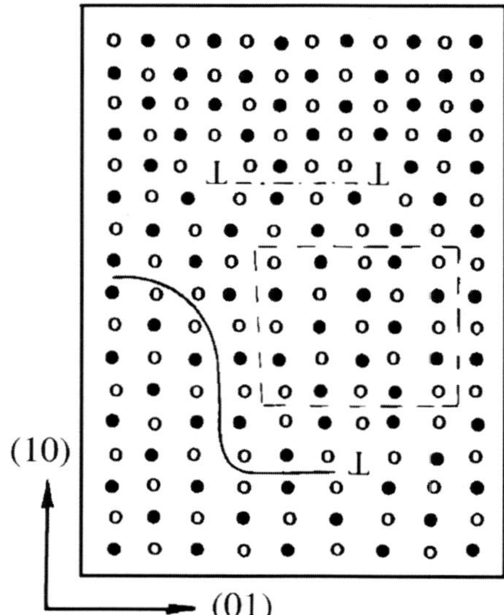


Fig. 2.9. Two-dimensional ordered crystal lattice of A (•) and B (○) atoms showing an antiphase domain boundary (APB) defect and a "superdislocation" (taken from [2.14])

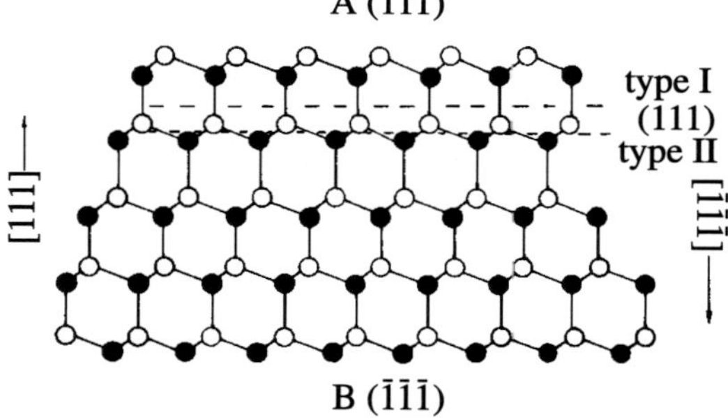


**Fig. 2.10.** Sphalerite structure in the ($1\bar{1}0$) projection showing two possible 111 surface types, the $(111)A$ and the $(\bar{1}\bar{1}\bar{1})B$ surfaces (taken from [2.14])

at which there are equal numbers of wrong *A*-*A* and *B*-*B* bonds (a Type I APB) is drawn in Fig. 2.11a, whereas in Figs. 2.11b and c, respectively, only wrong *B*-*B* and *A*-*A* bonds exist (Type II APBs). The latter type of APB is important in that it represents a two-dimensional excess of one class of atoms in the crystal; it may be viewed as a thin planar precipitate. In compounds grown under nonstoichiometric conditions, this type of defect may be prominent.

### 2.2.6 "Superdislocations"

Considering permissible glide displacements in the crystal lattice shown in Fig. 2.9, one may conclude that the passage of a dislocation with Burgers

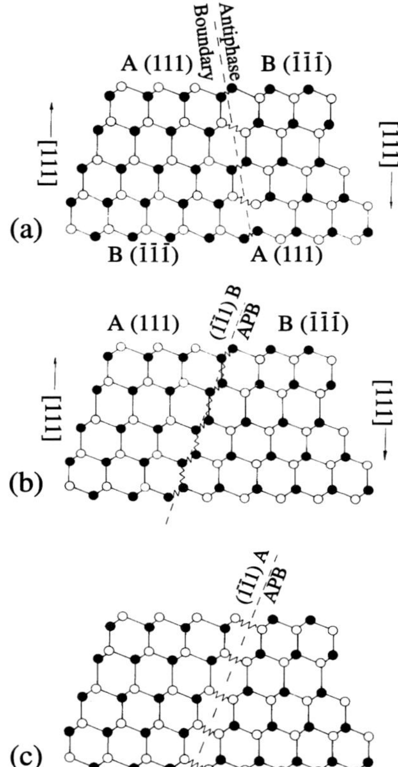


**Fig. 2.11.** Antiphase boundaries in the sphalerite structure [shown in the $(1\bar{1}0)$ projection]. Zig-zag lines denote wrong bonds. (**a**) Type I APB on the $(\bar{1}\bar{1}3)$ plane; (**b**) type II APB on the $(\bar{1}.1\bar{1})$ plane involving wrong $B$-$B$ bonds only and an excess of $B$ atoms; (**c**) type II APB involving wrong $A$-$A$ bonds with an excess of $A$ atoms (taken from [2.20])

vector $\frac{1}{2}[01]$ on a (01) plane creates an APB in its wake. In order not to create such a boundary, a dislocation must have a Burgers vector such as [01]; this is called a "superdislocation". A [01] superdislocation may dissociate into a $\frac{1}{2}[01]$ pair that bounds an APB, as in the upper part of Fig. 2.9. It is also evident from the lower part of this figure that, if an APB terminates within a single crystal, it does so at a dislocation.

### 2.2.7 Misfit Dislocations

In the interface area of lattice mismatched heteroepitaxial layers usually misfit dislocations (MD) [2.14, 15, 21, 22] occur. These dislocations are termed so because of their geometrical function in accommodating the misfit between "$e$" and "$s$" crystals. This is demonstrated in Fig. 2.12, where the lattice constant $a_o$ is less than $a_s$. Accordingly, there are unpaired atomic planes which terminate at the interface from above and which constitute the MDs. Geometrically, the misfit is said to be accommodated by MDs.

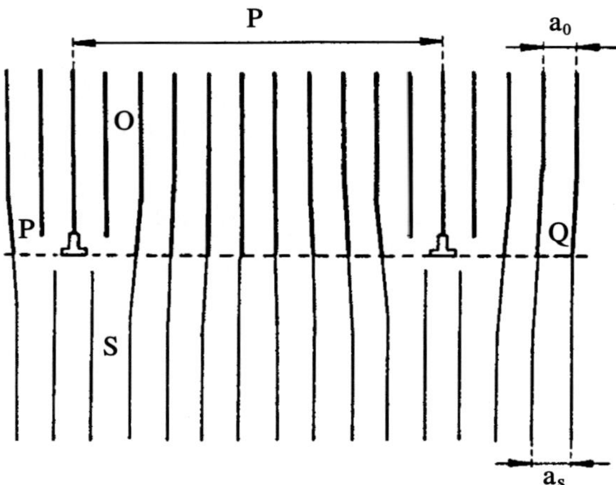


**Fig. 2.12.** A pure misfit dislocation geometry of edge type at interface $PQ$ of crystals $O$ (epilayer) and $S$ (substrate), with lattice constants $a_o < a_s$ (taken from [2.21])

In order to comply properly with the concept of a dislocation, significant strain should be localized in a small region around the dislocation. This is also the criterion that a sharp image of the dislocation can be formed in transmission electron microscopy. Basically, it requires the bonding across the interface to be strong enough. In the extreme case when the bonding is weak, there will simply be a misfit vernier and insufficient elastic strains to define MDs.

The dislocations shown in Fig. 2.12 are edge type, with their Burgers vector $\boldsymbol{b}$ in the interface and normal to the dislocation line. However, dislocation may also glide into the interface from either crystal on an oblique glide plane. The corresponding Burgers vector will then generally be at an angle both to the dislocation line and the interface (usually three types of such dislocations, namely, 45°, 60°, and 90° MDs may occur in the heteroepitaxy of cubic crystals [2.14, 15]). Only the projection on the interface of the edge component is available to accommodate misfit which originates from a difference in atomic spacings. When misfit includes symmetry differences, or when the crystals are twisted about a normal to the interface, MDs may be partially or completely of screw character. A MD can also dissociate into partials or react with other dislocations according to fixed rules. An example of how MDs penetrate heteroepitaxial layers is shown in Fig. 2.13 for the case of the GaAs-on-Si system [2.15].

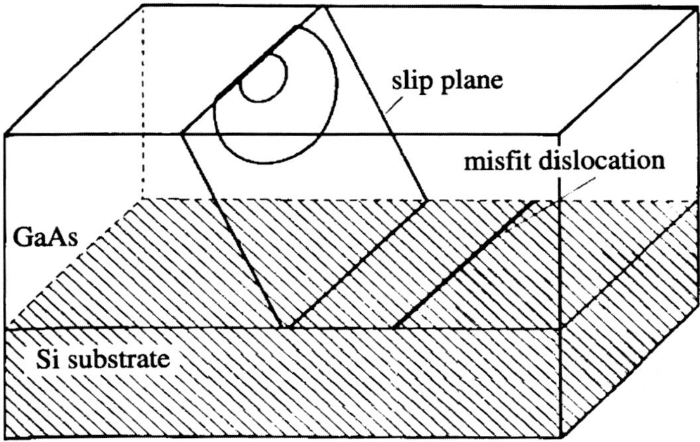


**Fig. 2.13.** Schematic representation of the generation of misfit dislocations in a semiconductor heterostructure: the dislocation loop which nucleates at the free (001) surface and glides in the {111} plane of the epilayer–substrate interface (taken from [2.15])

## 2.3 Peculiarities of Epitaxially Grown Layers

There exist significant differences in growth phenomena and in structural, electrical and optical properties between homoepitaxial layers and heteroepitaxial layers or layered structures. Let us discuss these properties of the two kinds of epilayers with the example of selected material systems; silicon homoepitaxial layers and silicon-based heteroepitaxial layers of the material system $Si_{1-x}Ge_x/Si$.

### 2.3.1 Homoepitaxial Layers

Epilayers grown in a single-component system (this means that the matrices of "$e$" and "$s$" phases are of the same chemical composition) are called homoepitaxial layers, even if they differ in doping with electrically active impurities from the substrate phase "$s$" or underlying epilayer "$e$" phase. Homoepitaxial layers or layer structures can be grown under conditions such that the crystallographic orientation of the layer is exactly determined by the substrate (or the already grown underlying epilayer) [2.23].

Most frequently, homoepitaxial layers are met in microelectronic semiconductor devices based on silicon structures. For example, integrated circuits utilize multilayer planar structures which are made by diffusion or implantation of dopant species into a single crystalline Si substrate, which is followed by epitaxial growth of one or more layers of silicon, appropriately doped and by further diffusion or implantation processes. Semiconductor varactors and avalanche diodes prepared as homoepitaxial structures constitute another example.

The technically important homoepitaxial structures are usually grown from the vapor phase [2.7] under two limiting conditions. In the first, nutrient particles arrive at the substrate surface from the vapor and stick wherever they land. In this case the growth rate of the layer is proportional to the difference between the actual impingement rate, controlled by the pressure in the vapor phase, and the equilibrium evaporation rate, defined by the equilibrium vapor pressure of the substrate crystal. This means that the nutrient particles (atoms or molecules of the reactant species) do not have an appreciable mobility on the surface and that they do not show a preference for attachment to the surface at steps and kinks. In order to grow an epilayer under this condition either the pressure (or supersaturation) in the vapor must be sufficiently high or the temperature of the substrate surface must be low enough (or both). The layer grows, however, continuously as fast as the nutrient particles are transported through the vapor phase towards the substrate surface. In this case the epilayers are usually highly defective crystals so that growth under the first condition is usually not used to prepare device-quality epilayers.

In the second limiting case, vicinal surfaces (that is terraces and steps) are intentionally created on the substrate surface. In order to grow epilayers now the nutrient particles must be transported through the vapor to a terrace, migrate there across to a step and then attach at a kink in the step (see Fig. 1.5 for the illustration of the SF-mode of epitaxial growth). Homoepitaxy performed under this condition allows in principle for the growth of structurally perfect single crystalline epilayers, with minimal interface energy and no mechanical strain [2.24]. However, if epitaxy:

(i) is performed on a contaminated substrate surface,
(ii) proceeds under non-optimal growth conditions, or
(iii) the homoepitaxial layers are heavily doped,

then even in this case of homoepitaxy different crystal lattice defects may occur in the epilayer. This holds for both the elemental growth (epitaxy of Si, Ge, Au, etc.) and the growth of compounds or alloys (e.g., GaAs, CdTe, Fe-alloys).

In the case of Si homoepitaxy, which serves here as an example, the growth of device-quality epilayers is performed on substrates with vicinal surfaces. It is important to note that at low growth temperature and high incoming flux of nutrient, the surface diffusion on the terraces is so slow that atoms do not have time to find low-energy sites before they are covered over. As a result of such growth conditions an amorphous layer is grown. At higher temperatures where the mean diffusion distance on the surface is still shorter than the distance between steps ($x_s < l$), the supersaturation at the surface builds up until two-dimensional (2D) nucleation occurs between the steps (on the terrace). This causes polycrystalline growth. Surface nucleation also affects the morphology of Si layers. When 2D nucleation occurs on the terrace, it may

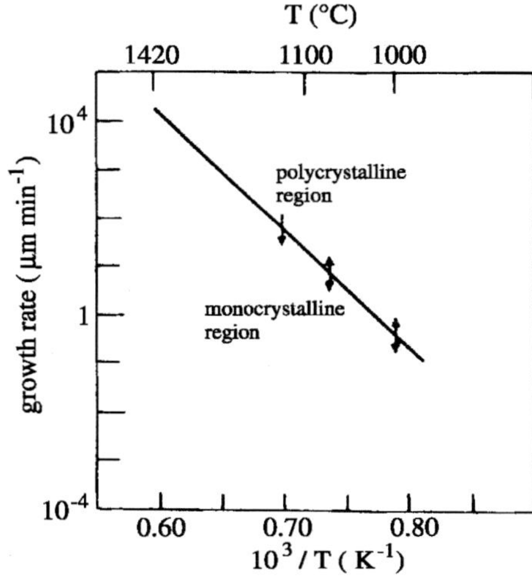


**Fig. 2.14.** Maximum growth rate for which monocrystalline Si epilayers can be obtained as a function of temperature (taken from [2.25])

produce multilayered islands or hillocks often observed in VPE layers. This is particularly true for growth on (111) faces where the surface is naturally smooth, necessitating steps for growth. Impurities on the surface also enhance 2D nucleation, and thus, all negative features of the epilayer which are caused by nucleation.

Perfect single crystalline layers are produced only, when growth occurs solely by the propagation of steps [2.7], that is, when the growth rates are sufficiently low, high temperatures ensuring that $x_s > l$ are used, and low impurity levels are retained. Therefore, from the point of view of application to technology of microelectronic devices Si homoepitaxy should be performed on (001) slightly misoriented Si substrates on which characteristic monoatomic steps and terraces occur (see Fig. 1.4). The relationship between growth rate and temperature required to grow single crystalline layers is shown in Fig. 2.14 [2.25]. The slope of the line separating the single and polycrystalline regions is 5 eV, which is equal to the activation energy for self-diffusion in Si. This lends itself to a simple interpretation. If the trapped, interstitial Si atom in the solid can diffuse to the surface and hence be incorporated into a lower energy site, the layer can "heal" itself during growth. It is well known [2.26] that Si epilayers are usually grown at temperatures higher than 1000°C. This temperature range is imperative if low defect densities and significant growth rates are required. In this regime the surface diffusion rate is sufficiently high and any chemical species that could interfere with Si perfect single crystal formation on the substrate surface can be removed by reactive desorption. Thus, the desired growth rate and crystal perfection can easily be obtained. However, with the continuing scaling of device structures to the submicrometer level two important questions arise when using epitaxial structures:

(i) how to minimize the transition region width (the so called autodoping region) between the substrate and the epitaxial layer, and

(ii) how to optimize the lateral and vertical isolation between the different active areas.

The immediate answer is to lower the deposition temperature. However, in such a case, adsorbed species such as oxygen, carbon, $H_2O$, chlorine and their related compounds interfere with the single crystal growth because of incomplete desorption. As a consequence the processing pressure should be reduced to facilitate desorption of impurities. Such considerations have led to new epitaxial processing techniques to achieve low temperature epitaxy in low pressure reactors which are now frequently used. Sharp epitaxial–substrate transition region widths have been obtained at temperatures as low as 550°C by using, for example, the MBE growth technique [2.27], or plasma enhanced chemical vapor deposition (between 700 and 800°C). In both cases, however, the epilayer quality was only moderate. The dislocation densities could be as high as $10^5$ cm$^{-2}$ [2.26].

An interesting report on experiments concerning formation of dislocations during growth of homoepitaxial Si by VPE at reduced temperatures is given in [2.28]. These experiments have shown that the bare Si substrate surface, once formed by oxide desorption during the pre-epitaxial bake (prebake), is able to accumulate carbon, either brought in on the wafer surface or from the vapor phase or both, which forms into immobile nuclei. These form subsequently into pyramidal features which grow in size during anneal. These features locally impede the desirable initial step flow growth, causing transient pits during the first $0.1-0.2$ μm deposition, a small proportion of which creates epilayer dislocations. Figure 2.15 shows two examples (wafers A and B) with images taken by laser scanning optical micrographs in differential phase contrast mode (SOM) and grazing incidence scanning electron microscope (SEM). The first of them (A) has been annealed at 900°C for 30 min in a 1 mbar $H_2$ stream, and on the second (B) a 0.1 μm thick Si layer was grown first and then the growth has been stopped. The sample A exhibits a pronounced “thermal roughening” effect, while sample (B) shows transient pits formed where contaminant particles, such as existing on the surface A, disrupted the lateral flow of surface misorientation steps during initial growth.

### 2.3.2 Heteroepitaxial Layers

Epitaxial growth of a layer with chemical composition and sometimes structural parameters different from those of the substrate leads to creation of a heteroepitaxial layer on the substrate surface. The growth process in heteroepitaxy depends strongly on whether the grown epilayer is coherent or incoherent with the substrate, i.e., on whether the interface between the epitaxial overgrowth and the substrate is crystallographically perfect or not [2.29]. If the overgrowth is incoherent with the substrate, then it is free to

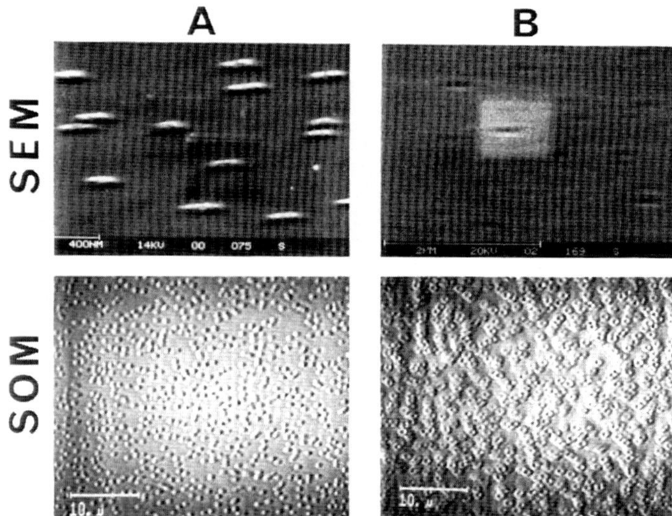


**Fig. 2.15.** Laser scanning optical micrographs (SOM) in differential phase contrast mode, and scanning electron images (SEM) taken from Si(001), slightly misoriented wafers A and B. Wafer A shows pronounced "thermal roughening" after prolonged prebake. Wafer B was prepared by stopping Si deposition after normal prebake, when a 0.1 μm thick epilayer was grown (taken from [2.28])

adopt any in-plane lattice constant that minimizes its free energy. If the overgrowth is coherent with the substrate, then energy minimization is achieved by adopting the in-plane lattice constant of the substrate, thus, creating a commensurable material system with the substrate. The resulting elastic-strain energy can then increase its overall free energy significantly [2.30]. Both of these growth processes are currently often applied for preparation of semiconductor devices, when using different epitaxial growth techniques.

Three factors are important from the point of view of the mutual relation between the substrate and the heteroepilayer. Lattice constant matching, or mismatch, is the first, crystallographic orientation of the substrate is the second, and its surface geometry, or surface reconstruction, is the third. The crucial problems of heteroepitaxy are, however, related to lattice mismatch.

Following van der Merwe [2.21], the term "misfit" is usually used to refer to the disregistry of the equilibrium interfacial atomic arrangements of the substrate and the unstrained epilayer. This disregistry results from differences in atomic spacings of lattice symmetries, which are characteristic of each of the two crystals in the absence of interfacial interaction between them. In the simplest case, where the epilayer is fairly thin in comparison to the thickness of the substrate crystal, the misfit $f_i$ may be quantitatively defined as [2.22]

*Definition 1* $$f_i = (a_{si} - a_{ei})/a_{ei}, \quad i = 1, 2 \qquad (2.1)$$

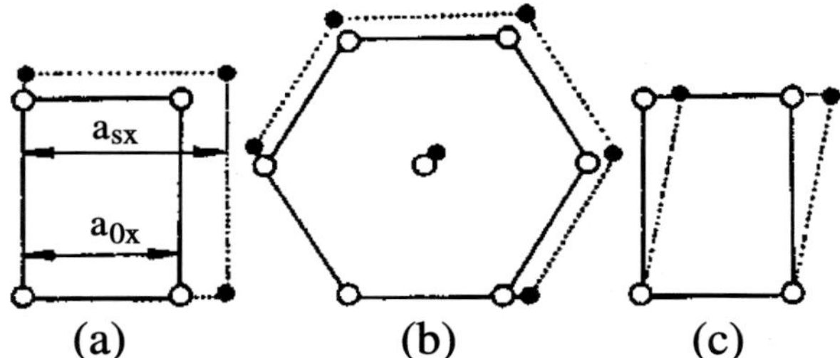


**Fig. 2.16.** Examples of interfacial atomic arrangements in the "$e$"/"$s$" systems of different crystallographic symmetry. Black dots represent the substrate atoms while empty circles stand for overgrowth atoms (taken from [2.21])

where the $a_i$ are the lattice constants perpendicular to the growth direction, and $s$ and $e$ designate the substrate and the epitaxial layer, respectively (Fig. 2.16). The convenient features of this definition are as follows. If the film is strained so that the lattices of film and substrate are in register at the interface, then the misfit strain, defined by

$$\textit{Definition 2} \qquad e_i = ({a_{ei}}^{\text{str}} - a_{ei})/a_{ei}, \quad i = 1, 2, \tag{2.2}$$

will be equal to $f_i$. In (2.2) ${a_{ei}}^{\text{str}}$ stays for the lateral atomic spacing in the strained epilayer. If, however, the misfit is shared between dislocations and strain, then

$$\textit{Definition 3} \qquad f_i = e_i + d_i, \quad i = 1, 2, \tag{2.3}$$

where $d_i$ is the part of the misfit accommodated by dislocations. A positive value for $f$ implies that the misfit strain is tensile and that the misfit dislocations are positive Taylor dislocations [2.31], i.e., extra atomic planes lie in the overgrowth (Fig. 2.12).

The above definition does not allow for reconstruction [2.32] or unevenness of the substrate surface. Therefore, for very thin epilayers (1 – 2 monolayers thick), for which the equilibrium interatomic spacing may differ from that of the bulk crystal by as much as 5 %, Frank and van der Merwe defined the misfit by [2.33]

$$\textit{Definition 4} \qquad f_i = (a_{ei} - a_{si})/a_{si}, \quad i = 1, 2. \tag{2.4}$$

By this definition misfit is related to the fixed substrate interatomic spacing. However, when the overlayer is thick enough (if its thickness becomes comparable to the thickness of the substrate), both crystals are usually treated on the same footing by defining the misfit as [2.22]

$$\textit{Definition 5} \qquad f_i = 2(a_{si} - a_{ei})/(a_{si} + a_{ei}), \quad i = 1, 2. \tag{2.5}$$

It is well known that if the misfit between a substrate and a growing layer is sufficiently small, the first atomic monolayers which are deposited will be strained to match the substrate and a coherent (perfectly matched)

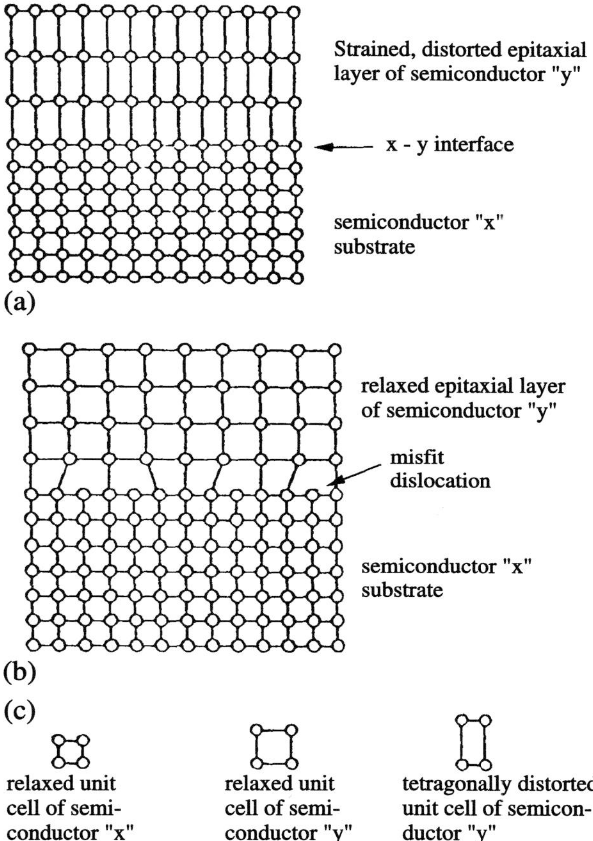


**Fig. 2.17.** Schematic illustrations of (**a**) strained and (**b**) relaxed epitaxial layers of a lattice mismatched heterostructure, as well as (**c**) strained and relaxed unit cells (taken from [2.35])

epilayer will be formed [2.29, 30]. For such a state of the "$e$"/"$s$" system the term "pseudomorphism" has been introduced by Finch and Quarrell [2.34]. However, as the layer thickness increases, the homogeneous strain energy $E_{\mathrm{H}}$ becomes so large that a thickness is reached when it is energetically favorable for misfit dislocations to be introduced (see Fig. 2.17). The overall strain will then be reduced but at the same time the dislocation energy $E_{\mathrm{D}}$ will increase from zero to a value determined by the misfits $f_1$ and $f_2$. These misfits are defined by (2.5) in which, however, $a_{ei}$ has been replaced by ${a_{ei}}^{\mathrm{str}}$, the lattice constant of the strained epilayer at the interface. The existence of the critical thickness was first indicated in the theoretical study by Frank and van der Merwe [2.33], then treated theoretically by others, and confirmed by various experimental observations [2.22].

The possibility of growing high-quality epitaxial layers of different materials on lattice mismatched substrates is a topic of considerable interest for the growth of heterostructures. The range of useful devices available with a given substrate is largely enhanced by this method. For example, GaAs and compounds related to it (AlGaAs, InGaAs, etc.) offer many advantages over Si in terms of increased speed and radiation resistance and its ability to process and transmit signals by light pulses. Si, on the other hand, is a well-established material for integrated circuits and exhibits superior mechanical and thermal characteristics. By growing, for example, epitaxial layers of GaAs on Si substrates, it would be possible to combine the advantages of both materials. However, these materials are not matched together, neither by lattice constants ($a_{\mathrm{Si}} = 0.543\,\mathrm{nm}$, $a_{\mathrm{GaAs}} = 0.565\,\mathrm{nm}$), nor by thermal expansion coefficients ($\alpha_{\mathrm{Si}} = 2.6\times10^{-6}\,\mathrm{K}^{-1}$, $\alpha_{\mathrm{GaAs}} = 6.8\times10^{-6}\,\mathrm{K}^{-1}$). Therefore, dislocations and other lattice defects are usually present in GaAs-on-Si heterostructures.

Let us now consider the peculiarities of heteroepilayers in more detail with the example of silicon-based heterostructures in the material system $\mathrm{Si}_{1-\mathrm{x}}\mathrm{Ge}_{\mathrm{x}}/\mathrm{Si}$. The main challenge in epitaxial growth of this material system is the fact that the lattice constants of Si and Ge differ by 4.17 %, and hence, it is quite difficult to grow dislocation-free $\mathrm{Si}_{1-\mathrm{x}}\mathrm{Ge}_{\mathrm{x}}$ layers on Si substrates. However, the lattice mismatch between these two materials can be accommo-

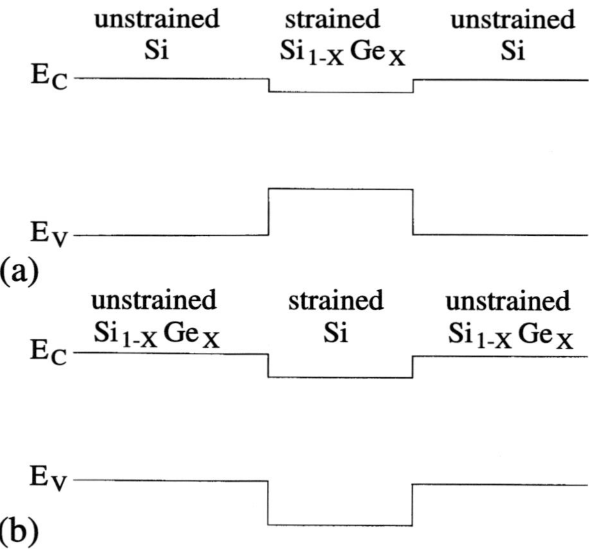


**Fig. 2.18.** Band edge lineup in Type I heterostructure (**a**) of a strained $\mathrm{Si}_{1-\mathrm{x}}\mathrm{Ge}_{\mathrm{x}}$ epilayer sandwiched between unstrained Si layers, and (**b**) in a Type II heterostructure of a strained Si epilayer sandwiched between unstrained $\mathrm{Si}_{1-\mathrm{x}}\mathrm{Ge}_{\mathrm{x}}$ layers (taken from [2.35])

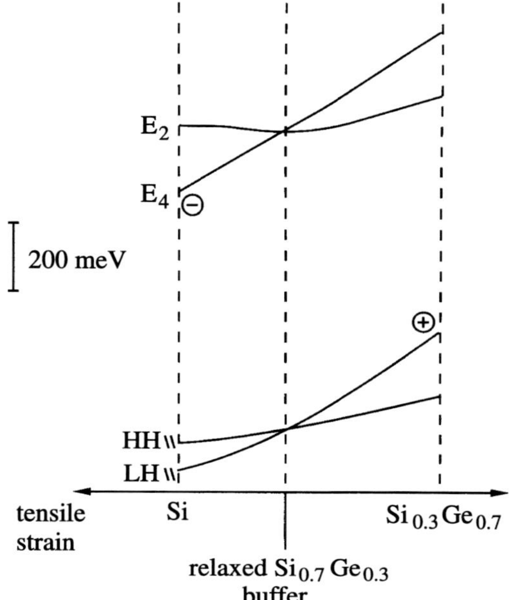


**Fig. 2.19.** Effect of strain on the conduction and valence band edges in the heterostructure $Si_{1-x}Ge_x/Si$. The substrate is assumed to be relaxed and to have 30 % Ge in the alloy. The negative $x$-axis represents tensile strain and the positive axis represents compressive strain with respect to the $Si_{0.7}Ge_{0.3}$ substrate (taken from [2.38])

dated by a finite degree of lattice distortion (pseudomorphic strained layers), after which planar film growth (FM-mode) either converts to 3D growth (SK-mode), or misfit dislocations nucleate at the heterointerface, thus relaxing the lattice strain (Fig. 2.17b). The last two decades have seen tremendous advances towards pseudomorphic epitaxial growth of $Si_{1-x}Ge_x/Si$ heterostructures [2.36]. Using such techniques as MBE and ultrahigh vacuum chemical vapor deposition (UHVCVD) one can currently grow heterostructures of this material system with very low densities of defects [2.37].

Strain plays a dominant role in determining the alignment of energy bands at heterointerfaces (see Fig. 2.19), thus determining the confinement energy of electrons (holes) in quantum wells [2.38]. Strain modifies also the crystal symmetry, and thus offers the possibility of influencing the recombination rates of electrons and holes in device structures. Consequently, understanding and manipulation of strain became the natural prerequisite for device applications of $Si_{1-x}Ge_x/Si$ heterostructures. In the $Si_{1-x}Ge_x$ heteroepilayer of the $Si_{1-x}Ge_x/Si$ structure a compressive strain occurs, which increases both for increasing Ge concentration in the alloy as well as for increasing the layer thickness.

Multi-layer strained structures of Si and $Si_{1-x}Ge_x$ alloys are used to create electron/hole quantum wells. The band alignment at the heterointerfaces of Si and $Si_{1-x}Ge_x$ is of type I, what means that the offset lies predominantly within the valence band (Fig. 2.18a). To give an idea of the scale of this effect, the offset may be estimated at about 200 meV for $Si_{0.7}Ge_{0.3}$ commensurate to an underlying Si substrate [2.38]. On the other hand, when strained Si is grown on relaxed $Si_{1-x}Ge_x$, a staggered band alignment of type II results at the heterointerface (Fig. 2.18b). In this case, the conduction band of $Si_{1-x}Ge_x$ is higher than that of Si (by about 200 meV for $x = 0.3$ in the relaxed alloy), and the valence band in $Si_{1-x}Ge_x$ is lower than that of Si (by about 150 meV for $x = 0.3$). The effect of strain on the energy of conduction and valence band edges may be seen in Fig. 2.19 on the example of $Si_{1-x}Ge_x$, for $0 \leq x \leq 0.7$, grown on a relaxed $Si_{0.7}Ge_{0.3}$ buffer layer. In addition to the band offsets, the electron/hole effective masses are greatly affected by strain. Due to the strong anisotropy of the effective masses, energy bands with heavy masses are split from those with light masses. For lateral transport it is desirable to have a light in-plane electron/hole effective mass in order to enhance the carrier mobility.

Concluding the considerations concerning the homo- and heteroepitaxial growth processes and the peculiarities of the homo- and heteroepitaxial layers it should be emphasized that while homoepitaxy and homoepitaxial layers (appropriately doped) became the basis for present day microelectronics, the heteroepitaxy and heteroepitaxial structures, much more difficult to be grown on high perfection level, have created present day semiconductor optoelectronics and opened the way to development of tomorrow's nanoscale electronics (see Sect. 3.1).

# 3. Application Areas of Epitaxially Grown Layer Structures

Epitaxial growth can be used for crystallization of thin film structures of different, metallic, semiconducting, and insulating material systems, including mixed systems like, e.g., metal–semiconductor or semiconductor-on-insulator device structures. The epitaxially grown structures may be of very high perfection, satisfying the so-called device-quality demands. They are suitable for application in many branches of the "high technology" part of electronics, mechanics and materials engineering. However, the main application areas of epilayer structures are defined by solid state electronics, optoelectronics, and photonics [3.1–5].

## 3.1 Low-Dimensional Heterostructures

Crystallization of low-dimensional heterostructures (LDH) by epitaxy belongs to the most fascinating problems related to crystal growth of multilayer structures. Single or multiple quantum wells, quantum wires, and quantum dots (also called quantum boxes) create the LDH family. These structures are the constituent elements of mesostructure electronics [3.6–8]. MBE and MOVPE make possible the fabrication of heterostructures with composition varying on a spatial scale of one to 100 crystal-lattice constants. LDHs exhibit distinctive physical properties resulting from the occurrence of an intermediate size between two physical domains. The first is the microscale, or quantum-effects domain of lattice constants, or the de Broglie wavelength of the charge carriers. The second is the macroscale domain associated with quasi-classical motion of the charge carrier. The motion of carriers in a direction where potential barriers created by heterointerfaces confine them to an area that has a size belonging to the microscale, is strongly quantized, while the carriers motion in directions where no such confining barriers exist, is quasi-classical. Figure 3.1 illustrates the LDH family and the variation of the density of states of electrons with the increase of the quantization dimension in the LDH structures.

The area of application of quantum well structures in semiconductor devices is already considerable, extending from high electron mobility transistors through double barrier resonant tunneling microwave diodes to quantum well lasers and quantum well intraband photodetectors [3.10]. The two other

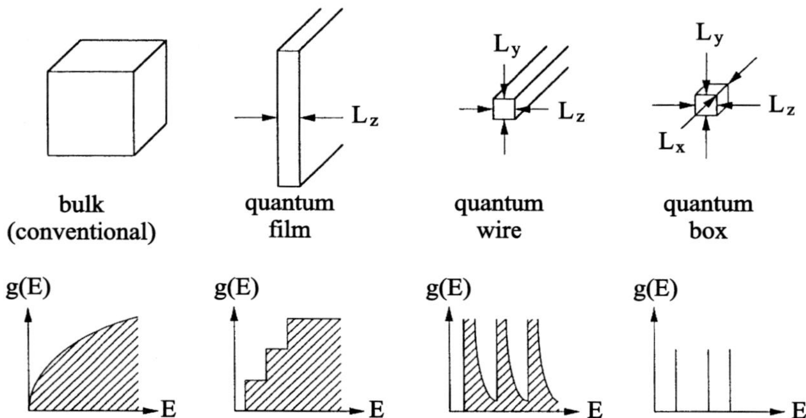


**Fig. 3.1.** The LDH structures (upper part) and the variation of the density of states of electrons (lower part) with the increase of the quantization dimension in LDH structures (taken from [3.9])

groups of LDH structures, namely, quantum wires (QWR) and quantum dots (QD) are still under development and need further intensive investigations. However, the first optoelectronic devices with QWR structures [3.11] and QD structures [3.12] have been demonstrated, and some examples are presented in the next section.

Here, we will describe an application of epitaxy to so-called band-gap engineering technology. In order to produce an artificially generated half-parabolic composition and potential energy profile in quantum well structures one may use the quantum size effect principle (the dependence of the position on the energy scale of the quantum well energy levels on the thickness of the well) to the growth of a multilayer heterostructure in which the thickness of the sequent quantum well layers is changed in a precisely controlled way according to the relevant algorithm ensuring the desired shape of the potential energy profile of the multilayer structure. This has been done by Miller et al. [3.13] with the GaAs–$Al_xGa_{1-x}As$ lattice matched material system.

The technique used was a computer-controlled long-period (24 Å) pulsed MBE in which the duty cycle of Al deposition with each pulse period was made proportional to the desired thickness of the $Al_xGa_{1-x}As$ barrier (or GaAs well) layers. The deposition sequence for one of the half-parabolic well samples is shown in Fig. 3.2. It contained five artificially made wells of thickness $L = 522{\pm}31$ Å separated by 299 Å thick $Al_{0.26}Ga_{0.74}As$ barriers, with each well comprising 43 alternate depositions of GaAs layers of different thicknesses. The algorithm for the deposition processes of GaAs layers was given by

$$(L/22)(1 - ((44 - i)/44))^2) \quad \text{with } i = 1, 3, 5, \ldots, 43 \tag{3.1}$$

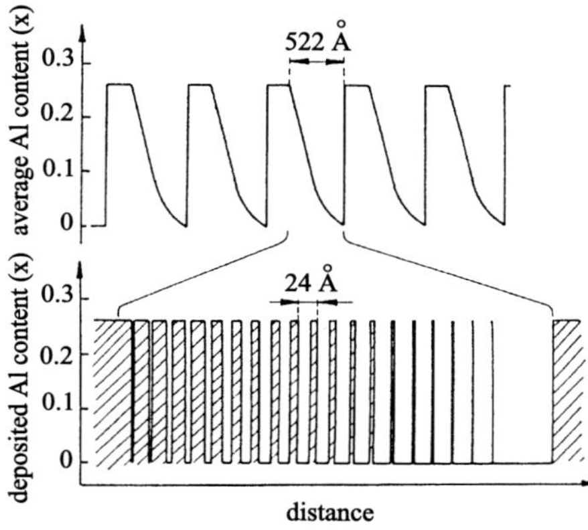


**Fig. 3.2.** Pulsed molecular beam deposition sequence used to produce half-parabolic compositions and potential energy profiles (taken from [3.13])

The analogous algorithm for deposition of the $Al_{0.26}Ga_{0.74}As$ layers has the form

$$(L/22)((43-i)/44)^2. \tag{3.2}$$

Deposition of layers less than 0.1 Å thick (layers 40 and 42) were omitted because of shutter timing considerations. Model calculations suggest that this produces no observable change in energy levels. The transmission electron microscope cross-section images of the grown structures confirmed the layer periodicity and its constancy to within 10% [3.13]

## 3.2 Device Structures with Epitaxial Layers

There is an increasing interest in growing single crystalline layered structures for device applications. This results from the unique electronic and optical properties exhibited by these structures. Up to now, nearly all kinds of electronic and optoelectronic devices have been prepared in the form, in which the most important functional elements of the device have been grown as epitaxial layers [3.2, 4, 5].

Device performance is critically dependent on the electrical and morphological quality of epilayers and interfaces. For example, high yield production of modulation-doped high electron mobility transistors, quantum well heterojunction lasers or superlattice avalanche photodiodes require pure materials, abrupt and smooth interfaces and defect-free surface morphology. A great impetus to development of epitaxially grown structures gave in the early 1960s the increasing attention paid by the semiconductor community to GaAs devices of high complexity. In particular, microwave and high speed component production has achieved remarkable increase in order to meet the high demands of the market for GaAs field effect transistors and integrated circuits. The development and production of the next generation of high speed discrete and integrated circuit devices was linked to the ability of growing epitaxially highly complex device structures. Present-day MBE and MOVPE have become the leading edge of this technology.

In the second half of the 1990s, the compound semiconductor industry was booming [3.14]. Compound semiconductor chips are found in numerous consumer products and used daily, often without people ever realizing it. With the development of blue and green as well as orange and red ultrahigh brightness light-emitting diodes, this form of lighting is fast replacing the conventional light bulb in every possible application from car lights and traffic signals, to outdoor advertising [3.14]. The revolution in lighting which we now face "is analogous to the replacement of the electronic vacuum tubes with the transistor" (*Financial Times*, 13 January 1998). Digital video disks using III–V compound lasers are replacing the CD-ROM as the next generation of mass memory storage. III–V solar cells are used to power huge numbers of space satellites, while heterostructure bipolar transistors are taking over as the standard chip used in mobile phones, television satellites and fiber-glass telecommunication.

Each material system which has to be grown epitaxially requires its own technological "know-how". Therefore, considering the application areas of epitaxially grown structures, one has to consider the problem of material-related peculiarities of epitaxial growth. This will be the subject of Chaps. 14 and 15 of this book. However, it is worth of mentioning that already in 1956 about 150 epitaxially grown materials systems could be listed [3.15]. Information on the intensive development of epitaxy, which was observed in the next 20 years (until 1975) is given in [3.16]. This enormous boom at that time was a result of both the development of important new diagnostic techniques for analyzing epitaxial growth phenomena, and the large interest in single-crystal layers for studies of their physical properties and possible applications. However, the most intensive expansion of epitaxial growth techniques in different application areas occurred in the last decade, when MBE and MOVPE achieved their mass-production maturity. [3.9, 14, 17].

We will not present here an exhaustive overview of the device applications of epitaxy. However, instead of this we will show how epitaxy may lead to most sophisticated device structures, namely, quantum wire semiconductor laser structures.

Two MBE growth procedures have been developed, so far, for preparing as-grown QWR structures. The first, realized on vicinal surfaces, concerns serpentine superlattices (SSL) with quantum wires as parts of these structures [3.18, 19]. The second is realized on (100) on-axis substrates by strain-induced lateral ordering (SILO) process [3.20]. SSL structures are produced with a cyclic-deposition growth technique on vicinal surfaces of slightly misoriented substrates (usually by 2°–4° from the (100) crystallographic plane toward the (111) plane) taking the relative coverage between successive deposition cycles as a variable, precisely controlled parameter [3.21]. The misoriented substrates provide a staircase of monolayer-height steps (most favorable when being parallel to the [110] direction) and plane-parallel terraces on which 2D nucleation may occur in epitaxial growth. During such growth on vicinal surfaces the shuttering of the molecular beams is under computer control and may be changed each cycle by a certain amount relative to the other cycles. Smoothly sweeping the per-cycle coverage back and forth through a range including monolayer coverage gives a structure with a continuously varying tilt. 2D electronic confinement (a typical feature of QWR structures) may be obtained in the winding wells, where the structure turns a corner.

Depositing fractional monolayers of $(GaAs)_m(AlAs)_n$ planar superlattice with $m+n=p\approx 1$, one may grow on vicinal surfaces a so-called tilted superlattice (TSL) [3.22]. In these superlattices, the interface planes are tilted with respect to the substrate surface plane, at an angle which can be controlled by adjusting the value of $p$ or the orientation of the substrate surface. The tilt angle and period of the TSL provide a sensitive measure of the growth kinetics and the influence of variations in the growth parameters during MBE growth. Crystallization proceeds by alternating depositions of between 1/2 and 1 monolayer of the column-III atoms, with deposition of the column-V atoms. An idealized representation of a TSL growth is shown together with the deposition algorithm in Fig. 3.3 [3.22].

TSLs have not been used successfully to form 2D confinement structures with good electrical and optical properties. One reason for poorly defined confinement is because of the thermodynamic difficulty of having fractional layers segregate fully on the vicinal terraces in the (Ga,Al)As system. Therefore, another cyclic deposition and epitaxial growth technique for vicinal surfaces has been presented [3.18]. This structure, i.e., the SSL, may be employed to obtain quantum wires for its purposefully meandering shape. Two-dimensional electronic confinement (characteristic to QWRs) may be obtained at the wide places in the winding wells, where the structure turns a corner, as is schematically represented in Fig. 3.4. The shape near these regions determines the energy spectrum of the confinement. Perhaps, the most interesting case occurs for a linear ramping of the per-cycle coverage through a range including vertical growth. At this instance an SSL results with a cross-section of periodically displaced parabolas having curvatures determined by the misorientation angle and the per-cycle coverage ramping rate $1/z_0$. The confinement energies are determined by $z_0$, the lateral well, the barrier widths, and the barrier height. The confinement in this case is insensitive to errors in the absolute growth rates. An error shifts the vertices on the parabolas up or down in the growth direction but does not change their shape. Thus a confinement may be obtained that is quite insensitive to growth rate variations across an entire substrate [3.20].

Quantum wire lasers with SSL have been grown and investigated by Hu et al. [3.19]. They have reported optical gain measurement from a serpentine superlattice nanowire-array laser sample, grown by MBE on a 2°-off (100)

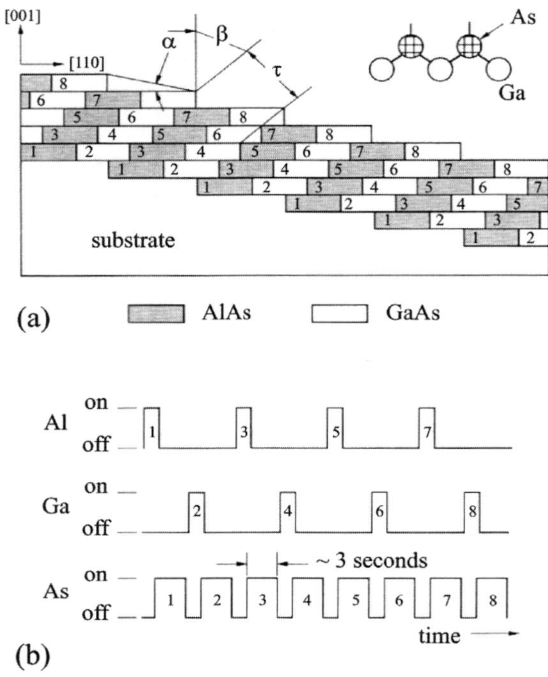


**Fig. 3.3.** Idealized representation of a tilted superlattice grown by alternating beam deposition. (**a**) Cross-section, perpendicular to the step edges of a superlattice with $p = 1.25$. The inset shows the arrangement of the As bonds with respect to the step edges. (**b**) Detail of growth, showing the order of atomic depositions. Numbers in (**a**) refer to partial monolayers deposited in (**b**) (taken from [3.22])

GaAs vicinal substrate. Gain spectra, obtained from in-plane ridge-waveguide lasers with stripes either parallel or perpendicular to the nanowire arrays at 1.4 K, showed that the optical gain for the TM mode became greater than that of the TE mode when the optical cavity was placed along the nanowire direction. This provides strong evidence that the lateral quantum confinement in the SSL is stronger than the vertical quantum confinement. A schematic diagram of the grown laser sample is shown in Fig. 3.5.

In the second MBE growth procedure, which uses the SILO process, it has been demonstrated that a strain-induced lateral periodic modulation of the composition along the [110] direction with a periodicity as small as 20 nm can be formed in vertical short period superlattices (SPS) of $(GaAs)_2/(InAs)_2$ and $(GaP)_2/(InP)_2$ grown on nominally (100) GaAs and InP, respectively [3.23]. This new path in growing sophisticated LDHs is often referred to as a self-organized growth of nanometer-scale features. It means spontaneous structural ordering, usually on a sublattice, of the deposits during MBE growth.

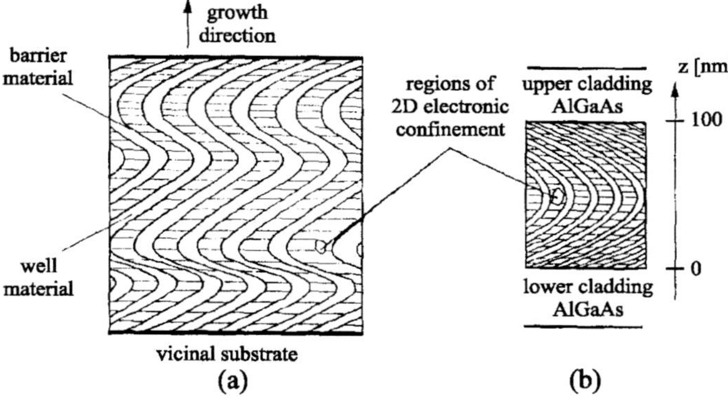


**Fig. 3.4.** An SSL cross-section is shown (**a**) that would result from sweeping the per-cycle coverage back and forth through a range that includes exact monolayer coverage. At such places, where the tangent to the structure is vertical, electronic states are confined to two dimensions. (**b**) Most of the grown structures have been single-crescent, parabolic geometries with barriers cladding the SSL region above and below (taken from [3.18])

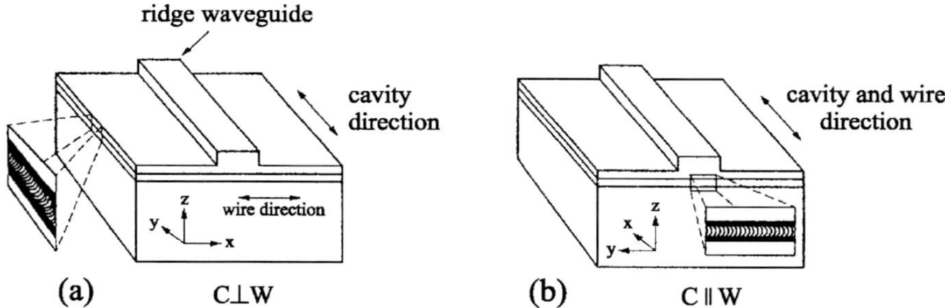


**Fig. 3.5.** Schematic diagram of the two device configurations of the SSL laser samples for the case (**a**) when the optical cavity is perpendicular to the nanowire array ($C \perp W$) and (**b**) when the optical cavity is parallel to the nanowire array ($C \parallel W$) (taken from [3.19])

$Ga_xIn_{1-x}P/Al_{0.15}Ga_{0.35}In_{0.5}P$ heterostructure graded-index separate-confinement visible laser structures with multiple quantum wire (MQWR) active regions have been formed *in situ* during MBE growth [3.11]. No regrowth or ex situ fabrication procedures were employed in the formation of the quantum wires. QWRs with cross-sectional dimensions of approximately $50\times120\,\text{Å}^2$ were routinely achieved with a linear density of $100\,\mu\text{m}^{-1}$. Broad area stripe geometry lasers with contact stripe oriented in the [110] and $[\bar{1}10]$ directions exhibited anisotropic threshold current densities varying in ratio by a factor of more than 3.75. Threshold current densities as low as $400\,\text{A}\,\text{cm}^{-2}$ were obtained for lasers with stripes in the [110] direction, perpendicular to the quantum wires. Strong dependence of electroluminescence polarization on

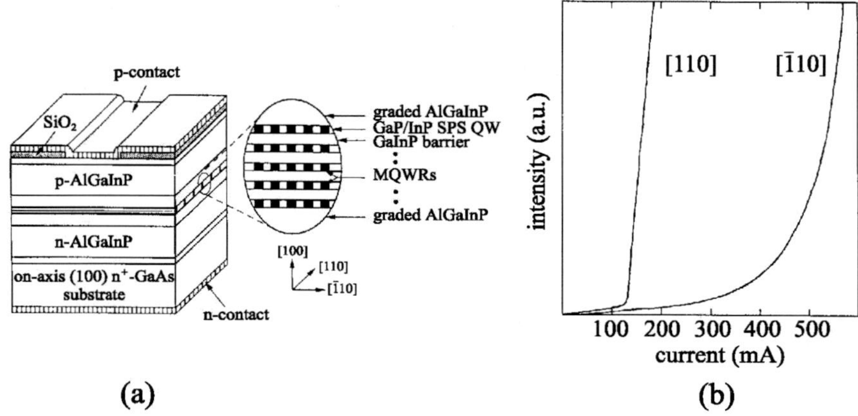


**Fig. 3.6.** (**a**) Schematic illustration of a stripe geometry $Ga_xIn_{1-x}P/Al_{0.15}Ga_{0.35}In_{0.5}P$ graded-index separate-confinement-heterostructure MQWR laser grown by MBE on (100) on-axis GaAs substrate. The MQWRs are aligned in the $[\bar{1}10]$ direction. (**b**) Light output power versus pulsed injection current at 77 K for AlGaInP MQWR laser diodes with contact stripes aligned in the [110] and in the $[\bar{1}10]$ directions. The spontaneously formed MQWRs are aligned in the $[\bar{1}10]$ direction in these lasers (taken from [3.11])

stripe direction was also observed. Figure 3.6 shows the laser structure and its characteristics.

A set of techniques has been developed and introduced into technological practice in order to produce by epitaxial crystallization highly perfect layers and layered structures, made of semiconductors, dielectrics, and metals, which are used as functional elements in different electronic devices. The most frequently used epitaxial growth techniques are: solid phase epitaxy (SPE), liquid phase epitaxy (LPE), vapor phase epitaxy (VPE), especially metalorganic vapor phase epitaxy (MOVPE), and molecular beam epitaxy (MBE) [3.24]. The basic concepts defining these techniques as well as the specific details of their different modifications are presented and discussed in Chaps. 4–8 on selected examples of material systems grown by epitaxy. The discussion will be sufficiently fundamental in the treatment of the subject. However, it should enable the reader to understand easily the future publications concerning different epitaxial growth techniques, when referring to the principles presented here.

# Part II

# Technical Implementation

# 4. Solid Phase Epitaxy

Solid phase epitaxy (SPE) occurs when a metastable amorphous layer in contact with a single crystal template crystallizes epitaxially in the solid state by the rearrangement of atoms at the interface between the two phases [4.1]. The regular array of atoms on the crystalline side of the interface serves as a template for the layer-by-layer addition of atoms from the disordered amorphous material to the ordered crystalline solid. The amorphous-to-crystal (a/c) transformation occurs solely in the solid phase and may be induced either by heating or by ion bombardment.

SPE occurs under conditions where the atomic mobility is comparatively low, and movement of atoms and changes in bonding configurations are largely limited to the region within a few bond lengths of the interface. Therefore, the structure of the amorphous phase, the extent to which the short-range order in the amorphous material approximates the structure of the crystal lattice, and the type of bonding, e.g., covalent or ionic, influence strongly the characteristics of the SPE process in a given material system. In materials that undergo SPE, the amorphous phase is metastable, and the free energy of the system changes abruptly upon crystallization.

Essentially all of the existing models of SPE have been developed within the framework of a model of the c/a interface that is derived from a continuous random network description of a-Si. A fundamentally different model for covalently bonded amorphous semiconductors, which involves paracrystalline clusters, has been proposed, and an alternative model for SPE involving rotational alignment of these clusters and their fusing to the crystalline substrate has been suggested, too [4.2].

Differences between the basic crystallization mechanisms in SPE of elemental and compound materials are not surprising in view of the fact that SPE in compound materials not only involves reorganization of atoms at the surface, but also requires that the cations and anions become located on appropriate lattice sites. Clearly this raises an additional complication in the atomic reconfiguration at the interface that does not exist for the elemental semiconductors.

It should be pointed out that progress toward understanding SPE has been, so far, limited almost exclusively to interpretation of experiments that probe the overall kinetics of the crystallization process. This has led to the development of empirical models which treat the fundamental steps in the SPE mechanism in only a semi-quantitative way. However, one may expect that the extension of molecular dynamics simulation methods to description of amorphous materials, together with development of advanced diagnostic techniques capable of probing the structure and properties of the c/a interface can have great potential for improving the current understanding of the fundamental solid phase crystallization mechanisms.

In discussing here SPE, we will not go into the theory of this phenomenon, referring the reader, instead, to more extended reviews on this subject [4.1, 2].

## 4.1 Technological Procedures

Figure 4.1 schematically depicts the SPE crystallization process in an amorphous layer in direct contact with a heated substrate crystal. Heating induces reordering of atoms at the c/a interface, with the resulting propagation of the interface towards the surface. Following SPE, the structure of the crystallized layer is indistinguishable from that of the underlying substrate.

SPE crystallization has been conducted using amorphous layers formed either by high-dose ion implantation of a crystalline substrate or by vapor phase deposition onto a single crystal surface. In both cases, the single crystal substrate serves as a pre-existing two-dimensional "nucleus" which grows in a layer-by-layer mode as the amorphous phase is consumed during SPE. The velocity of the moving interface (the growth rate) exhibits in this process an Arrhenius temperature dependence:

$$v = v_0 \exp(-E_\mathrm{a}/kT), \tag{4.1}$$

where $E_\mathrm{a}$ is the activation energy of the SPE process. Growth rates of about $1\,\text{Å}\mathrm{s}^{-1}$ are observed at temperatures of roughly one-half to one-third the melting point [4.2]. SPE can occur even when both participating phases are composed of different materials. This is often used for producing so-called buffer layers in heteroepitaxy of lattice mismatched "$e$"/"$s$" material systems (for more details see Sect. 4.3.2).

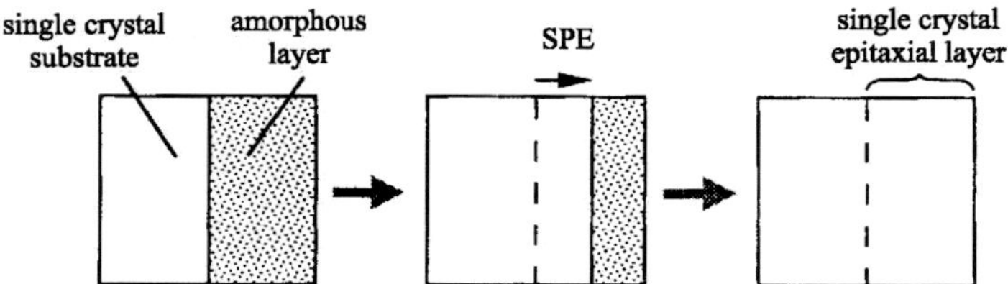


**Fig. 4.1.** Solid phase epitaxy of an amorphous thin layer on a single crystal substrate (taken from [4.2])

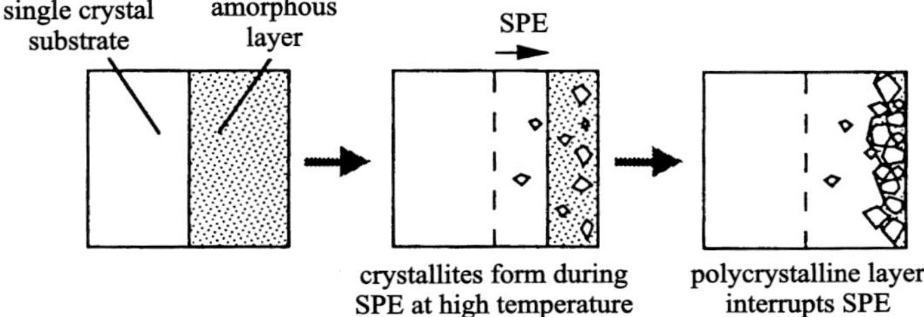


**Fig. 4.2.** Random nucleation and growth can interfere with SPE at high temperatures (taken from [4.2])

During the crystallization of an amorphous layer by SPE it is possible for nucleation and growth of randomly oriented crystallites to occur in the amorphous phase ahead of the advancing c/a interface, thereby interrupting the regular progression of the epitaxial growth front, as illustrated in Fig. 4.2. Random nucleation and growth occur when thermal fluctuations cause crystalline nuclei of various sizes to form within the amorphous matrix. The newly formed crystallites expand then by addition of atoms from the amorphous phase to the crystal. Since the kinetics of planar SPE and random nucleation (with subsequent growth) are different, the extent to which either component will dominate depends strongly on the crystallization temperature and thickness of the amorphous layer. However, in the presence of a single crystalline substrate, SPE is generally the dominant process at all but the highest temperatures.

Different experimental methods have been used for accomplishing SPE in a variety of material systems. An extended discussion of this problem has been presented in [4.2]. We will follow this treatment, when discussing here SPE.

The most basic, most commonly employed, configuration of the SPE growth system, i.e., the planar sample configuration, is shown in Fig. 4.3a. It

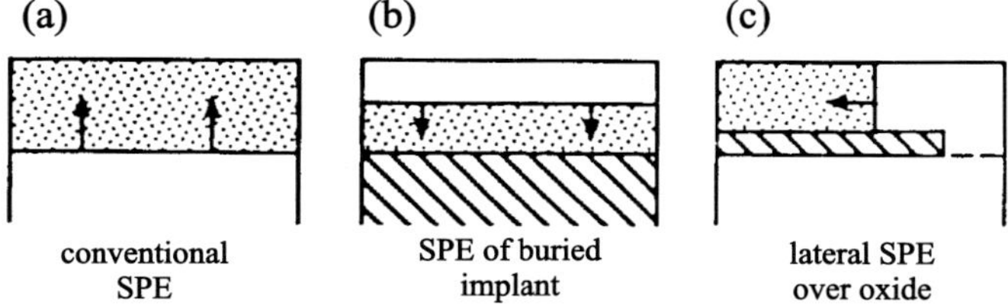


**Fig. 4.3.** Typical sample configurations employed for studies of solid phase epitaxy. (**a**) Amorphous film in direct contact with crystalline substrate – SPE growth toward surface; (**b**) buried amorphous layer formed by channeled ion implantation – SPE growth downward toward heterointerface; (**c**) amorphous layer deposited on patterned substrate – vertical SPE in seeded region followed by lateral SPE over insulator (taken from [4.2])

consists of a continuous amorphous deposit (most often a layer) in direct contact with an underlying single crystal (the substrate). Typical layer thickness, encountered in SPE, range from several thousand Å to several μm. The layer is uniform in thickness, and during growth its thickness decreases with time. Under nominal conditions the c/a interface remains flat and parallel to the surface. Cross-sectional TEM has shown that the transition from amorphous to crystal can occur in as few as 1–4 atomic layers, and in well-behaved specimens this planarity can be maintained through several microns of growth, as it occurs in a-Si on Si(100) [4.3]. Other sample configurations that are encountered in certain specialized embodiments of SPE growth are illustrated in Figs. 4.3b and c. The recrystallization of a buried amorphous layer formed by channeled ion implantation is shown in Fig. 4.3b. A thin single crystal layer remaining at the surface after channeled ion implantation serves as a template for downward regrowth of the buried amorphous layer toward the c/a interface. Use of SPE to recrystallize a buried amorphous layer formed by ion implantation has often been employed to improve the crystal quality of silicon-on-saphire (SOS) structures [4.4].

The sample configuration illustrated in Fig. 4.3c has been used when SPE is accomplished by lateral overgrowth. This technique, known as "epitaxial lateral overgrowth" (ELO), was applied for the first time applied in Si technology in the earlier 1980s [4.5–7]. An a-Si layer is vacuum deposited on a c-Si substrate containing a patterned $SiO_2$ overlayer. Holes in the surface oxide allow contact to be made between the deposited a-Si and the single crystal substrate. Vertical SPE growth originating at the c/a interface in the "seed" regions produces columns of single crystal material which then spread by lateral SPE growth over the surface of the oxide-covered region. This process, when repeated, leads to multilayer silicon-on-insulator (SOI) structures [4.2].

There are three important stages in SPE. The first is the formation of an amorphous phase (usually a layer) on the top of the crystalline substrate. The second is the programmed heating of the a/c system in order to get the SPE growth in a well controlled mode. The third phase, i.e., the measurement of the growth rate, is indispensable for gaining the required control over the SPE growth process. Let us discuss the three items in brief, on the example of the a-Si/c-Si material system.

### 4.1.1 Formation of the Amorphous Phase

As already mentioned, high dose ion implantation and vacuum deposition are the most commonly used methods to prepare the amorphous phase suitable for SPE. However, other methods like, e.g., rapid quenching of a molten layer, formed by ultrashort pulsed laser irradiation, have also been employed for formation of very thin amorphous layers [4.8]. For the sake of brevity, we will concentrate here only on the first two methods. Let us start with the subject of high dose ion implantation.

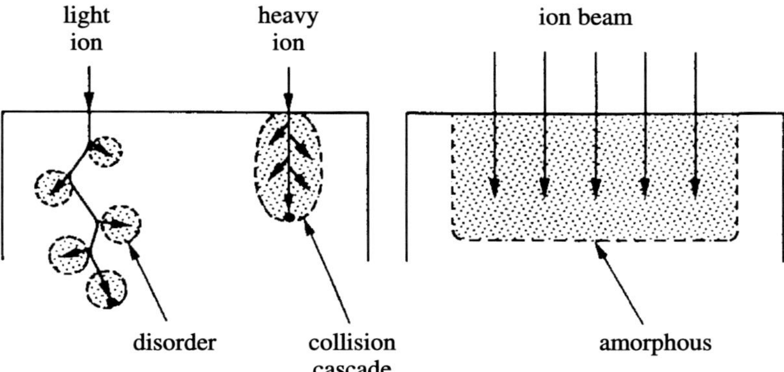


**Fig. 4.4.** Amorphous layer formation by ion implantation. Disorder is produced along the ion track. Overlapping damage zones produce a continuous amorphous layer (taken from [4.11])

Amorphous layer formation by ion implantation involves the interaction of energetic ions (typically 30 keV – 5 MeV) with the atoms of a single crystal substrate [4.9–11]. As depicted in Fig. 4.4, ion implantation creates disorder when the incident ion transfers sufficient energy to an atom in the crystal to displace that atom from its lattice site. Additional damage occurs as a result of the collision cascade created when the displaced atoms undergo collisions with atoms on lattice sites. When the mass and fluence of the incident ions are sufficiently high (e.g., $> 10\,\mathrm{amu}$ and $\approx 10^{14}\,\mathrm{cm}^{-2}$ for amorphization of Si) the collision cascades produce overlapping disordered zones which eventually result in the formation of a continuous amorphous layer in the implanted region.

Creation of a pure amorphous layer in elemental semiconductors is most conveniently accomplished using self-ion-implantation (e.g., $^{28}\mathrm{Si}^{+}$ for amorphization of Si) into single crystal substrates. Dynamic annealing of the amorphous layer due to heating of the sample by the ion beam is minimized by using reduced ion beam currents ($< 1\,\mu\mathrm{A\,cm}^{-2}$) and by maintaining the sample at cryogenic temperature (77 K). Using MV ion accelerators and high purity ion sources it is possible to produce continuous a-Si layers that are $>5\,\mu\mathrm{m}$ thick and which have impurity concentration below about $5\times10^{15}\,\mathrm{cm}^{-3}$ [4.12].

Ion implantation can also be used to form an amorphous layer suitable for SPE in III–V semiconductor compounds, as well as in metal oxides [4.2]. However, the formation of pure amorphous layers in compound semiconductors is complicated by the unwanted mixing of the atoms during ion implantation [4.9]. This can result in local regions of non-stoichiometry within the amorphous layer, and as a consequence, can cause breakdown of the planar c/a interface.

**Table 4.1.** $^{28}Si^+$ doses and ion energies used to form a 4.3 μm thick continuous amorphous layer in a Si(100) substrate held at 77 K during implantation (taken from [4.2])

| a-Si thickness (μm) | Energy of the $^{28}Si^+$ ions (MeV) | Ion dose ($10^{15}$ cm−2 ) |
|---|---|---|
| 0.9 | 0.5 | 5 |
| 1.4 | 1.0 | 5 |
| 2.2 | 2.0 | 5 |
| 3.0 | 3.5 | 6 |
| 3.7 | 5.0 | 7 |
| 4.3 | 6.5 | 8 |

The thickness of the amorphous layer created by ion implantation depends on the range (energy) of the ions in the specimen. In most semiconductors the formation of continuous amorphous layers thicker than a few thousand Å requires the use of several implants at different energies and doses to create a uniform damage distribution and to ensure that the damage density is above the threshold for amorphization in all portions of the layer. An example of a typical ion implantation schedule appropriate for the formation of a 4.3 μm thick continuous amorphous layer is given in Table 4.1.

Amorphous layers can also be formed by a variety of vacuum deposition methods, including evaporation, sputter deposition, radio frequency discharge and chemical vapor deposition [4.2]. In order to produce an amorphous (rather than polycrystalline) film, the substrate must be maintained at temperatures sufficiently low to prevent or inhibit the lateral migration of atoms to lattice sites on the surface. For the formation of amorphous layers appropriate for SPE a single crystal substrate is used, and great care must be taken to ensure that the surface of the substrate is free of contaminants and that unwanted impurities do not enter the amorphous film during the deposition process. This normally requires the use of ultrahigh vacuum (UHV) procedures and *in situ* substrate cleaning techniques [4.13].

Vacuum deposition is advantageous for generating atomically abrupt interfaces between the substrate and amorphous layer and for fabricating structures containing layers with different compositions. Examples of the latter include SOI structures like a-Si films deposited on an $SiO_2$ layer, and IV–IV and III–V multilayer structures.

There is a qualitative difference in the microstructure of amorphous films prepared by vacuum deposition and by ion implantation. With the exception of the damaged crystalline material produced at the end-of-range region, ion implanted layers are homogeneous and uniformly disordered throughout the implanted zone. In contrast, UHV deposition produces an amorphous layer that contains voids and low-density regions [4.13]. This is a natural consequence of the self-shadowing of atoms which occur when the surface-atom mobility is low. In addition, tensile stress can lead to cracking and delamination of the film from the underlying crystal. Amorphous Si films deposited onto atomically clean substrates by electron beam evaporation under UHV conditions delaminate when the amorphous layer thickness exceeds $\approx 2\,\mu m$ [4.2].

### 4.1.2 Programmed Heating of the a/c System

There are different heating methods which are usually applied for the SPE growth. These are: oven heating, resistively heated block, pulsed laser irradiation plus heated block, and pulsed electron beam. Let us discuss them on the example of an a-Si layer [4.2].

Oven heating enables the most accurate determination of annealing temperature and is the most uniform heating method. However for SPE growth its use is limited to low temperatures because relatively long times ($> 5\,\text{min}$) are required to reach steady state. As an alternative, placing the sample on a resistively heated block with a vacuum hold-down permits heating times on the order of 1 s to be achieved, allowing SPE with a-Si to be conducted up to approximately 700°C. Experiments at higher temperatures require the use of a pulsed, high power heating source, which can heat the sample at a much faster rate. Sources in this category include flashlamps, continuous wave and pulsed lasers, and electron beams. It should be noted that laser sources with pulse widths less than about 1 µs are not appropriate for SPE processes because the amount of growth that occurs during each pulse is negligible.

Regardless of the heating method that is employed, an accurate determination of temperature is imperative if meaningful information is to be obtained about the kinetics of the solid phase crystallization process. In the low temperature regime, temperature measurements are relatively straightforward and are usually performed using thermocouples or pyrometers calibrated against known temperature standards. Temperature measurements during SPE at high temperatures is more difficult since the heated area is small (typical laser spot diameter is 100 – 200 µm) and the annealing times are short ($< 1\,\text{s}$). In some cases the temperature never reaches steady state during the measurement, so it is necessary to explicitly monitor the variations of temperature with time during the experiment. One technique that has been used successfully for real-time temperature measurement during laser heating of a-Si is based on measurement of the surface reflectivity at the center of the laser heated spot [4.14]. Since the index of refraction of Si varies nearly linearly with temperature, the sample temperature can be derived from measurements of the reflectivity [4.2]. When knowing the temperature of the heated sample in real time the heating process can be easily programmed by electronic means.

## 4.2 Measurement of the Growth Rate

In SPE, the growth rate is defined to be the speed of the moving c/a interface. Therefore, a determination of the growth rate can be made from measurements of the thickness of the remaining a-Si layer as a function of annealing time. Ideally, during such a measurement, the temperature would be instantaneously raised to the desired value and maintained at that value for the duration of the measurement procedure. This condition can readily be achieved when the crystallization rate is much slower than the time to reach a steady-state temperature, as is the case for most oven annealing procedures. However, at high temperatures the crystallization times are often very short, and some variation of temperature during the measurement is unavoidable. This must be taken into account in interpreting the results of SPE growth kinetics studies. By using rapid heating techniques such as laser or electron beam irradiation coupled with real-time measurements of the c/a interface depth, it is possible to control SPE and competing processes such as random nucleation and growth at temperatures approaching the melting point [4.2].

For the real-time determination of the a-Si thickness during SPE, the use of time resolved reflectivity (TRR) [4.14], a technique based on reflection interferometry [4.15], has many advantages, including improved accuracy of the thickness measurement, ease of implementation, and adaptability to high temperature measurements. In addition to determining the growth rate with high accuracy, this method can be used to evaluate the extent to which processes such as nucleation and growth as well as impurity segregation compete with SPE. When coupled with laser based heating techniques, TRR can be used to measure the kinetics of both SPE and random nucleation and growth, up to temperatures approaching the melting point [4.2].

The use of TRR to monitor the growth rate in SPE is based on the interference between light reflected from the surface of the sample and light reflected from the moving c/a interface. Figure 4.5 schematically illustrates a simple embodiment of the TRR method in which SPE occurring in an a-

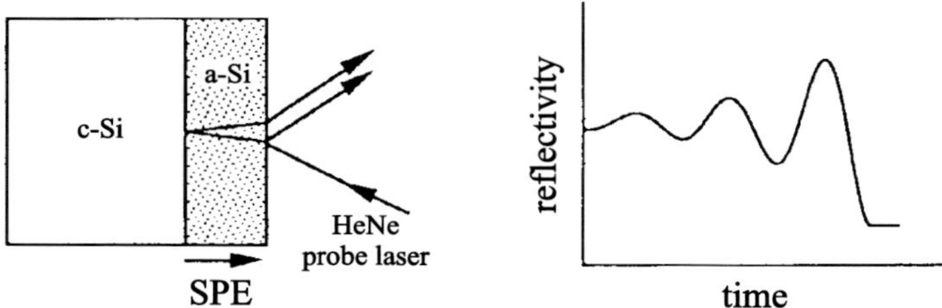


**Fig. 4.5.** Measurement of SPE growth rate by TRR. As the amorphous layer thickness changes, the reflectivity oscillates due to interference between beams reflected from the surface and from the c/a interface. Comparison with the theoretical expression yields the growth rate as a function of interface depth (taken from [4.2])

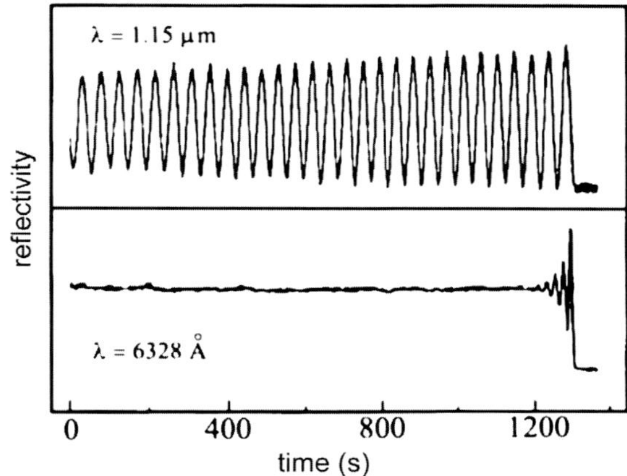


**Fig. 4.6.** Time resolved reflectivity data at $\lambda = 1.15$ μm and 6328 Å for SPE growth of 4.3 μm thick, ion implanted ($^{28}Si^+$) film at 625°C in vacuum (taken from [4.2])

Si film on a c-Si substrate is monitored using a low power HeNe laser. As the growth interface propagates toward the surface, interference produces an oscillation in the light intensity. Each time the a-Si thickness changes by $\lambda/2n$ ($n$: refractive index), the reflected light signal completes one interference cycle. Therefore, if the time between successive extrema is determined accurately, it is possible to directly calculate the speed of the interface motion. In principle, the depth resolution of this technique is on the order of 10 Å. It is also possible to obtain a continuous measure of the SPE rate as a function of c/a interface depth by comparing the measured reflectivity with the theoretical expression for reflectivity as a function of thickness given in Fig. 4.5. This method has been used extensively in studies on SPE.

Optical absorption in the a-Si layer causes the amplitude of interference oscillations to be reduced in thicker films. In a-Si, for example, absorption at the HeNe laser wavelength of 6328 Å limits the use of TRR to layers less than about 5000 Å. It is possible to employ a longer wavelength for which the amorphous layer is more transmissive, but there is an attendant sacrifice in depth resolution. Figure 4.6 shows TRR data simultaneously acquired at 6328 Å and 1.15 μm during SPE growth of a 4.3 μm thick a-Si layer at 625°C. Since there is a large difference in the absorption coefficients in a-Si at the two wavelengths ($1.2\times10^5$ cm$^{-1}$ at 6328 Å and $1.2\times10^3$ cm$^{-1}$ at 1.15 μm), the depth over which a usable TRR signal can be obtained is also different. For example, the TRR trace at the wavelength 6328 Å, exhibits oscillations in intensity only during the last 4000 Å of SPE growth. In contrast, the 1.15 μm data exhibit fully resolved oscillations over the entire 4.3 μm thick a-Si layer; the equal spacing between the interference peaks indicates that the SPE growth rate is constant throughout the layer.

## 4.3 Application Areas

The first reports on SPE in silicon [4.16] and gallium arsenide [4.17] stimulated a wide range of experimental studies devoted to the characterization of the SPE crystallization behavior in elemental and compound semiconductors. The ability to convert an amorphous layer to a highly ordered single crystal by SPE has been exploited in many technological applications. Some examples include:

(i) recovery of ion implantation damage produced during, (a) introduction of electrically active impurities into elemental and compound semiconductors, and (b) modification of the optoelectronic and tribological properties of metal oxides
(ii) preparation of semiconductor thin layers with dopant concentrations in excess of the solid solubility limit,
(iii) growth of buffer layers on single crystalline substrates for improving heteroepitaxy of lattice mismatched heterostructures,
(iv) low-temperature homoepitaxy and heteroepitaxy for high performance electronic and optoelectronic devices,
(v) formation of SOI structures for applications in multilevel and three-dimensional microelectronics,
(vi) growth of silicide layers for electrical contacts and Schottky barriers in Si-based devices [4.2].

All of the listed application areas of SPE are important from the point of view of the current technological practice in electronics. However, the items (ii) and (iii) seem to be most interesting for understanding the role of SPE in present day technology of electronic devices. Let us then discuss them in more detail.

### 4.3.1 Growth of Highly Doped Epilayers

The fundamental problems related to this subject concern the influence of impurities present in the amorphous layer on SPE growth rate, and the incorporation feasibility of the impurities into the crystalline phase. Following [4.2], we will discuss these problems on the example of the a-Si/c-Si materials system.

It is well established that doping impurities (substitutional atoms [4.18]), such as B (acceptor in Si), P and As (donors in Si), increase the SPE rate relative to the intrinsic value, i.e., the growth rate occurring in pure, intrinsic, amorphous phase, whereas non-doping impurities (interstitial atoms [4.18]), generally decrease the rate. However, the influence of impurities on the SPE process can be quite complex, since the rate-altering behavior is usually dependent on both the impurity concentration and the crystallization temperature. For example, impurities such as As and P that increase the rate of SPE when they are present at low concentrations can retard interface motion when they are present at high concentrations. Likewise, certain impurities cause the interruption of epitaxy at low temperatures but not at high temperatures. In other cases, increasing the temperature causes SPE to be disrupted by impurity-enhanced nucleation and growth. The behavior becomes even more complicated when two or more impurity species are present simultaneously in the amorphous layer. The large rate enhancements that are observed when doping impurities are present in an amorphous layer have been the subject of numerous studies. Basically, this effect can be discussed with respect to two separate concentration regimes. At low concentration, $\approx 10^{18} - 5\times10^{19}\,\mathrm{cm}^{-3}$, the SPE rate varies approximately linearly with dopant concentration. At higher concentrations the variation is non-linear and SPE eventually breaks down altogether.

Additional insight into the effect of doping impurities on the SPE rate was provided by the experimental fact that although n- and p-type impurities enhance the rate when acting separately, the simultaneous presence of both impurities at comparable concentrations in the amorphous layer acts to compensate the effect of either dopant acting alone and drives the SPE rate back to the intrinsic value [4.19, 20]. This dopant compensation effect has been shown to hold over a wide range of temperatures (550 – 900°C) in samples containing B and P and has been observed using other p- and n-type dopant combinations as well [4.2]. The observation of compensation provided the first evidence that the effect of group-III and group-V impurities on SPE in a-Si is primarily electronic in nature.

A large number of experimental studies have been conducted to understand the role of dopant atoms in the SPE process. Characterization of the rate enhancement occurring at concentrations for which the crystallization behavior is unaltered by competing effects such as impurity segregation is particularly important, since this information can provide unique insight into the fundamental mechanism of the SPE process. Virtually all early studies focused on the qualitative aspects of the rate enhancement induced by selected impurities at higher concentrations, and most of the available data were not sufficiently precise to allow the exact form of the concentration dependence to be established.

In extensive qualitative studies on impurity-induced rate enhancement, Walser and co-workers [4.21–23] showed that for concentrations less than about 0.1 at% the SPE rate increases linearly with dopant concentration. Both the pre-exponential factor and the apparent activation energy in (4.1) describing the SPE rate were found to be concentration dependent. They showed that in the low-concentration regime the SPE rate can be expressed in normalized form as

$$v/v_i(T) = 1 + N/N_i(T), \tag{4.2}$$

where $v$ is the SPE interface velocity, $v_i(T)$ is the intrinsic SPE velocity in undoped material, $N$ is the impurity concentration at the c/a interface, and

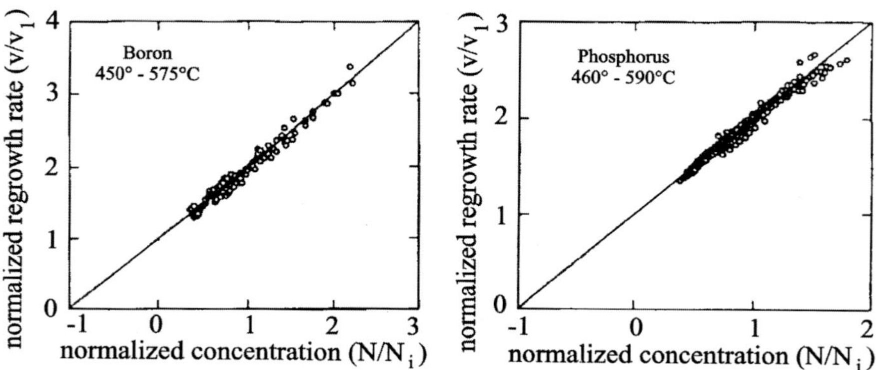


**Fig. 4.7.** Dependence of the normalized Si SPE rate on normalized impurity concentration for boron, and phosphorus. Straight lines are least squares fits of the data to equation (4.2) (taken from [4.23])

$T$ is the crystallization temperature. $N_i$ is a temperature dependent factor given by

$$N_{\mathrm{i}}(T) = N_0 \exp(-\Delta E/kT), \tag{4.3}$$

with $N_0 \approx 2.5 \times 10^{21}\,\mathrm{cm}^{-3}$ and $\Delta E = 0.25 - 0.35\,\mathrm{eV}$ for group-III and group-V impurities in a-Si. The form of this expression is significant. It suggests that impurity rate enhancement is an additive, not a multiplicative effect. Recognition of this point led to the development of an empirical model for impurity enhancement of the SPE rate [4.23] based on the participation of both neutral and charged defects, wherein the neutral defects are responsible for intrinsic SPE in undoped material, and the charged defects are responsible for the enhancement term $N/N_i$ that appears in (4.2). The factor $N_i$ contains an activation energy, $\Delta E$ (see (4.3)), which represents the reduction in the activation barrier for SPE due to the presence of charge on the defect.

Experimental data on the effect of doping impurities, taken from the work of Walser and Jeon [4.23] are presented in Fig. 4.7. These data were obtained during annealing in a vacuum oven at temperatures of $450-575°\mathrm{C}$, in $20-25°\mathrm{C}$ increments. Time resolved reflectivity was used for *in situ* determination of the SPE rate with a high degree of precision. Each sample contained a single energy implant of the species of interest into a layer pre-amorphized by $^{28}\mathrm{Si}^{+}$ implantation. The SPE rate as a function of the c/a interface depth was compared with secondary ion mass spectrometry (SIMS) measurements of the impurity concentration depth profile to arrive at the dependence of growth rate on concentration. For the data shown in Fig. 4.7, the lowest usable concentration was approximately $3\times10^{18}\,\mathrm{cm}^{-3}$ and the upper limit of the linear regime occurred at roughly $3-5\times10^{19}\,\mathrm{cm}^{-3}$ for the species studied.

The experiments on growth rate for the low-concentration regime are important for understanding of SPE peculiarities; however, their are not related

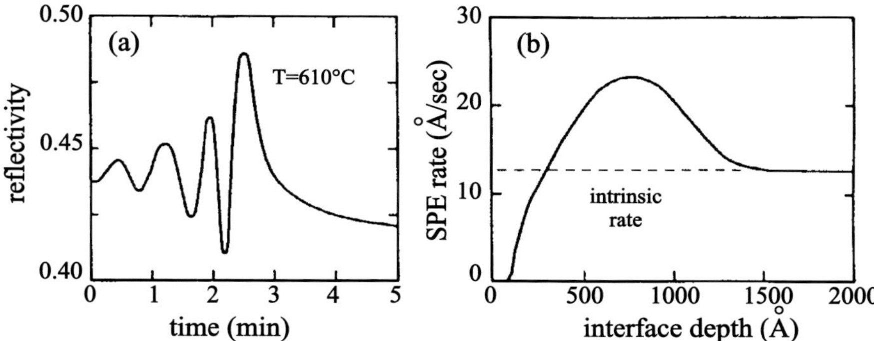


**Fig. 4.8.** Rate enhancement and retardation during SPE of a 2500 Å thick pre-amorphized a-Si layer implanted with indium ($C_{\text{peak}}$ =2.7×10$^{20}$ cm$^{-3}$). (**a**) TRR data obtained during SPE at 610°C. (**b**) Rate versus depth derived from data shown in (**a**). Dashed line gives the intrinsic SPE rate (taken from [4.2])

directly to understanding of the limits for SPE to occur in the high doping concentration regime.

The recrystallization behavior is markedly altered when the concentration approaches or exceeds the solid solubility limit. These changes are due to segregation of the impurity into the amorphous phase and the formation and propagation of defects at the c/a interface. Accumulation of impurity atoms at the interface retards SPE and eventually prevents further crystallization when the impurity concentration exceeds a critical value. This is illustrated by SPE in a-Si layers containing indium at concentration of about 3×10$^{20}$ cm$^{-3}$ (0.6 at %). Figure 4.8a shows TRR data obtained during crystallization of an $In^+$-implanted film at 610°C. The Si (100) substrate was previously amorphized by $Si^+$ implantation to a thickness of about 2500 Å. The $In^+$ was implanted at an energy of 125 keV and a fluence of 1×10$^{15}$ cm$^{-2}$. The peak of the In profile was about 550 Å from the surface, and the peak concentration $C_{\text{peak}}$ =2.7×10$^{20}$ cm$^{-3}$. By using a pre-amorphized layer the rate changes caused by the In could be compared directly with the intrinsic rate in the same specimen.

The data in Fig. 4.8b show that during crystallization at 610°C the interface moves through the intrinsic region at a nearly constant rate of 15 Ås$^{-1}$. As the interface moves into the region containing the highest In concentration, a rate increase is observed which is consistent with results obtained for other doping impurities in a-Si. However, a dramatic rate decrease occurs as the interface passes through the peak of the In profile and approaches the surface. Rate retardation in $In^+$-implanted layers has been studied in the low-temperature regime by numerous investigations and it has been shown that the rate reduction is caused by segregation of the impurity as the interface propagates through regions of high concentration [4.2]. The progressive accumulation of impurities at the c/a interface impedes interface motion to the extent that it can cause complete cessation of epitaxial growth. At In concentrations exceeding about 0.6 at % random nucleation and growth occurs, and a polycrystalline layer forms. It has been shown by Nygren et al. [4.24] that this amorphous-to-polycrystalline transformation can occur at temperatures as low as 200°C below the temperature for which random crystallization is observed in intrinsic material. This low-temperature a-to-p transformation has been described in terms of a melt-mediated process in which In diffuses and precipitates in a-Si to form an In-rich melt, which in turn promotes crystallization via dissolution and resolidification of a-Si [4.24].

Although epitaxy can be interrupted by segregation of impurities during low-temperature growth, the competition between SPE and impurity segregation at the c/a interface depends strongly on temperature, with SPE dominating during high temperature growth. It has been found that when samples of the type described in Fig. 4.8 are rapidly heated, using e.g., a cw laser, to temperatures approaching 900°C, essentially no retardation of epitaxial growth occurs as the interface approaches the surface of the sample. Microbeam Rutherford back scattering (RBS) analysis of the recrystallized laser irradiated spots indicates that dopant redistribution is minimal under these heating conditions [4.2]. The results show that high temperature annealing is much more effective than conventional furnace heating for removing ion implantation damage and activating dopant atoms in $In^+$-implanted layers.

During SPE, impurities can be incorporated at concentrations exceeding the equilibrium solid solubility limit. This important feature of SPE is summarized in Table 4.2, showing the data on the metastable solubility limits of dopant atoms achieved by SPE in ion-implanted a-Si. The equilibrium solid solubility limits for c-Si are compared with the metastable concentrations that have been achieved during SPE at $T < 600°C$.

Subsequent high temperature annealing of recrystallized samples containing doping impurities at the (metastable) concentrations shown in Table 4.2 results in a decrease of the substitutional atom fraction and attendant precipitation of the impurity species. The reader is referred to the review paper [4.2] and references cited therein for detailed information concerning the behavior of specific impurities and more complete exposition of the complex behavior that can occur during SPE in the presence of doping impurities at concentrations in excess of equilibrium solid solubility limits.

### 4.3.2 Growth of Buffer Layers

Dislocation densities in highly mismatched heterostructures may be considerably reduced by growing on the mismatched substrate a so-called buffer layer for the final epitaxial growth of the desired epilayaer. A buffer layer is, thus, an intermediate single- or multilayer material system which matches the substrate and the epilayer, which are highly lattice mismatched, one to the other. Recently, it has been shown that SPE is a useful technique for

**Table 4.2.** Equilibrium solid solubility limits and measured metastable solubility limits for selected impurities in Si, related to SPE processes at T $<$ 600°C (taken from [4.2])

| Impurity species | Maximum equlilbrium solubility (in $cm^{-3}$) | Metastable solubility limit (in $cm^{-3}$) |
|---|---|---|
| Ga | $4.5 \times 10^{19}$ | $2.5 \times 10^{20}$ |
| As | $1.5 \times 10^{21}$ | $9 \times 10^{21}$ |
| In | $8 \times 10^{17}$ | $5 \times 10^{19}$ |
| Sb | $7 \times 10^{19}$ | $1.3 \times 10^{21}$ |
| Tl | $< 5 \times 10^{17}$ | $4 \times 10^{19}$ |
| Pb | $< 5 \times 10^{17}$ | $8 \times 10^{19}$ |
| Bi | $8 \times 10^{17}$ | $9 \times 10^{19}$ |

growing buffer layers in some III–V [4.25] or II–VI [4.26, 27] heterostructure material systems.

Let us concentrate here only on the growth of an AlN buffer layer for heteroepitaxy of gallium nitride on sapphire ($\alpha$-$Al_2O_3$) [4.25]. GaN is one of the most promising semiconducting materials for optoelectronic devices in the region from blue to ultraviolet light, because it has a direct energy band gap of 3.39 eV at room temperature. It is extremely difficult to grow large scale bulk single crystals of GaN because of the high equilibrium pressure of nitrogen at growth temperatures over 1000°C. Therefore, vapor phase epitaxial methods, such as metalorganic vapor phase epitaxy (MOVPE) and hydride vapor phase epitaxy (HVPE) have been conducted using dissimilar substrates like sapphire or silicon for growing GaN single crystal films. However, it had been fairly difficult to grow high quality epitaxial films, in particular, with a smooth surface free from cracks, because of the large lattice mismatch and the large difference in the thermal expansion coefficients between GaN and those substrates as shown in Table 4.3.

To solve the problem, the deposition of a thin buffer layer before GaN growth was proposed [4.28]. In the case of sapphire substrates, surface morphology as well as electrical and optical properties of GaN films have been improved remarkably by preceding deposition of AlN [4.29] or GaN [4.30] buffer layer before MOVPE growth of GaN films or by preceding deposition of ZnO [4.27] buffer layer before HVPE of GaN. The improvement in matching of lattice constants, as well as thermal expansion coefficients gained by introducing a buffer layer into the heterostructure is evident from data shown in Table 4.3.

When growing GaN on sapphire (0001) surface by MOVPE a thin AlN buffer layer of about 50 nm thickness was deposited at about 600°C before starting the growth of GaN. Then, the substrate temperature was raised to a growth temperature of about 1040°C, and a GaN film was grown. The surface morphology as well as the electrical and optical properties of GaN film

**Table 4.3.** Lattice mismatch $\Delta a/a$ and differences in thermal expansion coefficients $\Delta\alpha/\alpha$ (taken from [4.25])

| GaN on substrate | $\Delta a/a$ (%) | $\Delta\alpha/\alpha$ (%) |
|---|---|---|
| GaN(0001)/Sapphire (0001) | +13.9 | –34.2 |
| GaN(0001)/Si(111) | –20.4 | +55.3 |
| **GaN on buffer layer** | $\Delta a/a$ (%) | $\Delta\alpha/\alpha$ (%) |
| GaN(0001)/AlN(0001) | +2.5 | +5.5 |
| GaN(0001)/ZnO(0001) | –1.9 | +1.6 |
| GaN(0001)/3C-SiC(111) | +3.4 | +48.1 |

have been remarkably improved by the preceding deposition of a thin AlN layer as a buffer [4.31]. From electron diffraction spots the crystallographic relations between GaN, AlN and $\alpha$-$Al_2O_3$ were found to be [0001]GaN $\|$ [0001]AlN $\|$ [0001] $Al_2O_3$ and $[1\bar{1}00]$GaN $\|$ $[1\bar{1}00]$AlN $\|$ $[1\bar{1}00]Al_2O_3$. The sharp spots of AlN indicate that this layer is crystallized epitaxially on the sapphire substrate during raising the temperature and/or the growth of GaN. The relations given above agree with those occurring in GaN grown directly on (0001) $Al_2O_3$ without the AlN buffer. Observation on initial growth stage and cross-sectional TEM images of GaN films revealed the growth process of GaN without and with an AlN buffer layer, which are shown schematically in Fig. 4.9. In the case of a GaN film without an AlN buffer layer the nucleation density of GaN on the sapphire substrate is low as shown in Fig. 4.9a. Many hexagonal GaN columns with different sizes and heights are formed and they grow three-dimensionally, resulting in rough surface and many pits at their boundaries. Furthermore, many crystalline defects generate near the boundaries between GaN grains, which is caused by misorientation of each island.

The growth process of GaN film with AlN buffer layer is shown in Fig. 4.9b. The AlN buffer layer has amorphous-like-structure at the deposition temperature of 600°C, but when the temperature is raised to the growth temperature of GaN (1040°C), AlN is crystallized by SPE and then it exhibits the columnar structure. Since the AlN films were single-crystal-like from electron diffraction spots, orientations of AlN columnar crystals were found to be crystallographically well oriented to each other. Each GaN column is grown from a GaN nucleus which has been generated on the top of each columnar fine AlN crystallite. Therefore, it is thought that high density nucleation of GaN occurs owing to the high density of the AlN columns, as shown in Fig. 4.9b, compared with the nucleation density of GaN grown directly on the sapphire substrate.

The columnar fine GaN crystals increase accordingly in size during the growth and the crystalline quality of GaN is improved in this stage. It is thought that geometric selection of the GaN fine crystals occurs. Each fine

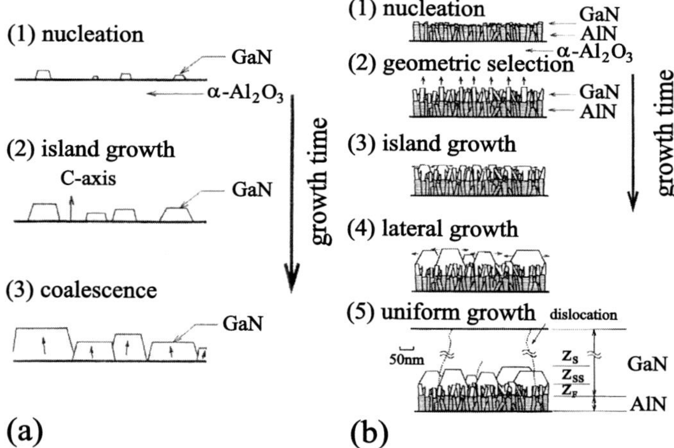


**Fig. 4.9.** Schematic diagrams showing the growth process of GaN as the cross-sectional views, (**a**) on a sapphire substrate, without an AlN buffer layer, (**b**) on an AlN buffer layer grown by SPE on the sapphire substrate (taken from [4.25])

crystal of GaN begins to grow along the c-axis, forming a columnar structure. Each column has a different random orientation and does not keep growing uniformly. The number of columns emerging at the front gradually decreases with the front area of each column increasing accordingly. Because only columns survive that grow along the fastest growth directions (i.e., the c-axis of each column is normal to the substrate surface), then all columns are arranged in the direction normal to the substrate surface, indicated by the arrows in Fig. 4.9b.

In the next stage the trapezoidal crystals are formed on the columnar crystal. As the front area of the column increases by the geometric selection, the c-face appears in the front of each column and trapezoid islands with the c-face are formed. These islands preferentially grow up to become larger trapezoidal crystals, which cover the minor islands nearby. The pyramidal trapezoidal crystals grow at a higher rate in a transverse direction, because the growth velocity of the c-face is much slower. After the lateral growth the islands repeat coalescence of each other very smoothly. Finally, since crystallographic directions of all islands agree well with each other, one can obtain a smooth GaN layer with a smaller number of defects as a result of the uniform coalescence. Thus, the uniform layer-by-layer growth occurs creating high quality GaN with low defect density and a smooth surface.

From the above results, the roles of the SPE-grown AlN buffer layer on GaN MOVPE growth are summarized as follows [4.25]:

(i) high density nucleation of GaN occurs on AlN columnar crystals,

(ii) geometric selection occurs among the GaN fine crystals which are able to arrange the crystallographic direction of GaN columnar crystals,

(iii) because of coalescence among GaN crystals which have been arranged in the crystallographic direction, crystalline defects near the interface between the grains are much reduced,

(iv) because of the higher lateral growth velocity of the pyramidal trapezoidal islands with c-face on the top, the surface is covered at an early growth stage and a smooth surface is easily obtained.

One may conclude that as a consequence of the above listed features of the GaN growth process, the SPE growth of the AlN buffer layer plays a crucial role for realization of uniform growth and for obtaining high quality GaN films with few defects even on a highly mismatched (13.8 %) substrate like $\alpha$-$Al_2O_3$.

# 5. Liquid Phase Epitaxy

Liquid phase epitaxy (LPE) is the deposition from a liquid phase (a solution or melt) of a thin single crystalline layer isostructural with the substrate crystal [5.1]. Usually LPE is performed using a solution as the liquid phase, because this is advantageous against the cases when a melt is used. In comparison to the growth from melts, growth from solutions:

(i) allows for epitaxy at lower temperatures,

(ii) enables better control of the amount of the crystalline phase grown or removed by dissolution,

(iii) leads to crystallization of layers with lower densities of defects, whether intrinsic (point defects) or extrinsic (e.g., impurities dissolved from crucible material),

(iv) makes it possible to grow layers sequentially from a series of solutions of differing composition, but with very similar and readily controlled liquidus temperature (liquidus is the border-line on the "temperature–concentration (composition)" phase diagram between the pure liquid region and the solid–liquid co-existence region (see Sect. 11.3)), and last but not least,

(v) allows for considerable reduction of the vapor pressure of volatile components of the compounds (e.g., P in the case of InP and As in the case of GaAs) by working at temperatures far below the melting point.

It is worth emphasizing that LPE operates near thermodynamic equilibrium, which results in low nutrient fluxes and makes mass transport and diffusion effects more important, while diminishing the importance of surface reaction and incorporation phenomena.

There are essentially two ways of promoting growth from a liquid solution on the solid substrate crystal: either the substrate may be immersed in the supersaturated solution, or this solution may be transported into the region of the crucible in which the substrate is located. There are, however, many variations of these basic processes differing more or less in the details [5.1–3]. The thermodynamic driving force for LPE is normally generated by cooling the substrate below the liquidus temperature on the relevant phase diagram. An LPE procedure consists, thus, of providing mechanisms for:

(a) creating supersaturation of the solution,

(b) introduction of the substrate platelet upon which the precipitation of the solid film from the solution will occur,
(c) controlling the morphology, uniformity and perfection of the epilayer, and
(d) removing from the melt the substrate with the epilayer grown on it.

Steps (b – d) are repeated for multiple layer growth, where layers of different composition and/or doping are grown sequentially on the same substrate.

## 5.1 Standard Techniques

Many of the growth systems used today were developed as a result of numerous experiments conducted with other designs. LPE is practiced in three major ways: tipping, dipping, and sliding. The first growth systems explored were the tipping device [5.4] and the dipping device [5.5]. Both systems have been found appropriate for growing single, nominally homogeneous layers of most of the semiconducting III–V and II–VI compounds important for

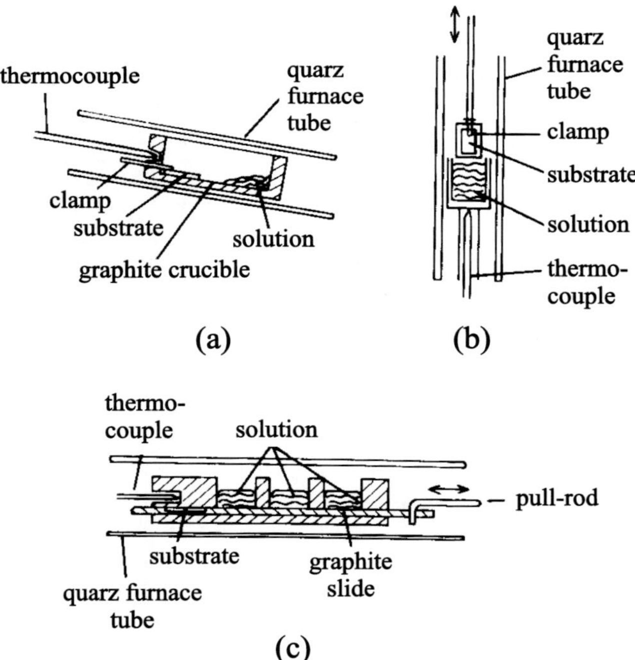


**Fig. 5.1.** Schematic illustrations of the basic LPE growth systems: (**a**) tipping furnace system [5.4], (**b**) vertical dipping system [5.5], (**c**) multicompartment slider boat system [5.6] (taken from [5.3])

optoelectronic devices. Application of these systems is obvious and straightforward (see Figs. 5.1a,b and 5.2). The original tipping furnace as well as the dipping apparatus have since been perfected into systems which permit simultaneous growth on many large-area substrates [5.3].

An important step in LPE technology was the introduction of multicompartment slider systems (Fig. 5.1c), which have been employed for the growth of multilayered structures [5.6]. These systems, and modified versions thereof, such as rotating drum devices [5.7] were developed mainly to prepare GaAs/AlGaAs heterostructures for biheterojunction lasers [5.8,9].

LPE is a fairly simple and quite flexible technique for the growth of semiconductor compounds and alloys. The quality of material obtained by LPE is excellent, often superior to the best obtained by MBE or VPE. This high material quality can be partially attributed to the stoichiometry of the solid,

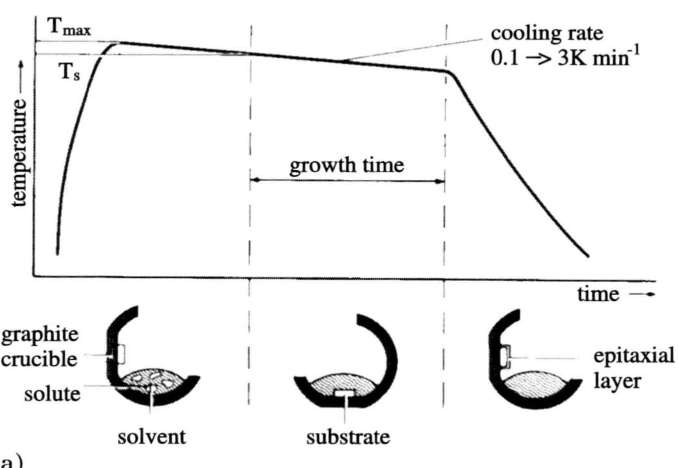

**Fig. 5.2.** (**a**) Temperature changes during one growth cycle in a tipping furnace (higher part); the substrate is exposed to the solution at temperature $T_S$. A schematic illustration of the rotated tipping crucible positions in the cycle is shown in the lower part. (**b**) Photograph of the rotated tipping crucible, supported by elastic quartz rods in the position when layer growth was terminated. A small droplet of solution residue is visible at the lower edge of the substrate (taken from [5.3])

which is always on the group III-rich boundary of the solidus [this is the border-line on the "temperature-concentration (composition)" phase diagram (see Sect. 11.2)], which causes reduction of the concentration of group III vacancies in the epilayer. In addition, the solution retains many harmful impurities because of their small distribution coefficients. For example, when Al is present in the solution, any oxygen present forms the very stable $Al_2O_3$ oxide, thus preventing oxygen incorporation into the solid. This "gettering" of impurities allows the routine growth of III–V semiconductors of very high purity with excellent electrical and optical properties. Thus, LPE has been the epitaxy technique of choice for producing material in the laboratory for research on new semiconductor devices.

However, LPE also shows serious drawbacks, especially from the point of view of producing abrupt interfaces. Large-scale production of materials e.g., the simultaneous growth on large numbers of substrates is technologically difficult by LPE, and the morphology is also notoriously difficult to control. Surface defects such as metal drops from incomplete liquid removal from the surface (see Fig. 5.2b), terraces, and so-called meniscus lines are commonly observed. In addition, thickness variation over several square centimeters have also been observed experimentally [5.10].

### 5.1.1 Transport Processes

From the basic point of view, LPE can be regarded as a form of solution growth (or dissolution) on a crystalline substrate that may consist of some of the elements that exist in the liquid phase. Therefore, one is concerned with the way in which atoms or growth units are transported through liquid and solid phases and how they attach to, or detach from the film surface. The process can be assisted by externally applied forces such as cooling programs, electric fields, thermal gradients, or convection. The case of near-stoichiometric binary compounds is easiest to understand, since then processes in the solid phase are insignificant. Of course, vacancy and impurity redistribution always occur, but presumably, they have only secondary effects on growth and dissolution kinetics.

The macroscopic theory of LPE growth and dissolution kinetics of binary compounds was fairly complete, at least for one-dimensional models, with the appearance of the work of Giess and Ghez [5.11]. To introduce proper terminology we will summarize the most important points of these models, here.

The left-hand part of Fig. 5.3 represents a part of the typical phase diagram for equilibrium states between a congruently melting solid $AB$ and a liquid phase consisting of $B$ atoms in a solvent $A$. Thus, at a given temperature $T$ a concentration $C_e$ of $B$ in the liquid is associated with a unique compound composition $C_s$. In other words, the solid–liquid equilibrium states are connected by the tie-line $S_e - L_e$. Points to the right of the liquidus, such

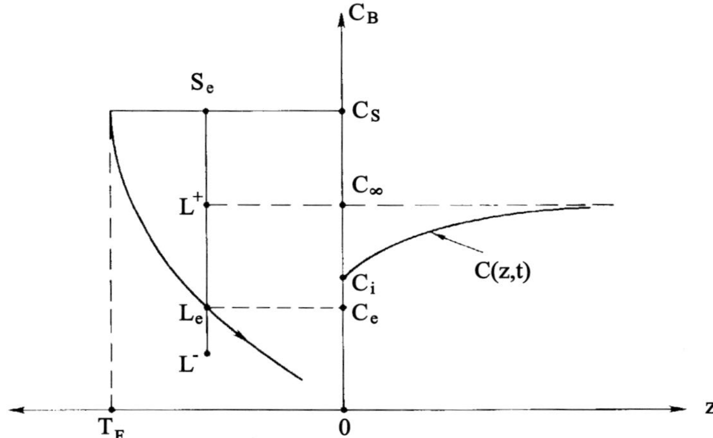


**Fig. 5.3.** Phase diagram (left-hand side) and solution's concentration distribution (right-hand side) in a binary system for which the solid (fusion temperature $T_F$) is nearly stoichiometric (taken from [5.1])

as $L^+$, correspond to metastable states that can be realized in supersaturated (or supercooled) solutions. Points to its left, labeled $L^-$, correspond to undersaturated solutions. Now, a solution whose bulk composition is $C_\infty$ is generally not in equilibrium with the solid at the processing temperature $T$. In this connection it is convenient to define the absolute supersaturation

$$\sigma = \frac{C_\infty - C_e}{C_s - C_e} \tag{5.1}$$

which is a dimensionless ratio of the previously defined quantities. If $L^+$ corresponds to a liquid composition in the two-phase region ($\sigma > 0$), as drawn in Fig. 5.3, then contact with the solid would tend to promote growth. This is demonstrated in more detail on the right-hand side of the diagram, where the concentration $C(z,t)$ of $B$ in the liquid decreases from its prepared value, far from the solid–liquid interface, to its interface value $C_i$. Surface rate limitations may cause $C_i$ to differ from $C_e$. An analogous picture holds under dissolution conditions that correspond to the initial state $L^-$ of the solution, i.e. when $C_\infty$ is less than $C_e$. In either case one can calculate the concentration distribution, hence the growth or dissolution rate $f(t)$. The time integral $h(t) = \int f(t)\mathrm{d}t$ represents the total amount of solid that has grown or dissolved during the processing time $t$ [5.1].

This picture of LPE describes the physical reality in the growth system only in the first approximation. Therefore, the following comments are required.

(i) One-dimensional diffusional mass flow is considered, while convection in the liquid phase is neglected. The only exceptions are estimates of edge effects in quiescent liquids and the influence of forced convection caused by substrates in a "moving" configuration (as in the case of a multicompartment slider boat system shown in Fig. 5.1c). In the latter case, the liquid is effectively bounded, and fresh growth units, at the concentration $C_\infty$, are fed to the boundary layer at its edge $z = \delta$. Even in the absence of forced convection, if the liquid is finite in extent (when it is, e.g., bounded by an inert wall), then a "no-flux" condition at the wall $z = L$, has to be imposed.

(ii) Surface rate limitations are handled by introducing a "surface–reaction" rate constant $k$. When $k$ is very large $C_i$ tends towards the equilibrium value $C_e$ [5.1].

(iii) Although LPE can be practiced at constant temperature, for some material systems (e.g., for compound semiconductors), there are many systems that require externally imposed driving forces such as negative-temperature ramps $T(t) = T_0 - r_c t$, where $r_c$ is the cooling rate. This drives $C_e$ downwards and causes an additional supersaturation. To first order in time, one finds that

$$C_e\,(T(t)) = C_0 \left(1 - \frac{r_c t}{m\,C_0}\right) \tag{5.2}$$

where $C_0$ and $m$ are, respectively, the equilibrium concentration and liquidus slope at the initial temperature $T_0$. Consequently, the system responds with a time constant $\tau = m\,C_0/|r_c|$. For this case, and $k$ being large, the time integral $h(t)$ is given in the form [5.12]

$$h(t) = \sqrt{D\tau}\left(\frac{2(C_\infty - C_0)}{\sqrt{2}(C_s - C_0)}\left(\frac{t}{\tau}\right)^{\frac{1}{2}} + \frac{\operatorname{sign}(r_c)\,4C_0}{3\sqrt{\pi}(C_\infty - C_0)}\left(\frac{t}{\tau}\right)^{\frac{3}{2}}\right). \tag{5.3}$$

(iv) When $k$ is not large (surface limitations are significant), then closed-form solutions are unavoidable. For the case of the first-order surface kinetics, one can reduce the problem to the solution of a rather simple integral equation, or to various perturbation series [5.1]. For short processing times, first-order kinetics implies that $h(t)$ be proportional to $t^2$, while $n$th-order kinetics implies that $h(t)$ initially varies as $t^{(3+n)/2}$.

### 5.1.2 Two-Dimensional Effects

Substrate wafers used in LPE processes are usually not large (seldom exceeding 25 mm in diameter) and are often patterned by masks or sometimes they develop morphological features, especially during dissolution. There is thus a need for two-dimensional models. The first two cases are almost completely understood in the sense that they represent cases of selective deposition in the vicinity of inert regions, either deposited masks or the wafer's edge. Then, flux lines converge towards those edges to cause enhanced growth. This has been successfully modeled analytically for LPE from quiescent, supersaturated solutions without temperature programming [5.1].

LPE is particularly rich in morphological features, and it is tempting to apply the morphological stability theory of Mullins and Sekerka [5.13] (see Sect. 11.6). Technically, this is incorrect because the unperturbed state is almost always time dependent, and therefore the perturbation amplitudes must depend on all the previous history, i.e., they must satisfy relevant integro-differential equations in time. However, when LPE occurs under quasi-steady-state conditions [5.14], Small and Potemski have shown [5.15] that morphological instabilities can be handled by standard methods.

### 5.1.3 LPE of Compound Semiconductors

The compound semiconductors have stimulated the development of technological applications of different implementations of the LPE technique by virtue of their feasibility to be used in light emitting devices like lasers and light emitting diodes (LEDs), as well as in photo detectors, especially designed for the far-infrared spectral regions [5.16]. Due to the binary nature of the simplest systems, there is a wide range of possible combinations, e.g., AlP to InSb, while combinations of binary pairs can be alloyed to make ternary or quaternary alloys with continuously variable properties. Thus, technological applications of LPE of compound semiconductors had been developed by the 1960s.

In implementation of the LPE technology, the design and construction of a graphite "boat" to allow the required sequence of layers to be grown, while also to reduce degradation of either the substrate prior to the growth or the completed structure following growth, is a basic factor. High-purity graphite is the material of almost universal choice for this function since it can be readily machined into parts of the necessary precision to make the boat. As already has been shown in Fig. 5.1c, the boat consists of a base, into which a recess is machined to accommodate a substrate, and a sliding part to contain the solutions. The two sliding parts are held together sufficiently firmly to prevent leakage of solution from their containers, while allowing free movement. Multiple-layer growth is normally practiced using a sequence of solutions ordered as the layers to be grown and traversed in one direction only, while for simpler structures the boat can be moved back and forth to reposition the same solution over the substrate. In addition to the parts listed, there may be graphite capping pieces, to cover and project into each well to define both the shape and the depth of the solutions, allowing different depths within one set (see Fig. 5.4).

An atmosphere of ultra-pure hydrogen is used to provide a clean environment in a simple closed-ended tube. A removable cap at one end allows access for removal and introduction of the boat. It also incorporates a sliding vacuum seal to couple mechanical movement to that of the slider. The glass parts should be of high-purity quartz and the sliding seal should be proof

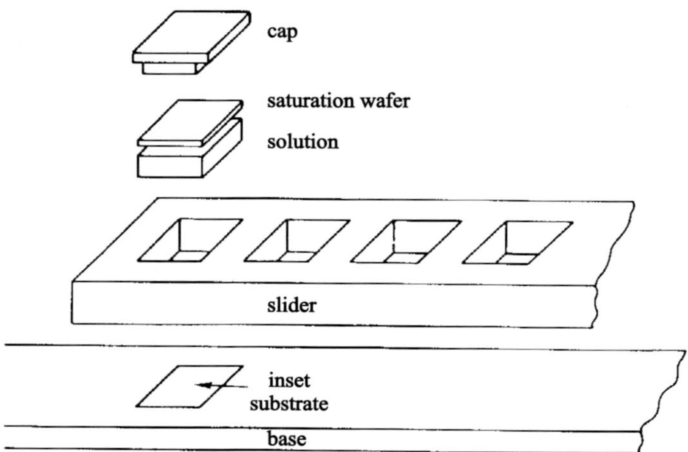


**Fig. 5.4.** Slider arrangement for growing multiple layers of compound semiconductors (taken from [5.1])

against the introduction of oxygen (the pressure inside the system is slightly above atmospheric).

Another major component is the tube furnace, whose function is generally to provide an environment with as small a temperature gradient as possible along the length of the boat. The relative temperatures of the parts of the boat should be reproducible to about 0.1°C while the absolute temperature control need not be better than about 1°C. Temperature gradients have been imposed along the direction of motion of the sliding part to cause solutions to be cooled as they are moved into the position where growth is to take place. When a series of layers is grown for different times, this results in supercoolings which are determined by the period of growth and the interface time integral $h(t)$. However, vertical gradients are in some growth processes more useful.

Loading the boat consists of the final preparation of the polished substrate and loading the measured constituents of each solution into the appropriate wells. A process of weighing the constituents of each solution with sufficient precision to achieve a small, known supersaturation at the time each is brought in contact with the substrate is impractical. An alternative is to deliberately introduce an excess of a major binary component, preferably in the form of wafer [5.17], which will float on the surface of each melt and maintain equilibrium with the surface layer of the solution. During cooling a concentration difference will develop between the top of the solution, in equilibrium with solid, and the bottom, which is in contact with graphite. Thin solutions, $\approx 3\,\mathrm{mm}$, will ensure that the supercooling will not be very large for the slow cooling rate used. By varying the depth of the solutions contained in the boat, different supercoolings can be achieved for different layers. The best geometrical uniformity can be obtained if the saturating solid material covers the entire surface of each melt. Ternary and quaternary materials can be grown using the described method. In the case of ternaries, such as (AlGa)As, the source substrate may be a GaAs wafer, the surface of which automatically attains the equilibrium alloy composition. For quaternary systems not all apparently suitable solid sources can be used, since some binaries will be dissolved.

In order to achieve a precise control over the LPE processes, experiments dedicated to fundamental understanding of the growth kinetics are required. Let us discuss some of these experiments.

Inatomi and Kuribayashi [5.18] have demonstrated how it is possible to view the interface of a III–V material using infrared radiation penetrating through the substrate towards the growing interface. Observations have revealed the existence of pits formed during dissolution, macrosteps with scalloped shapes between pinned points moving across vicinal surfaces on slightly off-oriented substrates, and hillocks formed at a density correlating with that of dislocations, growing and joining at larger off-orientations. The same technique has been applied to measure growth rate in real time [5.19].

Nishinaga et al. [5.20] have demonstrated how it is possible to measure the nitrogen concentration in GaP using the spatially resolved photoluminescence technique (SRPL) with sufficient resolution (5 μm) to show periodic variations across each terrace and lower concentrations at the terrace steps on vicinal surfaces of an off-oriented substrate, which observation is counter to conventional models of LPE. Zhang and Nishinaga [5.21] applied the same SRPL technique to GaP layers grown by epitaxial lateral overgrowth (ELO). Similarly to the already described ELO processes in SPE, also in LPE the epilayers grow first by restricted nucleation in small windows defined in $SiO_2$ on the substrate's surface, so reducing the number of dislocations which can propagate through the windowed areas. Growth emerges from the windows as star-like patterns advancing (growing) over the oxide plane, which reflects the strong orientation dependence of the lateral growth rate. Atomically flat surfaces are obtained in this way. Similar experiments with the ELO technique have been performed also with GaAs.

Let us conclude this survey of the main features of the standard version of LPE with the following statements taken from [5.1].

(i) LPE remains a relatively safe, cheap, flexible crystal-growth process with the potential for low processing temperatures compared to the melting point of the material in question. The epitaxial layers of compound semiconductors produced by this process exhibit very good opto-electronic properties which is largely due to low densities of point defects. This is partly due to the stoichiometry of LPE material. For example, the growth of GaAs from a Ga-rich melt always produces material with the most Ga-rich stoichiometry. Thus defects such as Ga vacancies and As atoms on Ga sites (the As antisite atoms) are virtually nonexistent in LPE material. The As antisite defects are believed to be related to the deep electron trap denoted EL2, which is known to have a deleterious effect on several materials properties [5.22]. Freedom from background elemental impurities is partly due to the availability of high-purity metals, which are typically used as solvents and the inherent purification process that occurs during the liquid to solid phase transition for solutes with distribution coefficients of less than unity. Very important for the LPE growth of Al-containing materials, such as AlGaAs, is the purification process where oxygen in the system forms highly stable $Al_2O_3$ on the surface of the liquid, thus preventing oxygen incorporation into the epitaxial layer. This allowed the early AlGaAs layers grown by LPE to be far superior to layers grown by any other technique.

(ii) The fact that the solution phase and crystalline phase are at equilibrium results in a number of serious disadvantages to the process, particularly that there can be intermixing between phases, and most importantly, this may happen during the formation of heterostructures. It would appear that the presence of vacancies at the interface results in fast interdiffusion, of dopants for example, while the interfaces of heterostructures are constitutionally unstable and render some systems either very difficult or impossible to grow.

(iii) The morphology of LPE material is determined by the process of introduction and removal of the solution phase to the surface, the crystal growth morphology, and also by the intersection of line defects with the surface. Morphology plays a role in the performance of devices and more work in this area would be rewarding.

(iv) Selective epitaxy through windows (ELO) has been shown to produce atomically smooth surfaces and this offers a route to improving the morphology where this is allowed by other constraints in the fabrication process.

(v) Modeling has been successful in treating the compositional interchange between solid and solution for ternary systems; however, there is no complete theory for quaternary and higher-order systems.

Without going into further details on standard LPE growth techniques, but instead referring the reader to [5.1–3] and [5.10] for more information concerning this subject, we will discuss now a special modification of LPE, known in the literature under the name liquid phase electroepitaxy (LPEE) [5.23–25].

## 5.2 Liquid Phase Electroepitaxy

Liquid phase electroepitaxy (LPEE) is a modification of LPE in which a direct electric current (DC) passing through the solution–substrate interface, and the phenomena related to this current (the Peltier effect and electromigration) are the driving forces for deposition of the epitaxial layer at a constant temperature (see [5.23, 26] for a review). LPEE has been found to be very effective for growth of bulk-like, very uniform, and nearly dislocation free crystals of multicomponent semiconductors. InGaAs crystals with diameters of 14 and 25 mm have been grown by this technique on InP [5.27] and GaAs [5.28] substrates, respectively, while AlGaSb crystals of 15 mm in diameter have been grown on GaSb substrates [5.29]. In all of these cases the crystals obtained were up to 3 mm thick and composition fluctuations in the growth direction as well as in the growth plane did not exceed 1%. The structural quality of the crystals was as high as that of thin epitaxial layers, i.e., much higher than the quality of the melt grown substrate crystals.

These facts have raised considerable interest in growing by LPEE bulk crystals which could be used as substrate platelets for other epitaxial growth processes. The thicknesses of such bulk-like single-crystalline platelets can exceed by many times the thicknesses of the substrates used for their growth by LPEE. Especially interesting is a combination of LPEE with heteroepitaxial lateral overgrowth (HELO) (see Sect. 14.1.6). When applying this hybrid technique high quality alloys can be grown on commercially available binary substrates, despite the alloy–substrate lattice mismatch. Thus, the growth of multicomponent III–V substrate crystals with a required value of the lattice constant becomes possible [5.30].

In discussing here the principles of LPEE, we will follow the pioneering work of Jastrzebski et al. [5.24]. It has been recognized for many years that precise control of the microscopic growth velocity (and thus control of segregation and defect structure) is a considerable problem in standard LPE growth processes. However, nearly perfect control of these parameters is much easier to be achieved through passage of electric current across the growth interface since it induces Peltier cooling (or heating) in the immediate vicinity of the interface and also causes electromigration of the solute to the growth interface. Thus, when the whole LPEE growth system is maintained at constant temperature, the growth occurs due to the Peltier and electromigration effects. Although approaches to perfect control of the LPEE growth process have been demonstrated on a laboratory scale, this technique has never been used on a production scale, because of unavoidable superimposed convective instabilities in the melt and other complexities introduced by Joule heating, which occurs when electric current is passing through a conductive medium (in LPEE, through the metallic solution contacting the semiconducting substrate crystal) [5.30].

## 5.3 The LPEE Process and Related Phenomena

Electroepitaxy utilizes the standard LPE configuration modified to permit passage of electric current through the solution–substrate interface. A schematic representation of a typical growth cell is shown schematically in Fig. 5.5. After thermal equilibration, growth is initiated and sustained by passing electric current through the substrate–solution interface in the appropriate direction (see below), while the overall temperature of the system is maintained constant.

Peltier cooling is, thus, the driving force for epitaxy in this case. The substrate and solution, being different electrical conductors, have different thermoelectric coefficients. Thus, flow of electric current across their interface is accompanied by absorption of heat or evolution of heat, depending on the current direction. The magnitude of this heat, $Q$, is proportional to the differences in Peltier coefficients, $\pi_p$, and the current density, $J$. One then gets the relation

$$Q = \Delta \pi_p J. \tag{5.4}$$

For III–V semiconductor compounds and the temperature range used in electroepitaxy, $Q$ is of the order of $1\,\mathrm{W\,cm^{-2}}$ at a current density of about $10\,\mathrm{A\,cm^{-2}}$. In an equilibrated isothermal LPE system the absorption of this heat decreases the interface temperature by $\Delta T_p$, which is typically of the order of 0.1 or 1°C and, thus, induces supersaturation leading to epitaxial growth.

Using the pertinent part of a schematic phase diagram of the III–V compound (see Fig. 5.6a), Peltier induced LPE can be illustrated by the vertical

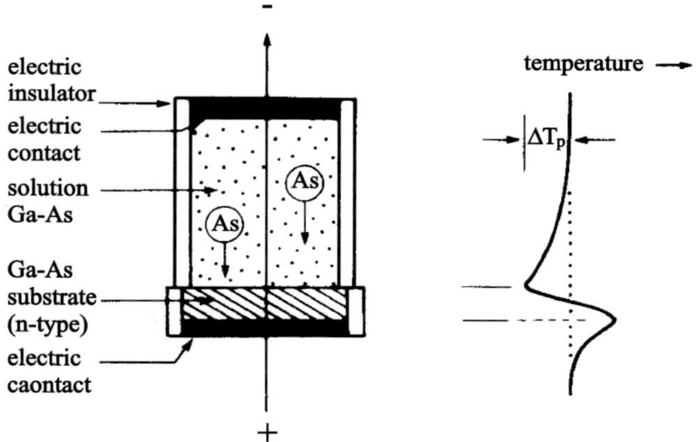


**Fig. 5.5.** Schematic representation of growth cell used for electroepitaxial growth of GaAs from Ga-As solution and of the temperature profile resulting from Peltier cooling at the substrate–solution interface and from Peltier heating at the substrate–electric contact interface (taken from [5.24])

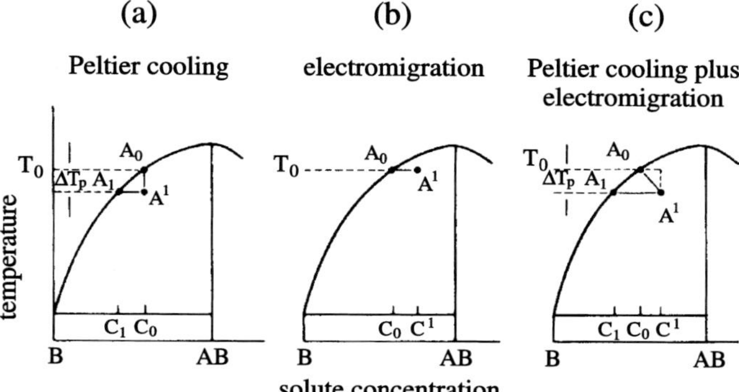


**Fig. 5.6.** Schematic representation of the pertinent part of the phase diagram of an III–V semiconductor compound system and transitions involved in electroepitaxial growth given for the case of (**a**) Peltier cooling (**b**) electromigration and (**c**) both (taken from [5.24])

transition $A_0 \to A^1$, i.e., supercooling, and the subsequent horizontal transition $A^1 \to A_1$ which leads to precipitation (solidification of the III–V compound). Using temperature–composition coordinates ($T$,$C$), the process can be expressed as $(T_0, C_0) \to (T_0 - \Delta T_p, C_0) \to (T_0 - \Delta T_p, C_1)$. Accordingly, in the following treatment of growth induced by Peltier cooling, the supersaturation $C_1 - C_0$ will be approximated as

$$C_1 - C_0 = \Delta T_p \left. \frac{\partial C}{\partial T} \right|_L^{T_0} \tag{5.5}$$

where $L$ designates the liquidus.

The solutions of the constituent elements of group III–V compounds used in electroepitaxy are metallic conductors; essentially they exhibit no ionic contributions to the current. However, in these solutions, as in other liquid metals, electromigration takes place due to electron-momentum exchange and/or electrostatic field forces. Under the electric field $\boldsymbol{E}$ induced by the current flow, species in solution migrate with a velocity $\boldsymbol{v} = \mu \boldsymbol{E}$ , where $\mu$ is the electron mobility. For example, in the Ga-As solutions, electromigration of As species is anode directed; thus, when the substrate has a positive polarity, the solution becomes supersaturated with As at the substrate–solution interface leading to epitaxial growth. As shown schematically in Fig. 5.6b, electromigration causes (in thermally equilibrated solution) a transition $A_0 \to A^1$, i.e., supersaturation, which in turn leads to a transition $A^1 \to A_0$, i.e., precipitation (solidification) of GaAs. In $T, C$ coordinates, these transitions can be expressed as $(T_0, C_0) \to (T_0, C_0 + \Delta C) \to (T_0, C_0)$. When both Peltier cooling and electromigration are significant, the following transitions are involved in the growth process (Fig. 5.6c): the diagonal supersaturation transition $A_0 \to A^1$, i.e., $(T_0, C_0) \rightarrow (T_0 - \Delta T_p, C_0 + \Delta C)$ and the horizontal precipitation (solidification) transition $A^1 \rightarrow A_1$, i.e., $(T_0 - \Delta T_p, C_0 + \Delta C) \rightarrow (T_0 - \Delta T_p, C_1)$.

In treating the solidification (precipitation) transitions as strictly isothermal, it must be assumed that the contribution of the heat of solidification is negligible. This assumption is justified for the commonly encountered rates of solidification in LPE which are of the order of 1 μm $\min^{-1}$.

### 5.3.1 Growth Kinetics in LPEE of GaAs

In the model of electroepitaxy presented above it is assumed that the solute transported to the advancing growth interface is removed from the solution only through epitaxial growth on the substrate. Transport of solute due to temperature gradients in the solution is neglected but solute transport due to convective flow in the solution is taken into consideration. In certain approximations and in calculations presented in this consideration, electroepitaxy of GaAs from Ga-As solution is used as a specific example [5.24].

The widely adopted isothermal diffusion treatment of LPE is here extended to include solute transport by electromigration which is controlled by the mobility $\mu$ of the migrating solute and the electric field $\boldsymbol{E}$ in the solution: thus,

$$D\left(\frac{\partial^2 C}{\partial x^2}\right) - v\left(\frac{\partial C}{\partial x}\right) - \mu|\boldsymbol{E}|\left(\frac{\partial C}{\partial x}\right) = \left(\frac{\partial C}{\partial t}\right), \tag{5.6}$$

where $D$ is the diffusion coefficient of the solute in the solution, $C$ is the concentration, $x$ is the distance from the advancing growth interface, $v$ is the growth velocity, and $t$ is the time. It should be noted that the sign of the electromigration term $\mu|\boldsymbol{E}|(\partial C/\partial x)$ is determined by the direction of the current flow.

The following boundary conditions are considered applicable in the present case (the notation is as in Fig. 5.6; the concentration in the solid is $C_s$ and the thickness of the solute boundary layer is $\delta$). In the absence of convection $(\delta = \infty)$:

(a) $D(\partial C/\partial x)|_0 - \mu|\boldsymbol{E}|C_l = v(C_l - C_s)$,
(b) at $t = 0, C = C_0$ for all $x$,
(c) at $t > 0, C = C_0$ for $x = \infty$ (absence of convection), or
(d) at $t > 0, C = C_0$ for $x > \delta$ (presence of convection),
(e) at $t > 0$ and $x = 0, C = C_l$ (growth follows the liquidus line).

By solving the transport equation (5.6), the following expressions are obtained for the growth velocity of electroepitaxy (the procedure for solving (5.6) is given in [5.24]):

$$v = \frac{\Delta T_p}{(C_s - C_l)} \cdot \left.\frac{\mathrm{d}C}{\mathrm{d}T}\right|_L \cdot \sqrt{\frac{D}{\pi t}} \cdot \frac{\exp\left(\frac{-(\mu|\boldsymbol{E}|t)^2}{4Dt}\right)}{\operatorname{erfc}\left(\frac{-\mu|\boldsymbol{E}|t}{2\sqrt{Dt}}\right)} - \mu|\boldsymbol{E}| \cdot \frac{C_l}{C_s - C_l} \tag{5.7}$$

in the presence of convection (finite boundary layer thickness)

$$v = \frac{\Delta T_p}{(C_s - C_l)} \cdot \left.\frac{\mathrm{d}C}{\mathrm{d}T}\right|_L \cdot \sqrt{\frac{D}{\pi t}} \cdot \frac{\exp\left(\frac{-(\mu|\boldsymbol{E}|t)^2}{4Dt}\right)}{\operatorname{erfc}\left(\frac{-\mu|\boldsymbol{E}|t}{2\sqrt{Dt}}\right) - \operatorname{erfc}\left(\frac{\delta-\mu|\boldsymbol{E}|t}{2t\sqrt{Dt}}\right)} - \mu|\boldsymbol{E}| \cdot \frac{C_l}{C_s - C_l}, \tag{5.8}$$

where $\Delta T_p$ is the temperature decrease at the interface due to Peltier cooling.

Equations (5.7) and (5.8) can be rewritten in a simplified general form where the contribution to the growth velocity by the temperature decrease at the interface (Peltier cooling) and by electromigration are designated $v_T$ and $v_E$, respectively:

$$v = v_T f_k(\boldsymbol{E}, \delta, t) + v_E, \tag{5.9}$$

where

$$v_T = \frac{\Delta T_p}{C_s - C_l} \cdot \left.\frac{\mathrm{d}C}{\mathrm{d}T}\right|_L \cdot \sqrt{\frac{D}{\pi t}}, \tag{5.10}$$

$$v_E = \mu|\boldsymbol{E}| \cdot \frac{C_l}{C_s - C_l} \tag{5.11}$$

in the absence of convection

$$f_k = f_1(\boldsymbol{E}, t) = \frac{\exp\left(-(\mu|\boldsymbol{E}|t)^2/4Dt\right)}{\operatorname{erfc}\left(-\mu|\boldsymbol{E}|t/2\sqrt{Dt}\right)} \tag{5.12}$$

and in the presence of convection

$$f_k = f_2(\boldsymbol{E}, \delta, t) = \frac{\exp\left(-\frac{(\mu|\boldsymbol{E}|t)^2}{4Dt}\right)}{\operatorname{erfc}\left(\frac{-\mu|\boldsymbol{E}|t}{2\sqrt{Dt}}\right) - \operatorname{erfc}\left(\frac{\delta-\mu|\boldsymbol{E}|t}{2t\sqrt{Dt}}\right)}. \tag{5.13}$$

Note that for $\delta \to \infty$, $f_2(\boldsymbol{E}, \delta, t) \to f_1(\boldsymbol{E}, t)$.

### 5.3.2 The Peltier Effect at the GaAs–substrate/(Ga-As)–Solution Interface

As already pointed out, knowledge of the temperature change $\Delta T_p$ at the interface, due to the Peltier effect, is essential for assessing the prevailing conditions in electroepitaxial growth. For this reason, the experimental determination of $\Delta T_p$ for the GaAs system as a function of the type of conduction and carrier concentration in the substrate, thickness of the substrate, and current density, is indispensable. For such measurements a standard electroepitaxy apparatus can be employed. In experiments, described in [5.24], a thermocouple was positioned in the solution about 0.3 mm from the substrate–solution interface. Measurements were carried out with p- and n-type substrates with

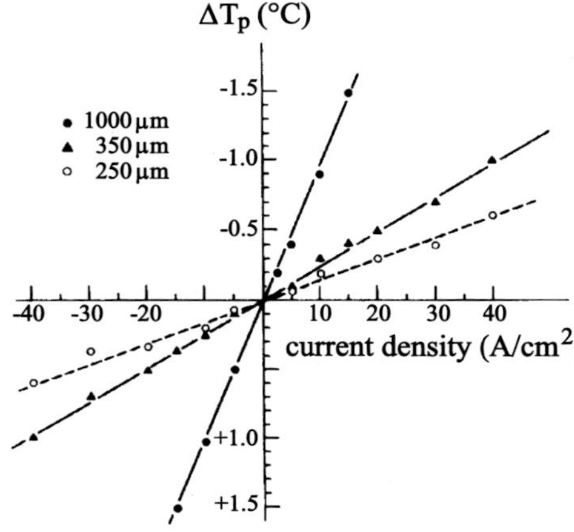


**Fig. 5.7.** Changes in temperature at the GaAs-solution interface due to the Peltier effect as a function of current density at 800°C; substrates were n-type $2 \times 10^{18}/\mathrm{cm}^3$ (taken from [5.24])

carrier concentrations from about $10^{16}$ to $10^{19}\,\mathrm{cm}^{-3}$, thickness ranging from 250 to 1500 μm, and under current densities up to $40\,\mathrm{A\,cm}^{-2}$ at temperatures of 800 and 900°C. The temperature changes $\Delta T_p$ were measured by applying DC. The contribution of Joule heating was determined by applying an equivalent amount of alternating current (AC) power. In most instances, Joule heating of the substrate was negligible in comparison with the Peltier effect, except when the substrates were semi-insulating (Cr doped), and their thickness exceeded 500 μm. The results are shown in Figs. 5.7–5.9. $\Delta T_p$ is plotted against current density for n-type substrate material of varying thickness in Fig. 5.7. The change in temperature is directly proportional to the current density and is negative for positive substrate polarity; for a given current density $|\Delta T_p|$ increases with increasing substrate thickness.

The values of $\Delta T_p$ are given as functions of substrate thickness in Fig. 5.8. The polarity of all substrates is positive; note that for the p-type substrates (having an opposite Peltier coefficient to that of the n-type substrate), $\Delta T_p$ is positive. It is seen that $|\Delta T_p|$ exhibits approximately a linear dependence on substrate thickness. A similar dependence of $\Delta T_p$ on current density and thickness to that in Figs. 5.7 and 5.8 has also been reported for n-type GaAs material [5.24]; differences in the absolute values of $\Delta T_p$ in the two studies are probably due to differences in the thermocouple location and/or differences in heat dissipation from the interface.

The values of $\Delta T_p$ as a function of carrier concentration are given in Fig. 5.9 for 300 μm thick substrates and current density of $10\,\mathrm{A\ cm}^{-2}$. It should be noted that the dependence seen in Fig. 5.9 is in qualitative agreement with the expected behavior of the GaAs Peltier coefficient [5.24].

Values of $\Delta T_p$ measured at 900°C were, within 20%, the same as those in Figs. 5.7–5.9. Thus, $\Delta T_p$ in GaAs electroepitaxy can be estimated for a given carrier concentration, substrate thickness and current density from the data given in Figs. 5.7–5.9, and by subsequent reasonable extrapolations.

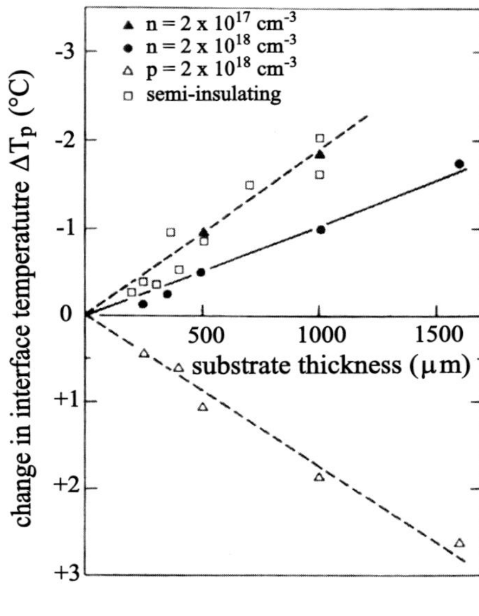


**Fig. 5.8.** Changes in temperature at the GaAs–solute interface due to the Peltier effect as a function of substrate thickness at 800°C under 10 A $cm^{-2}$ and for 300 μm thick substrates (taken from [5.24])

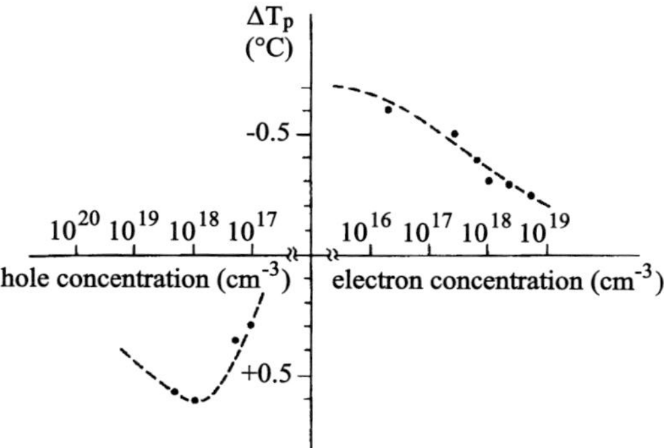


**Fig. 5.9.** Changes in temperature at the GaAs-solution interface due to the Peltier effect (800°C, 10 A $cm^{-2}$) as a function of substrate-carrier concentration. Substrates were 300 μm thick (taken from [5.24])

More advanced theory and additional experimental data concerning LPEE of GaAs can be found in [5.24].

Currently, LPEE is still only a laboratory technique. This is mainly because the size (diameter) of ingots produced with LPEE, as compared to sizes of commercially available, melt grown, binary III–V crystals is too small ($\Phi_{\mathrm{LPEE}} \approx 25$ mm while $\Phi_{\mathrm{melt}} \approx 100$ mm, or more). Although it has been proposed to simultaneously grow several crystals in a single growth run, the efficiency is still far too low for LPEE to be applied on an industrial scale. Therefore, the question of phenomena restricting the size of crystals obtainable by LPEE is of prime importance for further development of this growth technique.

This problem has been addressed in a series of experimental works by Zytkiewicz [5.30–32]. The technique used for *in situ* monitoring of disturbances occurring during LPEE growth of thick epilayers is presented in [5.32]. This technique, originally developed by Okamoto et al. [5.33], is based on the idea that the electrical resistance of the growing epilayer contributes to the total resistance $R(t)$ of the LPEE growth system. As it is the only component which is time dependent, it can be measured *in situ* informing about the progress of the growth process, because the time derivative of $R(t)$ is proportional to the growth rate $v_{\mathrm{LPEE}}$ (if the dissolution of the back surface of the substrate, by the melt, ensuring the electric contact to the substrate platelet, can be neglected). With the use of this technique the growth system instabilities have been detected and qualitatively explained for the LPEE growth of thick AlGaAs layers. An extension of these experiments to studies on the influence of the Joule effect in LPEE on the growth kinetics of thick AlGaAs and GaAs epilayers is presented in [5.31]. The complete theoretical as well as experimental treatment of the Joule effect as a barrier for unrestricted growth of bulk crystals by LPEE is given in [5.30]. With use of the one-dimensional mass transport model, the contribution of Joule heating in the crystal being grown to the LPEE growth kinetics, in the constant current mode, is discussed. The main results of this work can be summarized as follows:

(i) During the first period of electroepitaxy the growth rate is constant or slightly increases with time. Then, $v(t)$ exponentially decays to zero.

(ii) The crystal thickness $X_{\mathrm{cr}}(t)$ increases with time and then saturates when reaching the maximum thickness $X_{\mathrm{cr,max}}$. Such temporal dependence of $X_{\mathrm{cr,max}}$ agrees with experimental results published elsewhere [5.30].

(iii) The maximum crystal thickness $X_{\mathrm{cr,max}}$ feasible by LPEE decreases with increase of the electric current density $J$

(iv) For a thin source material the mass balance in the system limits the thickness of the crystal grown. When a sufficient amount of the source material is loaded to the growth cell ( i.e., for large enough $X_{\mathrm{so}}$ values) the Joule effect in the crystal limits its thickness.

(v) The constant growth velocity mode can be realized by changing the electric current density during the first part of the growth, only. When the crystal growth is thicker and the Joule effect becomes more pronounced, a constant value of the growth rate cannot be sustained.

These results are quite different from those reported earlier in [5.34], where unlimited, exponential increase of $v(t)$ and $X_{\mathrm{cr}}(t)$ with time, as well as unrestricted growth of bulk crystals by LPEE, are predicted for the case when Joule heating may be neglected.

# 6. Vapor Phase Epitaxy

The various techniques of growing epitaxial layers from the vapor phase can be divided roughly into two categories depending on whether the species are transported physically or chemically from the source to the substrate.

In physical transport techniques (referred to as physical vapor deposition (PVD)), the compound to be grown or its constituent elements are vaporized by evaporation, sputtering or laser ablation from polycrystalline or amorphous sources at high local temperatures, and subsequently transported through the relevant reactor toward the substrate in the form of vapor streams without any chemical change.

In chemical transport techniques (referred to as chemical vapor deposition (CVD)), volatile species containing the constituent elements of the layer to be grown are produced first in, or outside the reactor, which may be a closed (e.g., an ampoule) or open (e.g., a flow-through tube or barrel) system, and then transported as streams of vapor through the reactor toward the reaction zone near the substrate. These gaseous species subsequently undergo chemical reactions in the reaction zone, or dissociate thermally, to form the reactants which participate in the growth of the film on the surface of the substrate crystal.

In general, VPE may be divided into a number of subcategories, due to the availability of different transport gas species. The most frequently applied transport agents are:

(i) halides; used for growing metals ($WF_6 \rightarrow W$), elemental semiconductors ($SiCl_4 \rightarrow Si$), compound semiconductors ($GaCl + AsH_3 \rightarrow GaAs$), and rare earth substituted garnets ($YCl_3 + FeCl_2 + O_2 \rightarrow Y_3Fe_5O_{12}$),

(ii) oxides; for growing compound semiconductors ($Ga_2O + PH_4 \rightarrow GaP$),

(iii) hydrides; for elemental semiconductors ($SiH_4 \rightarrow Si$), silicon dioxide ($SiH_4 + H_2O \rightarrow SiO_2$), and silicon nitride ($SiH_4 + NH_3 \rightarrow Si_3N_4$), and

(iv) metalorganic compounds; for growing semiconductor compounds ($Ga(CH_3)_3 + AsH_3 \rightarrow GaAs$), and metals [$Al(C_4H_9)_3 \rightarrow Al$].

If no transport species is used, and the film constituents are propagating from their sources toward the substrate under vacuum conditions, the pure PVD process occurs. The best known, and currently most frequently used, representative of this subcategory of PVD is molecular beam epitaxy (MBE), especially in its version using solid sources of the constituent elements of the film to be grown, i.e., solid source-MBE (see Sect. 7.1).

Fluid flow, heat transfer, and chemical species transport are critical in determining both the access of film precursors to the growth interface and the degree of gas phase reactions occurring prior to the species participating in the surface growth reactions. A long residence time in the region near the heated substrate may lead to extensive gas phase reactions that alter the growth mode, in the worst case via the formation of particles or, more commonly, increased impurity incorporation from the gas phase reaction products. The movement of precursors to the growth surface and the amount of gas phase reactions are determined by the choice of reactor design and operating parameters. An inferior selection of design or operating parameters will lead to complex flow fields resulting in nonuniform deposition rates and increased impurity incorporation. It is therefore crucial to understand and control factors governing the transport processes determining crystal growth from the vapor phase [6.1].

In the case of a closed reactor, the ampoule includes a source (charge) material (or materials) and a transport gas and, sometimes, an inert gas. The transport gas reacts with the source material forming a volatile compound at (typically) higher temperatures, which decomposes at the surface of the growing crystal at a lower temperature. The classical example is the iodine transport of germanium when the source and the growing material is germanium. The transport gas, iodine, forms volatile $GeI_4$, which decomposes on the growing surface.

In open systems, the crystallizing species are brought to the system from outside as a gas or gases, which decompose or react with each other, depositing the material that should be grown. As examples of open growth systems common for VPE applications, four radio frequency (RF) heated systems are shown in Fig. 6.1. They are classified according to the direction of gas flow, i.e., horizontal or vertical, and whether the reactor walls are hot or cold [6.2].

The complex reactor geometries and thermal gradients characteristic of vapor phase crystal growth processes lead to a wide variety of flow structures affecting film thickness and composition uniformity. The horizontal cold wall and vertical pancake CVD reactors (Figs. 6.1a and 6.1b, respectively) are the most commonly used configurations in metalorganic VPE of compound semiconductors. Variations of the horizontal reactor have included substrate rotation and an inverted geometry with the susceptor facing downward. High-speed rotation of the susceptor ($\approx$ 500–1500 rpm) has been added to the vertical reactor to create a well-defined axisymmetric rotating disk flow, resulting in uniform deposition rates. Multi-wafer rotating disk reactors as well as reactors with multi-wafer planetary motion, have been used to achieve improved uniformity of thickness and doping [6.1]. The barrel (Fig. 6.1c) and pancake (Fig. 6.1b) reactors are examples of multi-wafer growth systems

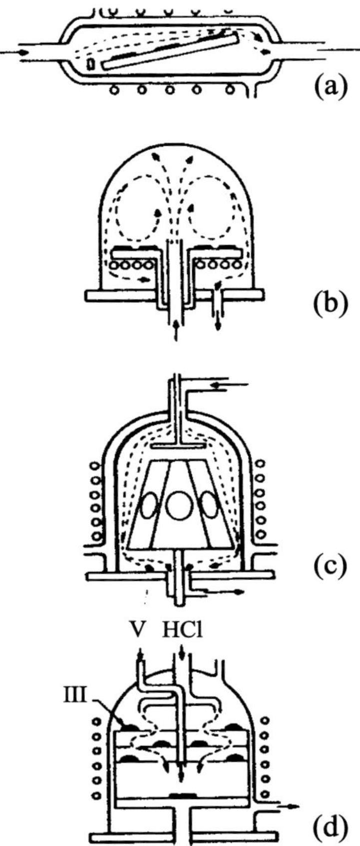


**Fig. 6.1.** Schematic illustrations of RF-heated reactor systems commonly used for VPE of semiconductors: (**a**) horizontal cold wall, (**b**) vertical ("pancake"), (**c**) barrel cold wall, and (**d**) vertical hot wall (taken from [6.2])

primarily used in Si technology. The next generation reactors are likely to resemble the classical horizontal and vertical systems, albeit modified to allow for automated substrate handling (see Fig. 8.14 in chap. 8, showing the AIX 2000/2400 multi-wafer planetary reactor)

Growth of thin films by epitaxy from the vapor phase involves the physical transport of gas phase precursors, or atomic/molecular constituents, to a heated solid crystalline surface (the substrate surface), where epitaxial growth occurs and gaseous byproducts are released. Physical approach to modeling of these growth processes consists of nonlinear, coupled partial differential equations that represent the conservation of momentum, energy, total mass, and individual species. This approach will be discussed in detail in Sect. 11.2.

## 6.1 Physical Vapor Deposition

Evaporation is the most fundamental technique which is used for vaporization of source materials in PVD processes. There exist a variety of evaporation phenomena, which are exhaustively described in [6.3]. However, the fundamental physics of these phenomena is connected mainly with the names of Hertz, Knudsen and Langmuir [6.4].

### 6.1.1 Evaporation Rates

The first systematic investigation of evaporation rates in a vacuum was conducted by Hertz [6.5] in 1882. He distilled mercury at reduced air pressure and observed the evaporation losses while simultaneously measuring the hydrostatic pressure exerted on the evaporating surface by the surrounding gas. From these observations, he drew the important and fundamental conclusion that a liquid has a specific ability to evaporate and cannot exceed a certain maximum evaporation rate at a given temperature, even if the supply of heat is unlimited. Furthermore, the theoretical maximum evaporation rates are obtained only if as many evaporant molecules leave the surface as would be required to exert the equilibrium pressure $p_{eq}$ on the same surface and none of them return. The latter condition means that a hydrostatic pressure of $p = 0$ must be maintained. Based on these considerations, the number of molecules $\mathrm{d}N_e$ evaporating from a surface area $A_e$ during the time $\mathrm{d}t$ is equal to

$$\frac{\mathrm{d}N_e}{A_e \mathrm{d}t} = (p_{eq} - p)\sqrt{\frac{N_A}{2\pi M k_B T}} \quad \left[\mathrm{m}^{-2}\mathrm{s}^{-1}\right], \tag{6.1}$$

where $M$ is the molecular weight of the evaporating species, $p_{eq}$ is the equilibrium pressure, $p$ is the hydrostatic pressure of the evaporant in the gas phase, and $k_B$ and $N_A$ are the Boltzmann and Avogadro constants, respectively.

The evaporation rates originally measured by Hertz were only about one tenth as high as the theoretical maximum rates. The latter were actually obtained by Knudsen in 1915 [6.6]. Knudsen argued that molecules impinging on the evaporating surface may be reflected back into the gas rather than incorporated into the liquid. Consequently, there is a certain fraction $(1 - a_v)$ of vapor molecules which contribute to the evaporant pressure but not to the net molecular flux from the condensed phase into the vapor phase. To account for this situation, he introduced the evaporation coefficient $a_v$, defined as the ratio of the observed evaporation rate in vacuum to the value theoretically possible according to (6.1). The most general form of the evaporation rate equation is then

$$\frac{\mathrm{d}N_e}{A_e \mathrm{d}t} = a_v (p_{eq} - p)\sqrt{\frac{N_A}{2\pi M k_B T}} \quad \left[\mathrm{m}^{-2}\mathrm{s}^{-1}\right], \tag{6.2}$$

which is commonly referred to as the Hertz–Knudsen equation.

Knudsen found the evaporation coefficient $a_v$ to be strongly dependent on the condition of the mercury surface. In his earlier experiments, where evaporation took place from the surface of a small quantity of mercury, he obtained values of $a_v$ as low as $5 \times 10^{-4}$. Concluding that the low rates were attributable to surface contamination, he allowed carefully purified mercury to evaporate from a series of droplets which were falling from a pipette and thus continually generated fresh, clean surfaces. This experiment yielded the maximum evaporation rate

$$\frac{\mathrm{d}N_e}{A_e \mathrm{d}t} = p_{eq}\sqrt{\frac{N_A}{2\pi M k_B T}} \quad \left[\mathrm{m}^{-2}\mathrm{s}^{-1}\right]. \tag{6.3}$$

### 6.1.2 Langmuir and Knudsen Modes of Evaporation

It was first shown by Langmuir in 1913 [6.7] that the Hertz–Knudsen equation also applies to evaporation from free solid surfaces. He investigated the evaporation of tungsten from filaments in evacuated glass bulbs and assumed that the evaporation rate of a material at pressures below 1 torr is the same as if the surface were in equilibrium with its vapor. Since recondensation of evaporated species was thereby excluded, he derived the maximum rate as stated by (6.3). Phase transitions of this type, which constitute evaporation from free surfaces, are commonly referred to as Langmuir or free evaporation.

An alternative evaporation technique was established by Knudsen [6.8] and is associated with his name. In Knudsen's technique, evaporation occurs as effusion from an isothermal enclosure with a small orifice (Knudsen cell). The evaporating surface within the enclosure is large compared with the orifice and maintains the equilibrium pressure $p_{eq}$ inside. The diameter of the orifice must be about one-tenth or less of the mean free path of the gas molecules at the equilibrium pressure, and the wall around the orifice must be vanishingly thin so that gas particles leaving the enclosure are not scattered or adsorbed and desorbed by the orifice wall. Under these conditions, the orifice constitutes an evaporating surface with the evaporant pressure $p_{eq}$ but without the ability to reflect vapor molecules; hence, $a_v = 1$. If $A_e$ is the orifice area, the total number of molecules effusing from the Knudsen cell into the vacuum per unit time, which will be called hereafter the total effusion rate $\Gamma_e$, is given by

$$\Gamma_e \equiv \frac{\mathrm{d}N_e}{\mathrm{d}t} = A_e(p_{eq} - p_v)\sqrt{\frac{N_A}{2\pi M k_B T}} \quad \left[\mathrm{molecules}^{-1}\right], \tag{6.4}$$

where $p_v$ is the pressure in the vacuum reservoir to which the molecules effuse from the cell orifice. This is the Knudsen effusion equation. It may be simplified by setting $p_v = 0$, which is reasonable for effusion into an UHV environment.

The Knudsen equation is often written in the form

$$\Gamma_e \equiv 3.51 \times 10^{22} \frac{p A_e}{\sqrt{MT}} \quad \left[\mathrm{molecules}^{-1}\right], \tag{6.5}$$

where $p$ is the pressure in the effusion cell in torrs, and all other quantities are in cgs units. Expressing these quantities in SI units, one has to replace the numerical factor in (6.5) by $8.33 \times 10^{22}$ [6.4].

Langmuir's as well as Knudsen's modes of evaporation have been employed in many experimental methods of determining the vapor pressure of materials and heats of vaporization. A critical examination of both techniques with their limitations has been published by Rutner [6.9]. Langmuir's method suffers from the uncertainty of whether or not an observed rate of weight loss truly reflects the equilibrium rate of evaporation. It is often used, however, to determine $a_v$ by comparing its results with independently known vapor-pressure data or with evaporation-rate measurements from Knudsen cells. The principal problem with Knudsen's technique is that an ideal cell with an infinitely thin-walled orifice yielding free molecular flow can only be approximated. In practice, orifices of finite thickness must be used (see Fig. 6.2), which necessitates the application of corrective terms in the effusion equation (see Sect. 11.2.3).

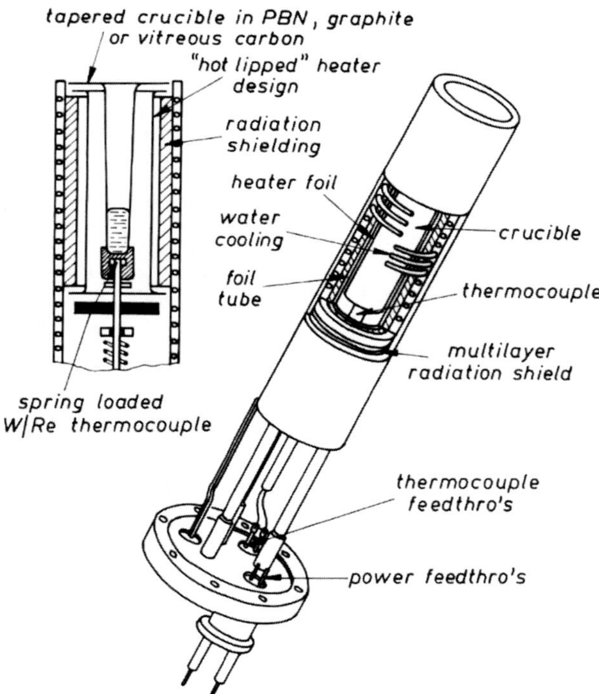


**Fig. 6.2.** Schematic illustration of a present-day effusion cell used in MBE systems. Section of the central part of the cell assembly (top left), and a cut-away diagram of a whole effusion cell assembly (bottom right) (taken from [6.4])

### 6.1.3 Principles of MBE

The most frequently used epitaxial growth technique based on the principle of PVD is solid source MBE. In this technique vaporization of source materials is realized by evaporation in the Knudsen mode (see Sect. 6.1.2). In this growth technique, thin epitaxial layers crystallize via reactions between thermal-energy molecular or atomic beams of the constituent elements and a substrate surface which is maintained at an elevated temperature in ultrahigh vacuum. The composition of the grown epilayer and its doping level depend on the relative arrival rates of the constituent elements and dopants, which in turn depend on the evaporation rates of the appropriate sources. The growth rate of typically $1\,\mu m\,h^{-1}$ (1 monolayer $s^{-1}$) is low enough that surface migration of the impinging species on the growing surface is ensured. Consequently, the surface of the grown layer is in general very smooth. Simple mechanical shutters in front of the beam sources are used to interrupt the beam fluxes, i.e., to start and to stop the deposition and doping (see Fig. 6.3). Changes in composition and doping can thus be abrupt on an atomic scale.

The characteristic features of MBE will be presented and discussed in detail in Chap. 7.

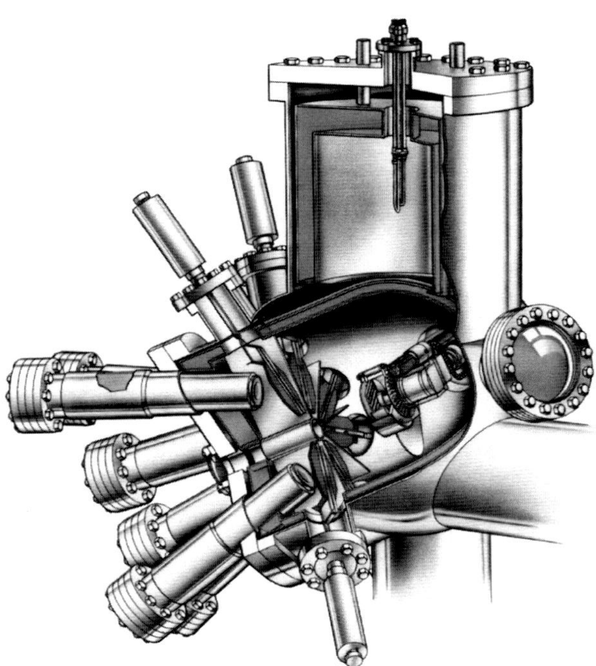

**Fig. 6.3.** Cutaway illustration of the deposition chamber of the V80H MBE system of VG Semicon company, showing the configuration and geometry of the evaporation sources, mechanical shutters, and substrate manipulator (taken from [6.4])

### 6.1.4 Sputtering

The second vaporization technique used in PVD processes is sputtering (sometimes called "cathode sputtering") [6.10–12]. Sputtering is similar to evaporation in that reduced pressures are required for each. The principal difference is that while thermal energy is used in evaporating the constituents of the epilayer or other deposited film, ion bombardment of the material, causing ejection of atoms, is used for sputtering. Thus, thin films of refractory materials may be deposited by sputtering without high source temperatures, such as required by evaporation. The ions are formed when a high electric field is applied to a low-pressure gas such as argon, creating a glow discharge in the deposition chamber. The positively charged argon ions are accelerated through the field to strike a cathode made of the material to be sputtered. The atoms of a cathode surface gain sufficient energy to leave the cathode and condense on the substrate (the anode electrode). The actual gas pressure during sputtering is of the order of $10^{-1}$–$10^{-3}$ torr. The described sputtering technique is called "diode sputtering".

If films of high purity are desired, it is necessary to evacuate the sputtering chamber to less than $10^{-4}$–$10^{-6}$ torr prior to sputtering so as to control the contamination of the deposited film. This is especially important with refractory metals normally deposited by sputtering since they are known for their gettering ability. This ability has also been put to advantage to produce films which are purposely doped by the addition of reactive gases to the system during sputtering. With involved reactive gases, this technique is known as "reactive sputtering".

An advantage of sputtering is the capability of depositing refractory metals onto relatively cool substrates at reasonable rates, and the capability of depositing compounds of these metals by addition of reactive gases. Control of the deposition parameters is somewhat more tenuous than that of evaporation due to the higher complexity of conditions in sputtering.

A large amount of work has been done on sputtering methods in the course of the past 40 years. New developments include "bias sputtering", for controlling film properties, "triode sputtering", in which much lower pressure may be used, "radio frequency sputtering", which is especially useful for depositing dielectric films, and "magnetron sputtering" [6.10–12]. Without going into details, let us define in short the listed sputtering techniques [6.13–15].

"Bias sputtering", called also "ion-plating", is a variant of diode sputtering in which the substrate is ion-bombarded during film deposition and prior to this in order to clean the substrate platelet. Ion bombardment during film deposition can produce some desirable effects, such as re-sputtering of loosely bonded film material (surface impurities), low-energy ion implantation, desorption of gases, conformal coverage of patterned surfaces, or modification of different film properties. The source material need not originate from a sputtering target, but can be an evaporation source, a reactive gas with condensable constituents, or a mixture of reactive gases with condensable constituents to form compounds.

With the aim of improving sputtering efficiency and of minimizing structural and chemical changes induced in films grown in a glow-discharge environment, several approaches have been developed in which ionization is enhanced at low gas pressures by "assisted discharge sputtering". A commonly accepted arrangement that employs both a thermionic electron source and an assisting magnetic field has been developed for use in bell-jar systems. The basic concept is based on employing a horizontal electron-gun system. Such a system has been used for epitaxial growth of a wide range of metals and oxides. Typical operating conditions for the epitaxial growth of Pt on $CaF_2$ are [6.14]: argon pressure $5\times10^{-3}$ torr; substrate temperature 530°C; anode potential 65 V; anode current 3 A; target potential 1200 V; target current $2\,mA/cm^2$; deposition rate $200\,Å\,min^{-1}$. One of the advantages of such a low-pressure system lies in the fact that once a few calibration runs have been made, the sputtering rate as a function of voltage and ion mass can be predicted for different metals [6.14].

It should be noted that all glow discharge processes involve sputtering in one form or another, since it is impossible to sustain a glow discharge without an electrode at which these processes occur. In "electrodeless" discharges radio frequency power (RF power) is capacitively coupled through the insulating wall of a tubular reactor. In this case, the inside wall of the tube is the main electrode of the discharge [6.13]. One of the biggest limitations experienced with DC sputtering has been the difficulty of applying this method to the deposition of insulating materials such as oxides and high-resistivity semiconductors. Application of a high negative potential to insulating targets leads to positive ion bombardment and, within a short time, to the buildup of a compensating positive surface charge that brings further ion bombardment to a half. Attempts to remove this charge using surface leakage grids or supplementary electron bombardment have proven somewhat cumbersome and unsatisfactory. It was not until the work of Anderson et al. [6.16], who used RF voltages successfully to clean the walls of glass discharge tubes, that an effective solution to this problem was found. Soon it was demonstrated that practical systems for the rapid deposition of insulators, based upon the use of RF voltages, so-called "radio frequency sputtering" systems could be built, and a new and possibly the most fruitful chapter in the field of sputtering began [6.14].

### 6.1.5 Film Deposition in a Glow Discharge

Although there are several techniques available for generating the positive ions necessary for sputtering, by far the simplest method is by establishing a glow discharge between two flat parallel electrodes in a low-pressure gas (the diode sputtering variant). The particular type of glow discharge which occurs between two electrodes depends upon several factors: the pressure of the gas;

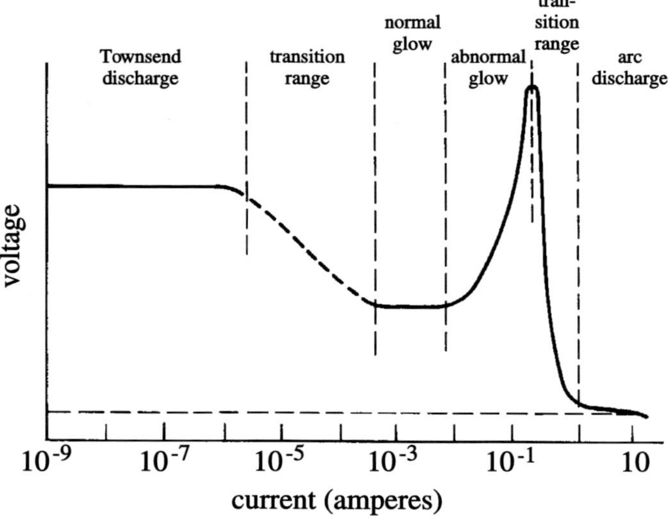


**Fig. 6.4.** Schematic data concerning a gas discharge. The voltage is represented by a linear scale while the current is a logarithmic scale (taken from [6.10])

the applied voltage; and the electrode configuration, which influences the path length of the discharge and its current density. Figure 6.4 shows the current–voltage characteristics of a discharge between two flat plates in a gas with a pressure in the range of $10^{-3}-10^{-1}$ torr. There is no appreciable current below some minimum voltage at which the gas "breaks down" abruptly. This region is known as the Townsend discharge, and the current may be increased within this region without a change in voltage. As the current is increased further, however, additional carriers are created and the discharge exhibits negative resistance. As the current is further increased, a second constant voltage region, known as the "normal" glow, is reached. If the current is increased beyond a certain level, the voltage rises with increasing current, and this region is known as the "abnormal" glow. It is in this region that most sputtering work is performed. If the current is further increased, the voltage drops abruptly and the discharge becomes an arc.

Figure 6.5 is a schematic illustration of the appearance of a glow discharge for a gas in the pressure region between $10^{-2}$ and $5\times10^{-1}$ torr. Below the diagram are shown plots of various parameters of the discharge along its length. The transport of current through a glow discharge occurs by the motion of electrons and positive ions parallel to the electric field. In order for the gas to be a conductor, however, some source of energy is necessary to continuously produce ions and electrons in the gas. An electron emitted from the cathode moves through the space in the deposition chamber. This electron is first accelerated by the strong field adjacent to the cathode, but it initially makes few, if any, ionizing collisions because its energy is not suffi-

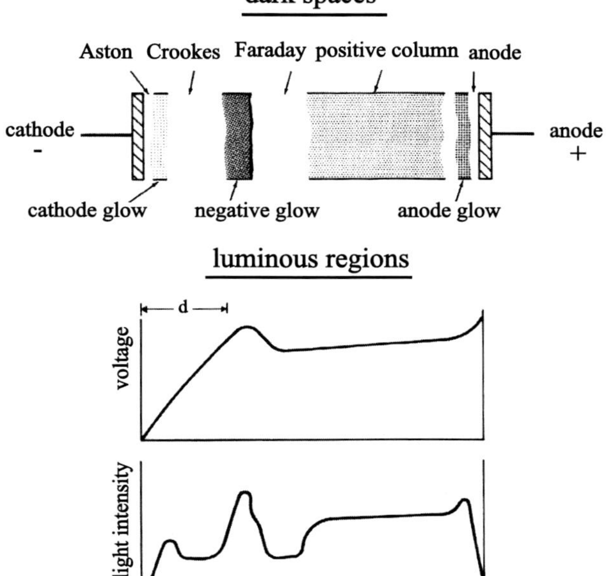


**Fig. 6.5.** Appearance of glow discharge at low pressure (taken from [6.10])

ciently above the ionization potential of the gas. Further from the cathode, however, the electron has gained sufficient energy to cause ionization when it collides with a gas molecule, and, in effect, gives rise to electron multiplication. In order to have a steady state, each electron emitted from the cathode must produce sufficient ionization to effect the release of one further electron from the cathode. Most of the ionization which is necessary to sustain the glow occurs within the Crookes' dark space region. If the anode is moved toward the cathode into the region of the Crookes' dark space, the discharge is extinguished since there is now insufficient ionization to sustain the glow.

The various luminous and dark regions of a glow discharge arise in the following manner: an electron usually leaves the cathode with very small initial velocity, such that its energy is of the order of 1 eV. It is not able to excite gas molecules until its energy is as great as the excitation potential of the gas, and this results in a region called "Aston's dark space". The cathode layer is the region in which the electrons reach an energy corresponding to the ionization potential, and this is the luminous region closest to the cathode. At distances beyond the cathode layer, i.e., in the "Crookes' dark space", the electron energies are mostly far above the maximum excitation potential, so that little visible light is emitted. By the time the negative glow is reached, the number of slow electrons (i.e., those produced by an ionizing collision) has become large and, while these electrons do not have sufficient energy to produce ionization, they possess sufficient energy to cause excitation and are the cause of the negative glow.

Since the mean free path of electrons is inversely proportional to the gas pressure, it follows that the distance required for an electron to travel before it has produced adequate ionization to sustain the glow would also be inversely proportional to the pressure. In actuality, the proportionality is to the gas density, so that the above statements apply only so long as the temperature remains constant. The thickness of the Crookes' dark space, then, increases as the pressure is decreased. If the pressure is made sufficiently low, the Crookes' dark space will expand until the plane of the anode is reached, and the discharge will become extinguished. In the region of the normal glow, the product of the thickness of the dark space and the pressure is independent of current. For argon, this product ($pd$) has a value of about 0.3 torr cm.

If very low pressures are desired for sputtering, a supplemental means for either producing a discharge or increasing the trajectory of the electrons must be provided. The former may be done by the use of radio-frequency excitation or by the use of a separate hot cathode to provide thermionic emission of electrons into the gas. The trajectory of the electrons may be increased by providing a magnetic field to cause the electrons to travel in spiral paths [6.10].

A schematic illustration of a simple sputtering apparatus is shown in Fig. 6.6. The apparatus consists of a large area cathode and an anode holding

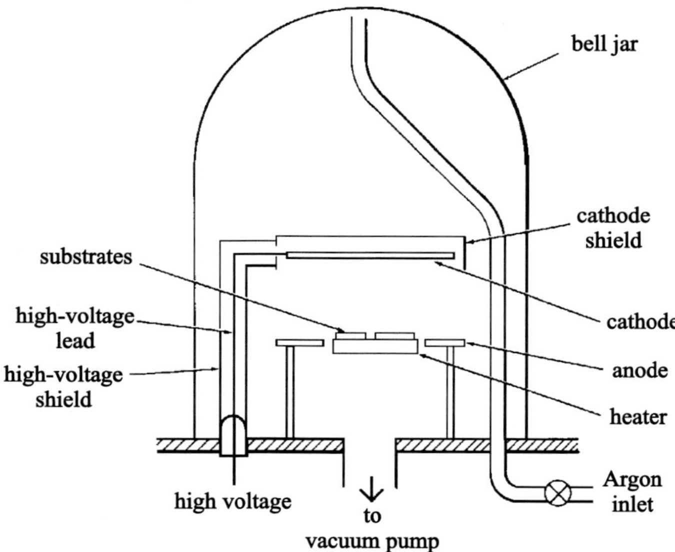


**Fig. 6.6.** Schematic illustration of a simple sputtering apparatus (taken from [6.10])

the substrates in a plane parallel configuration, and it is contained in a bell jar vacuum system. The gas to be used is admitted to the system to provide a pressure of about $10^{-2}-10^{-1}$ torr. The spacing of the electrodes is of the order of 1 to 10 cm, and the electrodes are typically from 5 to 50 cm in diameter. This type of system is usually operated with voltages of 1 to 10 kV.

The rate at which material is deposited at the substrate site is proportional to the rate at which it is removed from the cathode, and may be represented by

$$Q = CI\gamma, \tag{6.6}$$

where $Q$ is the deposition rate, $C$ is a constant which characterizes the sputtering apparatus, $I$ is the ion current, and $\gamma$ is the sputtering yield. It must be recalled that the yield is itself a function of the sputtering voltage and the particular ion being used. In most work, it is desirable to operate at the highest sputtering rate commensurate with the particular type of film to be deposited. The gas chosen as the medium for establishing the discharge should be chosen with the sputtering yield in mind. For example, argon is an excellent choice from this point of view as well as for its inertness. It would appear from (6.6) that the maximum possible current should be used if the highest deposition rate is to be achieved. While this is generally true, for most applications the situation is considerably more complex. The power available is not without limit, and the only way to increase current without increasing the power is to increase the pressure of the plasma. If the pressure is increased, however, there is a higher probability that sputtered atoms will return to the cathode by diffusion. As a matter of fact, at pressures in the range of $10^{-1}$ torr, only about 10 percent of the atoms sputtered from the cathode travel beyond the Crookes' dark space. This apparent drop in yield (as measured from the cathode) as pressure is increased leads to the conclusion that the pressure chosen from sputtering rate considerations alone would be the highest pressure for which the yield was still close to the maximum.

The position of the substrate relative to the cathode is also an important issue with respect to the deposition rate. The substrate should be as close as possible to the cathode without disturbing the glow discharge to collect a maximum of sputtered material. As the substrate approaches the cathode, however, the current will fall drastically, even before the edge of the dark space is reached.

### 6.1.6 Sputtering and Epitaxy

Let us now consider the problem of the interrelation of the sputtering parameters and the quality of epitaxial layers which are grown by using this vaporization technique of the constituent elements of the epilayers. In doing this, we will follow the experimental results and the concluding statements presented in [6.12] and [6.14].

Although epitaxial films grown through sputtering are generally indistinguishable from films grown by other methods, there is evidence that the way in which sputtered films grow may be quite different from the way in which films grow during, for example, vacuum evaporation. The first evidence for this is related to nucleation process in the growth of metallic films. Comparing sputtered and evaporated films of cadmium in a study of whether or not the critical nucleation density phenomenon found in evaporation was also present during sputtering, it could not be concluded that there exists a minimum accumulation rate below which cadmium films could not be grown by sputtering. This is quite different from what is observed for evaporation [6.12].

There are a number of reasons why one has to expect the growth conditions during sputtering to be different from those that prevail during evaporation and other epitaxial growth methods. For example, depending upon the method used to sustain the discharge during sputtering, the surface of the substrate and the growing epilayer will be subjected to a bombardment with a varying degree by electrons and neutral as well as charged gas atoms. Associated with these differences will be wide variations in the incident energy and directions of atoms arriving from the cathode. Therefore, the following phenomena should be taken into consideration when analyzing the growth of epilayers by sputtering [6.12]:

(i) the influence of inert-gas background pressure in the deposition chamber,

(ii) the role of energy of sputtered particles in the epitaxial growth phenomenon

(iii) the influence of charged-particle bombardment of the crystallizing phase at the substrate (growing epilayer) surface.

Let us illustrate this by considering the two cases of an unsupported (diode-type) glow discharge operated typically at pressures in the range $2 \times 10^{-2} - 10^{-1}$ torr, and thermionically and magnetically assisted discharge at pressures of $10^{-3}$ to $5 \times 10^{-3}$ torr. In the first case, the secondary electrons required to sustain the supply of positive ions are generated during ion bombardment of the target, and to maintain a supply of electrons sufficient to keep the discharge going, relatively high target voltages (2–5 kV) and gas pressures are needed. In the second case, the electrons are furnished by a thermionic (e.g., filament) source, and their ionizing effect is enhanced by the magnetic field, which increases their path length. In this situation the gas pressure can be relatively low, and a fairly high sputtering yield is achieved with modest target voltages (500–1000 V).

Under high-pressure, glow-discharge conditions, the sputtered atoms undergo many collisions with the gas atoms before reaching the substrate, and indeed a high proportion may be back-sputtered onto the target surface. These collisions attenuate the initial high ejection energy and cause the target atoms to arrive at the substrate over a wide range of incident angles. As a result of electron bombardment and energy dissipated at the substrate surface by ion–electron recombination, the surface temperature (especially of an insulating substrate) rises rapidly within a few seconds from the initiation of the discharge to reach values as high as several hundred degrees centigrade. Although the temperature rise may be suppressed by lowering the cathode potential, this leads to a reduction in sputtering rate to values where the relative rate of arrival of background impurities becomes significant. These effects have a significant bearing upon conditions chosen for epitaxial growth. Substrate temperatures chosen for epitaxy and the conditions of sputtering should be adjusted so that the discharge causes no further appreciable rise from the initial-growth temperature value.

Using low-pressure assisted discharge conditions of operation, the sputtered atoms, although possibly ejected at lower velocities (due to the smaller values of target voltage), undergo relatively few collisions during transfer to the substrate and consequently arrive with high energies and at more or less normal incidence. The high arrival energy is sufficient to cause significant lattice penetration and can induce surface damage and even alloying or chemical reactions with the substrate surface. Under the higher substrate temperature conditions used in epitaxial growth, the damage is usually minimized by annealing effects. An important difference produced by operating at low pressures is that the substrate heating effect due to the discharge is considerably reduced. In this case the temperature rises very slowly during sputtering and, on turning off the discharge, cooling occurs at a comparable rate. There seems little doubt, therefore, that the substrate heating under these conditions can be attributed primarily to radiation from the target. Water cooling of the target, as is done routinely in low-pressure RF sputtering, is found to reduce substrate heating considerably.

Several other sputtering techniques tend to introduce additional factors that are important in influencing the film nucleation and epitaxial growth processes. In particular, bias sputtering under both DC and RF conditions, through re-emission effects, can cause major changes in both the structure and electrical properties of the growing epilayer [6.12].

As a result of the complicated surface processes occurring in growth by sputter deposition, films obtained by this technique can grow as polycrystalline deposits. Figure 6.7 shows a microstructure zone diagram for metal films deposited by magnetron sputtering [6.17]. Different zones of different surface morphology and crystallinity of the deposited film occur on the diagram for different growth temperatures. There, also the zone T is shown, which is regarded as a transition region, between zones 1 and 2 (low temperature and high temperature growth, respectively) which appears characteristically only in sputtered films.

The structure named zone 1, caused by migration of incident atoms on a substrate surface, is affected by adsorbed atoms. This structure is constructed

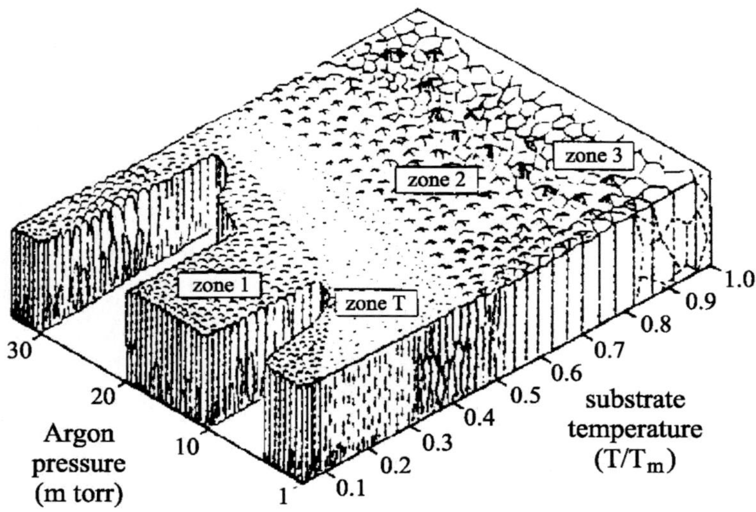


**Fig. 6.7.** Microstructure zone diagram of Thornton [6.17] for metal films deposited by magnetron sputtering ($T$ is the substrate temperature; $T_m$ is the temperature of the melting point of the deposited metal) (taken from [6.15])

from tapered crystallites with domed heads and contains voids in the grain boundaries. In the transition region (zone T) the film reveals fibrous structure in which crystallites grow perpendicularly to the surface plane of the substrate. Since the crystallites develop close to each other, the density of this type of film is nearly equal to that of the bulk material. The surface is relatively smooth and the film has large tensile strength and hardness values. In zone 2, the migration of atoms on the substrate surface becomes active. The structure is constructed of columnar grains. These grains increase their size with increasing growth temperature $T$ (in relation to the melting point temperature $T_m$).

Finally, zone 3 is a region where interdiffusion of atoms in the film controls the final film structure, thus, the film surface becomes smooth. Recrystallization progresses in the film during film formation, and the film becomes, therefore, polycrystalline and isotropic, built of randomly oriented polycrystals.

Because in sputter deposition, the energy of sputtered atoms incident upon the substrate is large, intermixing and mutual diffusion between incoming atoms and substrate atoms tend to occur easily. Therefore, the adhesion of film to substrate obtained by sputter deposition is stronger than that by evaporation or plasma assisted CVD.

Having grown a polycrystalline film by sputter deposition, one may get a single crystalline film by subsequent SPE procedure. This means a relevant heat treatment of the deposit under suitable growth conditions, i.e., in vacuum or in inert gas environment (see Sect. 4.1).

A lot of different material systems have been grown epitaxially by sputtering [6.14]. Among them the following groups are of largest technical importance:

(i) Metals and alloys, including: nonreactive metals (Au, Ag, and Pt), refractory metals (Ti, Zr, Nb, Ta, Mo, W), magnetic metals and alloys (Fe, $NiFe_2$), non-magnetic alloys (Au-Ni, Ag-Mg, Ag-Al).

(ii) Elemental semiconductors, mainly Ge and Si.

(iii) Compound semiconductors, including: tellurides ($Bi_2Te_3$, HgTe, CdTe, PbTe, PbSnTe), sulfides (CdS, ZnS), III–V compounds (InSb, GaAs, AlSb, AlN).

(iv) Oxides, including: magnetic materials ($MgFe_2O_4$, $Gd_3Fe_5O_{12}$, Gd-$FeO_3$), simple oxides ($Ta_2O_5$, $Al_2O_3$, $ZrO_2$, ZnO, $VO_2$), mixed oxides (GdI-garnet, $TbFeO_3$, $Bi_4Ti_3O_{12}$).

An exhaustive collection of experimental data concerning the epitaxial growth by sputtering of the listed materials is given in [6.14].

Concluding the considerations on growth of epitaxial films by means of sputtering one has to emphasize that epitaxy, often leading to films of excellent single crystal quality, has been achieved in a wide variety of metals, alloys, semiconductors, and oxides. It is clear that the cases studied to date represent only a small part of the available materials for exploration. The sputter technique is immediately applicable to the epitaxial growth of any metal or alloy, providing the requisite conditions of cleanliness and target cooling are met. The problem of producing epitaxial semiconductors that are technologically useful may be somewhat more complex, and requires a more complete understanding of the role of electron and ion bombardment in influencing the crystallographic perfection and carrier concentration of the grown film. The studies concerning epitaxial films of Ge and Si suggest that electrical properties can be optimized by means of growth in a plasma-free region or by careful adjustment of bias-sputtering conditions.

One of the most challenging segments of this field involves the further exploration of semiconductor and insulator compounds of potential use in electronic devices. In addition to such advanced and widely used methods as MBE and MOVPE, sputtering, with its large area capability, offers an attractive supplement for the sophisticated metal-oxide-semiconductor (MOS) or silicon-on-insulator (SOI) technologies of electronic devices [6.14].

### 6.1.7 Pulsed Laser Deposition

Conceptually and experimentally, PLD is extremely simple, probably the simplest among all thin film growth techniques. Figure 6.8 shows a schematic

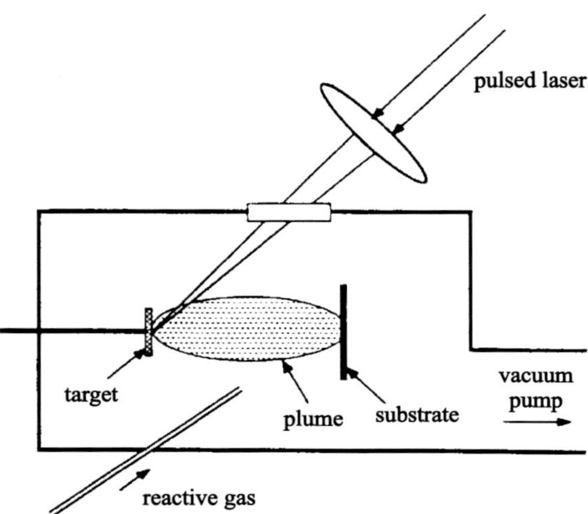


**Fig. 6.8.** Schematic diagram of a common PLD system. In special applications variations of this general arrangement can be introduced (taken from [6.21])

illustration of an experimental setup. It consists of a target holder and a substrate holder housed in a vacuum chamber. A high-power laser is used as an external energy source to vaporize materials and to deposit thin films. A set of optical components is used to focus and raster the laser beam over the target surface. The decoupling of the vacuum hardware and the evaporation power source makes this technique so flexible that it is easily adaptable to different operational modes without the constraints imposed by the use of internally powered evaporation sources. Film growth can be carried out in a reactive environment containing any kind of gas with or without plasma excitation. It can also be operated in conjunction with other types of evaporation sources in a hybrid approach.

Early work in PLD ventured into the studies of organic crystals, semiconductors, dielectric materials, and refractory metals. Most work was carried out without a persistent effort to make a long-lasting impact. The diversity is reflected in the fact that this technique has been without a common name and acronym for a long time. Names such as laser sputtering, laser assisted deposition and annealing (LADA), pulsed laser evaporation (PLE), laser molecular beam epitaxy (LMBE), laser-induced flash evaporation (LIFE), and many others were used. The present name, pulsed laser deposition (PLD), was designated by the first Material Research Society Symposium on Pulsed Laser Ablation held in San Francisco in April 1989 [6.23]. Since that time the PLD technique has become one of the most versatile and powerful methods for producing multicomponent thin films.

During the interaction of a laser beam with a solid target, complex physical processes occur. For acquiring a better understanding of the basic phenomena involved in laser–solid interaction, several models have been proposed, which have served as the first approximation to the physical reality of this interaction.

The general view is that material removal from a solid impacted by a laser beam involves a two-step process. As the laser beam hits the target surface, photons are absorbed and their energy is coupled into the material, which results in the formation of a molten layer that vaporizes. The vaporization process creates a recoil pressure on the liquid layer, which contributes to expel the molten material. Thus, the material removed is a combination of vapor and liquid. This model gives a simplified view of a more complicated process.

The real mechanism that leads to material ablation depends on both the laser characteristics, and the optical, topological, and thermodynamic properties of the target. When the laser radiation is absorbed by a solid surface, electromagnetic energy is converted first into electronic excitation and then into thermal, chemical, and even mechanical energy to cause evaporation, ablation, excitation, plasma formation, and exfoliation. Evaporants form a so-called "plume" consisting of a mixture of energetic species including atoms, molecules, electrons, ions, clusters, micron-sized solid particulates, and molten globules. The collision mean free path length inside the dense plume is very short. As a result, immediately after the laser irradiation, the plume rapidly expands into the vacuum from the target surface to form a nozzle jet with hydrodynamic flow characteristics. This process has many advantages as well as disadvantages. The advantages are: flexibility, fast response, energetic evaporants, and congruent evaporation. The disadvantages are: the presence of micron-sized particulates, and the narrow forward angular distribution that makes large-area scale-up a very difficult task (see Fig. 6.8 below).

Another important aspect is the secondary interaction of the laser radiation with the target and the plume ejected from it. For example, photons from the laser can be absorbed in at least three different ways:

(i) Volume absorption by electrons and phonons in the lattice of the solid,
(ii) Surface absorption by free carriers in the molten layer, and
(iii) Absorption by an emitted plume (formation of a plasma).

The second and the third mechanisms mentioned are higher-order perturbations, but they directly affect the film deposition by inducing changes in the plasma properties.

Efforts to understand the laser–solid interaction during the deposition of multicomponent oxide films have been largely related to high temperature superconductor (HTSC) materials and have concentrated mainly on characterizing the laser-generated plume. The interaction of a short-pulse, high-power ($10^7$–$10^8$ W/cm$^2$) laser with a ceramic target, under conditions appropriate for the growth of high-quality multicomponent oxide thin films, can be viewed as occurring in several stages, as shown in Fig. 6.9 [6.22].

Optical irradiation of the target causes surface heating. The increase in the surface temperature depends on the optical penetration depth of the material, the thermal diffusivity of the target, and the rate at which energy is deposited into the system (laser pulse width). For metallic and semiconductor narrow-gap materials the thermal diffusion distance is long compared to the optical absorption distance. Generally, the electronic structure (i.e. electronic states and band gap) of the target is not important since the high electric fields generated by the laser result in dielectric breakdown (i.e. at these power densities all materials absorb a significant fraction of the radiation, resulting in surface heating). Under these conditions the surface temperature can exceed the melting temperature of a ceramic material, like for example the multicomponent oxide HTSC material $YBa_2Cu_3O_x$ written in abbreviated form as YBCO, which has the melting point $T_m \approx 1400°C$.

For the laser-power levels mentioned above, surface heating is followed by melting and evaporation. Laser-etching studies have shown that there is a threshold energy density for removal of material from the surface. For YBCO impacted by a pulsed laser beam of 248 nm, for example, this threshold has been measured to be $\approx 0.11\,J/cm^2$[6.22]. Above the threshold, material re-

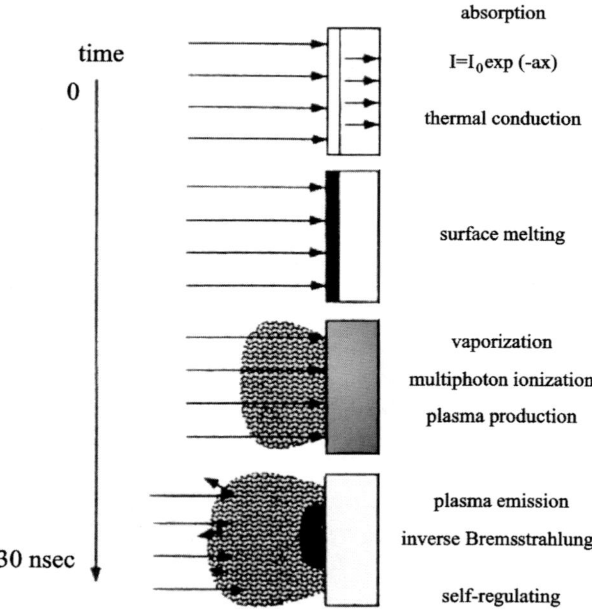


**Fig. 6.9.** Schematic representation of the laser–solid target interaction stages (taken from [6.22])

moval varies linearly with laser fluence [6.23]. However, for efficient material removal and congruent evaporation of multicomponent targets, short wavelengths (e.g., ultraviolet radiation) and short pulse widths ($\approx$ 30 ns) yield superior film quality, i.e., smooth and stoichiometric epilayers [6.21]. Presumably this is due to a change in the optical penetration depth, which decreases with decreasing wavelength. The decreased volume of material with which the laser can interact results in more efficient coupling of the optical energy to the target. From etch-depth measurements, the optical penetration depth of a 248 nm laser beam in YBCO, for example, is $\approx$ 500 Å [6.22]. Long optical pulses (μs) and/or infrared radiation results in a more thermal-like heating of the target material. With IR sources, differences in the equilibrium vapor pressures of the target constituents are more likely to affect the composition of the laser-produced vapor and lead to the production of non-stoichiometric films.

The PLD has advantages and disadvantages that need to be appropriately considered. Major advantages of the PLD technique include:

(i) relatively high deposition rates,

(ii) the possibility of producing multicomponent thin films with the same stoichiometry of the multicomponent materials used as targets, provided that appropriate deposition conditions are used,

(iii) the possibility of producing as-deposited films with the appropriate microstructure at relatively low substrate temperatures,

(iv) the potential for producing multilayered films of many different materials, since their parent targets can be efficiently ablated with virtually similar conditions of laser energy density, and

(v) the possibility of time-sharing the laser beam among several deposition systems (the steering of the beam from one chamber to the other can be easily accomplished by automated mirror technology).

Major disadvantages of the PLD method include:

(i) the formation of particulates during the target ablation, which are subsequently deposited on the film (however, various methods are being explored to eliminate this problem),

(ii) the highly focused nature of the ablated plume makes it difficult to scale-up the PLD method to cover large areas (various schemes are being explored to extend the PLD method to cover large-area substrates (at least 100 mm in diameter), but this will not be as easy as with other techniques, such as plasma- and ion-beam-sputter deposition, CVD, or sol-gel techniques),

(iii) the need for polishing the surface of the targets for each new film deposition, due to the extensive erosion of the target by the laser beam (this problem would add undesirable extra time in an industrial thin-film processes), although new target-holder hardware recently developed allows an operator to change targets through a load-lock system (as it usually is in the case of MBE systems) without breaking vacuum, and

(iv) the need for protecting the special quartz UHV windows, used to introduce the laser beam into the deposition chamber, from deposition of the material ablated from the target (this problem may be serious since film deposition on the window requires frequent cleaning to avoid laser-light absorption, and thus reduction of the energy available at the target for ablation).

Obviously, future research and development directions should include work on the main disadvantages of the PLD technique.

## 6.2 Chemical Vapor Deposition

The chemical systems involved in CVD processes are to a large extent defined by the volatile transporting agents. The most frequently used agents are: halide, oxide, hydride and metalorganic compounds. Accordingly, the CVD technique is often divided into different epitaxial growth methods, like for example: halogen transport epitaxy [6.24], hydride vapor phase epitaxy [6.25], or metalorganic vapor phase epitaxy (MOVPE) [6.3]. We will not go into details of each mentioned growth method, instead, some general considerations concerning epitaxy by CVD (the main features of CVD processes) will be presented here, because of the particular importance of this technique for growth on an industrial scale of many kinds of materials used in electronic devices and integrated circuits [6.24]. The MOVPE method, the most frequently used CVD method in technological research laboratories, but also widely introduced to mass scale production of semiconductor heterostructure devices will then be discussed more in detail in subsequent chapters. this method occurs in the literature also with other abbreviations like MOCVD, OMVPE or OMCVD. We will stick throughout this book to MOVPE, used in the international conferences and indicating that epitaxy "E" is a relevant aspect.

### 6.2.1 Principles of CVD Processes

Following [6.26] we will define epitaxy by CVD as any process where at least one of the component elements of a crystal to be grown is supplied as a halide (oxide, hydride, metalorganic compound) gas to the substrate, where a halogen (oxide, hydride, metalorganic compound)-elimination reaction occurs, resulting in epitaxial growth. This method is particularly important when a component element of the crystal has a very low vapor pressure. By changing the element to its volatile transport agent, the delivery of this element to a growth region becomes possible. The transport agents are chlorine or metal organic compounds in most of the present growth systems because of easier purification and moderate vapor pressures of these gases, suitable to control the grown film's properties.

The transport agent's reactions are divided into several categories: disproportionation, hydrogen reduction, and thermal decomposition reactions. The disproportionation reaction proceeds with changing the valency of the chemical bond of the transporting gas, such as from $GaCl_3$ to GaCl, in the case of halides. This method is, however, seldom used at present. The hydrogen reduction method is currently more popular, since the presence of hydrogen is very effective in preventing the epitaxial surface from oxidizing.

The characteristics of epitaxy by CVD have been reported by many researchers. The discussions were based on well-developed theories of thermodynamics and fluid dynamics and various parameters were used. Partial pressure calculations have been frequently performed for various transport systems. In order to know the potential for crystal growth from a given gas, several chemical parameters have been defined such as the ratio Si/Cl, Ga/Cl, or Cl/H, in the case of halogen transporting agents. The supersaturation has also been derived, by combining the calculated partial pressures of the gas phase components. The chemical potential is known to be a good parameter to understand the chemical situation of a given system. It will also give information as to whether crystal growth would or would not occur. Therefore, we will begin the considerations on epitaxy by CVD with a short introduction to the relevant definitions and relations concerning this parameter. More general information on this subject can be found in Sect. 11.1.

The chemical potential of a given phase is defined as the Gibbs free energy increase when one mole of substance is added in that phase at constant temperature and pressure:

$$\mu = \left.\frac{\partial G}{\partial n}\right|_{T,p} \tag{6.7}$$

If we use the molar Gibbs energy, $\mu = \Delta G$, the Gibbs energy is expressed in terms of enthalpy $\Delta H$ and entropy $\Delta S$, which are determined from the specific heat $C_p$.

$$\mu = \Delta G = \Delta H - T\Delta S = \Delta H_f^0 + \int_{298}^{T} C_p \mathrm{d}T - T\left(\Delta S^0 + \int_{298}^{T} \frac{C_p}{T}\mathrm{d}T\right). \tag{6.8}$$

When the standard heat of formation $\Delta H_f^0$, the standard entropy $\Delta S^0$, and the specific heat of the substance $C_p$ are known, one can obtain the chemical potential of the substance in any given phase [6.24]. Suppose now that we have a vapor transport system as shown in Fig. 6.10 which is typical of a CVD process. This system consists of a solid $M$ and gaseous species $A$, $B$, $C$, and $D$. The component that occurs in one of the gaseous species is transported to the solid phase through the halogen transport reaction given by

$$M_{\mathrm{solid}} + aA_{\mathrm{gas}} \rightleftharpoons bB_{\mathrm{gas}} + cC_{\mathrm{gas}} + dD_{\mathrm{gas}}. \tag{6.9}$$

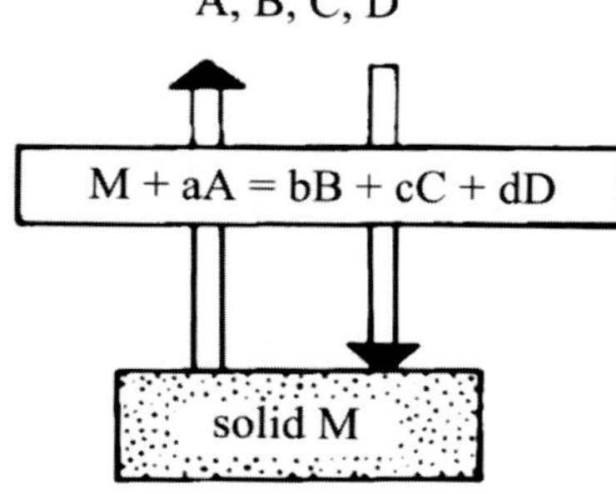


**Fig. 6.10.** Vapor transport system comprising vapor and solid phases and the transport reaction between them (taken from [6.24])

Halogen atoms are included only in two gaseous species, namely, in $A$ and in one of $B$, $C$, or $D$. When the system is nearly at equilibrium, the total chemical potential change caused by a slight perturbation must be zero

$$\sum_i \mu_i \mathrm{d}n_i = 0 \tag{6.10}$$

where $\mathrm{d}n_i$ is the molar increase of species $i$ due to the perturbation. Since no material is added to the system from outside, a material flow balance is established between the existing species

$$\mathrm{d}n_M = \frac{\mathrm{d}n_A}{a} = -\frac{\mathrm{d}n_B}{b} = -\frac{\mathrm{d}n_C}{c} = -\frac{\mathrm{d}n_D}{d}. \tag{6.11}$$

The constants $a, b, c$, and $d$ are the reaction coefficients of reaction (6.9). In equilibrium

$$\mu_{M,\mathrm{solid}} + a\mu_a = b\mu_b + c\mu_c + d\mu_d. \tag{6.12}$$

If we use the ideal-gas approximation, we can write

$$\mu_{M,\mathrm{solid}} = -a\mu_A^0 + b\mu_B^0 + c\mu_C^0 + d\mu_D^0 + RT\log(p_A^{-a}\, p_B^b\, p_C^c\, p_D^d). \tag{6.13}$$

The equilibrium constant of reaction (6.9) is given by

$$K(T) = p_A^{-a}\, p_B^b\, p_C^c\, p_D^d. \tag{6.14}$$

We thus obtain

$$\mu_{M,\mathrm{solid}} = -a\mu_A^0 + b\mu_B^0 + c\mu_C^0 + d\mu_D^0 + RT\log(K_1(T)). \tag{6.15}$$

Next, we obtain the chemical potential under non equilibrium conditions. In doing this we assume a hypothetic chemical potential of the solid $M$, i.e., $\mu_{M,\mathrm{gas}}$, which is expressed in the partial pressures of the gaseous species $A$, $B$, $C$, and $D$ as

$$\mu_{M,\mathrm{gas}} = -a\mu_A^0 + b\mu_B^0 + c\mu_C^0 + d\mu_D^0 + RT\log(K(T)). \tag{6.16}$$

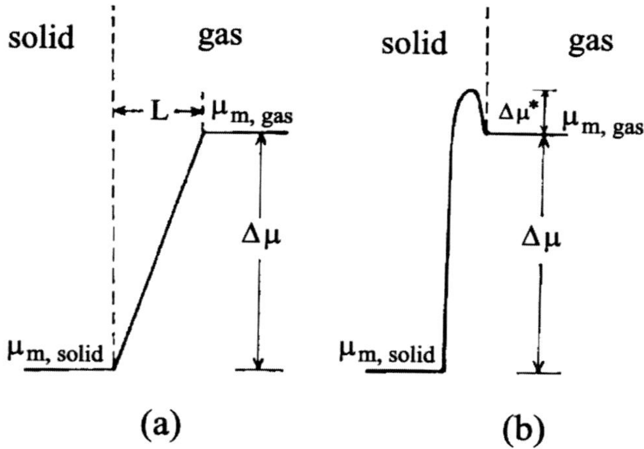


**Fig. 6.11.** Chemical potential profiles in vapor and solid phases during the CVD process (taken from [6.24])

Here $\mu_{M,\text{gas}}$ can be regarded as a vapor chemical potential of the solid $M$, as if the solid $M$ is existing in the vapor phase.

The following relations are important for understanding the phase transitions at the solid surface of the substrate during the CVD process:

(i) $\mu_{M,\text{gas}} = \mu_{M,\text{solid}}$ means equilibrium between the two phases

(ii) $\mu_{M,\text{gas}} > \mu_{M,\text{solid}}$ means that growth of the solid film occurs

(iii) $\mu_{M,\text{gas}} < \mu_{M,\text{solid}}$ means that etching (vaporization) of the solid surface occurs.

The difference between the chemical potentials of the vapor and solid phases, $\Delta\mu = \mu_{M,\text{gas}} - \mu_{M,\text{solid}}$, is the driving force for crystal growth. The evolution of the chemical reaction will result in $\Delta\mu$ becoming zero. However, in an actual growth system, gas having a chemical potential higher than that of the substrate is continuously supplied, and hence $\Delta\mu$ remains greater than zero in the system. If we know all partial pressures of the gaseous species supplied to the solid, the hypothetical chemical potential defined by (6.16) is obtained by substituting those pressures:

$$\Delta\mu = RT \log\left(\frac{p_A^{-a}\, p_B^b\, p_C^c\, p_D^d}{K_1(T)}\right). \tag{6.17}$$

$K_1(T)$ is here the equilibrium constant of the deposition reaction given by (6.9) at the temperature $T$. We can calculate $\Delta\mu$ if $K(T)$ is known. Figure 6.11 shows chemical potential profiles for two cases. The case shown in Fig. 6.11a expresses a situation where the chemical reaction at the substrate surface is very fast compared with the diffusion speed of the gas species. In this case

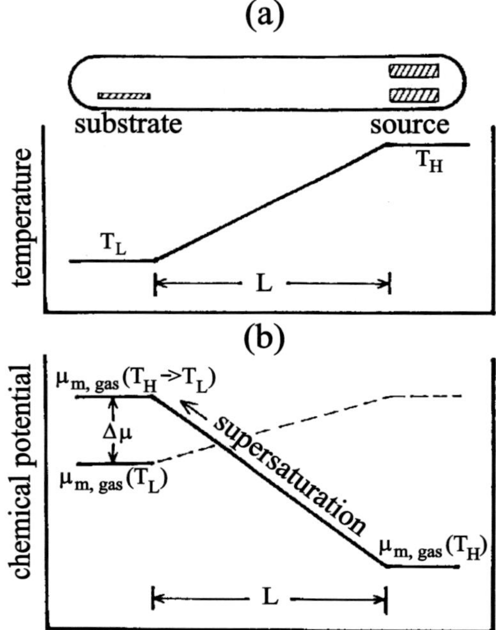


**Fig. 6.12.** Crystal growth in a closed-tube system and the chemical potential profiles (taken from [6.24])

the chemical potential difference is developed in the vapor phase, resulting in a chemical potential gradient $\Delta\mu/L$. The distance $L$ is dependent on the diffusion velocity of the gaseous species. On the other hand, as shown in Fig. 6.11b, when the reaction velocity is very low, the chemical potential difference does not exist in the vapor phase but instead it occurs at the interface between the vapor and the solid phases, namely on the substrate surface, resulting in a chemical potential step $\Delta\mu$.

Let us now consider the two possible cases of CVD growth systems, namely, the closed-tube system (see Fig. 6.12) and the open-tube system (see Fig. 6.13), on the example of halogen transport epitaxy. In the closed-tube growth system the source and the substrate are positioned at opposite ends of the tube maintained at temperatures $T_H$ and $T_L$. A controlled amount of halogen gas such as $I_2$ or HCl is introduced, which gives rise to a disproportionation reaction. For instance, this reaction in the case of the GaAs-Cl system is given by [6.24]

$$\mathrm{GaAs} + \frac{1}{2}\mathrm{GaCl_3} \rightleftharpoons \frac{3}{2}\mathrm{GaCl} + \frac{1}{4}\mathrm{As_4}. \tag{6.18}$$

The vapor phase at both ends is assumed to be in equilibrium with the respective solid. We shall write, using the definition (6.16),

$$\mu_{M,\mathrm{gas}} = \mu_M^0(T) + RT\log(K(T)). \tag{6.19}$$

However, at $T = T_H$ and $T = T_L$

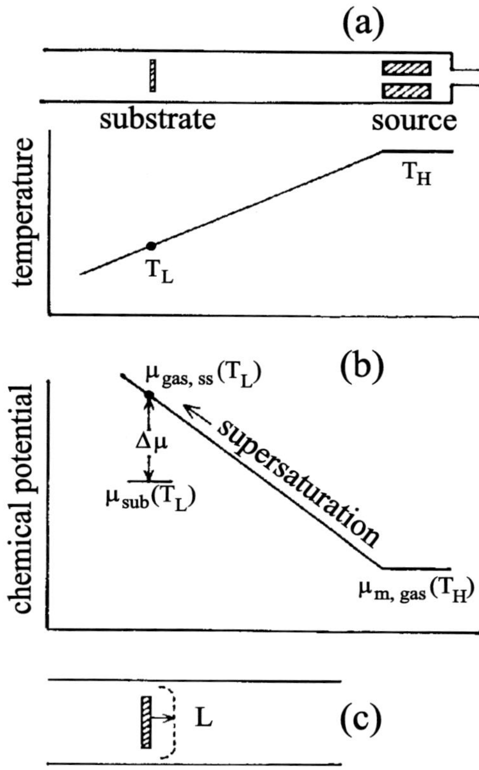


**Fig. 6.13.** Crystal growth in an open-tube reactor and the chemical potential profile (taken from [6.24])

$$\mu_{M,\mathrm{gas}} = \mu_{M,\mathrm{solid}}. \tag{6.20}$$

Here, $K(T)$ is the equilibrium constant of reaction given by (6.18). The driving force for crystal growth might be supposed to be the difference between the chemical potentials at $T_H$ and $T_L$, namely $\Delta\mu = \mu_{M,\mathrm{gas}}(T_H) - \mu_{M,\mathrm{gas}}(T_L)$. However, this is not true, because the chemical potentials in the difference should be defined at an identical temperature, which should be the substrate temperature $T_L$. Therefore, we write

$$\Delta\mu = \mu_{M,\mathrm{gas}}(T_H \rightarrow T_L) - \mu_{M,\mathrm{gas}}(T_L). \tag{6.21}$$

where $\mu_{M,\mathrm{gas}}(T_H \rightarrow T_L) = \mu_M^0(T_L) + RT_L \log(K(T_H))$. Thus, we obtain

$$\Delta\mu = RT_L \log\left(\frac{K(T_H)}{K(T_L)}\right). \tag{6.22}$$

The advantage of the expression using the chemical potential is that we can understand the thermodynamic situation of the gas mixture of interest also in the case of growth in an open tube CVD system. It is not necessary to calculate partial pressures to know whether a given gas mixture will exert a driving force for crystal growth. The chemical potential of a gas mixture must change as it flows down the reactor. A model case is shown in Fig. 6.13. The source and the substrate, kept at $T_H$ and $T_L$, respectively, are the same material. Halogen gas is introduced and reacted with the source, generating several gas species. The chemical potential along the flow direction is shown in Fig. 6.13b. The vapor phase is assumed to be in equilibrium with the source, i.e., the chemical potential of the source and the gas mixture are equal, $\mu_{\text{source}}(T_H) = \mu_{\text{gas}}(T_H)$. The gas mixture is flowing toward the substrate without any deposition in the middle region. It becomes supersaturated when it reaches the substrate. The gas temperature is decreased from $T_H$ to $T_L$, but the gas composition is unchanged. The chemical potential of the supersaturated gas, $\mu_{\text{gas,ss}}$ can be expressed as

$$\mu_{\text{gas,ss}} = \mu^0{}_{\text{gas}}(T_L) + RT_L \sum_i \vartheta_i \log p_i . \tag{6.23}$$

Since the source region is in equilibrium, $\sum \vartheta_i \log p_i = \log K(T_H)$, and the chemical potential of the supersaturated gas can be expressed as

$$\mu_{\text{gas,ss}} = \mu^0{}_{\text{gas}}(T_L) + RT_L \log K(T_H) \tag{6.24}$$

and its chemical potential becomes higher than that of the substrate

$$\mu_{\text{gas,ss}}(T_L) = \mu_{\text{sub}}(T_L) + \Delta\mu . \tag{6.25}$$

The chemical potential difference thus generated exists over some distance. The boundary-layer theory [6.3] is well known in flow dynamics for the calculation of the transition length, $L_{BL}$, near the substrate. It seems that $L_{BL}$ corresponds to the length of the chemical potential difference. However, $L_{BL}$ is defined as the transition region thickness in the velocity profile of the flowing gases, and is not always identical to the thickness of the concentration profile. Figure 6.13c shows a rough sketch of $L$. The gas species entering the region defined by $L$ contribute to the growth reaction. Within $L$, the gas composition is changed due to the diffusion flow of the reactant species. If we define $L$ as the boundary layer thickness of the concentration profile, it determines the chemical potential gradient $\Delta\mu/L$. In the case of the closed-tube reactor, $L$ is equal to the distance between the source and the substrate. However, in the open-tube reactor, the separation between the source and the substrate has little meaning. We use a forced flow in the open-tube reactor, where the flow pattern is strongly dependent on the carrier-gas flow rate and the substrate holder geometry. So far, no relation has been reported between the boundary layer thickness $L_{BL}$ and the length of the chemical potential difference $L$. Since actual reactors have a complicated structure in the substrate region, it is very difficult to obtain the exact profiles of gas velocity and concentrations. In order to overcome such difficulty, Secrest et al. [6.27] have proposed a procedure based on the finite-element method involving the balance of mass, balance of momentum, and balance of energy. They calculated the GaAs growth rate with the assumption that the growing surface is in equilibrium and the growth rate is limited by the transport rate of $As_4$ in the transition layer. They obtained a velocity profile near the substrate in which the boundary layer is not found since the distance between the substrate surface and the tube wall is too short to develop a wall-formed boundary layer as predicted by the conventional theory. A good agreement is obtained for the GaAs growth rate. It is particularly noteworthy that the rates were calculated without empirical determination of parameters or use of the boundary layer approximation, so far used in many papers. Although this kind of procedure needs a very fast computer, it became at present the major technique for the analysis of vapor transport growth, since not only the velocity profile but also many thermodynamic characteristics, such as the chemical potential gradient, can be readily visualized in figures.

### 6.2.2 Mass Transport and Heat Transfer in CVD Reactors

In a VPE environment the mass transport, i.e. the flux $\boldsymbol{j_i}$ of reactant $\boldsymbol{i}$ to the gas–solid interface is given by convective flow, by diffusion in a concentration gradient and by thermodiffusion (temperature gradient driven) [6.29]:

$$\boldsymbol{j}_i = \frac{\boldsymbol{p_i} v}{kT} - \frac{D_i}{kT}\left[\nabla p_i + \frac{\alpha_i}{T} p_i \nabla T\right] \tag{6.26}$$

where $\boldsymbol{p_i}$ is the partial pressure of the species $i$, $D_i$ is the diffusion constant of species $i$ within the carrier gas and $\alpha_i$ is the thermodiffusion coefficient. Clearly inhomogeneous velocity profiles $v$, distributions of growth determining species $p_i$ as well as inhomogeneous temperature profiles can create very inhomogeneous deposition according to (6.26). Much effort therefore has been undertaken in the past to study and, as a consequence, to control the flow and heat transfer in a CVD reactor in order to achieve uniform deposition. This is especially needed for semiconductor technology. The range of CVD reactor models include analytical boundary layer inspired models, two-dimensional boundary layer models, two-dimensional elliptic descriptions, and three-dimensional models [6.30]. Because of the difficulties in obtaining experimental flow rate data, these models have typically been verified through comparison of measured and predicted deposition rates. However, the deposition rate is a cumulative function of chemical reaction and transport effects and cannot give a detailed picture of the actual gas phase transport phenomena occurring above the susceptor.

The most accurate modeling calculations have been performed on the mathematical basis of the nonlinear differential equations describing conservation of mass, momentum and energy in the reactor (11.15)–(11.22). Experimental verifications have been done via smoke visualization, velocity and temperature measurements. The most complete studies have been performed for horizontal reactors which will be used here as an example to present the principal features with respect to external macroscopic parameter such as flow rates, susceptor temperature or pressure [6.28, 31].

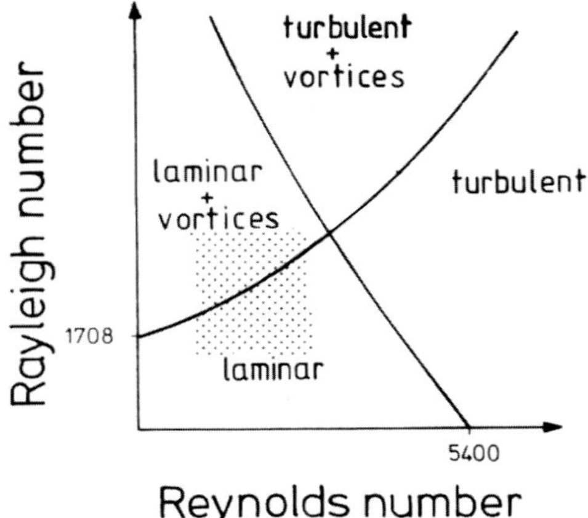


**Fig. 6.14.** Schematic illustration of the different hydrodynamical flow modes occurring as a function of Rayleigh and Reynolds numbers. CVD is desired to operate in the laminar area but because of necessary temperature gradients is quite often also near the border line where thermally driven vortices (rolls) occur in the reactor. (taken from [6.28])

All known CVD reactors operate in or near the laminar flow regime. This hydrodynamical regime is specified by the magnitudes of the Reynolds number

$$R_e = \frac{<v>h}{\nu} \tag{6.27}$$

and the Rayleigh number:

$$R_a = \frac{g\beta h^3 \Delta T}{\nu\alpha} \tag{6.28}$$

where $<v>$ is the average velocity, $h$ the height of the channel, $\nu$ the kinematic viscosity, $g$ the gravitational constant, $\beta$ the thermal expansion coefficient, $\Delta T$ the temperature difference and $\alpha$ the thermal diffusivity.

The former describes the momentum flux by convection in relation to that by diffusion and the latter the ratio of the buoyancy force to the viscous force [6.32]. Depending on the actual values the flow pattern can be laminar, show vortices (rolls) or may be turbulent (Fig. 6.14). In the region where vortices are occurring transport is difficult to control by external parameters and may lead to unexpected transport features of the reactants. The turbulent region, in principle, might be useful for a homogeneous deposition too. It would require, however, inconvenient high flow rates. To the authors' knowledge epitaxial growth under such conditions has not yet been reported. The computational solution of (11.15)–(11.22) in general assumes steady state conditions, i.e. the time derivatives are set equal to zero with the consequence that dynamical effects like gas switching or rapid thermal changes cannot be treated and also the turbulent regime of Fig. 6.14 is excluded. Beside the steady state approximations in general one assumes also low concentrations of the reactants, e.g., that the total pressure is dominated by a carrier gas. Consequently volume changes due to a change of the number of moles during gas phase reactions are neglected. Similarly, energy contributions from the heat of reactions are not considered. Therefore, the flow and energy solution can be decoupled from the mass transfer analysis. These simplifications might not always be possible (for example in systems involving mercury compounds) [6.33], but are nearly always fulfilled in MOVPE reactors.

The hydrodynamical boundary conditions include the standard hydrodynamic conditions that the velocities are all zero at the reactor walls (no slip condition) or in case of a susceptor rotating with frequency $\omega$ that the azimuthal velocity $v_\theta = \omega \cdot r$ ($r$ = radial distance from the axis) is zero. Moreover, the inlet flow profile is specified for the only component in the axial direction as either parabolic (Hagen–Poiseuille) or as a constant (plug flow).

The development of vortices in a CVD reactor is also a phenomenon usually not wanted in the light of a very homogeneous deposition. Vortices (rolls) appear at large Rayleigh numbers, e.g. when the buoyancy force is large (large density of the gas, large dimensions of the reactor, large temperature gradients). Vortices may lead to uncontrolled mass transfer within the reactor and may increase also the residence time of the gas mixture considerably. As a consequence of the former thickness and composition uniformity may be adversely affected while the latter will increase gas switching times which might become rather long and might make the growth of sharp heterointerfaces impossible.

Besides affecting the flow pattern the temperature distribution is critical since the chemical reactions are temperature activated. The actual heat transfer within CVD reactors, however, is complex involving radiation heat transfer between solid surfaces, convection in the gas phase as well as heat conduction in the gas, in the reactor walls and in the susceptor. Thus, the more critical boundary conditions are those for the temperature. In a first guess one might assume the wall and the susceptor temperatures to be isotherms. However, measurements show that this is by no means the case and one has to model the full thermal flow by heat conduction in the solid parts of the reactor (walls, susceptor), heat conduction by the gas and by radiation [6.28]. As an example we show in Fig. 6.15 a result for the temperature field above the susceptor. The fixed temperature condition at the top wall (Fig. 6.15a) in principle can be realized with a cooling system. It leads to a quite nicely developed classical boundary layer (see Sect. 11.2.2). The adiabatic wall case leads to an increase of the top wall temperature along the length of the reactor and the isotherms will terminate perpendicular on the top wall. On the

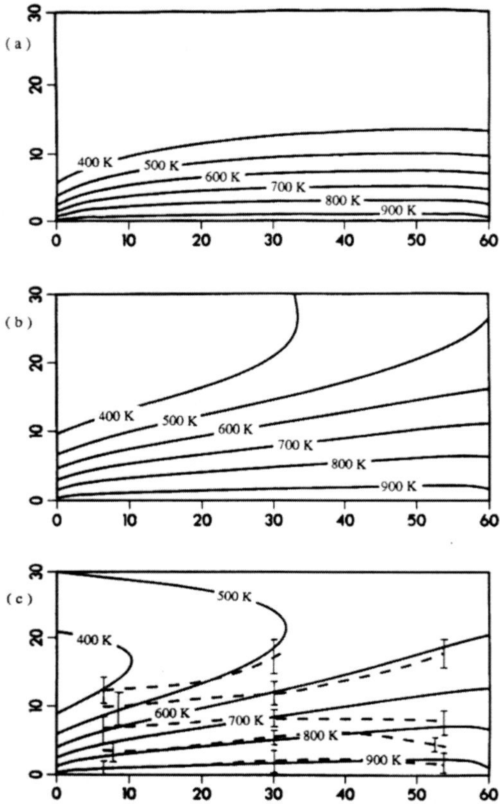


**Fig. 6.15.** Effect of temperature boundary conditions on predicted isotherms in a horizontal reactor heated at the bottom with gas inlet at the left: (**a**) the temperature of the upper wall is fixed at 300 K; (**b**) the upper wall is adiabatic; (**c**) detailed boundary conditions together with experimental data (dashed lines). The x (y) axis of the graphs is the flow direction (height above the susceptor) respectively. Units are in cm (see geometry in Fig. 6.16) (taken from [6.28])

other hand, the agreement between calculations performed with very specific boundary conditions and experiments (Fig. 6.15c) shows that thermal boundary conditions are very important and have to be specified in detail. Only when detailed wall balances, including heat conduction in the wall, heat transfer to the surroundings and radiative heat transfer (using Planck's radiation law with correct emissivities) were taken into account was it possible to reproduce the experimental results.

These and other quite detailed theoretical studies as well as experiments have been performed for horizontal reactors such as the one displayed in Fig. 6.16 and we will discuss in the following examples obtained with this

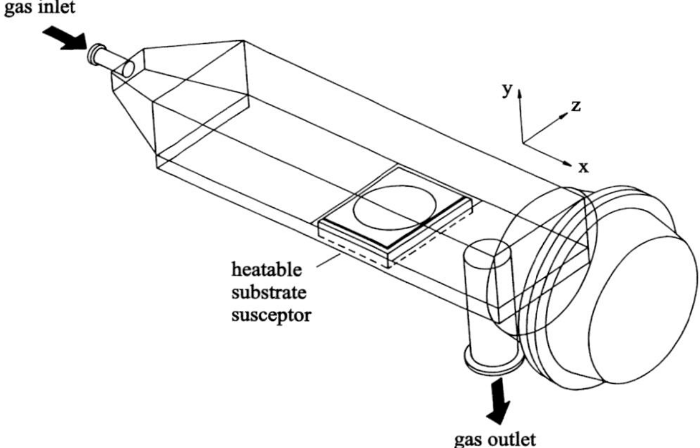


**Fig. 6.16.** Schematic illustration of a horizontal reactor cell used in studies of CVD growth (taken from [6.28])

geometry in more detail here. Equations (11.15)–(11.22) are then solved numerically by the Galerkin finite element technique [6.34]. An appropriate mesh is generated which in general will be not rectangular but adapted to the shape of the reactor walls. Figure 6.17 gives examples for a horizontal reactor with a horizontal and a tilted susceptor. In the modeling procedures performed in [6.28] the reactor is considered to be symmetric about the

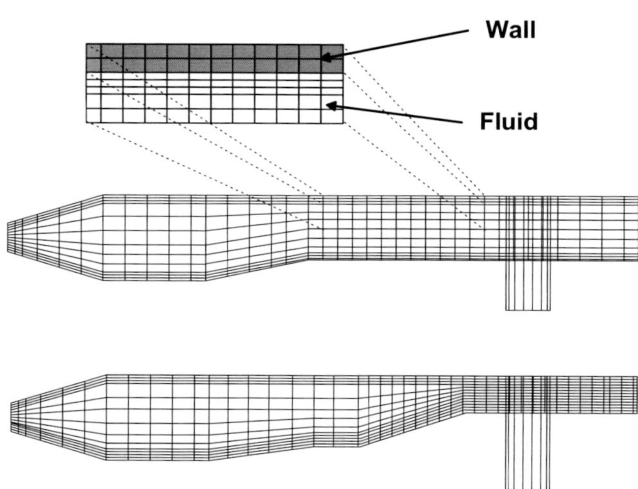


**Fig. 6.17.** Finite element mesh used in computations: (**a**) horizontal susceptor; (**b**) tilted susceptor (taken from [6.28])

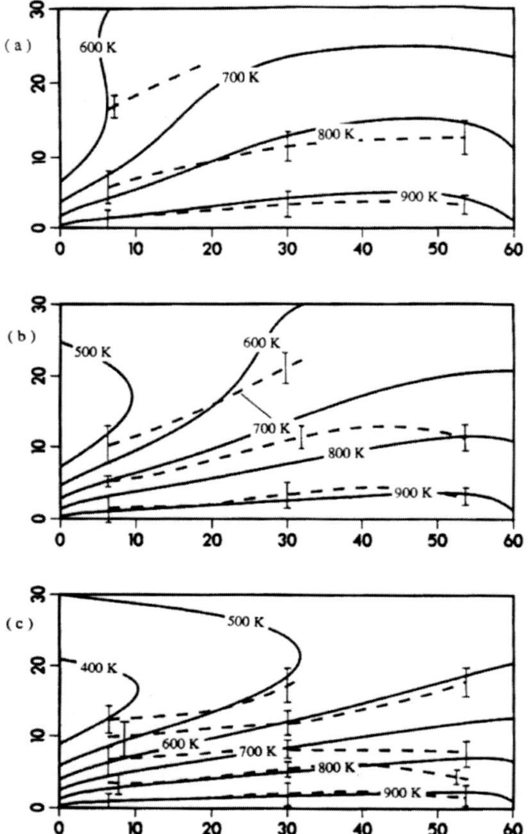


**Fig. 6.18.** Measured (dashed lines) and predicted (solid lines) isotherms above the susceptor for different inlet flow rates of hydrogen at 1 atm pressure: (**a**) 2 SLM, (**b**) 4 SLM and (**c**) 8 SLM (SLM: standard liters per minute) The x (y) axis of the graphs is the flow direction (height above the susceptor) respectively. Units are in cm (see geometry in Fig. 6.16) (taken from [6.28])

midplane, and this plane is then taken as the two dimensional domain for the solutions of the basic equations describing the carrier gas flow and the temperature distribution in the reactor.

The solution vector was approximated by sets of piecewise continuous polynomials formulated on the finite element mesh and multiplied by unknown scalar coefficients. The velocity and temperature components were represented by biquadratic polynomials and the pressure by bilinear polynomials. This corresponds to the standard mixed finite element representation for an incompressible fluid [6.35]. Details of the finite element solution can be found in [6.36].

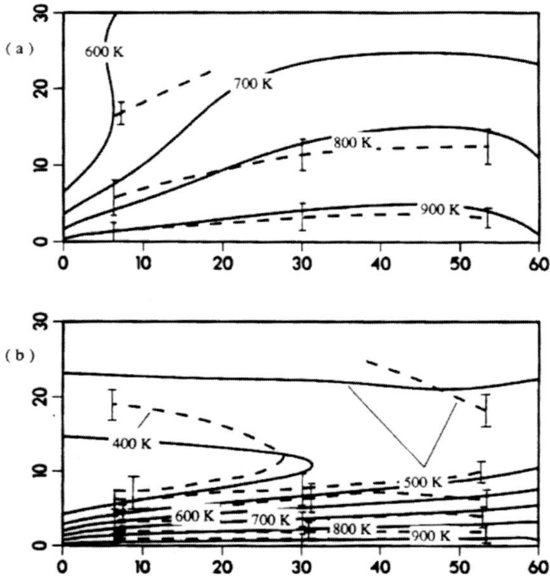


**Fig. 6.19.** Measured (dashed lines) and predicted (solid lines) isotherms above the susceptor for (**a**) hydrogen and (**b**) nitrogen at 2 SLM inlet flow at 1000 mbar The x (y) axis of the graphs is the flow direction (height above the susceptor) respectively. Units are in cm (see geometry in Fig. 6.16) (taken from [6.28])

In the following we will discuss results obtained from the variation of common CVD reactor parameters such as inlet flow rate, type of carrier gas, susceptor tilt, thermal boundary conditions and the development of rolls.

The variation of flow rate (Fig. 6.18) is discussed in terms of standard liters per minute (SLM) fed to the reactor. At low inlet flow rates heat conduction is sufficiently fast relative to convection to establish a situation where the gas temperature decreases nearly linear with vertical distance from the susceptor corresponding to a conduction dominated case. At higher flow rates convection becomes important and a boundary layer develops as may be observed from the position of the 700 K isotherm for the three flow rates shown in Fig. 6.18. The "cold finger" is the combined result of the radiative heating of the top quartz wall by the hot susceptor and the heat transfer from the top wall to the incoming colder gas (top wall boundary layer). It becomes most pronounced at the highest velocity.

Variations of the total pressure, which have also been considered, lead only to minor changes in the temperature field [6.28], a fact which is expected, since the heat transfer characteristics at constant mass flow should be independent of pressure. However, the kind of carrier gas has a strong influence as can be seen from the hydrogen vs nitrogen comparison in Fig. 6.19.

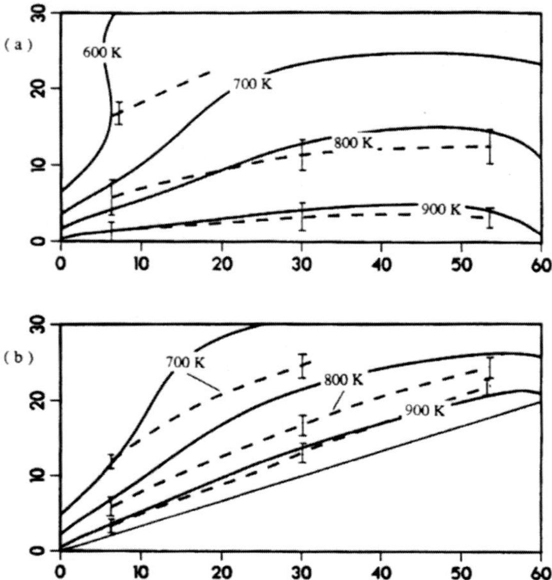


**Fig. 6.20.** Predicted (solid lines) and measured (dashed lines) isotherms above the susceptor for 2 SLM inlet hydrogen flow at 1000 mbar and two different susceptor tilt angles: (**a**) no tilt and (**b**) 18.4° tilt angle The x (y) axis of the graphs is the flow direction (height above the susceptor) respectively. Units are in cm (see geometry in Fig. 6.16) (taken from [6.28])

Since nitrogen has a thermal conductivity lower by a factor of approximately seven compared to hydrogen, convection dominates in nitrogen. As a result a definite and uniform boundary layer develops above the susceptor with the presence of a very long "cold finger". Consequently, in growth studies with nitrogen as carrier gas very uniform deposition has been observed [6.37, 38]. Thus from the viewpoint of uniformity of growth it would be desirable to use nitrogen instead of hydrogen as the carrier gas. The fact that hydrogen is still the carrier gas of choice is related to the ease of purification.

Figure 6.20 illustrates the effect of tilting the susceptor. Clearly sharper and more uniform temperature gradients are obtained which will lead to more uniform growth. The effect of changing the susceptor orientation relative to gravity (buoyancy force) has been explored too by rotating the reactor cell from the conventional horizontal with the susceptor at the bottom of the reactor (Fig. 6.16), to an aided flow case (vertical) where the buoyancy force acts in the same direction as the main flow (chimney configuration) and to an inverted geometry with the susceptor and sample mounted at the top of the reactor. Both these configurations ("chimney-reactor", "upside-down-reactor") have been shown to produce uniform deposition rates [6.39, 40]. The calculations and measurements presented in [6.28, 41, 42] show, however,

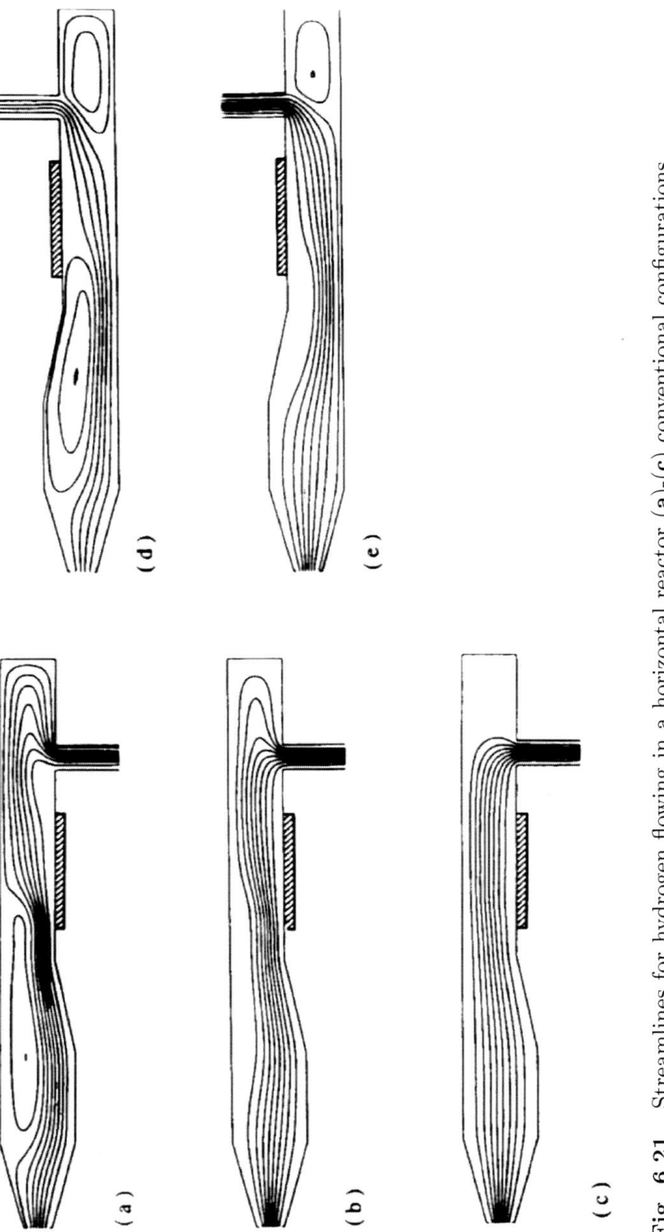


**Fig. 6.21.** Streamlines for hydrogen flowing in a horizontal reactor (**a**)-(**c**) conventional configurations, (**d**) and (**e**) with susceptor placed upside-down. Conditions: (**a**) 2 SLM, 1000 mbar; (**b**) 4 SLM, 1000 mbar; (**c**) 2 SLM, 100 mbar, (**d**) 2 SLM, 1000 mbar; (**e**) 4 SLM, 1000 mbar (taken from [6.28])

that the increase in uniformity is only rather slight and thus in reality may be offset by the larger effort to mount a substrate vertically or upside down at the top of the reactor wall.

The previous examples all gave excellent agreement between theoretical predictions and experiments. This demonstrates that with today's computational possibilities the temperature fields within reactors can be modeled quite accurately such that detailed measurements seem to be no longer necessary. The effect of the thermal boundary layers on the deposition rates can be further explored by using the analogy between heat and mass transfer [6.43]. According to this analogy the variation in heat transfer along the substrate surface will be equivalent to the variation in mass transfer if the deposition is governed by a single mass transfer limited reaction and the reactant is dilute in the carrier gas. The Nusselt number $N_u$

$$N_u = \frac{H \, \partial T/\partial y|_{\text{susceptor}}}{\Delta T}, \tag{6.29}$$

which is a measure of the heat transfer rate, may then readily be taken in order to discuss the temperature profiles, given in the last figures, in terms of deposition uniformity. Isotherms being parallel to the substrate have the same temperature gradient everywhere and thus lead to uniform growth. A larger (smaller) temperature gradient normal to the surface means larger (smaller) deposition rates.

The model simulations are also able to produce the occurrence of rolls in the flow pattern. Figure 6.21 give examples for a horizontal reactor under different operation conditions and mountings. The recirculation cells occurring in Figs. 6.21a, d, e are of course unwanted because they make gas switching processes slow and thus make it difficult to grow sharp interfaces. They can be avoided as shown in Figs. 6.21b, c by increasing the flow rate (larger Reynolds number) or by reducing the total pressure (smaller Rayleigh number, see Fig. 6.14). These and other calculations are usually performed for computational reasons, as mentioned, in two dimensions only in a reactor midplane containing the flow direction. This plane is assumed to be symmetric with respect to all physical quantities in the transverse direction ($z$ in Fig. 6.16). This seems to be the case indeed for all practical reactors. But unfortunately, no statement can be made of course about the magnitude of transverse mass transport in such a two-dimensional calculation. However, depending on the geometries and conditions used transverse transport might quite severely affect the deposition. Such transport my be caused by the development of transverse rolls in the reactor. They can be simulated only by three-dimensional solutions of (11.15)–(11.22).

Figure 6.22 gives one of the few examples for a three-dimensional calculation showing the development of rolls along the length of the reactor [6.31]. The calculation is performed for atmospheric pressure and in a reactor with a large aspect ratio (height divided by width). Both facts lead to large Rayleigh numbers (or buoyancy force) and thus stimulate the occurrence of rolls. The

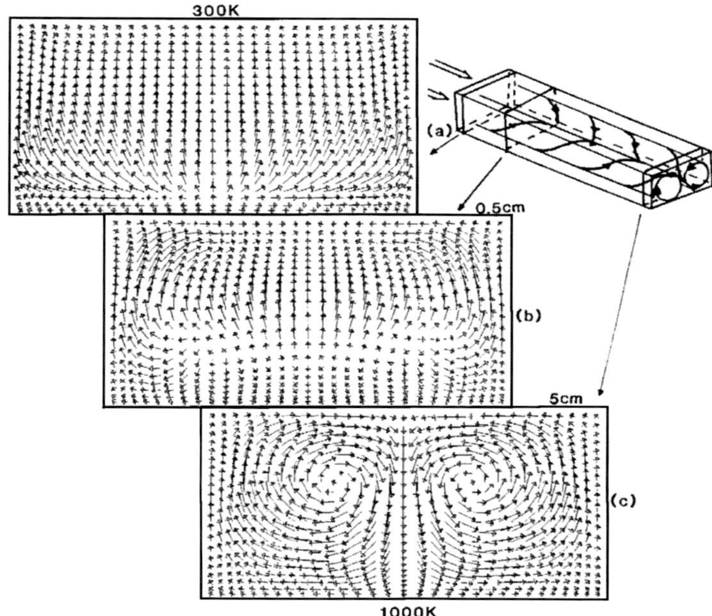


**Fig. 6.22.** Transverse velocities at three axial reactor positions calculated for a case with adiabatic side walls. Reactor height 30 mm and width 60 mm (taken from [6.31])

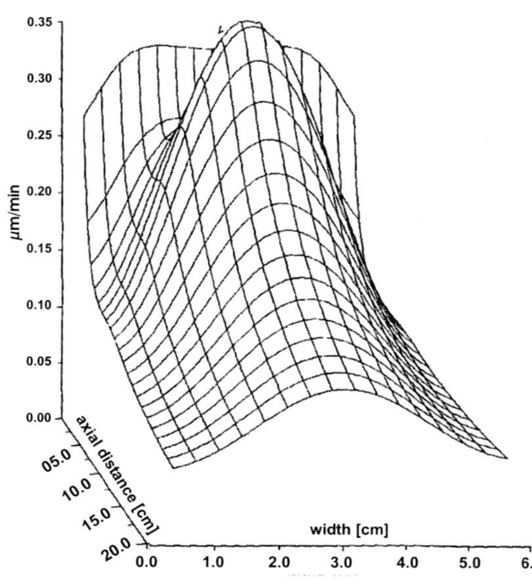


**Fig. 6.23.** Growth rate of GaAs from TMGa and $AsH_3$ in a horizontal reactor corresponding to the flow simulation in Fig. 6.22 (taken from [6.31])

corresponding growth rate shown in Fig. 6.23 (calculated with adiabatic side walls) is by no means uniform any more. The rolls deplete the growth determining precursor in the side regions of the reactor and the growth rate decreases there correspondingly.

As a summary of this section on gas phase transport one can state that in general high flow rates and lower pressures are promoting uniform boundary layers and circumvent the appearance of rolls. Thus uniform deposition is much easier to achieve.

### 6.2.3 Principles of the MOVPE Process

The practical demand to decrease the growth temperature in order to make the heterojunctions in epitaxial layers as sharp as possible generated an intensive development trend of CVD processes based on metalorganic compounds (MOCVD, or MOVPE) decomposing at lower temperatures [6.44] than those at which the reactions in the Cl-H systems occur [6.24]. The classical example is the growth of GaAs from trimethylgallium and arsine:

$$Ga(CH_3)_3 + AsH_3 \rightarrow GaAs + 3CH_4. \tag{6.30}$$

The process is also often referred as metalorganic chemical vapor deposition (MOCVD) or following the chemical terminology "organometallic" as organo metallic vapor phase epitaxy (OMVPE) or organo metallic CVD (OMCVD).

The first reports about MOVPE growth of III–V semiconductors have been given in patent files as early as in 1954 (InSb) [6.45] and 1962/63 (GaAs) [6.46, 47]. This work was, however, not published in the standard literature, and therefore the pioneering epitaxial work of Manasevit and Simpson [6.48] is usually taken erroneously as the first report on MOVPE growth.

The main concern of MOVPE development in the first few decades has been the complex and unknown chemistry of the metalorganic compounds and the resulting purity of the epitaxial layers specifically in comparison to MBE. These problems have been solved by intensive analytical and chemical research and MOVPE has become a versatile and broadly applied technique for deposition of high-quality epilayers of virtually all III/V and II/VI semiconductor in a mass-production scale [6.49]. In addition, nearly every device structure that has been made by other epitaxial growth technique has now been duplicated using MOVPE, in many cases also with superior properties. Moreover, phosphorus containing semiconductors (Ga-In-P) are exclusively deposited with MOVPE because of the inherent difficulties of solid phosphorus sources in MBE. Similarly group III-nitrides are mainly grown by MOVPE. Because of its technical importance, we will discuss the MOVPE technology in more detail in chapter 8.

## 6.3 Atomic Layer Epitaxy

Atomic layer epitaxy (ALE) is not so much a new technique for preparation of thin films as a modification to existing methods of VPE, either PVD with MBE at one limit, or CVD with halogen VPE or MOVPE at the other limit. It is a self-regulatory process which, in its simplest form, produces one complete molecular layer of the grown compound per operational cycle with a greater thickness being obtained by repeated cycling. There is no growth rate in ALE as in other crystal growth processes [6.50]. Originally invented by Suntola and Antson [6.51] in 1974, ALE has been applied since then to epitaxial growth of a whole variety of compound semiconductors, with greatest successes achieved for II–VI and III–V compounds. Especially useful is ALE for preparation of large-area thin-film electroluminescent displays, based on Mn doped ZnS [6.50]. This growth technique offers particular advantages for preparation of ultrathin films of precisely controlled thickness in the nanometer range, and thus has a special value for growing low-dimensional structures.

Presenting here the basic features of ALE, in its CVD-related variant, we will follow the review by Suntola [6.52] but the reader is also refered to [6.53]; for information on the MBE-related variant, i.e., on UHV ALE, the reader is referred to Sect. 7.2.

### 6.3.1 Principles of the ALE Process

ALE is a surface controlled process for thin-film manufacturing and for epitaxial growth of single crystal layers. It is primarily developed for compound materials such as II–VI and III–V semiconductors, oxides and nitrides but it can also be extended to covalent materials [6.54].

In ALE, epitaxial growth is obtained by sequentially controlled surface conditions, which not only control the structure of the growing layer but also the rate of the growth, to one atomic layer (monolayer) in one reaction sequence. The inherent structural control of the growing layer, and the "digital" rate control of the growth process, makes ALE attractive for the growth of crystalline compound layers, complex layered structures, superlattices and layered alloys.

The sequential control of the growth in ALE is based on saturating surface reactions between the substrate and each of the reactants needed for the compound to be grown. Each surface reaction adds one monolayer of the material on the surface. A monolayer obtained in a reaction sequence may be a "full monolayer" corresponding to the density of atoms in the corresponding crystal plane of a bulk crystal or it may be a "partial monolayer" due to preferred surface reconstructions or steric-hindrance effects related to the reactant used [6.52].

If the reactants used are the constituent elements of the material to be grown, the surface reactions are additive and the necessary condition for saturation is that condensation of the reactant does not occur (see Fig. 6.24,

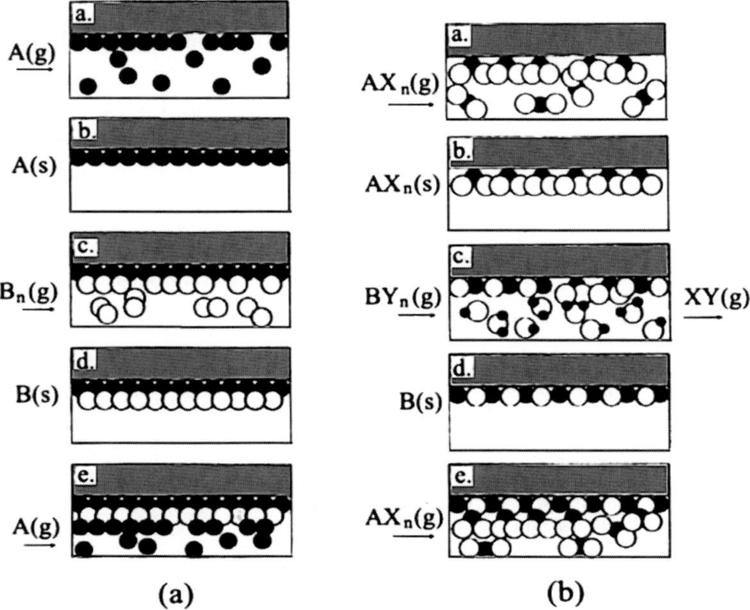


**Fig. 6.24.** Basic sequences of ALE for compound AB. "A" denotes a group-I, II, III element, while "B" denotes a group V, VI, VII element. "$X_n$" and "$Y_n$" denote ligands of the reactants $AX_n$ and $BY_n$, respectively. In the ALE process shown in part (**a**), the sequence of reactions for elemental reactants A(g) and $B_n$(g) is indicated. In the ALE process shown in part (**b**), it is required that the ligands $X_n$ and $Y_n$ are able to form volatile compounds in a surface exchange reaction.
In part (**a**): step (a.) illustrates the introduction of A(g) onto the substrate surface; step (b.) shows the formation of an A(s) monolayer surface; step (c.) illustrates the introduction of $B_n$(g) onto the A(s) surface, while step (d.) indicates the formation of a B(s) monolayer surface. Step (e.) is just a repetition of the cycle shown in (a.).
In part (**b**) the sequence of steps of ALE with compound reactants are shown: step (a.) illustrates the introduction of $AX_n$(g) onto the substrate surface; step (b.) shows the formation of an $AX_n$(s) monolayer surface; step (c.) illustrates the introduction of $BY_n$(g) onto the $AX_n$(s) surface; step (d.) indicates the formation of a B(s) monolayer surface with simultaneous release of XY(g). Step (e.) is the repetition of the cycle from step (a.). The release of ligands $X_n$ and $Y_n$ may also take place partially in each reaction sequence (taken from [6.52])

part A). In the case of compound reactants there are several mechanisms of saturation, depending on the nature of the surface reactions in question (see Fig. 6.24, part B). Saturation of each surface reaction is the characteristic feature of ALE; sequencing alone does not result in the surface control. Saturation makes the growth rate proportional to the number of reaction cycles and not to the intensity of the reactant flux. Instead of monitoring thickness during the growth process, in ALE the thickness can be determined by counting the number of reaction steps. For ALE of covalent (elemental) materials a compound reactant or a pair of compound reactants is needed. Two approaches to saturated reactions can be used in this case. The first approach comprises a reaction adding a monolayer of the covalent material to be grown and a separate reduction sequence activated with extra energy which re-establishes the surface for a new monolayer (see Fig. 6.25). The reduction sequence may also be a chemical reaction resulting in release of the surface ligands (see Fig. 6.26). In the best case the release of the surface ligand is performed with a reactant also adding a monolayer of the material grown (see Fig. 6.27). A pair comprised of a halide and a hydride of a group-IV material could satisfy such a requirement. The saturation mechanisms in this approach are basically similar to the saturation mechanisms of ALE of compound materials using compound reactants.

Advantages obtainable with ALE are dependent on the material to be processed and the type of applications in question. In thin film epitaxy ALE may be a way of obtaining lower growth temperatures. It is also a method for making precise (very sharp) interfaces and epilayers needed in superlattice structures. In applications for non-epitaxial thin-film growth ALE makes it possible to obtain excellent thickness uniformity over large areas. The surface control of ALE results also in reproducible physical characteristics with good homogeneity of the grown films.

In its simplest mode of operation ALE is performed using the elements of the compound to be grown as the reactants. In this mode, usually realized by UHV ALE, the compound grows one full or partial monolayer in

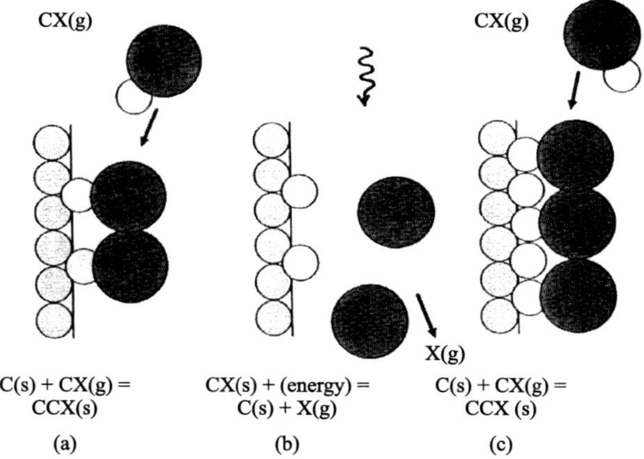


**Fig. 6.25.** ALE of covalent material "C" (group-IV element) comprising: (**a**) a sequence of formation of a monolayer using reactant $CX_n$ and (**b**) a sequence for re-activating the surface for the next monolayer by extra energy, heat, or photons. (**c**) presents a sequence for growing the next monolayer (taken from [6.52])

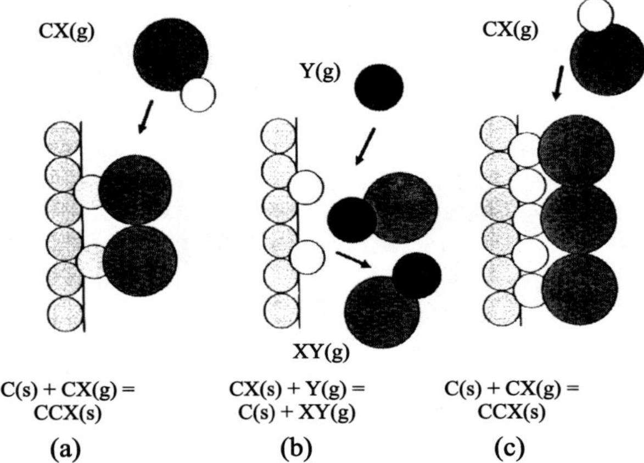


**Fig. 6.26.** ALE of covalent material "C" (group-IV element) comprising: (**a**) a sequence for a monolayer formation using reactant $CX_n$ and (**b**) a sequence for releasing the surface ligands by a reaction with $Y_n(g)$ gas. (**c**) shows the sequence for the next monolayer to be grown (taken from [6.52])

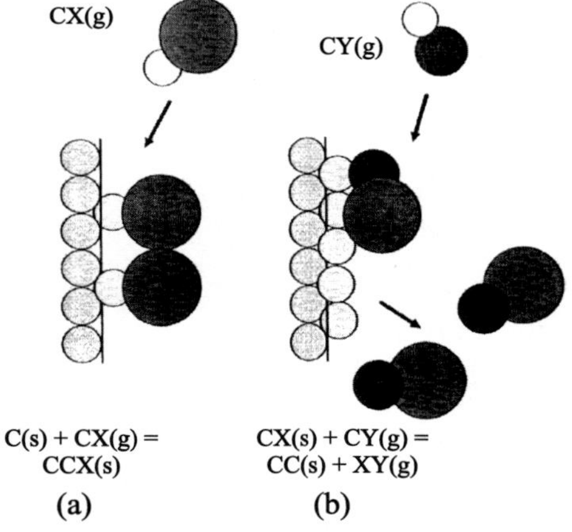


**Fig. 6.27.** ALE of elemental materials by a pair of reactants $CX_n$ and $CY_n$ each adding a monolayer of the material: (**a**) the reactant $CX_n(g)$ adds a monolayer CX(s) onto the C(s) surface, (**b**) the reactant $CY_n(g)$ adds a C(s) monolayer while releasing the $X_n(s)$ surface ligand as nXY(g) vapor. The ligand exchange may also be divided between the $CX_n(g)$ and $CY_n(g)$ reactions. Then the surface can be expressed as $CX_{x\text{-}k}(s)$ and $CY_k(s)$, respectively (taken from [6.52])

each reaction step as a result of an additive chemisorption in each cycle. For monoatomic elemental reactants additive surface reaction may occur directly, without a precursor state or an activation energy. For such a reaction one must also assume that there are bonding sites directly available on the surface. If a modification of the surface structure such as breaking of a surface reconstruction is needed to create a bonding site, an activation energy of the surface reaction has to be assumed. The activation energy can be described with a precursor state connected to the final bonding site. A precursor state and an activation energy can also be used to describe the surface reaction condition when modification of the reactant molecule, such as dissociation or structural modification, is needed before chemisorption.

The use of compounds as the reactants extends the use of the ALE process to materials where the vapor pressures of the elemental components are too low to give rise to the additive mode of the ALE process. A basic requirement of a reactant is a reasonable vapor pressure at the processing temperature. Low vapor pressure of a reactant leads to slow processing, especially with large-area substrates. Other requirements of the reactant are a good reactivity with the surface it is reacting with and also that the surface formed after the reaction is reactive with the second reactant used.

Full advantage of the saturation of the surface reactions in ALE is obtained when the process is performed in thermal equilibrium. This also means that the reactant should be chemically stable at the processing temperature. In the case of additive ALE processes it was stated that the saturation densities of the monolayers are mainly determined by the surface reconstructions formed in each reaction sequence. In the case of compound reactants, the saturation densities are also subject to the effects of surface reconstructions, but they may also be determined by the physical sizes of the ligands of the reactant molecules. Ligands may suppress the formation of surface reconstructions. This helps in obtaining a "full monolayer per cycle" growth condition. The ligands have an important role in the surface reactions; accordingly, the choice of the compound reactants has a major effect on the ALE process [6.52].

The characteristic feature of ALE in all modes of operation is the saturation of each surface reaction, which causes the growth to proceed incrementally, one reaction step at a time. Accordingly, the surface reactions control the growth of the material. The structure and the density of a monolayer formed in an ALE reaction sequence is strongly dependent on the characteristic surface reconstruction of the material and the crystalline face in question. For some III–V materials the saturation density has been observed to correspond to the density of atoms in a crystalline face of the bulk material This case is often referred to as "a full monolayer per cycle" growth. In most cases the saturation density has been observed to be less than a "full monolayer", which refers to the important role of surface reconstruction in ALE growth steps. Use of large reactant molecules may also reduce the saturation density of a monolayer.

In contrast to the requirements for reactants in CVD processes, optimal reactants in ALE should react aggressively with each other, i.e., the activation energy of the surface reaction should be low. In CVD a reaction threshold is needed for uniformity of thickness. A reaction threshold is also needed to avoid gas phase reactions. In ALE, gas phase reactions between reactants are automatically eliminated by the sequencing of the reactants. Thickness uniformity results from the saturation mechanism. High reactivity assures efficient saturation and a high material utilization factor. For a uniform saturation it is also important that the reactants used are stable at the processing temperature. Undesirable decomposition or partially completed exchange reactions may be reasons for incomplete saturation or sources of impurities in the resulting material. Unstable ligands formed in the surface reactions may make the process sensitive to the gas flow dynamics of the system (they cause the loss of the self-regulation mechanism, characteristic of perfect ALE growth) [6.52].

### 6.3.2 Growth Systems for CVD-like ALE

The main building blocks of an ALE reactor are presented in Fig. 6.28. ALE processes in the CVD-like mode of operation are usually carried out in an inert gas atmosphere. Therefore, the standard CVD reactors can be operated in ALE mode, provided that there are means for independent valving of the sources.

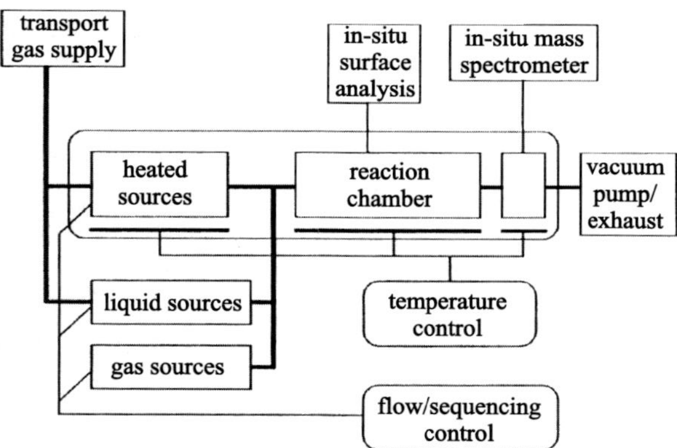


**Fig. 6.28.** Block diagram of an ALE reactor. The reaction zone may be in a vacuum or in an inert gas system at low pressure or at atmospheric pressure (taken from [6.52])

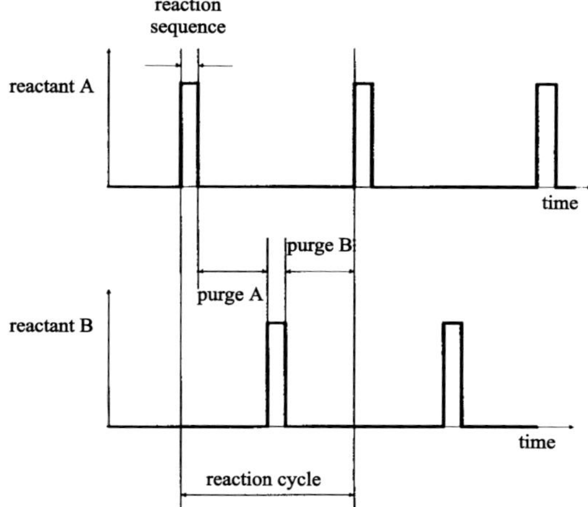


**Fig. 6.29.** Definition of reaction and purge sequences (taken from [6.52])

Most CVD reactors, especially MOVPE reactors, are designed for reactants which are volatile at room temperature. They usually have a cold-wall design with inductive or infrared heating of the substrates. This minimizes the hot surface area and reduces undesirable growth on the reactor walls. In order to minimize gas phase reactions in a CVD process, reactants which are too aggressive are avoided. In standard MOVPE processes it is essential that, for uniform thickness, the reactant flux onto the substrate surface is uniform over the whole substrate area. In ALE, the rate of growth is controlled by the surface saturation mechanism, thus, it is proportional to the speed of sequencing of the reactants and to the saturation density of the reactants. A full ALE cycle consists of a reactant supply sequence and a purge sequence after each reactant sequence (see Fig. 6.29). A uniform thickness in a material layer is achieved, provided that the supply of each reactant is high enough for the saturation of every part of the substrate surface. This gives extra freedom and different boundary conditions for the design of ALE reactors.

In the traveling-wave ALE reactor (see Fig. 6.30d), the reactants are fed through a cassette of substrates by an inert gas flow. A traveling-wave reactor utilizes multiple hits of reactant molecules on the substrate, which results in a high material utilization efficiency. The saturation mechanism of ALE ensures uniform thickness over the whole length of the substrate. Figure 6.30 summarizes some reactor constructions used for ALE in its CVD-like mode.

### 6.3.3 Specific Features and Application Areas

ALE is a surface-controlled method of producing atomically smooth epitaxial layers and thin films. In spite of the principal simplicity of the process, the

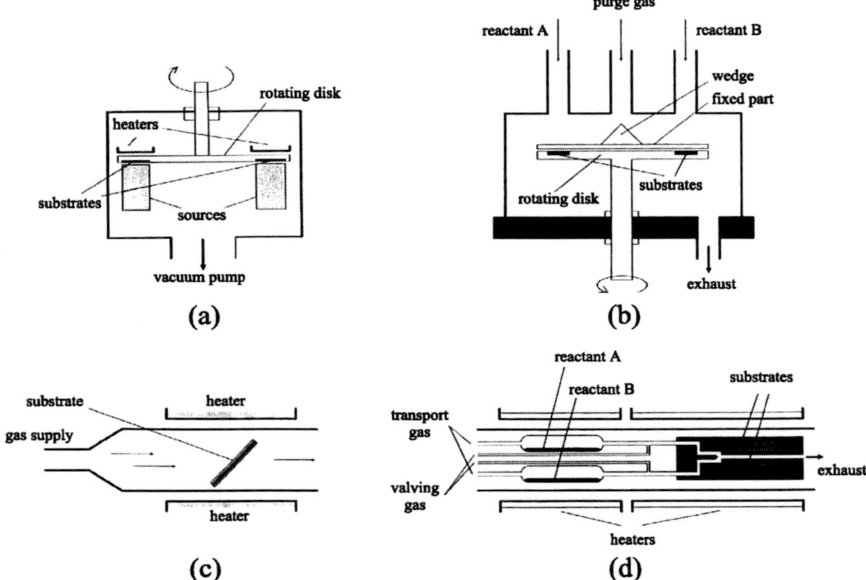


**Fig. 6.30.** Schematic illustrations of reactors used for realization of the CVD-like mode of ALE. (**a**) High-vacuum reactor with a rotating substrate holder. (**b**) CVD-type reactor utilizing a rotating disk for sequencing. Purging and separation of the different reactants is enhanced by a purge gas supplied to the center of the rotating disk. (**c**) Flow-tube-type reactor often used in atmospheric-pressure CVD. For cold-wall operation inductive heating of the substrate is used. (**d**) A "traveling-wave" ALE reactor comprising hot-wall design, fast flow-dynamics and inert-gas-valved hot sources (taken from [6.52])

surface chemistry of ALE is not simple, and is still not very well understood. The surface chemistry is strictly individual for each material to be processed and for each combination of reactants used for the process (for an exhaustive review on materials-related growth peculiarities in ALE the reader is referred to the review of Suntola [6.52]).

In ALE, the chemical reactions involved are divided into well-defined sub-reactions. This helps, by penetrating deeper into the reaction mechanisms and reaction pathways in ALE, to obtain a link between theoretical and experimental chemistry of this process.

Application-oriented investigations concerning ALE are directed to areas like:

(i) epitaxy of III–V and II–VI materials for optoelectronic devices and low-dimensional structures,

(ii) layer-by-layer controlled epitaxy of covalent (elemental) materials, their conformal growth and application in advanced semiconductor device structures,
(iii) extremely sharp interfaces, thin dielectric layers, controlled superalloys,
(iv) large-area thin-film growth optimization for applications in electroluminescent displays and thin-film solar cells,
(v) tailored molecular surfaces, including heterogeneous catalysts.

ALE also has shown its productive capability in electroluminescent flat-panel display technology. The cost efficiency of the process is based on the inherent self-regulation property of ALE, which makes handling of large batches in a small volume possible. ALE also makes it possible to produce different types of materials in one process. For electroluminescent thin-film structures the whole stack of dielectric-semiconductor-dielectric thin films can be made in one process by using ALE. While ALE was originally developed to meet the qualitative requirements of the material, it has turned out that the inherent self-regulation of ALE is also a key to productive and effective processing [6.52].

# 7. Molecular Beam Epitaxy

MBE is a versatile technique for growing thin epitaxial structures made of semiconductors, metals or insulators [7.1] (see also the definition given in Sect. 6.1.3). What distinguishes MBE from previous vacuum deposition techniques is its significantly more precise control of the beam fluxes and growth conditions. Because of vacuum deposition, MBE is carried out under conditions far from thermodynamic equilibrium and is governed mainly by the kinetics of the surface processes occurring when the impinging beams react with the outermost atomic layers of the substrate crystal. This is in contrast to other epitaxial growth techniques, such as LPE or atmospheric pressure VPE, which proceed at conditions near thermodynamic equilibrium and are most frequently controlled by diffusion processes occurring in the crystallizing phase surrounding the substrate crystal.

In comparison to other epitaxial growth techniques, MBE has a unique advantage. Being realized in an ultrahigh vacuum environment (the pressure in the growth reactor is less than $10^{-7}$ Pa), it may be controlled *in situ* by surface diagnostic methods such as reflection high energy electron diffraction (RHEED), reflection mass spectrometry (REMS) or optical techniques involving reflectance anisotropy spectroscopy (RAS), spectroscopic ellipsometry (SE), and laser interferometry (LI). These powerful facilities for control and analysis eliminate much of the guesswork in MBE, and enable the fabrication of sophisticated device structures using this growth technique. For detailed information about these characterization techniques, the reader is referred to Chaps. 9 and 10 of this book.

The characteristic feature of MBE is the beam nature of the mass flow towards the substrate. From this point of view it is important to consider the admissible value of the total pressure of the residual gas in the vacuum reactor, which has to be ensured in order to preserve the beam nature of the mass transport. The highest value of the residual gas pressure may be estimated from the condition that the mean free path $L_b$ of the molecules of the reactant beam penetrating the environment of the residual gas has to be larger than the distance from the outlet orifice of the beam source to the crystal surface $L_{os}$ (usually $L_b > L_{os} \geq 0.2\,\mathrm{m}$). For simplicity, assuming that the beam and the residual gas create a mixture of two gases, and that the average velocities of the residual gas molecules are much smaller than the velocities of the beam molecules, one may use the following formula to calculate the admissible partial pressure of the residual gas $p_g$ [7.1].

$$p_g = k_B T \frac{L_b^{-1} - \sqrt{2}\pi n_b d_b^2}{\frac{\pi}{4}(d_b + d_g)^2}, \tag{7.1}$$

where $n_b$ and $d_b$ are the concentration and diameter of the molecules in the molecular beam, respectively, while $d_g$ is the concentration of the molecules of the residual gas in the reaction chamber. In the case of MBE growth of GaAs (the first ever compound grown by MBE), the maximum value of the residual gas pressure is $p_{g,\max} = 7.7 \times 10^{-2}$ Pa [7.2]. It is evident that the beam nature of the mass transport is already preserved at high vacuum conditions ($1.33 \times 10^{-1}\,\text{Pa} \geq p \geq 1.33 \times 10^{-7}\,\text{Pa}$). However, the low growth rates typical for conventional MBE techniques (about $1\,\mu\text{mh}^{-1}$, or $1\,\text{monolayers}^{-1}$) coupled with the obvious requirement of negligible unintentional impurity levels in the crystallized epitaxial layer lead to much more rigorous limitations for the total pressure of the residual gas in the MBE reactor [7.2].

The condition for growing a sufficiently clean epilayer can be expressed by the relation between the monolayer deposition times of the beam $t_1(b)$ and the background vapor $t_1(v)$. The time $t_1(v)$ during which one monolayer of contaminants is deposited on the substrate surface from the residual gas contained in the vacuum reactor should be at least $10^5$ times longer than the time $t_1(b)$ necessary for the deposition of a film of one monolayer thickness from the molecular beams, ($t_1(b) \leq 10^{-5} \cdot t_1(v)$). For a beam flux of $10^{19}\,\text{atoms}\,\text{m}^{-2}\,\text{s}^{-1}$ reaching the substrate surface (this is a value typical for MBE growth of GaAs) and an equilibrium surface density of the substrate crystal atoms equal to $10^{19}$ atoms $\text{m}^{-2}$ one gets for $t_1(b)$ the value of 1 s. Consequently, the above condition will be fulfilled if the time $t_1(v)$ is not shorter than $10^5\,\text{s} \approx 28\,\text{h}$ [7.2]. From this requirement the pressure of the residual gas in the vacuum reactor $p_{g,\max}$ should be less than $1.7 \times 10^{-9}$ Pa. This numerical result indicates that MBE growth at low rates ($1\,\mu\text{mh}^{-1}$) should be carried out in an ultrahigh vacuum (UHV) environment.

Usually the residual gas pressure routinely achieved in MBE vacuum systems is in the range of low $10^{-8}$ Pa to mid $10^{-9}$ Pa [7.1]. These values are larger than the calculated value for $p_{g,\max}$; however, despite this, very clean thin epilayers may be grown by MBE. This positive effect should be ascribed to another feature of the MBE growth process, according to which the concentrations of unintentional contaminants incorporated into a growing film depend also on their sticking coefficients on the substrate at the growth temperature. The present success of MBE technology stems, thus, more from the small sticking coefficients of the unintentional impurities on the heated substrate than from the attainable low residual gas pressure [7.2].

## 7.1 Solid Source MBE

When the sources of the reactants used in MBE are solids, then the MBE growth technique is commonly called solid source MBE (SSMBE). In this case the Knudsen mode of evaporation (see Sect. 6.1.2. for definition) is the fundamental phenomenon on which the operation principle of effusion cells, used as reactant sources in SSMBE, is based. It should, however, be mentioned that electron beam and laser radiation heated sources have also been successfully applied for growing certain materials by SSMBE. For example, for growing mercury-containing compounds (characteristic of a very high vapor pressure of Hg) laser heated sources have been used, while electron beam heating is usually used for growing Si/GeSi heterostructures. Dissociation, or cracker cells are used in SS MBE, too. These cells are modified versions of conventional effusion cells in which the effusion beams are directed from a conventional Knudsen-type crucible via a higher-temperature (cracker) region onto the substrate. The cracker cells are mainly used for the group V materials As and P, which produce the tetramers $As_4$ and $P_4$ when they are directly evaporated from elemental charges at lower temperature. The cracking region provides an elevated temperature multiple collision path for the beams and thus enables effective dissociation of the tetrameric molecules into dimers $As_2$ and $P_2$, respectively [7.1]. The essential elements of a MBE system are shown schematically in Fig. 7.1. It is apparent from this illustration that each MBE arrangement may be divided into three zones where different physical

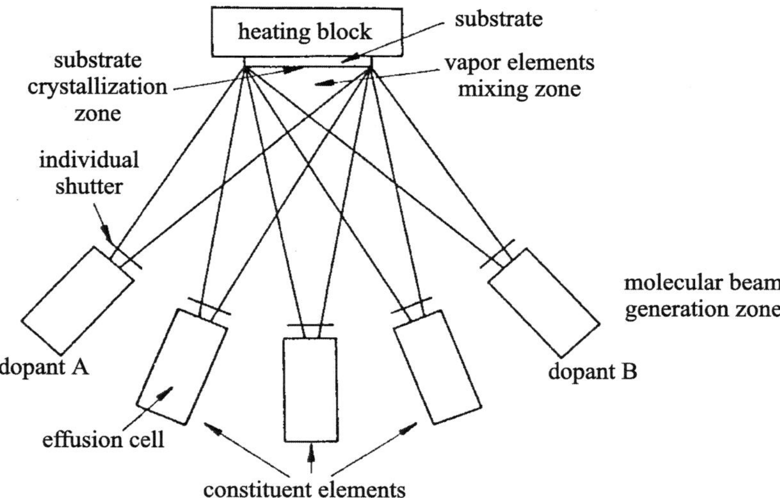


**Fig. 7.1.** Schematic illustration of the essential parts of a MBE growth system. Three zones where the basic processes of MBE take place are indicated (taken from [7.1])

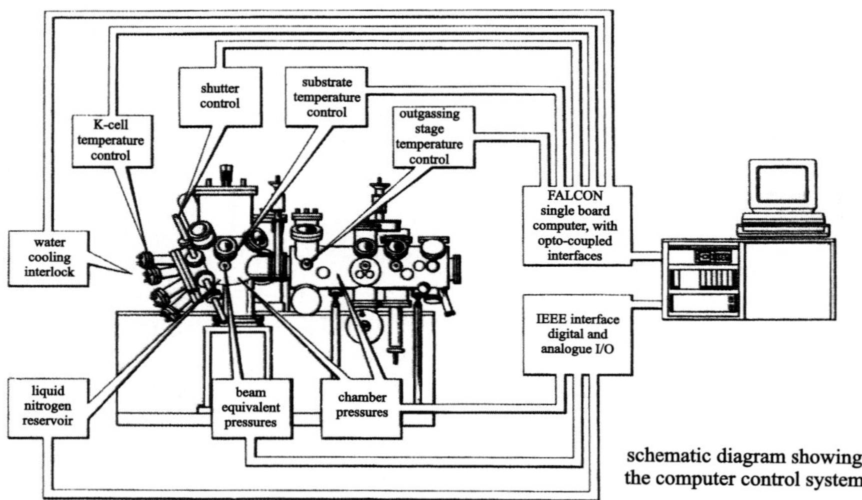


**Fig. 7.2.** Schematic diagram showing the computer control system of the V80H MBE machine of VG Semicon company (taken from [7.1])

phenomena take place. The first zone is the generation zone of the molecular beams. Next is the mixing zone where the beams from different sources intersect each other and the vaporized elements mix together creating a very special, rarefied gas phase contacting the substrate area. This area, where the crystallization processes take place, can be regarded as the third zone of the MBE physical system. The maturity which the MBE technique has now achieved is reflected in the demand for high throughput, high yield MBE machines. A whole set of companies currently manufacture MBE-growth and MBE-related equipment that is sophisticated in design and reliable in application. The modular approach to system design has been established as the most effective and flexible one. Modularity means providing the system in "building block" units, combining the advantages of standardization and flexibility. The enormous potential of MBE to produce structures of unequaled complexity can be fully realized only by utilizing effective automatic control using computer systems operating in real-time control mode. Real-time control ensures precise synchronization of events during the growth process. For example, shutter status and molecular beam fluxes must be assessed and adjusted simultaneously every one or two seconds in many growth procedures, especially when complicated device structures are to be grown with MBE. This means, however, that a computer control system with appropriate software (computer programs) is an indispensable part of the MBE equipment. An example of the computer-control system for a MBE machine (the V80H

**Fig. 7.3.** The V80H MBE system of VG Semicon installed in a clean room (taken from [7.1])

MBE system of the VG Semicon company) is depicted in Fig. 7.2. The photograph of the V80H system, installed in a clean room, is shown in Fig. 7.3.

More information on the construction blocks and functional elements of modern MBE production systems available commercially are presented and described in detail in Chap. 3 of [7.1], taking as examples the ISA Riber and the VG Semicon MBE systems.

### 7.1.1 Basic Phenomena

The molecular beams are generated in the first zone of the SSMBE growth system, under UHV conditions, from sources of the Knudsen effusion-cell type (see Fig. 6.2), whose temperatures are accurately controlled. By choosing appropriate cell and substrate temperatures, epitaxial layers of the desired chemical composition can be obtained.

The uniformity in thickness as well as in the composition of the epilayers grown by MBE depends on the uniformities of the molecular beam fluxes, and thus also on the geometrical relationship between the configuration of the sources and the substrate. Optimization of these parameters is very often an extremely difficult task. Therefore, if possible, the substrate is rotated during the growth with a constant angular velocity around the axis perpendicular to its surface. The substrate rotation causes a considerable enhancement in thickness and composition homogeneity of the grown epilayers but, unfortunately, excludes the possibility of growing patterned structures, which is often required in the technology of present-day electronic devices.

The second zone in the MBE vacuum reactor is the mixing zone, where the molecular beams intersect each other. Little is known at present about the physical phenomena occurring in this zone. This results from the fact that usually the mean free path of the molecules belonging to the intersecting beams is so long that collisions and other interaction between the molecules of different species do not occur there.

Epitaxial growth in MBE is realized in the third zone, i.e., on the substrate surface. A series of surface processes are involved in MBE growth; however, the following are the most important:

(i) adsorption of the constituent atoms or molecules impinging on the substrate surface,

(ii) surface migration and dissociation of the adsorbed molecules,

(iii) incorporation of the constituent atoms into the crystal lattice of the substrate or the epilayer already grown,

(iv) thermal desorption of the species not incorporated into the crystal lattice.

These processes are schematically illustrated in Fig. 7.4. The substrate crystal surface is divided there into so-called crystal sites with which the impinging molecules or atoms may interact. Each crystal site is a small part

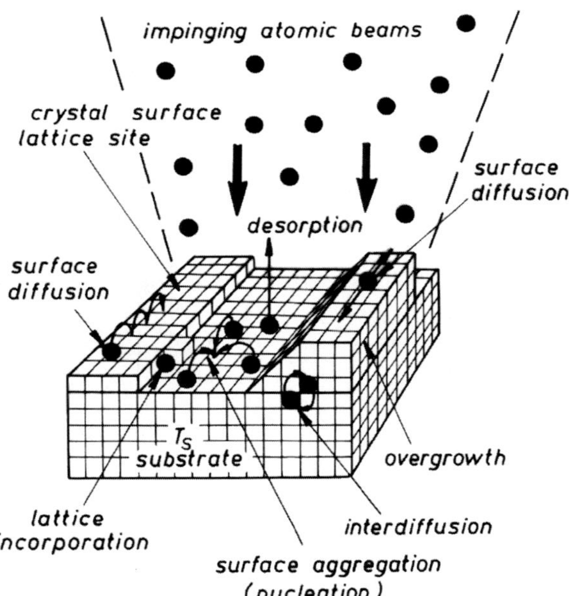


**Fig. 7.4.** Schematic illustration of the surface processes occurring during epilayer growth by MBE (taken from [7.1])

of the crystal surface characterized by its individual chemical activity. A site may be created by a dangling bond, vacancy, step edge, etc. [7.3].

The surface processes occurring during MBE growth are characterized by a set of relevant kinetic parameters that describe them quantitatively. The arrival rate is described by the flux of the arriving species, which gives the number of atoms impinging on the unit area of the surface per second. For MBE growth of compound semiconductors, the required fluxes are typically between $10^{18}$ and $10^{20}$ atoms $\mathrm{m}^{-2}\,\mathrm{s}^{-1}$ [7.2].

The atoms arriving at the substrate surface will generally have an energy distribution appropriate to the temperature of their place of origin. In the case of MBE being considered this will most frequently be the effusion cell temperature $T_i$. After arrival at the substrate surface, which has a temperature $T_s$, usually lower than $T_i$, the atom may reevaporate immediately, carrying with it an energy corresponding to temperature $T_e$. The impinging atoms may, however, also exchange energy with the atoms of the substrate until they are in thermodynamic equilibrium at $T_s$. The quantitative description of this process is possible by defining the thermal accommodation coefficient as

$$\alpha = \frac{(T_i - T_e)}{(T_i - T_s)}. \tag{7.2}$$

Clearly, when $T_e$ is equal to $T_s$ the accommodation coefficient is unity. Thus, it emerges as a measure of the extent to which the arriving atoms reach thermal equilibrium with the substrate [7.4]. It is important to differentiate between the accommodation coefficient defined above and the sticking coefficient $s$. The latter is defined as the ratio of the number of atoms adhering to the substrate surface $N_{\mathrm{adh}}$ to the number of atoms arriving there $N_{\mathrm{tot}}$:

$$s = \frac{N_{\mathrm{adh}}}{N_{\mathrm{tot}}}. \tag{7.3}$$

There are two types of adsorption. The first is physical adsorption, often called physisorption, which refers to the case where there is no electron transfer between the adsorbate and the adsorbent, and the attractive forces are van der Waals type. The second is chemisorption, which refers to the case when electron transfer, i.e., chemical reaction, takes place between the adsorbate and the adsorbent. The forces are then of the type occurring in the appropriate chemical bond. In general, adsorption energies for physical adsorption are smaller than for chemical adsorption.

It is obvious that for these two kinds of adsorption different sticking coefficients can be defined. According to the definition of the chemisorption process the sticking coefficient $s_c$ for the chemisorbed phase may be dependent on crystallographic orientation of the substrate surface, as well as on the nature and spatial distribution of atoms already adsorbed on this surface. Physisorption, on the other hand, generally shows little or no dependence on the surface site arrangement. Consequently, the sticking coefficient $s_p$ for the physisorbed phase may be taken to be independent of the local environment, i.e., of the orientation and coverage of the substrate surface.

It is experimentally well documented [7.3] that crystal growth in MBE in many cases in which molecular species are involved, proceeds through a two-step condensation process in which the molecular species reach the chemisorbed state via a precursor physisorbed phase. A possible version of the condensation process via the precursor state may be sketched as

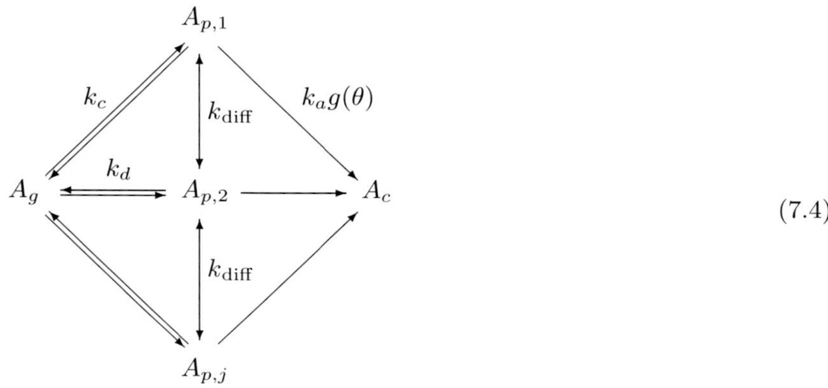


(7.4)

where $A_g$, $A_p$, and $A_c$ represent the adsorbate in the gas, precursor and chemisorbed states, respectively. The rate constants $k_c$ and $k_d$ apply to condensation and desorption of the adsorbate in the precursor state, and $k_a g(\theta)$ is the rate constant for conversion from the precursor state to the chemisorbed state. The conversion rate is assumed to be proportional to $g(\theta)$, the probability of having suitable vacant sites in the chemisorbed state, where $\theta$ is the surface coverage counted in monolayers.

This model implies that if the site visited by the molecule in the precursor state is occupied by a chemisorbed species, the molecule is allowed to diffuse over the surface with a rate constant $k_{\text{diff}}$ to another site.

The interaction potentials due to the surface, seen by an incoming molecule perpendicular to the surface, for chemisorbed and precursor states, are shown schematically in Fig. 7.5. It is evident from this figure that the molecule adsorbed in the precursor state has to overcome a lower barrier when it is subsequently chemisorbed at the surface, than in the case when it re-evaporates into the vacuum, because $E_a < E_{dp}$ [7.1]. The illustrated situation is a special example; a variety of other potential configurations can also exist.

The most important task for the MBE crystal grower is a proper adjustment of the growth temperature. The low-temperature limit for MBE to occur is defined by surface migration processes. Below some limiting temperature $T_L$ the deposited film will not longer be crystalline. MBE occurs near

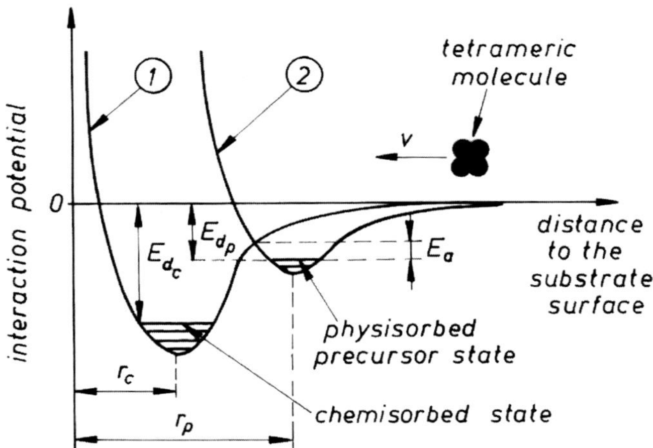


**Fig. 7.5.** The interaction potential due to the surface, as seen by a molecule impinging perpendicularly to the surface for chemisorbed (curve 1) and physisorbed (curve 2) states (taken from [7.1])

$T_L$ when the surface migration rate, multiplied by some weighting factor relevant for the density of appropriate lattice sites on the surface (of the substrate or the already grown epilayer), exceeds the deposition rate [7.5]. The high-temperature limit of MBE, $T_H$, is defined by the balance between adsorption and desorption processes, which results from thermodynamic phase equilibrium relations [7.6].

For an example of experimental procedures leading to the determination of the MBE temperature window, the reader may be referred to the data given in [7.7] for $Cd_{1-x}Zn_xTe$ ($0 \leq x \leq 1$) compounds grown on GaAs substrates. Simultaneous reflection mass spectrometry (REMS), laser interferometry (LI), and reflection high-energy electron diffraction (RHEED) measurements of the sticking parameters for the constituent elements of these II-VI semiconductor compounds are reported there. By these measurements the reactivity of the surface to incoming cation elements (Zn, Cd) both in the absence and in the presence of the anion element (Te) at the initial growth stages of CdTe on ZnTe/GaAs and ZnTe on CdTe/ZnTe(3 monolayers)/GaAs substrates as well as CdZnTe on these substrates have been evaluated. These studies have allowed the determination of the MBE window and the understanding of the effects of strain and surface composition at different growth temperatures lying in the range between $T_L$ and $T_H$. It has been found that the MBE temperature window for the investigated compounds was equal to $\Delta T = T_H - T_L = 370°\text{C} - 300°\text{C} = 70°\text{C}$. Only in the case of the growth of CdZnTe on ZnTe, we do have $\Delta T = 90°\text{C}$, with $T_H = 410°\text{C}$.

### 7.1.2 Evaporation Sources

In effusion cells used in SSMBE, the source material is held in condensed phase in an inert crucible which is heated by radiation from a resistance heated source. A thermocouple is used to provide temperature control (see Fig. 6.2) [7.1].

Conventionally the heater is a refractory metal wire wound noninductively either spirally around the crucible or from end to end and is supported on insulators or inside insulating tubing. Care is taken to place the thermocouple in a position to give realistic measurement of the cell temperature. This is either as a band around a midposition on the crucible or spring loaded to the base of the crucible. Experience has shown that Ta is the best refractory metal both for the heater and for the radiation shields, principally because it is relatively easy to outgas thoroughly, is not fragile after heat cycling, can be welded, and has a reasonable resistivity [7.8]. Where temperatures exceed 150°C refractory materials should be used, as metallic impurities (Mn, Fe, Cr, Mg) have commonly been observed in layers grown in systems where stainless steel is heated. It is also important that the refractory metal is of high purity ($> 99.9\%$) and has a low oxygen content. The insulator material has proven to be even more critical. Sintered alumina ($Al_2O_3$) proved to be unsuitable for this task; degraded optical and transport properties in grown layers have been associated with the use of this material. In a series of modulated beam mass spectrometry experiments it was clearly shown that at temperatures as low as 850°C, $Al_2O_3$ was reduced when in contact with a refractory metal, and at temperatures above 1100°C significant dissociation of the ceramic occurred. In addition, metallic impurities were often found in layers when this material was present in the source, presumably associated with a volatile impurity species.

The preferred insulating material now used for sources is pyrolytic boron nitride (PBN), which can be obtained with impurity levels $< 10$ ppm. Although dissociation of this material does occur above 1400°C, the nitrogen produced has not yet been shown to have a deleterious effect on the grown layers.

The preferred crucible material is also PBN, although other materials have been successfully employed (notably ultrapure graphite [7.9] or sapphire [7.10] for high-temperature evaporation and quartz for temperatures below 500°C), and there is now some interest in the use of nonporous vitreous carbon, which is available with $< 1$ ppm total impurity level. Graphite, however, is difficult to outgas thoroughly and is by nature porous, with a large surface area, thus making it susceptible to gas adsorption when exposed to the atmosphere. Also, Al reacts with graphite at the usual Al evaporation temperature ($> 1100$°C) and so PBN is normally used as the crucible for this metal, but not without some problems [7.9]. Aluminum tends to "wet" the crucible internal surface and also moves by capillary action through the growth lamellae of the PBN crucible. When the liquid Al solidifies on cooling, the subsequent contraction can often crack the PBN crucible. To overcome this limitation in crucible lifetime, it has been found that if only small quantities of Al are loaded into the crucible (viz., one-tenth the total crucible capacity) wetting is incomplete and the crucible does not break. A more practical solution is to position a second Al crucible inside the main one and replace this crucible when it is damaged or before significant leakage has occurred. It is also possible to fabricate a PBN crucible with a double skin of differing expansion coefficients specifically for Al evaporation, which then operates in a similar manner to the two separate crucibles but with obviously improved thermal properties.

The standard thermocouple material employed for the sources is W-Re (5% and 26% Re). These refractory alloys are suitable for operation at elevated temperatures and are inert to the reactive environment present. Multiple radiation shields surround each furnace to improve both temperature stability and thermal efficiency. Radiation heat transfer, however, is still not negligible at high temperatures and the shields themselves can become secondary centers for the generation of extraneous gases. To this end, gas-tight radiation heat shielding is also used around the cell orifice to restrict the range of angles over which these gases can reach the substrate by a straight-line path.

Effusion cells designed as specified above have an operating temperature range of up to 1400°C, which makes possible short out-gassing sequences at 1600°C. Although these temperatures are more than adequate for the common III–V materials, in practice most cells are limited to operating temperatures $< 1200°$C, which is only just within the range of that required for Ga, Al, and Si (as a dopant source) evaporation. This is because of out-gassing problems caused principally by the large temperature difference between the heater and the actual temperature of the charge in the cell (up to 300°C at 1200°C). In order to reduce this effect recent cell designs employing a much larger radiation surface heater area have been developed (see Fig. 6.2). An important feature of this design is the application of high efficiency, self-supporting tantalum radiant heater elements. Together, these elements make a large area radiation heater assembly capable of high temperature operation. Thus, the crucible runs at a temperature close to that of the foils. The need of ceramic insulators is minimized because the heater foils are almost entirely self-supporting. This means less outgassing and reduced contamination during the MBE growth. The low thermal mass results in quick response to cell temperature adjustments. A spring-loaded W-Re thermocouple provides temperature monitoring and control. The whole cell has an integral water-cooled surround, effectively thermally isolating it from the surroundings.

Water cooling is considerably more efficient at dissipating heat than liquid nitrogen, which boils if heat is directly radiated onto the cryopanel surface. Intermittent hot spots can occur where the nitrogen gas forms an insulating barrier. Avoiding these hot areas of stainless steel of which the cryopanel is made prevents irregular outgassing, which can cause contamination of the films to be grown. For crucible materials high-purity graphite, vitreous carbon or PBN have been used in the VG Semicon cell (Fig. 6.2). The exit orifices of the crucibles are large, so that the cell is a nonequilibrium effusion source.

One consequence of the large exit orifices now employed in the source cells used in SSMBE is that the radiation heat loss can be significant when the beam shutter is opened. This can result in a temperature drop at or near the orifice. In the worst situation the source material condenses at the end of the cell and reduces the exit orifice dimensions, thus changing the flux intensity.

Flux transients in MBE growth limit film reproducibility and are an inconvenience to the MBE user. These transients can be reduced by the use of partially filled crucibles so that the influence of the charges on the radiative shielding provided by the shutter are reduced. This approach, however, is impractical as the reduced charge limits machine operation between cell recharging. An interesting crucible arrangement with little or no flux transient has been proposed by Maki et al. [7.11, 12]. It involves a large volume crucible (40 $cm^3$) with a conical insert. Figure 7.6 shows the proposed crucible assembly and a geometrical construction for constant melt area as seen from the substrate position. Within the cone angle a nearly constant melt area is obtained, while outside this angle poor beam uniformity results from

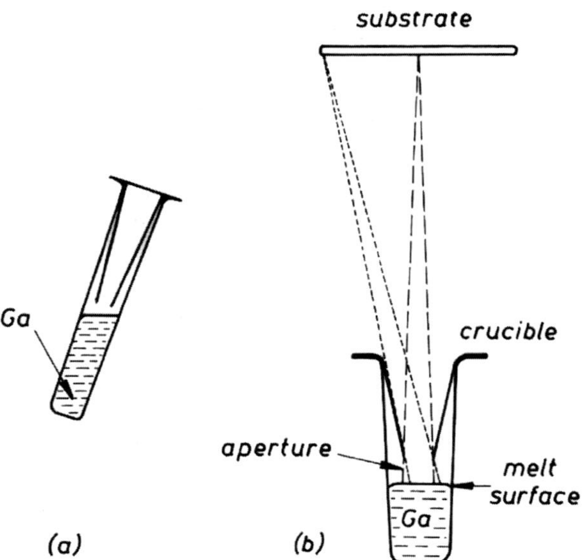


**Fig. 7.6.** (**a**) Diagram of the crucible assembly of Maki et al. [7.11], designed as a Ga source and used in the Varian GEN II MBE system. (**b**) The geometric construction for constant melt area as seen at the substrate position. The diagram is not drawn to scale and is in a vertical orientation for clarity (taken from [7.1])

a shadowing by the crucible lip. The small flux transient observed from this cell reflects a more stable melt temperature. This is attained both as a result of the melt location deep in the furnace and through the additional radiation shielding provided by the crucible insert.

The described effusion cells have been used mainly for growing III–V compounds. When materials with high vapor pressure have to be grown, the cell construction is usually changed. For growing CdTe and related compounds, a fused quartz cell with a special collimating tube has been designed [7.13], while for growing Cd and Zn II–VI compounds a high-rate nozzle effusion cell was proposed. The most difficult among the II–VI compounds, from the MBE point of view, are, however, the Hg-based compounds (HgTe, $Hg_{1-x}Cd_xTe$ or $Hg_{1-x}Zn_xTe$) [7.1], which are very important for infrared techniques [7.14]. This is based on the fact that the evaporation of these compounds is highly noncongruent. The effusion cell needs a special design, because a very high Hg atom flux is necessary to maintain the growth conditions for these compounds.

The equilibrium vapor pressure of Hg is $2\times10^{-3}$ torr at 300 K, so the Hg cell cannot be left in the growth chamber either during bakeout or prior to the growth, and consequently a transferable Hg effusion cell which can be loaded into the growth chamber through a vacuum interlock should be used. Figure 7.7 shows an example of a Hg cell used in SSMBE. This is the cell of Harris et al. [7.15], which provides an extremely stable flux of Hg during

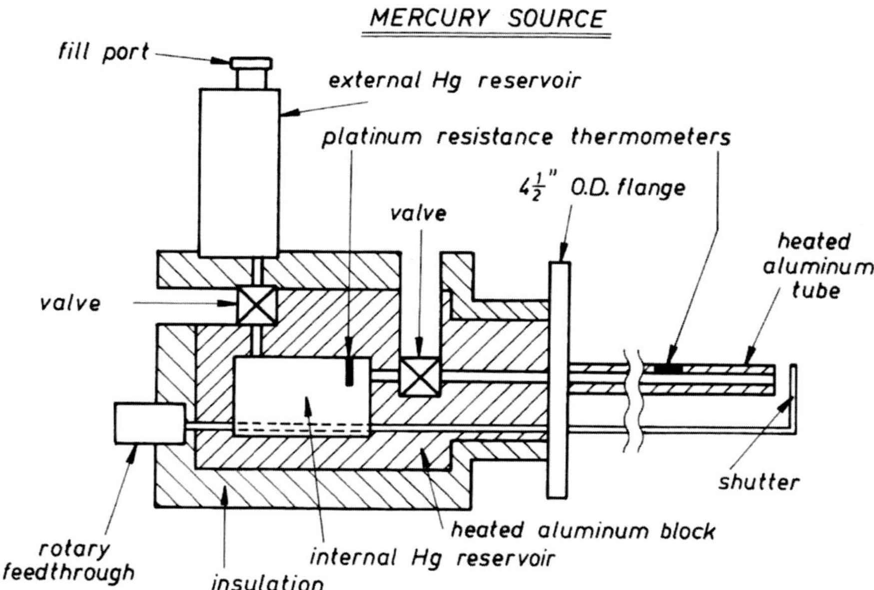


**Fig. 7.7.** Mercury vapor source used in SSMBE (taken from [7.15])

film growth over long periods of use. This source is refillable without disturbing the UHV ambient of the growth chamber, and is capable of generating vapor flux densities as large as $10^{-3}$ torr (this is the so-called beam equivalent pressure, BEP) at the substrate during film growth. The source consists of two Hg reservoirs and a heated tube separated by two UHV valves. The Hg source is attached to a UHV Conflat flange which mounts to the MBE growth chamber source flange. All internal surfaces of the Hg source are constructed of stainless steel. These surfaces are surrounded by solid aluminum heat sink material to provide good temperature stability. The temperature of the heated tube and the internal Hg reservoir are measured with platinum resistance thermometers and controlled with precise temperature controllers (e.g., made by the Eurotherm company). Platinum resistance thermometers are used rather than thermocouples because of their increased sensitivity over the temperature range employed ($T$ = 100–200°C). To fill (or refill) the Hg source, the internal Hg reservoir is first evacuated to UHV and then the valve between the reservoir and MBE chamber is closed. Next, Hg is introduced into the larger external reservoir through the fill port. The valve between the two reservoirs is then momentarily opened so that Hg fills the internal reservoir. In this way, Hg is transferred to the internal reservoir with minimum oxide contamination. As can be seen in Fig. 7.7, the modified Hg source is a one-channel circular effusion cell.

Cracker cells are used in SSMBE mainly for producing dimeric arsenic molecules ($As_2$) from elemental charges. The phosphorus-containing compounds are at present most frequently grown by gas source MBE, where

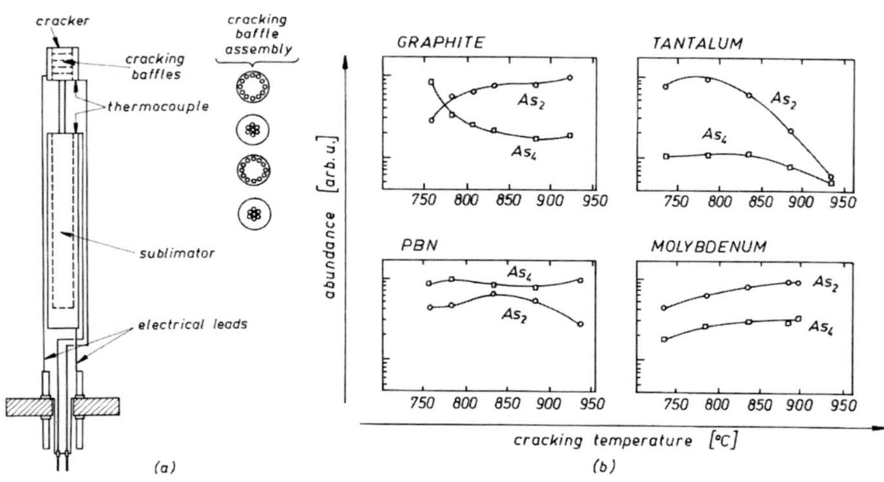


**Fig. 7.8.** (**a**) Schematic sketch of a typical arsenic cracking cell used in SSMBE, and (**b**) measured relative abundances of $As_2$ and $As_4$ molecular species in the beams generated by the cell shown in (**a**) (taken from [7.1])

cracking processes are performed in gas leak sources [7.1] (see Sect. 7.2.1). The temperatures for efficient cracking of the $As_4$ or $P_4$ beams to dimers have been calibrated by modulated beam mass spectroscopy experiments and are typically within the range 800 – 1000°C for 100% dissociation of both $As_4$ and $P_4$. When designing the cracker region of the dissociation cell, similar rigorous precautions concerning the choice of refractory materials and heater construction should be taken as with conventional effusion cells. Any significant heat transfer between the hot zone (the cracking zone) and the effusion cell region should also be avoided, otherwise uncontrolled evaporation could occur. A schematic illustration of a cracking cell for $As_4$ is shown in Fig. 7.8, together with some characteristics concerning the abundances of the dimeric and tetrameric species, respectively.

Instead of supplying energy by resistance heating, as in the case of the effusion cells, described so far, the gas particle beams for MBE may also be generated with sources heated by electron bombardment or by laser irradiation. This is favorable because the interactions between the evaporant and the source walls may be greatly reduced in this way. Two examples of such sources are presented in Figs. 7.9 and 7.10.

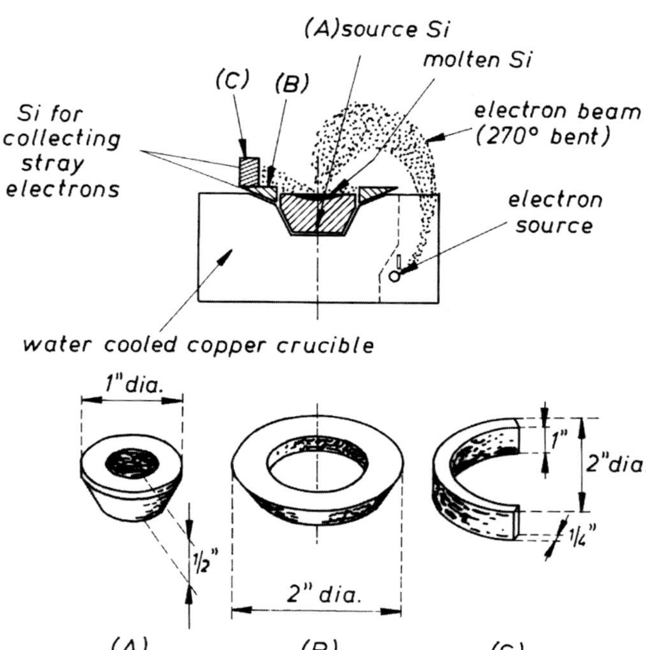


**Fig. 7.9.** Electron bent-beam heated silicon source used in SSMBE of silicon and silicon related compounds (taken from [7.16])

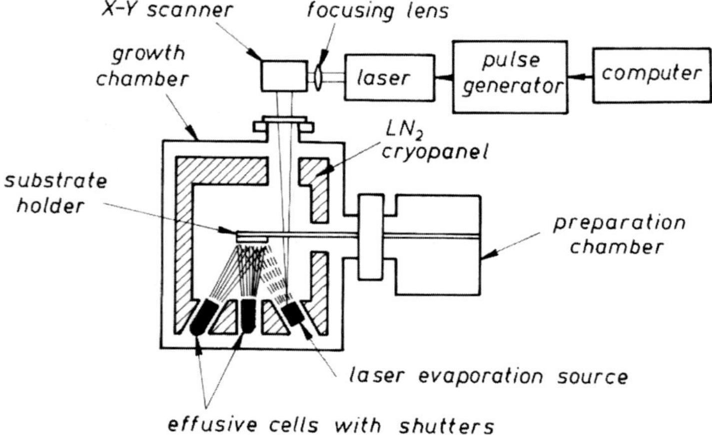


**Fig. 7.10.** Schematic drawing of a laser heated Hg source applied in SSMBE for crystallization of HgCdTe compound epilayers with complex composition profiles (taken from [7.17])

## 7.2 Gas Source MBE

Introduction of gas sources into the arsenal of tools available for MBE technology was stimulated by considerable difficulties met in growing semiconductors with P and As anions as constituents of ternary or quaternary compounds like, e.g., $Ga_xIn_{1-x}As_yP_{1-y}$. The difficulty of As/P beam intensities ratio control, especially at high intensity levels [7.15], and considerable technical problems with rapid change-over of As and P beams, have led to the introduction of gaseous sources, $AsH_3$ and $PH_3$, to MBE technology. The hydrides were first used in MBE in 1980 by Panish [7.18], and then by Calawa [7.19] for the growth of GaAs and GaInAsP.

Since the initial work on gas source MBE (GSMBE) by Panish et al. [7.18–21], Tsang [7.22, 23] and Tokumitsu et al. [7.24, 25], it has become clear that the use of gas sources for cracking of $AsH_3$ and $PH_3$ does in fact permit a degree of control that makes possible the epitaxy of ternary and quaternary compounds on InP substrates, with dimensional precision similar to that achieved by conventional SSMBE for $Al_xGa_{1-x}As/GaAs$ heterostructures [7.26].

In the infancy state of the GSMBE using hydride sources, standard effusion cells have been used for generation of elemental Al, Ga, and In beams. Thus, this kind of GSMBE, also called hydride source MBE (HSMBE) [7.27], still retained the disadvantages of SSMBE. The change of the elemental group–III flux intensity with time and its distribution over the cell depending on the consumption of source charge inside the effusion cell, the surface defects generated from the use of elemental Ga source, and the need to open the UHV growth chamber to air when solid source materials are depleted, belong to the most serious disadvantages of this growth technique.

Modifications of the GSMBE process [7.18], mostly by involving the use of organometallic gas sources [7.23, 28], considerably enhanced the usefulness of the gas source method, not only for growing III–V compounds but also for growing II–VI compounds [7.29]. For the modification of GSMBE that uses organometallic gas sources the names chemical beam epitaxy (CBE) [7.22, 23] and Metal Organic MBE (MOMBE) [7.24, 25] have been introduced. In what follows we will use exclusively the name MOMBE [7.1].

### 7.2.1 Beam Sources Used in GSMBE

Gas sources were introduced to the UHV thin film growth technique as early as in 1974 by Morris and Fukui [7.30]. They used arsine ($AsH_3$) and phosphine ($PH_3$) cracked in a boron nitride cracker tube at 800°C as arsenic and phosphorus sources for growing GaAs and GaP films on Si substrates.

Two types of crackers may be distinguished [7.26, 31–33]; one that cracks the hydrides at pressures in the range of hundreds of torr ($\sim 10^4$ Pa), called high pressure gas sources (HPGS), and the second that cracks them at pressures that are in the range of several torrs ($\sim 10^2$ Pa), the low pressure gas sources (LPGS). These are illustrated in Fig. 7.11.

The HPGS has two stages of cracking. At the usual operating temperature of 900°C, $AsH_3$ and $PH_3$ are expected, on the basis of thermodynamic equilibrium, to decompose to form the tetramers $As_4$ and $P_4$ plus $H_2$ under

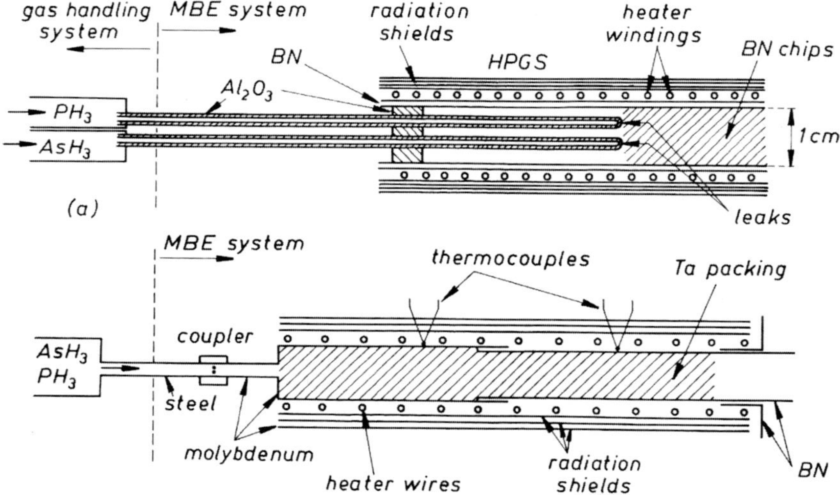


**Fig. 7.11.** Arsine and phosphine gas crackers: (**a**) high pressure gas source, and (**b**) low pressure gas source (taken from [7.26])

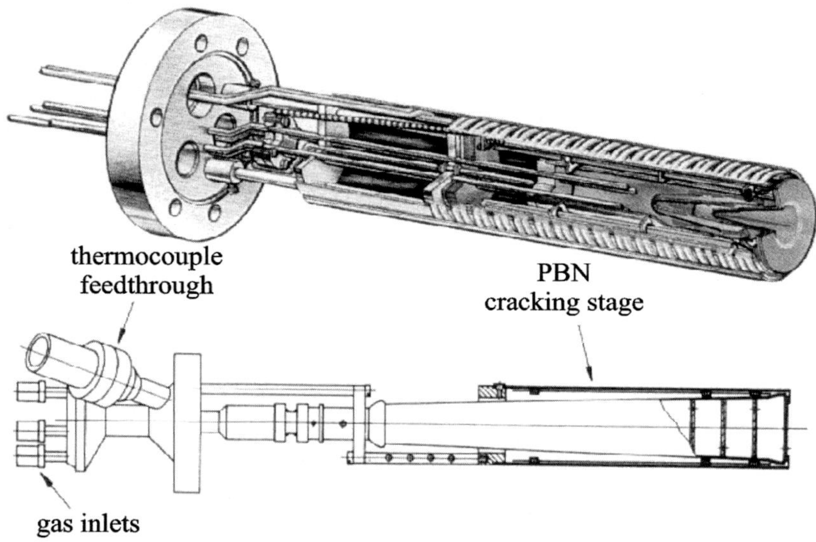


**Fig. 7.12.** Cutaway illustration of the hydride gas source used in the V80H MBE deposition chamber. Four inlet lines and the pyrolytic boron nitride cracker assembly are shown [Courtesy of VG Semicon (top) and ISA Riber (bottom)] (taken from [7.1])

the pressure conditions in the alumina cracking tube illustrated in Fig. 7.11a. These species then diffuse through a small leak in the alumina tube into the low pressure portion of the cracker where further thermal cracking to dimers occurs. Because it is presumed that the HPGS will yield species as expected from thermodynamic equilibrium, that cracker has been referred to as a thermodynamic cracker [7.31].

The LPGS is a catalytic cracker (Fig. 7.11b), used mostly when group III organometallics are involved in the GSMBE processes. What is significant is that both types of crackers can yield primarily dimeric species under proper experimental conditions [7.1].

Gaseous sources designed for the GSMBE systems by VG Semicon Company and by ISA Riber, respectively, are shown in cutaway illustrations in Fig. 7.12 [7.1]. The group V hydrides source has a cracking zone held at 900 – 1000°C. It has several introduction lines, which permit several gases to be introduced into the same cell, thus giving high compositional uniformity, both in space at the output of the cell and in time, due to precise flow regulation control. The VG Semicon source has four independent inlet lines, while the ISA Riber source has five inlets.

Organometallic gas sources are gas injectors and do not require any cracking zone as shown in Fig. 7.13. The VG Semicon source has four independent gas inlet lines with micromachined orifices. The gas flows into a PBN conical

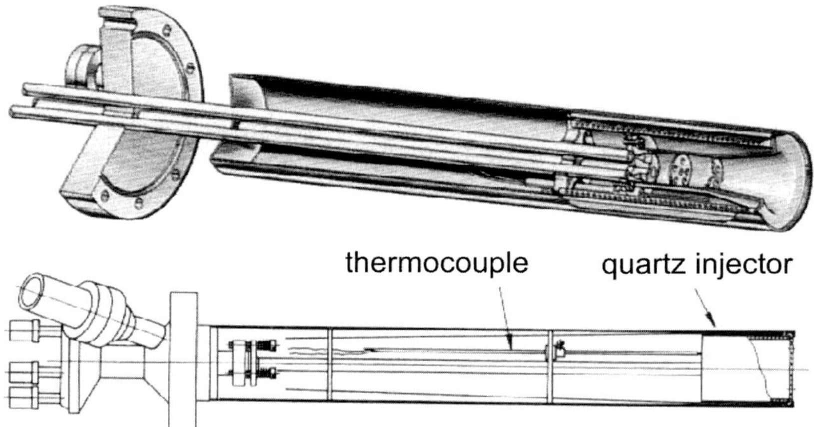


**Fig. 7.13.** Cutaway illustration of the organometallic gas source showing three of the four independent gas lines, the pyrolytic boron nitride mixing zone and the integral heater [Courtesy of VG Semicon (top) and ISA Riber (bottom)] (taken from [7.2])

crucible with perforated baffle plates to ensure thickness and compositional uniformity. An integral heater ($< 100°$C) ensures that the gas does not condense in the cell. The gas is then decomposed at the substrate wafer surface.

The ISA Riber source has five independent low pressure gas inlet lines. The gases are mixed in the cell flange for compositional uniformity. They then flow through a quartz crucible and are directed towards the substrate through a quartz diffuser, ensuring high flux uniformity at the substrate surface. The source has an outgassing facility (750°C) and preheating capability (50°C) with sufficient precision to avoid organometallic condensation and decomposition.

To ensure the precision and stability of the gas flow demanded by the GS MBE process, the gas handling system of a MBE machine using gas sources, instead of effusion sources, is specially designed. Each gas inlet line has an associated independent control system comprising a number of isolation valves which control gas inlet, gas outlet, gas exhaust, bypass and isolation. All control units are mounted in a gas cabinet which is easily connected to the laboratory vent system and scrubber. A dedicated exhaust pumping system is configured for fast, safe gas exhaust and ease of servicing. A vacuum trapped rotary pump forms the basis of the system with the addition of a dedicated turbomolecular pump.

### 7.2.2 Metal Organic MBE

MOMBE combines many important advantages of MBE and MOVPE [7.34] and advances epitaxial technology beyond both these techniques. In MOVPE, the gas pressure in the reactor is between $10^5$ and 1 Pa. As a result, the transport of the reactant gases is by viscous flow. If the pressure is further reduced down to $10^{-2}$ Pa, the gas transport becomes a molecular beam. Hence the process evolves from vapor deposition to beam deposition. If the thin film deposited is an epitaxial layer, the process is called MOMBE, which can use exclusively metal organic gas sources.

Unlike MBE, which employs beams (e.g., Al, Ga and In) evaporated at high temperature from elemental sources, in MOMBE the sources are gaseous or they are charged with reactants (solid or liquid) that have sufficiently high vapor pressure that they produce gaseous beams, without significant heating above room temperature. Al, Ga and In are derived by the pyrolysis of their organometallic compounds, e.g., trimethylaluminum (TMAl), trimethylgallium (TMGa), and trimethylindium (TMIn), at the heated substrate surface. $As_2$ and $P_2$ are obtained by thermal decomposition of their hydrides passing through a heated baffled cell. Unlike MOVPE, in which the chemicals reach the substrate surface by diffusing through a stagnant carrier gas boundary layer above the substrate, the chemicals in MOMBE are admitted into the high vacuum growth chamber in the form of a beam.

Comparing with conventional MBE, the main advantages of MOMBE include [7.22]:

(i) the use of group III organometallic sources that are solids or liquids which exhibit sufficiently high vapor pressure that they do not need to be heated significantly above room temperature, which simplifies multiwafer scale-up;
(ii) semi-infinite source supply and precision electronic flow control with instant flux response (which is suitable for the production environment);
(iii) a single group III beam that guarantees material composition uniformity;
(iv) no oval defects even at high growth rates (important for integrated circuit applications).

Comparing with MOVPE, the advantages of MOMBE include:

(i) no flow pattern problem encountered in multiwafer scale-up;
(ii) beam nature produces very abrupt heterointerfaces and ultrathin layers conveniently;
(iii) clean growth environment;
(iv) compatibility with other high vacuum thin-film processing methods, e.g., metal evaporation, ion beam milling, and ion implantation.

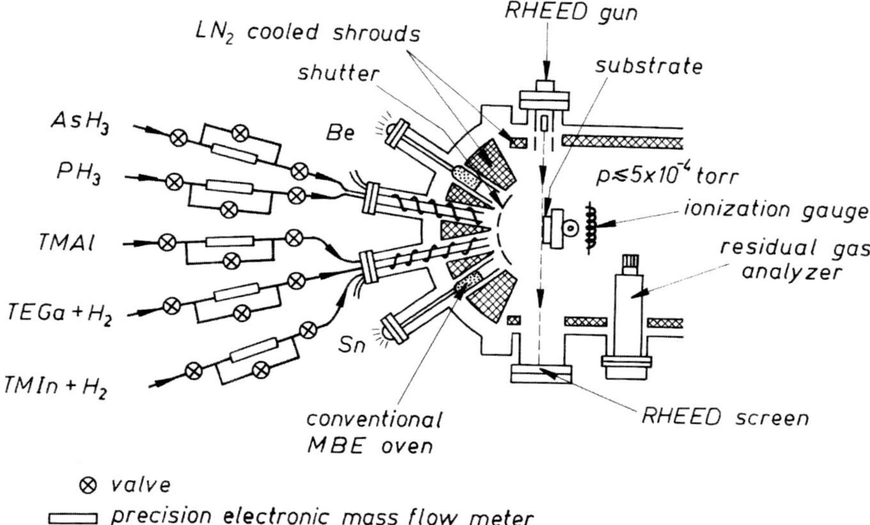


**Fig. 7.14.** Gas handling system and growth chamber with *in situ* surface diagnostic capabilities incorporated into a MOMBE system (taken from [7.23])

A gas handling system with precision electronic mass flow controllers is used in MOMBE for controlling the flow rates of the various gases admitted into the growth chamber as shown in Fig. 7.14. Hydrogen is used here as the carrier gas for transporting the low vapor pressure group-III organometallic compounds. Separate gas inlets are used for group-III organometallics and group-V hydrides. A low-pressure arsine ($AsH_3$) and phosphine ($PH_3$) cracker with a reduced input pressure of $3 \times 10^4$ Pa is maintained on the high pressure side of the electronic mass flow controller. To avoid material condensation, the manifold is warmed up to $\approx 40°C$. The cracking temperature is $\approx 920°C$. Decomposition of arsine and phosphine into arsenic, phosphorus, and hydrogen may be routinely achieved as observed with an *in situ* residue gas analyzer [7.23, 28]. In earlier studies on MOMBE of III–V compounds [7.35] group V alkyls were thermally decomposed after mixing with $H_2$. Though these alkyls are safer than hydrides, their purity at present is rather insufficient. Hence, hydrides are preferred.

TEGa maintained at 30°C, TMIn at 25°C, and TMAl at 25°C are used. The TEGa, TMIn and TMAl flows are combined to form a single emerging beam impinging line-of-sight onto the heated substrate surface. This automatically guarantees composition uniformity [7.35]. The typical growth rates are $2-3\,\mu m\,h^{-1}$ for GaAs. $4-6\,\mu m\ h^{-1}$ for AlGaAs, $3-5\,\mu m\,h^{-1}$ for GaInAs, and $1.5-2.5\,\mu m\,h^{-1}$ for InP. Such growth rates are higher than those typically used in conventional MBE growth processes.

### 7.2.3 Hydride Source MBE

Following the evident success of MOMBE in growth of III–V and II–VI compound heterostructures, hydride source MBE (HSMBE) has been demonstrated to be very promising for epitaxial growth of heterostructures made of the SiGe/Si material system [7.36]. For the application to Si large-scale-integrated (LSI) device fabrication processes, the conventional SSMBE [7.1] exhibits several disadvantages. Micron-size oval defects, which come from the use of a molten elemental Si source belong to the most fatal effect for LSI application, which is characteristic for this technology. The low throughput due to source exchange and the difficulty of selective-area epitaxial growth are also problems.

The remedy for these problems has been found by introducing to the Si-LSI technology the HSMBE with Si-hydride gaseous sources. Moreover, using phosphine and/or arsine gas source crackers, the difficulty of the rate control of P and As doping of Si has become considerably diminished [7.27].

In the epitaxial growth techniques using gaseous sources, e.g., MOVPE, two kinds of reactions should be considered. One is the reaction in the vapor phase and the other is the reaction on the substrate surface. In the case of HS MBE, and in general in all of the modifications of the GSMBE technique, only the latter reaction contributes to the growth process. This is so because the pressure in the growth chamber is low enough ($p \leq 10^{-3}$ Pa) for excluding vapor phase reactions. In high vacuum, which is the environment for the GSMBE process, the mean free path of gaseous reactants exceeds the effective length of the growth chamber. In this case, for gaseous molecules, the probability of collisions with the substrate is much larger than that of collisions with other gaseous molecules in the gas phase. In this respect, GSMBE, and thus also HSMBE, is defined as a growth process which proceeds only by surface reactions.

In HSMBE the source gas molecules move in the growth chamber from the gas inlet along the line-of-sight toward the substrate surface. In the case of Si-MBE, usually the source molecules consist of Si and hydrogen atoms. For example, $SiH_4$, $Si_2H_6$, and $Si_3H_8$ are candidates for the source gas for Si epitaxial growth [7.27]. Because the thermal decomposition of these molecules does not occur in the gas phase, the source molecules should decompose on the substrate surface leaving there free Si atoms, in order to produce epitaxial growth of the Si film. Finally, the remaining parts of the source molecules (hydrogen atoms) should desorb from the surface. This means that dissociative adsorption followed by hydrogen desorption are the key growth steps in group IV HSMBE [7.37].

A typical apparatus for group IV HSMBE consists of a main growth chamber, a sample loading chamber, and a subchamber with the gas mixing facility for source gas flow control [7.27]. The base pressure in the growth chamber is $\approx 10^{-8}$ Pa. For Si and Ge growth, $Si_2H_6$ and $GeH_4$ are used as source gases, respectively. This is because $Si_2H_6$ has a much larger sticking

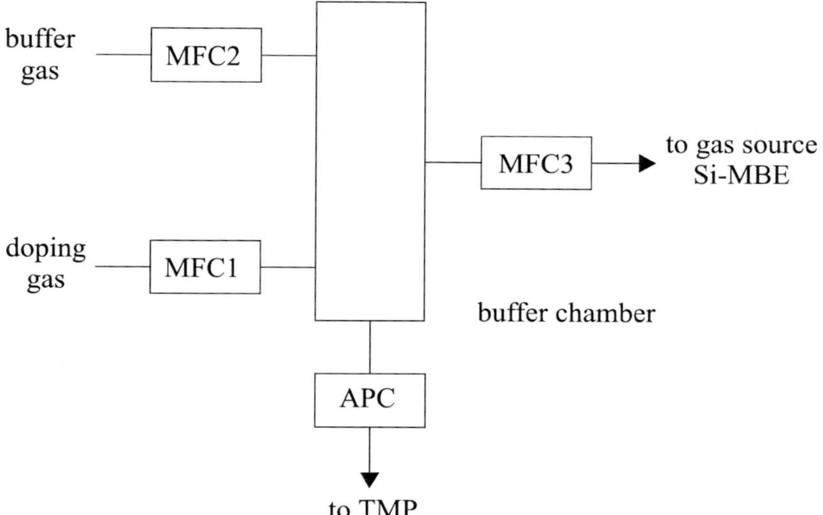


**Fig. 7.15.** A schematic illustration of the principle of the gas mixing system for dopant-gas flow control in the Si-HSMBE (taken from [7.27])

coefficient on Si substrates than $SiH_4$ [7.38]. Hence, a larger growth rate is achieved with $Si_2H_6$ [7.39]. B and P doping are achieved by using 5% $B_2H_6$ (diluted by $H_2$) and 5% $PH_3$ (diluted by $H_2$) gas sources, respectively.

Basically, the growth rate and doping concentrations are in HSMBE proportional to the flow rates. Hence, precise control of the flow rate is required. Especially for doping, a wide range of carrier-concentration control between $10^{15}$ and $10^{20}\,cm^{-3}$ is required for device fabrication. This means that the doping gas flow should be controllable over a five-decade range [7.27]. For this purpose, a specially designed gas mixing system [7.40] has to be used in the MBE growth apparatus. The system is illustrated in Fig. 7.15. Basically, it consists of three mass flow controllers (MFC) and a buffer chamber evacuated by a turbomolecular pump (TMP). In the figure, MFC1 controls the doping-gas flow rate and MFC2 controls the buffering gas. Because each MFC controls the flow rate over a two-decade range, the doping-gas concentration at the buffer chamber can be changed over a four-decade range. From the buffer chamber, a part of the buffered doping gas is introduced into the growth chamber through MFC3, and the remainder is evacuated by the TMP. The evacuation of the remaining gas keeps the absolute doping gas concentration in the buffer chamber constant. With this system, control of the boron doping concentration over a five-decade range has been achieved [7.40].

One of the great advantages offered by Si-HSMBE is the selective-area epitaxial growth of Si on $SiO_2$ patterned Si substrates. This growth mode could be achieved due to the different sticking coefficients of $Si_2H_6$ on Si and

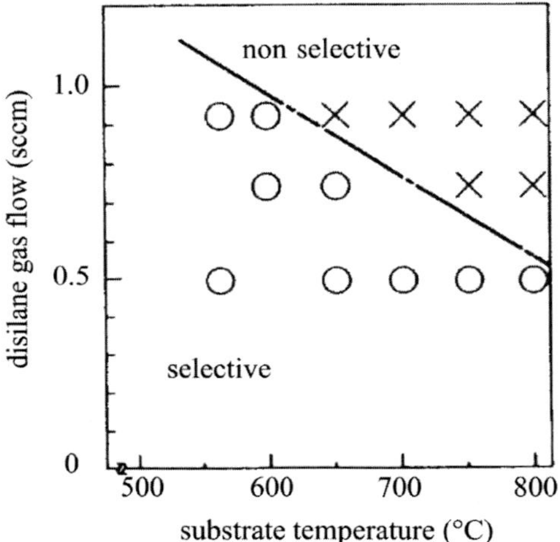


**Fig. 7.16.** Selective-growth condition in Si-HSMBE, with disilane as the Si gas source, as functions of the growth temperature and the molecular beam intensity. Open circles indicate no polycrystalline-Si grows on the $SiO_2$ surfaces, while at conditions indicated by crosses, polycrystalline-Si growth does occur on the $SiO_2$ surfaces (taken from [7.27])

$SiO_2$. Si growth epitaxially only on the bare Si areas of the substrate, while no Si growth is observed on $SiO_2$. As usual in MBE, this growth mode is achieved only in a definite temperature window. In Fig. 7.16, an example of these favorable conditions is illustrated as a function of substrate temperature and the $Si_2H_6$ flow rate [7.27]. One may recognize that high temperature and large flow rate tend to be unfavorable for selective-area growth.

The transition from selective to non-selective growth occurs suddenly. This abrupt transition suggests that it is caused by critical-nuclei formation on the $SiO_2$ surface. $SiH_n$ adsorbates migrate on the oxide surface. They occasionally collide with each other and nucleate. From the balancing of the surface and volume energies, the total energy of the nucleus has a maximum at a critical size [7.41]. Hence, when the size of a nucleus exceeds the critical one, it rapidly grows and the selectivity of the Si growth disappears. In fact, all the unfavorable conditions for selectivity are explained by this argument. The large flow rate enhances the probability of critical-nuclei formation on the oxide surface, because more Si atoms impinge on this surface in unit time. The high substrate temperature also enhances the probability of critical nucleation through the increase of the surface migration length of the adsorbed Si atoms. Therefore, to keep the selectivity of the Si growth, nuclei on the oxide surface should always be removed before their size exceeds the critical size. For this purpose, addition of halogen etching gas to the source gas is effective [7.27]. One more problem related to selective-area growth of Si by HSMBE is facet formation. Si(111), Si(311) and Si (511) faceting surfaces are observed at the $SiO_2$/epitaxial–Si(100) grown film boundary. This effect causes severe problems when HSMBE has to be applied to device fabrication. The key to solve the faceting problem is to select special growth conditions in the so-called reaction-limited regime [7.27].

Finally, in this section, the $Si_{1-x}Ge_x$ alloy growth using HSMBE is considered. This alloy is an attractive material for band-gap engineering with Si, because its band gap is smaller than that of Si. However, Ge has a 4 % mismatch in lattice constant to Si. Hence, commensurate $Si_{1-x}Ge_x$ alloy growth by HSMBE is indispensable for device application of the SiGe/Si heterostructures.

$GeH_4$ and $Si_2H_6$ are used in this case of HSMBE as source gases for Ge and Si, respectively. The Ge concentration can be controlled by changing the flow rate ratio of both of the reactants. However, at smaller Ge concentrations, below 30 %, the concentration is approximately proportional to the $GeH_4$ flow rate, at fixed $Si_2H_6$ flow rate [7.42].

Selective epitaxial growth of $Si_{1-x}Ge_x$ can also be achieved with HSMBE growth on $SiO_2$ patterned Si substrates. The reason for the successful selective epitaxial growth is the same as that for Si homoepitaxial-layer selective growth. However, faceting at the SiGe/$SiO_2$ boundary is due to its low surface energy. The facet at the Si/$SiO_2$ boundary disappears during the growth in the reaction-limited regime, whereas the facet at a SiGe/$SiO_2$ boundary still remains even during growth in this regime. The suppression of the faceting effect at the SiGe/$SiO_2$ boundary requires a microscopic understanding of the mechanism by which the alloy faceting surface energy can be controlled [7.27].

## 7.3 Growth Techniques Using Modulated Beams

Among the presently known modifications of MBE, the group of growth techniques based on the pulsed mode of supplying the reactant species to the substrate or the growing epilayer surface (modulated-beams operation mode of MBE) play a special role. This is so because all of these techniques lead to perfect layer-by-layer growth through surface controlled kinetic processes. Thus the growth is not source controlled (like in conventional MBE) but should be treated as surface controlled [7.1].

Atomic layer epitaxy carried out in an ultrahigh vacuum environment (UHV ALE) became the first representative of this group [7.43]. Later on, phase-locked epitaxy (PLE) [7.44], Molecular Layer Epitaxy (MLE) [7.45], and migration enhanced Epitaxy (MEE) [7.46] have been demonstrated as other growth techniques belonging to the same group of MBE modifications [7.47]. We will shortly discuss here the basic features of these growth techniques with modulated beams, which are most frequently used at present (see also Sect. 6.3.1).

### 7.3.1 Ultrahigh Vacuum Atomic Layer Epitaxy

In discussing UHV ALE we will take as an example the growth peculiarities of CdTe, the first ever compound grown by ALE epitaxially [7.43, 48]. The used CdTe substrate surfaces were the polar (111)A and (111)B, Cd and Te rich surfaces, respectively. The deposition timing scheme for this growth procedure is shown in Fig. 7.17, and the key conceptual steps associated with the growth by UHV ALE are illustrated in Fig. 7.18, following a model proposed by Herman et al. [7.49]. This model is indeed for the growth of CdTe on CdTe(111) substrates however, it may be considered as quite general for UHV ALE because CdTe may be treated as nearly ideal for studying the mechanism of this growth process [7.50].

The model includes (i) the existence of transition layers of both Cd and Te species intermediate between the crystalline (well-ordered) substrate or epilayer phase and the gaseous, non-ordered phase created by the impinging beams (or the residual gas of the UHV chamber in the dead time of the ALE timing scheme), and (ii) partial reevaporation of the first chemisorbed monolayer of the deposited constituent elements [7.51]. The transition layers, also called near surface transition Layers (NSTL), create reaction zones several monolayers thick near the substrate surface, where the atoms or molecules are weakly bound to the solid surface, so that they may readily migrate over the surface, become incorporated into the crystal lattice of the growing epilayer, or thermally desorb into the UHV environment [7.48, 51]. It should be emphasized that in the NSTL relevant to the UHV ALE (they are created only by one of the constituent elements of the binary compound to be grown), two

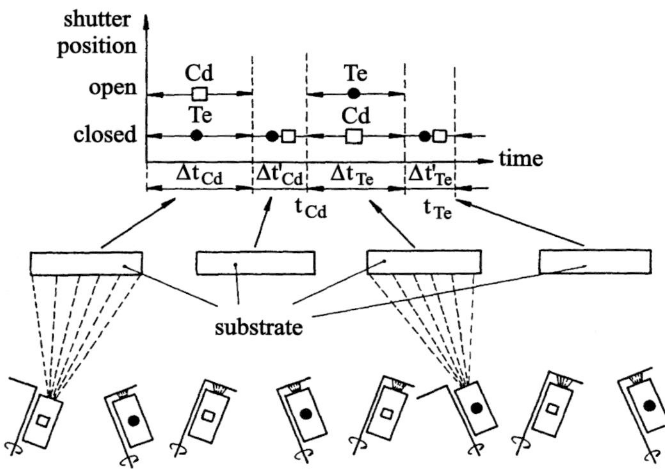


**Fig. 7.17.** Deposition timing scheme of ALE growth of CdTe in aUHV environment (taken from [7.1])

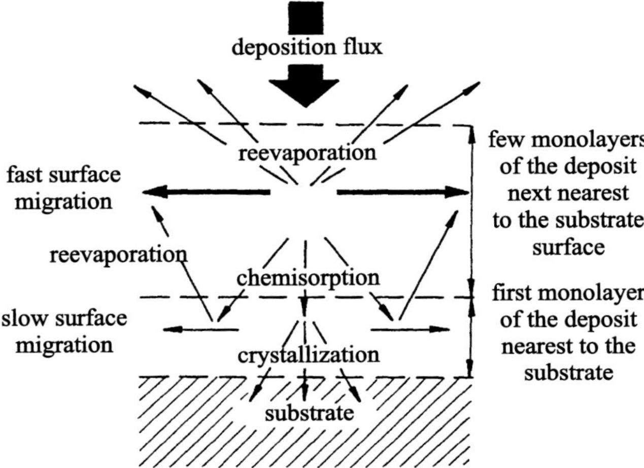


**Fig. 7.18.** Schematic illustration of the key conceptual steps associated with the growth of CdTe epilayers by UHV ALE. Here "substrate" means the single crystalline bulk substrate wafer and the already-grown epilayer (taken from [7.1])

different regions may be distinguished, as in the NSTL occurring in SSMBE (the transition layer is created by all constituents of the compound to be grown). Adparticles belonging to different regions of the ALE-relevant transition layer are bound to the crystalline surface with different bonds. The first monolayer, chemisorbed with covalent bonds, creates the first region, nearest to the substrate (Fig. 7.18). Atoms of this monolayer are much more strongly attached to the surface than the adparticles of the other monolayers of the deposited element [7.52]. Those monolayers next nearest to the substrate create the second region of the transition layer. The adparticles of this region are attracted by Van der Waals forces (this corresponds, for example, to physisorption of anion molecules), or by relatively weak chemical interactions (this corresponds to multilayer chemisorption by bonds characteristic of the bulk phase of the pure constituent element of the compound to be grown).

The principal technological parameter in UHV ALE is the substrate temperature during the growth process. Its value should be high enough to break all bonds attracting the adparticles belonging to the second region of the ALE-relevant transition layer, and thus to cause the thermal desorption of these adparticles during the dead time shown in the deposition timing scheme in Fig. 7.17. However, this temperature should be low enough to preserve the chemical bonds of the first monolayer, and thus to cause the growth of the compound as "layer of the first element-by-layer of the next element" [7.48]. Thus, again a temperature window exists for the MBE growth, here in its UHV ALE modification.

A characteristic feature of UHV ALE is the possibility of partial desorption of the first chemisorbed monolayer if the dead time is too long. This desorption, indicated on the scheme shown in Fig. 7.18, is a consequence of the UHV environment in which the growth process is performed (see, for example, the Langmuir adsorption isotherms relating the surface coverage of a given solid to the gas pressure at a given temperature [7.53]). In conclusion, one has to consider the UHV ALE as not a self-regulatory process, which is opposite to atmospheric-pressure ALE (see Sect. 6.3.1). This means that timing and dosing in the deposition cycle play a crucial role in this process. Thus, UHV ALE has to be precisely designed when smooth and flat layers in an atomic scale have to be gown. The smoothest surface morphology is obtained in UHV ALE when the growth proceeds in an exactly two-dimensional layer-by-layer mode. This happens, however, only if the deposited atoms of the constituent elements build up exactly one complete monolayer by the end of the respective dead times in the deposition cycle. If this so-called "one-monolayer coverage criterion" [7.54, 55] is not satisfied during the UHV ALE process, three-dimensional growth may occur, causing surface roughness. Consequently, an understanding of the nature of the NSTL occurring in sequential deposition processes of the constituent elements of the grown epilayers seems to be fundamental for successful handling of UHV ALE [7.56, 57].

The main body of the papers published on UHV ALE concerns binary semiconductor compounds [7.57], which results from the fact that the growth process becomes much more complicated when a ternary or quaternary semiconductor compound has to be grown [7.58]. One of the crucial problems in the growth of ternary compounds by UHV ALE is the selection of such a growth procedure (timing and dosing in the ALE cycle) which assures the desired composition of the grown compound. Especially important is the case of compounds consisting of incongruently evaporating elements. An example of such a compound is $Cd_{1-x}Mn_xTe$ ($0 \leq x \leq 1$), a diluted magnetic semiconductor. In this material system the vapor pressures of the constituents are equal to: 1.33 Pa for Cd, $8.34 \times 10^{-2}$ Pa for Te and $1.76 \times 10^{-15}$ Pa for Mn, at the temperatures of 270°C, typical for UHV ALE of CdTe [7.59]. In other words, during a cycle of evaporation in the ALE process excess Cd and Te atoms reevaporate from the substrate surface while Mn tends to stick still on this surface. Therefore a special mode of evaporation is needed when growing $Cd_{1-x}Mn_xTe$.

The first successful attempt of growing $Cd_{1-x}Mn_xTe$ by UHV ALE [7.59] concerned epilayers with fairly low content of Mn ($x \leq 0.3$) grown on p-type CdTe(111)B single crystal substrates. Later on, after this report, the results of UHV ALE growth of $Cd_{1-x}Mn_xTe$ with ($0 \leq x \leq 0.9$) were published [7.60].

Another interesting example of ternary compounds concerns $Cd_{1-x}Zn_xTe$, which exhibits nearly equal equilibrium pressures of the constituent elements. The first ever UHV ALE growth of this compound was reported in 1997

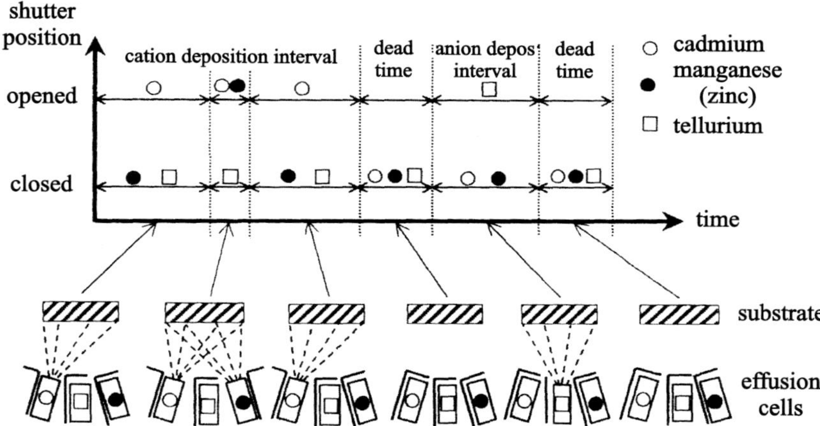


**Fig. 7.19.** Deposition timing scheme in UHV ALE of $Cd_{1-x}A_xTe$ (A = Zn, Mn) used for studies on the growth processes of these compounds (taken from [7.58])

[7.61]. The deposition timing scheme used in the performed UHV ALE growth experiments of the ternaries $Cd_{1-x}A_xTe$ (A = Zn, Mn) is shown schematically in Fig. 7.19.

The fundamental data concerning the growth kinetics in UHV ALE of the ternaries $Cd_{1-x}A_xTe$ (A = Zn, Mn) studied by reflection mass spectrometry (REMS) and reflection high energy electron diffraction (RHEED) are presented and discussed in [7.61, 62].

### 7.3.2 Migration Enhanced Epitaxy

A slight modification of UHV ALE, in which no dead-time in the deposition timing was introduced, has been demonstrated for low-temperature growth of GaAs and GaAs-AlAs quantum well structures by Horikoshi et al. [7.46]. They named this modification migration enhanced epitaxy (MEE).

The idea of the pulsed mode of supplying reactant species of the growing epitaxial film to the substrate has been introduced to the MBE technology of III–V semiconductor compounds by Kawabe et al. [7.63, 64], as a new method of composition control for the $Al_xGa_{1-x}As$ compounds. This idea was later applied to GaAs in the growth techniques named molecular layer epitaxy [7.45].

In MEE the interruption of arsenic supply to the growing surface during Ga or Al supply is essential to the surface migration enhancement of the metallic surface adatoms. Therefore, Ga or Al atoms and arsenic were alternately supplied to the GaAs substrate surface in order to obtain metal-stabilized surfaces. The typical switching behavior of $As_4$ and Ga beam intensity, as measured by an ionization gauge placed at the substrate holder position in the growth chamber, is shown in Fig. 7.20. Bearing in mind the

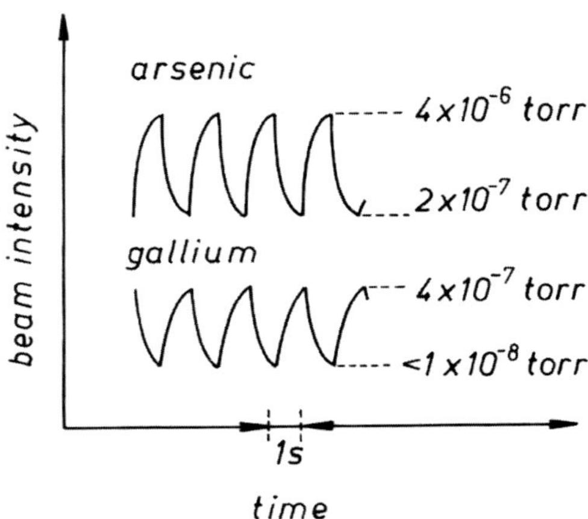


**Fig. 7.20.** Beam intensity changes for Ga and $As_4$ species caused by shutter operation characteristic of MEE (taken from [7.46])

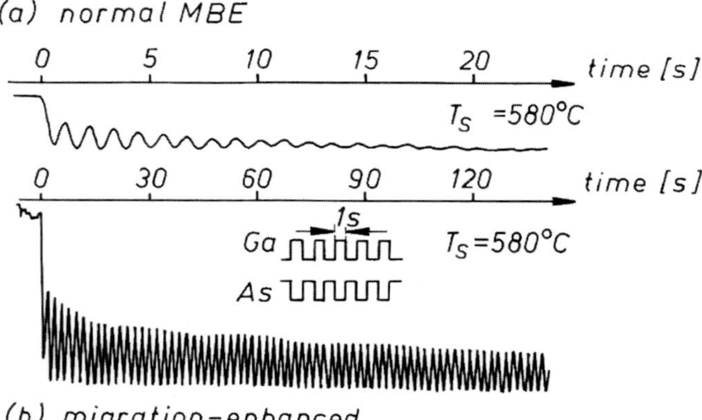


**Fig. 7.21.** RHEED specular beam intensity oscillation during growth of GaAs for conventional MBE growth (**a**) and for MEE growth (**b**) (taken from [7.46])

response time of the ionization gauge amplifier, the observed response indicates that the beam intensities change very quickly following the shutter operation.

It has been observed experimentally that the MEE growth process proceeds in a layer-by-layer mode [7.65]. This can be seen when recording the RHEED specular beam intensity oscillation during MEE. In Fig. 7.21, a comparison of the MEE results with those obtained for conventional MBE growth of GaAs on GaAs(001) substrates at the temperature of 580°C, is presented. In both cases the same beam flux intensities of Ga and of $As_4$, respectively, have been used. In conventional MBE growth, the RHEED oscillations almost completely disappear after about 20 periods because the surface smoothness on the atomic scale deteriorates. In contrast, the oscillations continue during the entire layer growth process when the MEE method is applied. This result indicates that a better surface smoothness is maintained during MEE even after the growth of thousands of layers. Most probably this is caused by the rapid migration of Ga atoms on the substrate surface in a very low arsenic pressure environment.

The amplitude of the RHEED pattern intensity oscillation in MEE is closely related to the amount of Ga atoms supplied to the growing surface per unit cycle, i.e., the amplitude is maximum when the number of Ga atoms per cycle is equal to the number of surface sites available for the migrating Ga atoms. This fact is consistent with the "one-monolayer coverage criterion" found earlier in UHV ALE of CdTe [7.54]. It also means, that low-temperature growth of GaAs by MEE is possible if the growth conditions are selected for MEE, which satisfy the demand of this criterion [7.65]. The high quality of GaAs epilayers grown at 200°C, and of an AlAs-GaAs double quantum well structure grown at 300°C with MEE was confirmed by photoluminescence measurements during the first demonstration of this growth technique [7.46].

MEE was later successfully applied by Salokatve et al [7.66] for reduction of surface oval defects in GaAs grown in a two-step process. They have shown that if a thin GaAs buffer layer is deposited by supplying alternately Ga atoms and $As_4$ molecules to a GaAs substrate, prior to further growth by MBE, the density of oval defects in the final layer is reduced reproducibly by a factor of 7, from about 490 to $70\,cm^{-2}$, when compared with that obtained using MBE alone under closely similar conditions. The improved surface morphology produced by the pulsed beam method (the MEE growth mode) was ascribed to the two-dimensional layer-by-layer growth characteristic of MEE.

The same group has also demonstrated that MEE may cause beneficial effects in MBE growth of GaAs on Si substrates [7.67]. They grew GaAs films on Si(100) substrates using a two-step growth process of a 300°C GaAs buffer layer (by MEE) followed by a layer grown at 600°C by conventional MBE. The films were examined by Rutherford backscattering and X-ray diffraction methods. A significant reduction in the defect density near the GaAs/Si interface and in the bulk of these films was observed when the buffer layer was deposited by supplying alternately Ga atoms and $As_4$ molecules to the substrate, rather than applying conventional MBE. More information concerning the growth mechanism of MEE may be found in [7.68–74].

### 7.3.3 Molecular Layer Epitaxy

Molecular layer epitaxy (MLE) is a thin film growth method using the chemical reaction of adsorbates on the semiconductor surface, where gas molecules containing one element of the compound semiconductor are introduced alternately not continuously as in gas source MBE into the growth chamber

[7.45, 75–77]. GaAs MLE using $AsH_3$ as an As source and TMGa as a Ga source was first demonstrated by Nishizawa et al. [7.45]. Later, instead of TMGa, TEGa was used as the Ga source, resulting in higher quality of epitaxial layers at lower growth temperature. An excimer laser which operated with ArF ($\lambda$ = 193 nm), KrCl ($\lambda$ = 222 nm), KrF ($\lambda$ = 249 nm), XeCl ($\lambda$ = 308 nm), or XeF ($\lambda$ = 350 nm) or a high pressure Hg lamp was used as the light source for substrate (growing layer) irradiation during growth with MLE. A schematic diagram of the experimental setup used for MLE with photostimulation facilities is shown in Fig. 7.22, and the changes of the gas pressure in the deposition chamber during the gas admittance cycles are shown in Fig. 7.23.

In the first ever experiments on MLE [7.45] the growth conditions, defined by admittance pressure of the reactants and the growth temperature, have been optimized experimentally by measuring the film thickness per deposition cycle as a function of reactant gas pressure (Fig. 7.24) and substrate temperature, and determining the conditions which guarantee one monolayer coverage in each cycle. The lower growth temperature obtained when GaAs is grown from TEGa and $AsH_3$ results from the fact that the binding energy of $C_2H_5$–Ga in TEGa is lower than the binding energy of $CH_3$–Ga in TMGa.

The MLE process may be considerably improved when using other As-sources, but not arsine. It has been demonstrated [7.79] that the tempera-

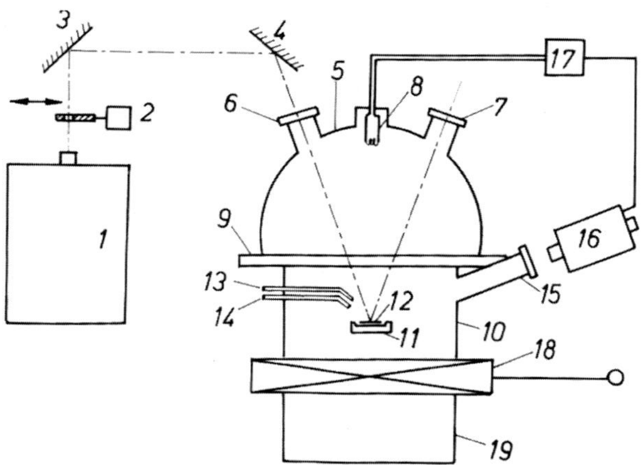


**Fig. 7.22.** Schematic diagram of the experimental setup used in the early MLE studies with photostimulation of the surface growth reactions: (1) excimer laser, (2) light beam chopper, (3,4) mirrors, (5) lamp housing, (6,7) quartz windows, (8) high pressure Hg lamp, (9) quartz plate, (10) growth chamber, (11) quartz sample holder, (12) substrate, (13,14) gas admittance nozzles, (15) sapphire window, (16) pyroscope, (17) temperature controller, (18) gate valve, (19) pumping unit (taken from [7.78])

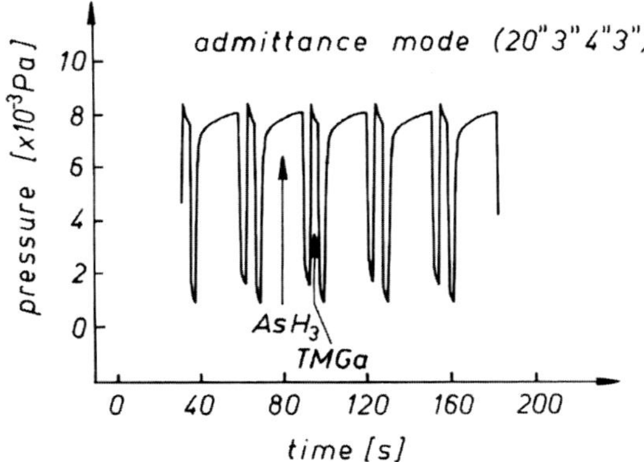


**Fig. 7.23.** The pressure in the deposition chamber during the gas admittance cycles of MLE growth of GaAs from TMGa and $AsH_3$ sources. This periodic admittance mode is repeated up to the desired film thickness (taken from [7.45])

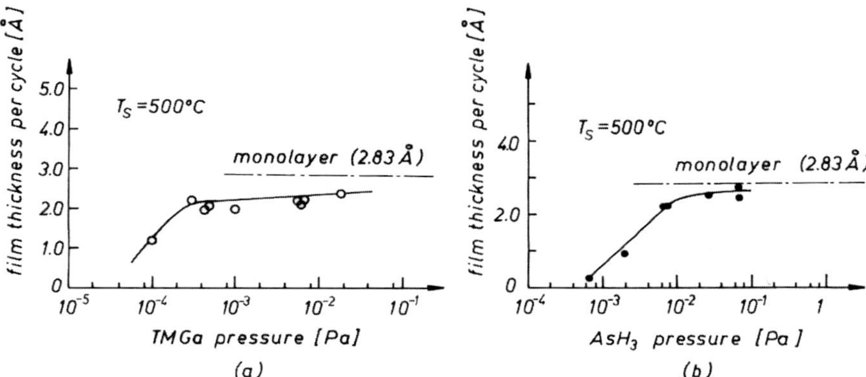


**Fig. 7.24.** Experimental determination of the growth conditions for MLE of GaAs from TMGa and $AsH_3$ by measuring the film thickness per deposition cycle as a function of (**a**) TMGa, and (**b**) $AsH_3$ pressures at a substrate temperature of 500°C with the admittance mode shown in Fig. 7.23 (taken from [7.78])

ture range for MLE using TEGa is substantially expanded when arsine is replaced with an alkyl-As, namely, tris-dimethylamino-arsenic ($As(N(CH_3)_2)_3$). In MLE using TMGa, tertiarybutylarsine has been utilized as the As-source, which has resulted in more effective surface reaction kinetics in the self-limiting growth regime of GaAs.

## 7.4 Externally Assisted MBE

In many cases of MBE, external agents are used in order to improve the growth processes. This concerns mainly material systems which require special growth conditions different from conventional MBE conditions, so far described. Here the silicon-based and nitride-based semiconductor compounds as well as organometallics can be mentioned as examples. As external agents, which influence the surface kinetic processes leading to epitaxial growth in MBE, the following are most frequently applied: (i) irradiation with light of different spectral ranges, (ii) activation of reactants by plasma , and (iii) ion beam assisted growth and doping. All of the externally assisted MBE growth techniques have been described and discussed in detail in [7.1]. Here, for the sake of brevity, we will consider only a few examples.

### 7.4.1 Irradiation with UV Light in MLE of GaAs

The effect of substrate irradiation with UV light during films growth by MLE has been studied by Nishizawa et al. in 1985 [7.45, 78]. The results may be summarized in general by stating that UV irradiation during MLE greatly improves the surface morphology (Fig. 7.25) as well as the electrical

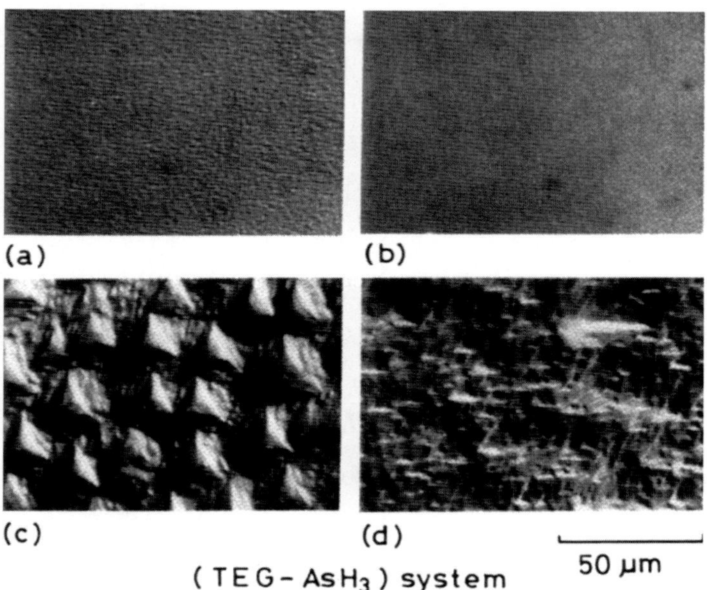


**Fig. 7.25.** Surface morphology of GaAs epitaxial films prepared by MLE on the GaAs(100) surface without irradiation (**a**) and with KrF (249 nm) irradiation (**b**) and on the GaAs(111)B surface without irradiation (**c**) and with Hg lamp irradiation (**d**) (taken from [7.78])

**Table 7.1.** Electrical properties of GaAs films prepared by MLE and photoassisted MLE from TMGa and $AsH_3$ under the same gas admittance conditions (taken from [7.45])

| $T_s$ (°C) | Light source | $\mu_h$ ($cm^2\,V^{-1}\,s^{-1}$) | Carrier concentration ($cm^{-3}$) | Conductivity type |
|---|---|---|---|---|
| 600 | no illumination | 56 | $1.9 \times 10^{19}$ | p |
| 600 | Hg lamp | 103 | $1.3 \times 10^{18}$ | p |
| 700 | no illumination | 68 | $1.2 \times 10^{19}$ | p |
| 500 | Hg lamp | 103 | $3.1 \times 10^{18}$ | p |
| 500 | Laser ($\lambda = 257.3\,nm$) | 84 | $1.6 \times 10^{18}$ | p |
| 500 | Laser ($\lambda = 257.3\,nm$ + $\lambda = 514.5\,nm$) | 110 | $2.4 \times 10^{18}$ | p |

properties of the films (Table 7.1). These effects have been ascribed to enhanced chemical decomposition [7.81] and enhanced surface migration [7.78] of adsorbates caused by UV light irradiation. It has been suggested that the essential migrating species is a Ga complex adsorbate like $Ga(CH_3)_x$ but not the As complex. Experimental evidence has been presented that the irradiation effects do not result from surface reactions promoted by increased electron–hole pair generation during the irradiation process.

The mechanisms of energy transfer from the photons to the crystal surface during the growth processes are difficult to identify. One possibility is to measure the electrical properties of the grown films, namely the Hall mobilities and carrier concentrations (Table 7.1). The following experimental results have been obtained:

(i) Irradiation ($\lambda = 249\,nm$) synchronized with TMGa admittance (see Fig. 7.23) increases the impurity level in the GaAs layer grown with MLE at optimal growth conditions, while synchronization of the irradiation with $AsH_3$ admittance causes a decrease in the impurity level, which is, however, still higher than that obtained without irradiation.

(ii) Irradiation synchronized with admittance pauses (dead times) causes an intermediate impurity level lower than in the case of TMGa synchronization, but higher than in the case of $AsH_3$ synchronization.

These results demonstrate that some of the reaction or migration steps were indeed separately enhanced by irradiation, which is evidence that the light irradiation effect is not only that of annealing, even if there are some heating effects [7.82].

### 7.4.2 Ion-Assisted Doping in Si-MBE

In conventional MBE technology intentional doping is achieved by co-deposition, which means that the dopant species are generated thermally in effusion cells and delivered to the growing film surface together with the beams of constituent elements of the film. The rate of doping is defined by suitable selected impingement rate of the neutral atomic or molecular doping species. In many cases, however, doping of MBE grown films may be achieved more effectively with better control of the incorporation rate of the dopant atoms, by applying ionized particle beams.

Doping using ionized particle beams may be performed either as a post-growth process, realized usually with focused high energy (about 100 keV) ion beams in a separate processing chamber of the MBE growth system (doping by ion beam implantation), or as an in-growth process, realized with low energy ion beams (0.1 – 3 keV). Here, we will discuss only the latter doping procedure.

Two modes of in-growth doping may be distinguished, i.e., the direct implantation mode and the secondary implantation mode. In the case of direct implantation, the low energy ion beam of dopant atoms is deposited on the substrate surface simultaneously with the neutral beams of the constituent elements. Thus, the dopant atoms are continuously incorporated into the growing crystal lattice of the film. Neither surface reevaporation nor surface segregation of dopant atoms is present during this growth-and-doping process [7.83]. The secondary implantation mode is based on a recoil process. The dopant-atom beam is generated thermally in an effusion cell and deposited on the substrate prior to the beginning of epitaxial growth of the layer to be doped. After a submonolayer of adsorbed dopant atoms has been formed on the substrate surface, the constituent element beams, of which at least one consists of ionized atoms, are allowed to impinge on the substrate, which causes the start of MBE growth. The substrate is electrically biased during growth by application of a potential of 0.1 – 1 kV, negative in relation to the electrically grounded beam sources. Consequently, the ionized atoms of the constituent element beam are accelerated towards the growing film surface. At this surface the ions collide with the accumulated dopant adatoms, knocking a fraction of them a short distance into the layer. Buried below the surface of the epilayer the dopants can no longer segregate or re-evaporate [7.83]. There are two reasons why in-growth ion implantation is employed to introduce intentionally impurities into the MBE-growing films. The first is an accurate control of the doping levels of those dopant species which exhibit strong surface segregation, e.g., antimony in silicon. The second reason is that in order to obtain doping levels higher than $10^{18}\,\mathrm{cm}^{-3}$, the sticking probability of dopants needs to be increased. However, in the case of Zn in GaAs there is no other way of introducing the doping element atoms into the MBE growing film than ion implantation, because the sticking coefficient of neutral Zn atoms is zero at MBE growth temperatures of GaAs [7.84–86]. The use of direct implantation for in-growth doping has been extended to other III–V compounds like, e.g., $GaP_xAs_{x-1}$ doped with $N_2^+$ [7.87]. However, it has become especially useful in MBE of Si and related compounds [7.83]

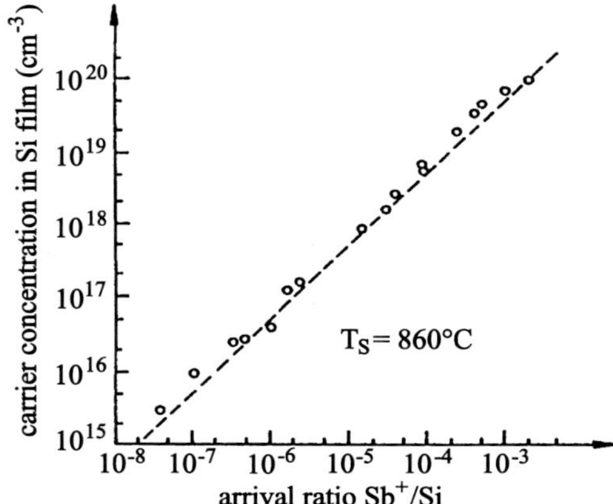


**Fig. 7.26.** Carrier concentration in Sb-ion-doped Si films as a function of $Sb^+$/Si flux ratio. Dashed line indicates 100% Sb-doping efficiency (taken from [7.89])

doped with $As^+$ [7.88] or $Sb^+$ [7.89]. An example of accurate control of the Sb doping level over the range $10^{16} - 10^{20}\,cm^{-3}$ by using an Si–MBE growth procedure with $Sb^+$ ion in-growth implantation is shown in Fig. 7.26. This figure presents the experimental data concerning the relationship between the carrier concentration in the Si MBE-grown films and the $Sb^+$/Si flux ratio at a substrate temperature of 860°C. It can be seen that the number of incorporated Sb dopant atoms is proportional to the $Sb^+$/Si flux ratio over four decades. The dashed line in this figure represents 100% doping efficiency. The doping efficiency achieved with $Sb^+$ by in-growth ion implantation is two orders of magnitude greater then that for evaporation doping by co-deposition.

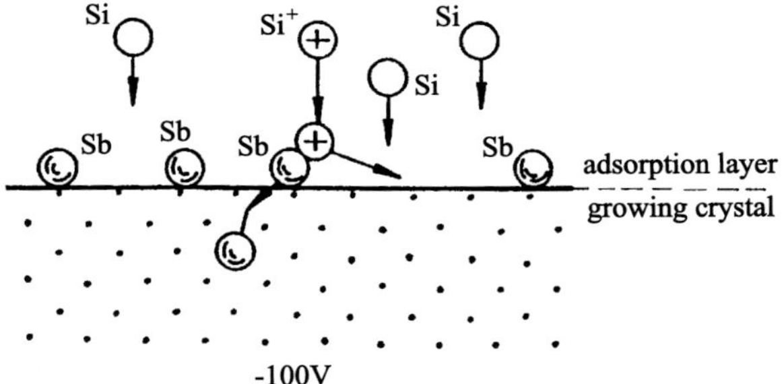


**Fig. 7.27.** Schematic illustration of the principle of doping by secondary implantation. The adsorbed Sb neutral atoms are implanted by recoil momentum from $Si^+$ ions impinging on the growing film surface. The incorporation depth is only some atomic distances in the film lattice (taken from [7.92])

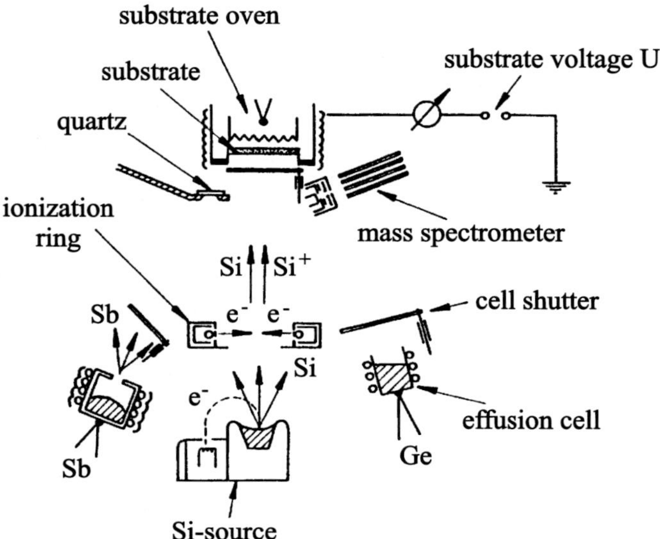


**Fig. 7.28.** Schematic illustration of the Si-MBE growth system used for growth of modulation doped Si/SiGe strained layer superlattices with in-growth secondary implantation of Sb dopants (taken from [7.92]),

This increase in doping efficiency is most probably caused by the relative high kinetic energy of the accelerated dopant ions (130 – 1000 eV) [7.89].

The secondary implantation technique [7.90–92] has been introduced in order to diminish the effect of surface segregation tendency of the doping Sb atoms during Si MBE growth. The idea of this technique, described above, is illustrated in Fig 7.27, while the schematic drawing of the Si-MBE growth system in which secondary implantation has been achieved is shown in Fig. 7.28.

After growing a nonintentionally doped buffer layer of Si, Sb atoms were deposited during MBE growth by opening the Sb cell shutter. This establishes an Sb adsorption layer, leading to doping in the $10^{16}\,\mathrm{cm}^{-3}$ range by spontaneous incorporation. After switching-on the substrate voltage $\mathrm{Si}^{+}$ ions are accelerated towards the substrate resulting in a considerable increase of dopant concentration by three orders of magnitude from about $10^{16}\,\mathrm{cm}^{-3}$ up to $10^{19}\,\mathrm{cm}^{-3}$. A carrier concentration profile in the grown film, as evaluated by spreading resistance measurements, is shown in Fig. 7.29. The observed exponential decrease of the carrier concentration to $10^{18}\ \mathrm{cm}^{-3}$ is caused by the loss of dopant adatom density by incorporation into the bulk of the film.

A slightly modified version of the secondary implantation technique (without any pre-growth deposition of dopant atoms) has been introduced by Kubiak et al. to Si MBE technology under the name potential-enhanced doping (PED) [7.93].

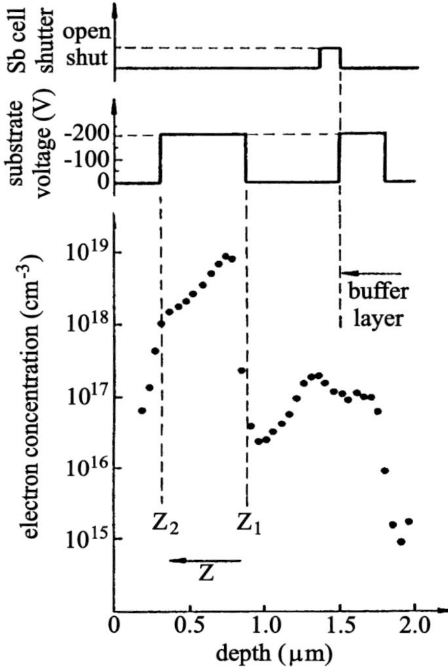


**Fig. 7.29.** Carrier concentration versus layer depth evaluated by spreading resistance. The growth temperature was 650°C. The program of the Sb cell shutter operation and the applied substrate voltage is given in the upper part of the figure (taken from [7.90])

### 7.4.3 Plasma-Assisted MBE Growth of GaN and Related Compounds

Band-gap engineering in the range of 3.4 – 6.2 eV can be achieved either by solid solutions or by layered structures of GaN and AlN [7.93]. The lattice constant mismatch between AlN and GaN is 3.5%, thus layered structures, e.g., multiquantum well structures of these two materials, may be used for producing low-dislocation density GaN-and-AlN devices.

AlN/GaN layered structures with layer periods between 1.5 and 40 nm have been grown on (0001) oriented sapphire and on 6H-SiC substrates by the plasma-assisted MBE technique [7.93]. Standard effusion cells were used as sources of Al and Ga, and a small, MBE compatible electron cyclotron resonance (ECR) plasma source was used to activate the nitrogen gas prior to deposition. The design and performance of such an ECR plasma source has been described in [7.94].

ECR plasma sources provide a denser plasma compared to radio frequency (RF) or conventional microwave sources and operate over a much larger pressure range. These sources use a magnetic field that simultaneously causes

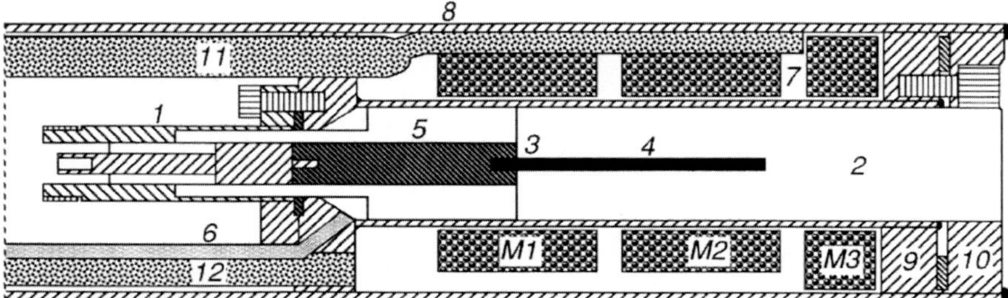


**Fig. 7.30.** Cross-sectional view of a small ECR plasma source used for deposition of III–V nitrides. The numbered elements are: 1 – 50 $\Omega$ coaxial feedthrough; 2 – plasma cavity; 3 – 50 $\Omega$ coaxial line; 4 – microwave antenna; 5 – boron nitride insulator; 6 – gas input; 7 – electromagnets M1, M2, and M3; 8 – stainless-steel shall; 9,10 – front conflat flanges, and 11,12 – outlet and inlet for cooling water. Black points represent welded joints (taken from [7.94])

electrons to move in circular orbits and confines the plasma. When the orbital frequency of the electrons, $f = eB/(2\pi m_e)$, equals the frequency of 2.45 GHz, resonance occurs when the magnetic field density equals 87.5 mT (875 Gs) [7.94]. The cross-sectional view of the small ECR plasma source used for deposition of III–V nitrides at low temperatures (400– 600°C) in a MBE environment, using molecular nitrogen, is shown schematically in Fig. 7.30. For another design of a MBE compatible ECR plasma source, see [7.95].

The MBE growth experiments on GaN/AlN layered structures [7.93] have shown that one of the most important parameters in III–V nitride epitaxy is the state of the nitrogen species generated by the ECR source. High ECR microwave powers coupled with large nitrogen background pressures result in films with inferior electrical and structural properties, caused by energetic ion damage, as determined by Hall and X-ray diffraction measurements, respectively. On the other hand low nitrogen background pressure leads to films with Ga droplets, indicative of reactive nitrogen deficiency. Under optimum, experimentally determined growth conditions encompassing nitrogen over-pressure, growth temperature, magnetic field intensity, and ECR power, GaN films of 1 μm thickness with n-type electrical conductivity and quality comparable to the best films obtained by MOVPE can be grown by the described plasma-assisted MBE technique [7.93]. A lot of other investigations on plasma-assisted MBE growth process of III–V nitrides have been performed [7.96–99] (see also Sect. 15.2.1). An exhaustive review on GaN based III–V nitride MBE plasma-assisted growth is given in [7.100, 101]. Commercialization of light-emitting optoelectronic devices for the blue and UV spectral ranges still stimulate considerable interest in III–V nitrides, which has caused an enormous publication activity on this subject by many research laboratories all around the world.

# 8. Metal Organic Vapor Phase Epitaxy

MBE and MOVPE are the main epitaxial methods used today. For a long time MOVPE was thought to be more of a production instrument while MBE was the scientific tool. This happened because MBE could be dealt with straightforwardly with the available surface science know–how while MOVPE was subject to the poor knowledge of the underlying chemistry, hydrodynamics and moreover surface analytical tools were not available. This situation has changed and both methods can be utilized for growing structures well defined on the atomic level. There are differences however which give preference to one of the methods for certain problems. Because of the large variety of chemical compounds for example MOVPE is essentially capable of depositing all solids while MBE with some elements (phosphorus, nitrogen) is problematic. On the other hand diffusion in MOVPE seems to be in general faster with the consequence of larger growth rates and thus the growth of very small structures (quantum dots) requires larger effort in MOVPE than in MBE. The larger possible growth rates together with the quasi-infinite supply of precursors makes MOVPE of course the ideal method for very thick layers in the micrometer regime and also for industrial production. Therefore the question of which of the methods should be used should be answered problem oriented.

There are many excellent reviews on MOVPE and we specifically recommend also looking through the recent proceedings of the International Conference on MOVPE (ICMOVPE) [8.1–5] which give an excellent overview on the state of art of MOVPE. In this chapter we will follow the reviews of Stringfellow [8.6], Kisker and Kuech [8.7] as well as Richter and Zahn [8.8].

## 8.1 Basic Concepts

The MOVPE process starts with a gas mixture which contains the molecular compounds, termed precursors, necessary for growth, and a carrier gas (Fig. 8.1). The latter is usually hydrogen (for special reasons also nitrogen), with an operating pressure between $10^3$ Pa and atmospheric pressure ($10^5$ Pa). For reasons of uniform deposition usually nowadays low pressures are preferred. This is a consequence of the more laminar flow pattern and the more homogeneous temperature field within the reactor (Sect. 6). The choice

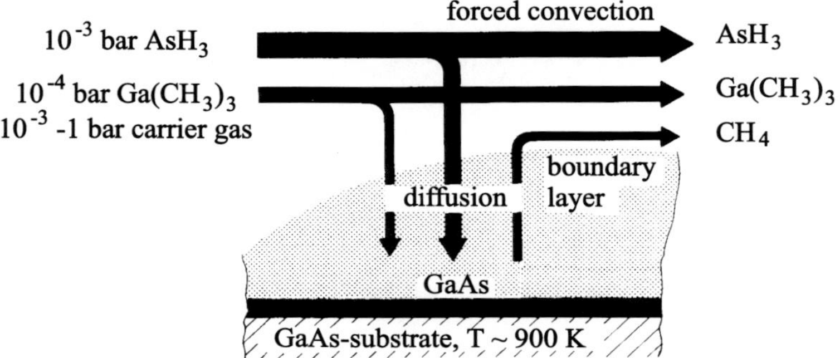


**Fig. 8.1.** Schematic picture of a MOVPE epitaxial growth process using the example of GaAs growth (taken from [8.8])

of precursors depends on the material to be deposited. For III–V semiconductors the standard precursors are, for example, metalorganic compounds of the group III elements, like trimethylgallium ($Ga(CH_3)_3$) (in abbreviated form TMGa), and hydrides of the group V elements, e.g., arsine ($AsH_3$). Thus, the general chemical reaction in the case of MOVPE of III–V semiconductor compounds is of the type:

$$A^{III}R_3 + B^{V}X_3 \rightarrow A^{III}B^{V} + \text{organic by-products}, \tag{8.1}$$

where R is an alkyl or other organic group, X is usually a hydrogen atom, but can also be an organic radical or even a halogen atom, and $A^{III}$ and $B^{V}$ are the cation and anion, respectively [8.7]. Depending on R and X, the by-products can be simple alkanes or more complicated species. In fact, for most MOVPE reactions, relatively little is known about the reaction details, so that (8.1) only schematically represents the actual process. Because of many undesired properties of the alkyls and hydrides, however, new organic compounds ("alternative precursors") are tested today for replacement [8.8].

The gas mixture is fed into an open reactor where the heated substrate is placed as shown in Fig. 8.1. Many designs for MOVPE reactors have been proposed, tested, and are in use. The goal of such development work is the homogeneity of properties such as thickness, stoichiometry, and carrier concentration across the whole substrate surface.

One common design, which is mostly used in research laboratories, is the horizontal, cold wall reactor displayed in Fig. 8.2, where, however, quite often the substrate is rotating and heating is either done by an RF coil around the reactor or radiation heated from the bottom. Typical gas-dynamic operating conditions in such a reactor are a laminar flow and gas velocities in the range from 0.1 to $1\,\mathrm{m\,s^{-1}}$. In growth processes which operate far from thermal equilibrium, as is the case for the decomposition of metalorganic compounds, only the substrate but not the reactor walls are heated in order to avoid decomposition and subsequent deposition onto walls. As a consequence only

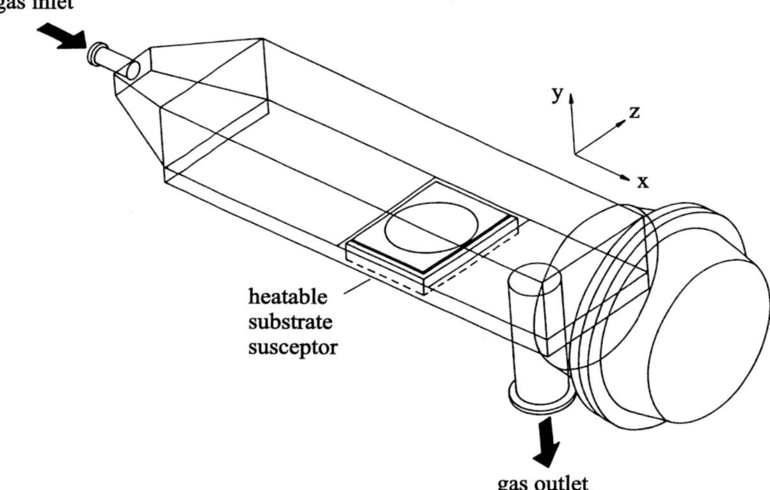


**Fig. 8.2.** Simplest realization of a horizontal cold wall reactor for gas phase epitaxial growth (taken from [8.8])

the gas in a region close to the substrate is heated (thermal boundary layer) and large temperature gradients are created in the reactor (see Fig. 6.18). Within this region gas phase reactions (homogeneous reactions) and reaction at the gas–solid interface (heterogeneous reactions) take place leading under appropriate conditions to epitaxial growth of semiconductor layers.

Two main factors determine the growth rate, as can be seen from Fig. 8.3 which outlines the various steps leading to growth. One is given by the gas phase transport (mass transport) of the precursors or their reaction prod-

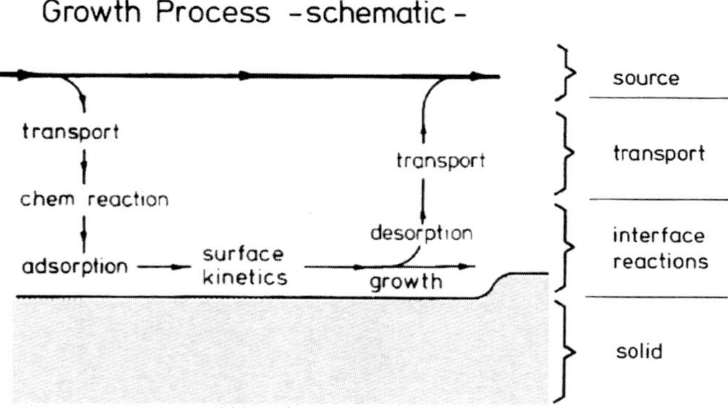


**Fig. 8.3.** Schematic diagram of sequential steps relevant to VPE growth processes, and thus also to the MOVPE process (taken from [8.8])

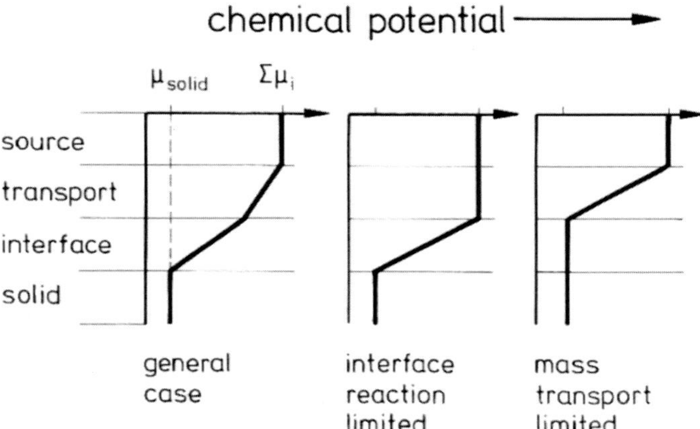


**Fig. 8.4.** Schematic plot of the chemical potential within the four regions defined in Fig. 8.3. The chemical potential difference between the source and the solid drives the transport and the interface reactions (taken from [8.8])

ucts, containing the growth relevant atoms, to the interface. The other is determined by the reactions near or at the interface converting the precursor molecules into group III or V atoms, and their incorporation into the lattice structure of the solid. Both can be recognized by their very different temperature dependencies. In the pressure range of $10^3$ to $10^5$ Pa mass transport can be described within the gasodynamic formalism and is governed by convection and diffusion. Near the substrate the convective flow is essentially zero and diffusion will be dominant. Since the diffusion constant is only weakly temperature dependent in the range of typical growth temperatures (800 to 1100 K) a weak temperature dependence is expected also for the transport contribution to the growth rate. For the chemical reactions taking place near or at the interface the reaction rates with their exponential factors should yield a strong exponential temperature dependence. These two temperature dependencies are observed when either one of the two contributions is the rate limiting step.

Rate limitation is explained in Fig. 8.4 where the change of chemical potential between source and the solid is shown schematically. The total difference in chemical potential is the driving force for the growth. In the general case the chemical potential is partly consumed for the transport as well as for the reactions at the interface. In case the transport is fast the potential will drop mainly at the interface and growth is called reaction or kinetically limited. This situation occurs at low temperatures and large partial pressures of the precursor. Just the temperature then determines the growth rate which displays exponential behavior following from the reaction rates (region 1 in Fig. 8.5). At high temperatures and low partial pressures the reactions are fast and the mass transport is the limiting factor.

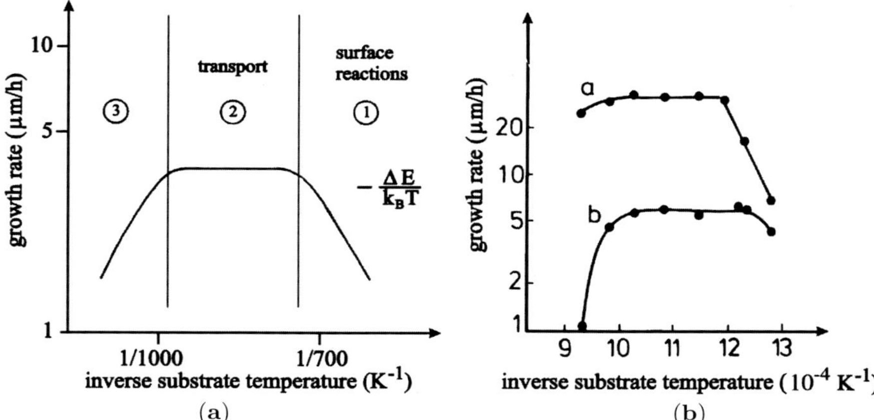


**Fig. 8.5.** Schematic diagram showing (**a**) the three main regimes of temperature dependence of the growth rate in MOVPE and (**b**) the MOVPE growth rate of GaAs versus temperature for two different partial pressures (taken from [8.8])

The corresponding temperature independent growth rate is shown in region 2 of Fig. 8.5. A decreasing growth rate is finally observed generally at very high temperatures (region 3). This could be expected from desorption of the growing species from the surface. In reality it might be caused also from an enhanced pre-deposition on the reactor walls or on the susceptor because of an extended temperature increase throughout the reactor. These temperature dependencies have been indeed observed for GaAs in temperature ranges where either one of these factors was the limiting one, i.e., the slowest step (Fig. 8.5). Exponential behavior is found at low temperatures and high pressures (reaction limited or kinetically controlled growth); temperature independence is found at higher temperatures and low pressures (mass transport controlled growth). At the highest temperatures finally either desorption or pre-deposition might decrease the growth rate. The difference between curves a and b in Fig. 8.5 displays the dependence of the diffusive mass transport on total pressure. At lower pressure the mean free path is larger and thus an increase of the growth rate with decreasing total pressure is observed.

In the mass transport limited case the growth rate shows a linear dependence on group III element supply to the reactor [8.8, 9] (case 2 in Fig. 8.6). Nearly all III–V semiconductor compounds are usually grown under these conditions [8.7]. However, at very high partial pressures and/or at lower temperatures the growth rate may saturate because the supply of reactants to the surface is larger then what can be incorporated by the interface reaction processes (case 1 in Fig. 8.6).

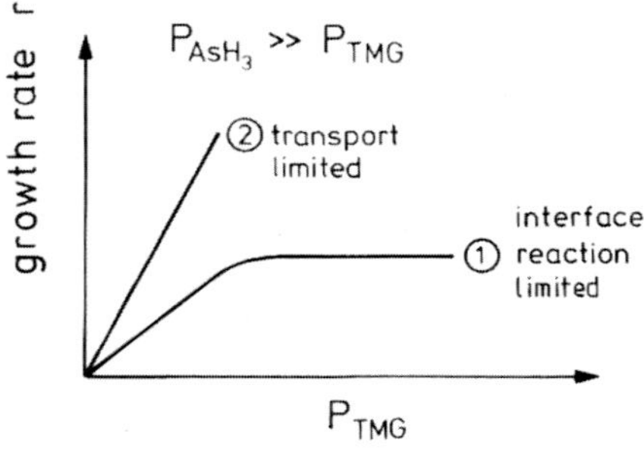


**Fig. 8.6.** Schematic MOVPE growth rate versus TMG partial pressure in the reaction limited (low T) and in the transport limited (high T) regime (taken from [8.9])

## 8.2 Growth Equipment

In order to compete with MBE growth, much effort has been made in the last few decades to control the hydrodynamics of MOVPE reactors in order to match the homogeneity and control found in MBE growth. One of the main problems found was the creation of convective flows driven by the large temperature gradients present in cold wall reactors. As a consequence unpredictable flow pattern and transport occurred. The main reason for this, besides nonfavorable geometries, were the high total gas pressures which increased the buoyancy force (proportional to the density) and caused for example vortices in the flow. Many reactor designs were tested with respect to their flow patterns, temperature distributions and homogeneity of deposition. All these problems had been essentially solved or were strongly reduced when the operating pressure was decreased from atmospheric pressure to values below 100 mbar because of the lower density. Moreover, care was taken to avoid abrupt changes of reactor cross–section which may stimulate non laminar flow as well as decreasing the reactor height which reduces the Rayleigh number, the figure of merit for thermally driven convection.

### 8.2.1 Commercial MOVPE Reactors

Three reactor types are commonly in use today. One is the horizontal reactor already shown in Fig. 8.2 which is used mainly in research or small scale growth and is either RF or radiation heated. For larger scale growth and in production the vertical rotating disk reactor [8.10] and the planetary reactor [8.11] are the ones used nowadays nearly exclusively. Both reactors employ substrate rotation in order to achieve uniform growth. The general hydrodynamic features discussed in Sect. 6.2.2 with the aim of a uniform boundary layer hold for both these as well. This means a total pressure considerably below 1 bar (100 mbar), sufficiently large flow rates and a carrier gas with low thermal conductivity. Similarly only small advantages for upside down mounting are obtained which might not compensate for the more difficult sample mounting. We will discuss both reactors here briefly and show especially the effects of substrate rotation.

**Vertical Rotating Disk Reactor.** This reactor type (Figs. 8.7–8.10) consists of a disk rotating around a vertical axis with a speed in the order of 1000 rpm. The disk holds the substrate(s) on its upper surface. The gas inlet is from the top. As can be seen from the streamline visualization (Fig. 8.8) the rotation forces the streamlines parallel to the susceptor surface. The effect of different rotation frequencies on streamlines and temperature profiles can be seen in Fig. 8.7.

The realization of such a reactor system with a preparation chamber and a load lock is shown in schematic and in total view in Figs. 8.10, 8.11.

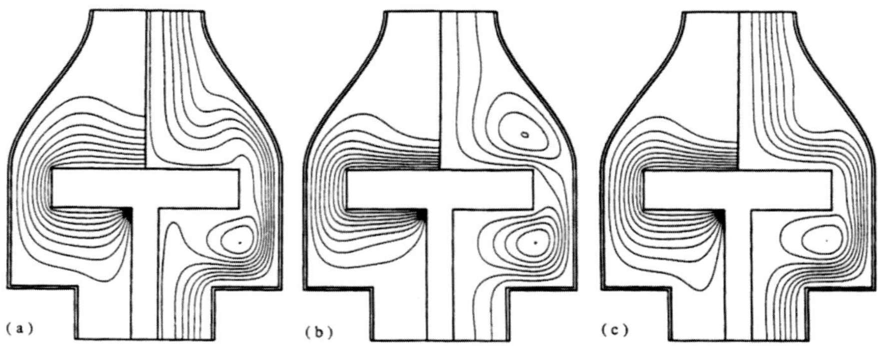


**Fig. 8.7.** Vertical rotating disk reactor: Effect of susceptor rotation on streamlines (right side) and isotherms (left side) (**a**) 600 rpm, 4 SLM, (**b**) 2400 rpm, 4 SLM, (**c**) 1200 rpm, 8 SLM. Centerline ($r = 0$) growth rates (**a**) 1.8 μm h$^{-1}$, (**b**) 3.5 μm h$^{-1}$ (**c**) 3.1 μm h$^{-1}$. Carrier gas hydrogen, total pressure 100 mbar, susceptor temperature 900 K (SLM: standard liters per minute, rpm: rotation per minute) (taken from [8.10])

**Fig. 8.8.** Smoke visualization of flow lines in a vertical rotating disk reactor (with courtesy from Emcore)

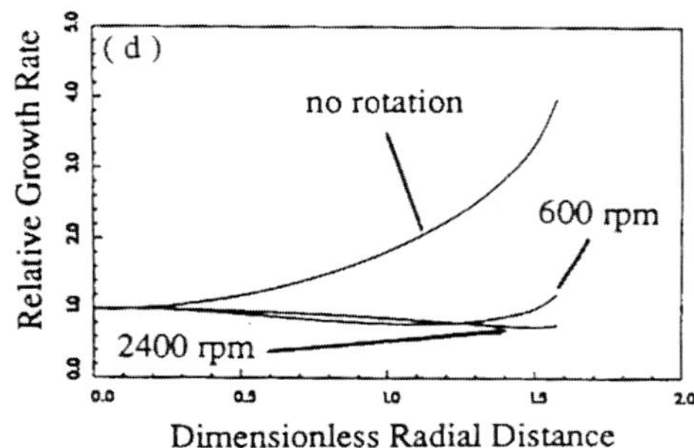


**Fig. 8.9.** Relative growth rate as a function of radial position for an inlet flow rate 4 SLM and different speeds of rotation (taken from [8.10])

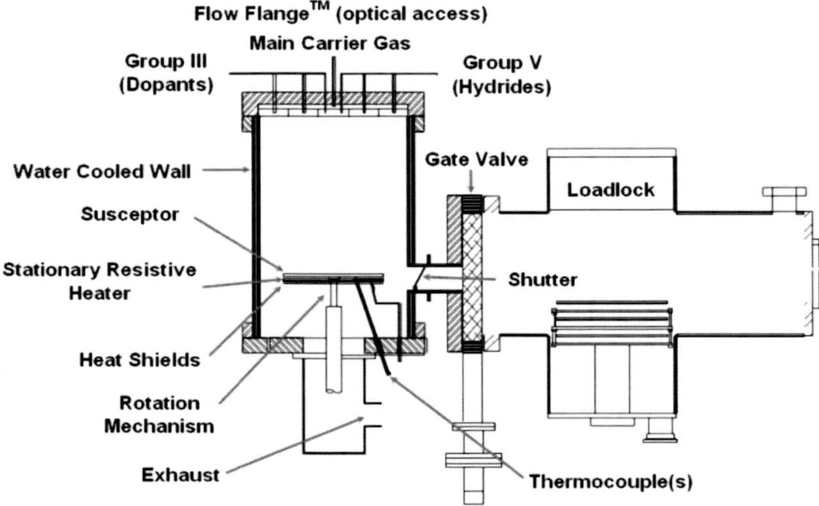


**Fig. 8.10.** Construction of a commercial vertical rotating disk reactor (with courtesy from Emcore)

**Fig. 8.11.** Total view of a commercial vertical rotating disk MOVPE reactor (with courtesy from Emcore)

**Planetary Reactor.** The alternative route to very uniform deposition taken in commercial MOVPE reactors is the planetary reactor developed by Frijlink [8.11]. The design in the reactor (Fig. 8.12) consists of a rotating substrate holder (2) and rotating substrates (4). They are lifted and rotated by a gas foil. The rotational frequencies are around 0.5 Hz and 5 Hz for the main plate and the substrates, respectively. The reactants enter from the top (9,10), their flow directions given by the arrows. Rolls are suppressed even at atmospheric pressure by proper design of the geometry of the ring where the substrates are placed and a cylindrical entrance grating (12) which provides a uniform stable distribution of gases.

The radial distributions of various relevant physical parameters are given in Fig. 8.13. The streamlines (**b**), the isotherms (**c**) and the Ga isobars (**d**) show excellent uniformity above the substrate region, a fact which is re-

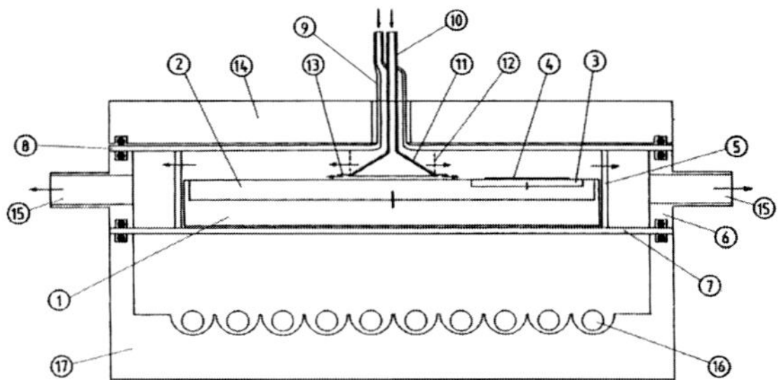

**Fig. 8.12.** Construction scheme of the planetary reactor. The in (9, 10) and out (15) flow of the reactants is indicated by arrows. The substrates in off-center position (4) on the substrate platform (2) as well as the substrate platform rotate. Both are supported by a gas foil which also provides the rotation (taken from [8.11])

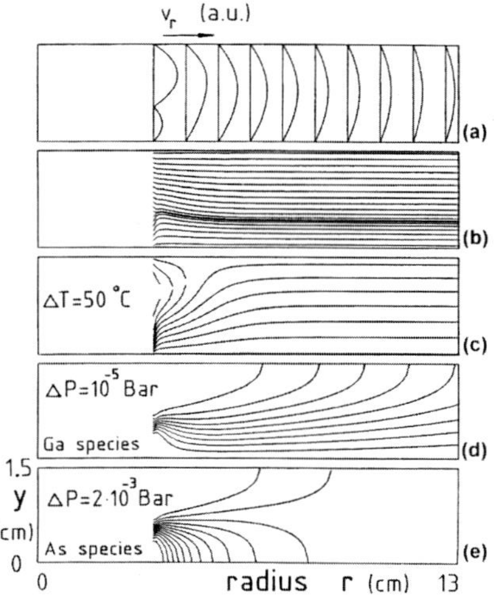


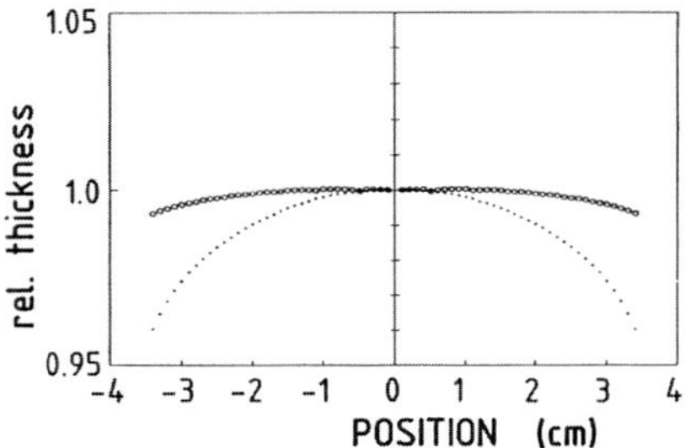


**Fig. 8.13.** Top: calculated radial dependence of thickness for two different carrier gas flows. Left (top to bottom): calculated cross-sectional maps for velocity, streamlines, temperature and partial pressures for Ga and As, respectively. (taken from [8.11])

sponsible for the high uniformity in growth rate (Fig. 8.13). Levels in the sub-percent range can be reached.

An example of a mass-production compatible growth system produced by the Aixtron Company is shown in Figs. 8.14 and 8.15.

**Fig. 8.14.** AIX 2000/2400 multi wafer planetary reactor consisting of three blocks, as seen from left to right: reaction chamber, gas mixing unit, and computerized control unit (with courtesy from Aixtron)

**Fig. 8.15.** AIX 24×2 MOCVD AIXTRON company planetary reactor. Wafer capacity: 35×2″ (GaAs), 24×2″ (nitrides) or 7×6″/8×4″ (optional) (with courtesy from Aixtron)

### 8.2.2 Gas–Vapor Delivery Systems in MOVPE

The sources in MOVPE partially exist as gases ($AsH_3$, $PH_3$, $NH_3$) but nearly all organometallic source materials (precursors) are provided in liquid form only a few are solids. The amount of gaseous reactants as well as that of the carrier gas are easily controlled in conjunction with a standard thermal mass flow controller (MFC) as shown in Fig. 8.16. There a thermal conductivity sensor in a well-defined bypass controls a valve which regulates the flow to a preset mass flow value. The partial pressure of the gaseous precursor $P_{\text{gp}}$ is then given in terms of the total pressure $P_{\text{reactor}}$ as

$$P_{\text{gp}} = \frac{Q_{\text{gp}}}{Q_{\text{tot}}} \cdot P_{\text{reactor}} \tag{8.2}$$

where $Q_{\text{gp}}$ is the flow of the gaseous precursor and $Q_{\text{tot}}$ is the total flow into the reactor.

The amount of liquid reactants fed to the reactor is controlled via the vapor pressure of the liquid given by its temperature. By using well-regulated temperature baths (thermostats) the vapor pressure has a well defined value which is then picked up by the carrier gas flowing ("bubbling") through the liquid and the vapor above. An example of such a so-called bubbler is given schematically in Fig. 8.17. The partial pressure $P_{\text{gp}}$ of the then gaseous precursor in the reactor is then given by

$$P_{\text{gp}} = \frac{Q_{\text{lp}}}{Q_{\text{tot}}} \cdot \frac{P_{\text{reactor}}}{P_b} \cdot P_{\text{vap}}(T) \tag{8.3}$$

where $P_{\text{vap}}(T)$ is the equilibrium vapor pressure of the liquid, $P_b$ the pressure in the bubbler, and $Q_{\text{lp}}$ the flow through the bubbler. $Q_{\text{tot}}$ and $P_{\text{reactor}}$ are

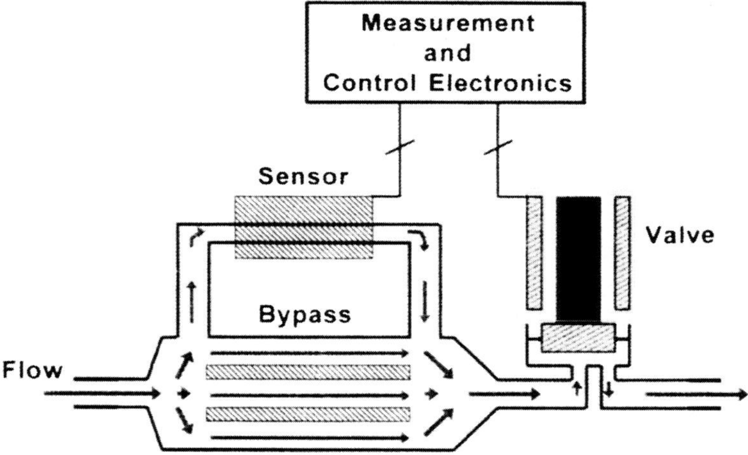


**Fig. 8.16.** Thermal mass flow controller (taken from [8.15])

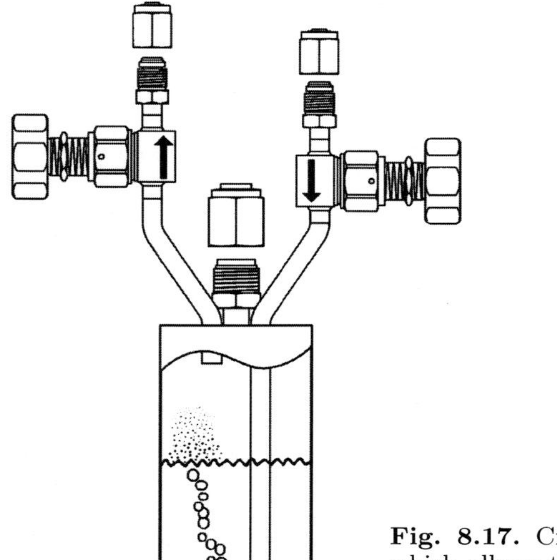

**Fig. 8.17.** Cross–section of a so–called bubbler which allows the carrier gas to pick up the vapor pressure of the liquid precursors.

defined above. Bubblers are nearly exclusively used in MOVPE as delivery systems for liquids. Their operation assumes a reasonable vapor pressure and that the carrier gas remains sufficiently long in contact with the vapor to reach the equilibrium vapor pressure. There are few problems with the use of modestly volatile liquids (vapor pressures in the range of 5–30 mbar). There are, however, problems when using materials with very low (e.g., TMIn) or very high vapor pressures (e.g., DMZn).

In the case of low vapor pressures one can of course increase the temperature of the bubbler. However, this will cause problems when the bath temperature becomes too high. The vapor might then condense in the colder tubing behind the bubbler. In such a case it might become necessary to heat the relevant tubes up to the reactor. Cooling high vapor pressure liquids might give on the other hand stability problems since the vapor pressure depends exponentially on $T$ and therefore the concentrations might easily fluctuate. For the low pressure case, especially when the precursor is partially a solid (e.g., TMI) it is common practice to connect two bubblers in series in order to reach better equilibrium pressures by doubling the interaction time with the precursor. Alternatively a solution system (Fig. 8.18) initially developed in [8.13, 14] can provide constant concentration in the gas phase. In this kind of delivery system the excess solid concentration (TMIn) will be in equilibrium with the solution, which in turn then has a better chance to equilibrate the gas phase.

Direct liquid delivery of precursors might also be useful especially when larger amounts of precursors are needed and the accurate value of the vapor pressure is not as critical. The solution is then carried most often via a syringe

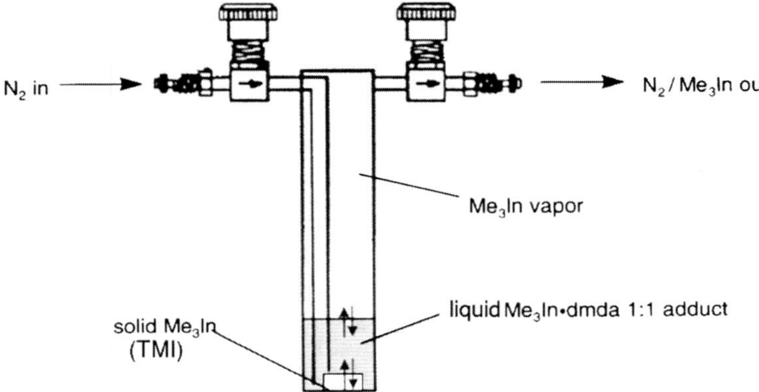


**Fig. 8.18.** Construction of a bubbler which allows a solid precursor (TMI) into the gas phase to be brought more easily via a saturated solution (in N,N-dimethyldodecylamine: dmda) (taken from [8.15])

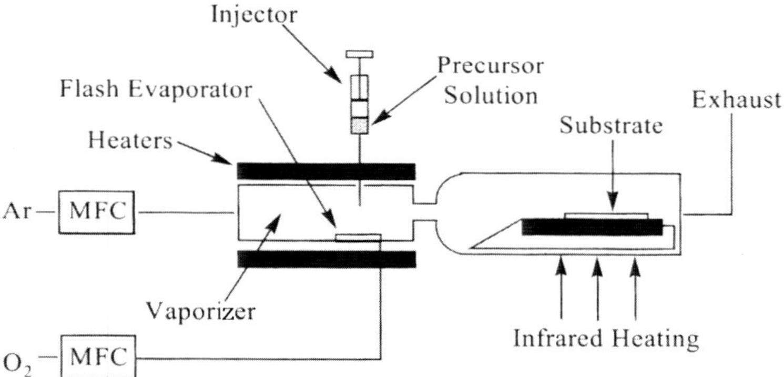


**Fig. 8.19.** Schematic diagram of a liquid injection system for liquid precursors (taken from [8.15])

to a vaporization chamber directly adjacent to the reactor [8.16] (Fig. 8.19). Such a system also has the advantage that the precursor stays most of the time, except when evaporated, at room temperature.

All the gases are then fed together with the gaseous precursor (arsine, phosphine, ammonia) and dopant gases into the reactor ("Run" in Fig. 8.20) for epitaxial growth. Before doing so the bubblers and the gas flows are allowed to stabilize and leave the system through an exhaust exit ("Vent" in Fig. 8.20 ). Growth is initiated by switching the relevant valves from the "Vent" to the "Run" line. Care has also to be taken for a mixing zone in the beginning of the reactor in such a way that a homogeneous gas mixture appears in the growth region of the reactor. The temperature of the bubbler, the pressures and the switching of the valves is generally controlled today

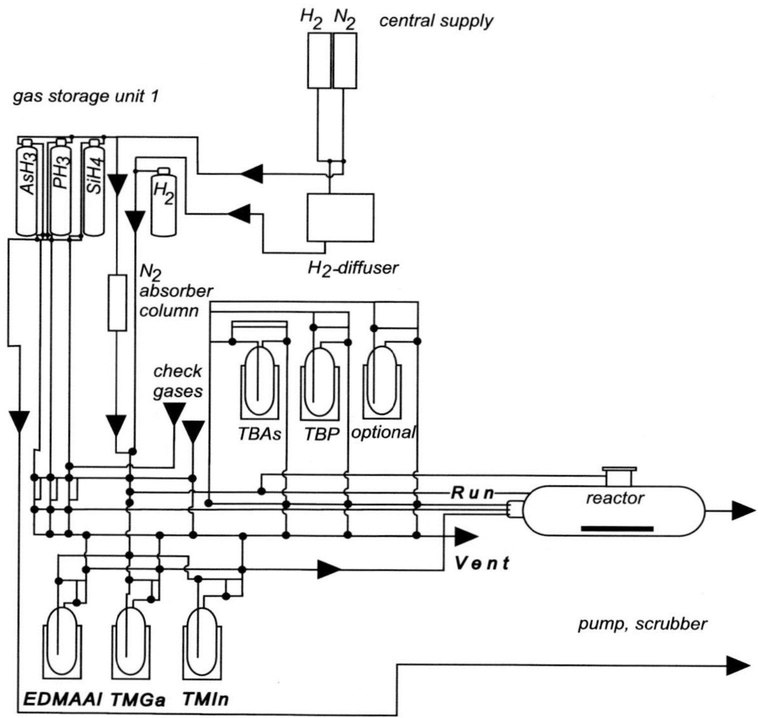


**Fig. 8.20.** Schematic outline of a MOVPE epitaxial growth apparatus (taken from [8.17])

by computers. This allows for depositions in the monolayer regime with a similar reproducibility as in MBE.

One of the main drawbacks in MOVPE as compared to MBE had always been the fact that *in situ* analysis tools were not all standard or not available. This was in contrast to MBE were from the beginning nearly always the full arsenal of surface science techniques relying on vacuum conditions was available either in the growth chamber or in a connected analysis chamber. With the help of these tools fluxes to the substrate, substrate structure and growth rate could be determined before and during the growth procedure and if necessary the growth parameters could be changed or the growth stopped. This was until recently not possible in VPE growth and only post growth analysis was performed. Delay times of days until the results were fed back to the epitaxial grower were standard. This situation has changed considerably in the last decade.

The most convenient techniques for analysis of epitaxial growth in the case of MOVPE are optical techniques [8.8]. These techniques in the growth environment typical for MOVPE separate into two groups, namely those methods which are suitable for gas phase analysis, and those which provide information from the growing surface. Optical techniques are especially ideally suited to perform analysis in the quite hostile environments met in MOVPE. For epitaxial growth however surface analysis is the most important task. Optical methods for surface analysis have been quite far developed for the epitaxial needs in the last decade. Optical equipment is commercially available now for the determination of surface reconstruction, surface morphology, layer thickness, surface temperature and growth rate even on the monolayer scale. This equipment is applied to the growth apparatus from outside through windows and thus can be easily added or removed and used also for comparison to the vacuum growth techniques like CBE or MBE. For more information on the optical characterization methods of epitaxial growth processes see Chaps. 9 and 10.

## 8.3 Precursor Materials

The last decade of development of the MOVPE technology has shown that the design of the source molecules (see the left side of (8.1)), called also precursor molecules, became an integral part of the design of the overall MOVPE process [8.6–8].

It has been discovered that designing molecules with specific properties is possible, including such parameters like pyrolysis temperature, vapor pressure, toxicity, etc. They have quite difficult notations (for a crystal grower, who is not a metalorganic chemist), thus, it is useful to explain in brief the existing rules. The metalorganic precursors are commonly designated by the MOVPE community using M, E, NP, IP, NB, IB, TB, and A to denote the radicals methyl, ethyl, n-propyl, i-propyl, n-butyl, i-butyl, t-butyl, and alkyl, respectively. Another common radical, cyclopentadienal ($C_5H_5$), is commonly denoted Cp; M (mono), D (di), and T (tri) are used to denote the number of specific radicals. Thus, DMZn represents dimethylzinc while DNPTe is used to denote di-n-propyltellurium.

New precursor designs include, among others, Te precursors such as MATe and DTBTe, designed to have lower pyrolysis temperature than the commonly available DETe and DMTe. Group VI molecules such as $C_4H_4S$ and MSH are designed to pyrolyze at higher temperatures for usage together with DMZn and DMCd to retard homogeneous nucleation of the solid upstream from the substrate ("parasitic reactions"). The triethyl-group III molecules suffer from prereaction difficulties, while TMIn has the disadvantage of being a solid at room temperature. The combination EDMIn, with a single methyl group replaced by an ethyl radical, is a liquid that has been discovered to behave chemically in the MOVPE reactor similarly to TMIn. It is now available in electronic-grade purity and has been used to grow extremely high-purity InP. Perhaps most dramatic has been the development of less hazardous group V metal organic molecules to replace the highly toxic hydrides of As and P [8.6]. The strong recent interest in III-nitrides for optical devices has pushed similarly development of N-precursors applicable at lower growth temperatures than necessary for the N-hydride ($NH_3$: $T_{\text{growth}} \sim 1300\,\text{K}$). A reduction in growth temperature would be desired in order to reduce the large number of N-vacancies created by the high volatility of the nitrogen. However, the results obtained with hydrazine (Hy)-derivates (dimethylhydrazin -e DMHy) or with butyl-derivates (tertiarybutylamine-TBAm) have until now not been convincing enough to replace ammonia. One should note, however, that the high purity required for precursors is a major problem or at least a time factor in development. The usual chemical purification ($10^{-3}$) is by far not sufficient for semiconductor purposes. Electronic grade precursors are needed which represent first of all a problem in analyzing the precursor material for the relevant impurities and possibly developing an appropriate purification method or sometimes also designing a new route for synthesizing. This is usually a time consuming process which relies also strongly on the interaction of epitaxial growers and precursor manufacturers.

Selection of the optimum source molecules is most frequently unrelated to the reactor pressure and geometry. In this case, the aspect of the process design related to the choice of source molecules is based on criteria such as toxicity, convenience (mainly vapor pressure, melting point, etc.), availability in high-purity form, pyrolysis temperature, and compatibility (as in the case of growth of HgCdTe compounds). The other major fundamental consideration is freedom from C contamination.

The group II molecules of the type $MR_2$, such as dimethylzinc and diethylcadmium, are linear, as shown schematically in Fig. 8.21a. This can be understood from the valence bond theory of hybrydized covalent bonding [8.19]. The group II elements have two s electrons in the outer shell. To form two covalent bonds a hybridization occurs, resulting in two sp orbitals, with which the ligands are bonded. The sp hybridized orbitals are linear. Thus, the two ligands are separated at an angle of 180°, producing a linear molecule. The molecules are electron acceptors, or Lewis acids, due to the unfilled p-orbitals.

An alternative approach to understanding the molecular configurations is the valence-shell, electron-pair repulsion (VSEPR) model [8.20]. The electrons in the valence shell of the central atom are brought into spin-paired couples by interactions with the ligands, and these paired electrons repel each other to form the geometries of lowest energy. For molecules of the group II alkyls, this leads to a linear molecule with bond angles of 180°.

The bonding in group III molecules is similar. The incomplete electron shell of the atoms contains one p and two s electrons. The three covalent bonds are formed with a hybridized $sp^2$ bonding configuration. Thus, a planar, trigonal molecule is formed with the three ligands separated by angles of 120°, as shown in Fig. 8.21b. Important is that an unfilled p-orbital remains after the three covalent bonds are formed. This unfilled p-orbital, lying perpendicular

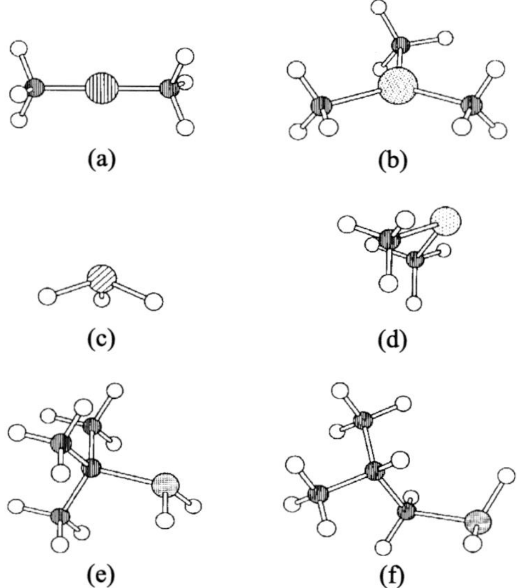


**Fig. 8.21.** Schematic diagrams of simple precursor molecules used in MOVPE growth of II–VI and III–V semiconductor compounds: **(a)** DMZn, **(b)** TMGa, **(c)** $AsH_3$, **(d)** DMTe, **(e)** TBP, and **(f)** IBP. Hydrogen atoms are represented by the small open circles, carbon by the intermediate-sized circles with the vertical hatch pattern. The largest circles represent the metal atoms (taken from [8.18])

to the plane of the molecule, makes it electrophilic, i.e., it attracts electrons. This makes the group III alkyls electron acceptors, or so-called Lewis acids.

The group V atoms each have three p electrons and two s electrons in the unfilled shell. The formation of three covalent bonds satisfies the bonding requirements. In this case the $sp^3$ hybridization gives a tetragonal bonding configuration, shown schematically in Fig. 8.21. The three R groups, if identical, form a structure resembling a three-legged stool, with bond angles of approximately 109.5°. Atop this "stool" is a pair of electrons forming no covalent bond. This so-called lone pair is important for interactions of the group V precursor molecules with surfaces and with other molecules. The molecules behave as electron donors, or so-called Lewis bases in these interactions.

The group VI elements have two s electrons and four p electrons in the outer shell. The two covalent $sp^3$ bonds are separated by approximately the tetrahedral angle, as seen in Fig. 8.21d. In this case, two lone pairs occupy the other two tetrahedral positions. The molecules, which resemble $H_2O$, are also Lewis bases.

An important parameter for MOVPE is the energy of the metal–carbon bond in the precursor molecule. This particular bond energy is significant because it determines the stability of the molecule against decomposition by free-radical homolysis. The metal–carbon bond strength is less important

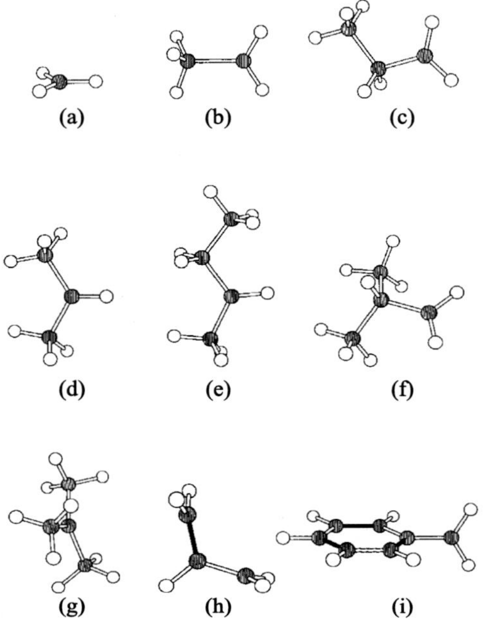


**Fig. 8.22.** Schematic diagrams of simple alkyl radicals: **(a)** methyl, **(b)** ethyl, **(c)** n-propyl, **(d)** i-propyl, **(e)** sec-butyl, **(f)** i-butyl, **(g)** t-butyl, **(h)** alkyl, and **(i)** benzyl. Hydrogen atoms are represented by the small open circles, carbon by the larger-sized circles with vertical hatch pattern. Double bonds are shaded (taken from [8.6])

for other pyrolytic mechanisms, such as $\beta$-elimination, which also occur for certain metalorganic precursor molecules. In general the metal–carbon bond strength depends both on the nature of the metal, i.e., the electronegativity, and the size and configuration of the radical.

The common ligands encountered in MOVPE are shown schematically in Fig. 8.22. The simplest, methyl and ethyl radicals, have only a single configuration, excluding the rotational conformations. The larger radicals have several configurations or isomers. In general, the metal–carbon bond strength is decreased as the number of carbons bonded to the metal bonded carbon (central carbon), is increased [8.6]. For example, the strongest metal–radical bond will involve the methyl radical, since the central carbon is bonded only to H atoms. The bond strength is slightly less than for the metal–H bond. For the ethyl, n-propyl, and i-butyl radicals this rule indicates the metal–carbon bond strengths to be equal, since in each case the central carbon atom is bonded to one additional carbon. For the i-propyl and t-butyl radicals, the metal–radical bond strength should be considerably reduced since the central carbon atom is bonded to two and three carbon atoms, respectively, in addition to the metal atom.

Weakening of the carbon–metal bond in this manner is attributed to delocalization of free-radical electronic charge. Even weaker carbon–metal bonds are formed for the alkyl radical, where a double bond is formed, and the benzyl radical, where the $C_{rad}$ atom is bonded to a benzene ring, as shown in Fig. 8.22i. A corollary to the rule described above is that the more stable the radical, the more rapidly it is formed. Experimentally determined bond strengths for several group II, III, V, and VI alkyls, where data are available, are summarized in Table 8.1. $D_1$ represents the energy required to break the first carbon–metal bond, which is typically the activation energy for pyrolysis when radical mechanisms dominate. Also listed is the average bond strength ($D_{avg}$) determined from combined thermochemical and kinetic data.

Another important property that varies systematically with the alkyl group is the vapor pressure. In general, the vapor pressures are highest for

**Table 8.1.** Bond strengths (energy required to break the first carbon–metal bond) of common precursors and related molecules (taken from [8.6])

| Precursor | $D_1$ (kcal mol$^{-1}$) | $D_{avg}$(kcal mol$^{-1}$) |
|---|---|---|
| DMZn | 51 (54) | 42 |
| DEZn | | 35 |
| DMCd | 53 | 33 |
| DECd | | 26 |
| DMHg | 58 | 29 |
| DEHg | 42.5 | 24 |
| DNPHg | 47 | |
| DIPHg | 41 | |
| DNBHg | 48 | |
| TMAl | 65 | 67 |
| TEAl | | 58 |
| TMGa | 60 | 59 |
| TEGa | | 57 |
| TMIn | 47 | |
| TMTl | 27 | |
| $PH_3$ | | 77 |
| TMP | | 66 |
| TEP | | 62 |
| $AsH_3$ | | 59 |
| TMAs | 62.8 | 55 |
| TMSb | 57 | 52 |
| TMBi | 44 | 34 |
| $H_2S$ | | 83 |
| DMS | | 65 |
| DES | | 65 |
| $H_2Se$ | | 66 |
| DESe | | 58 |
| $H_2Te$ | | 57 |

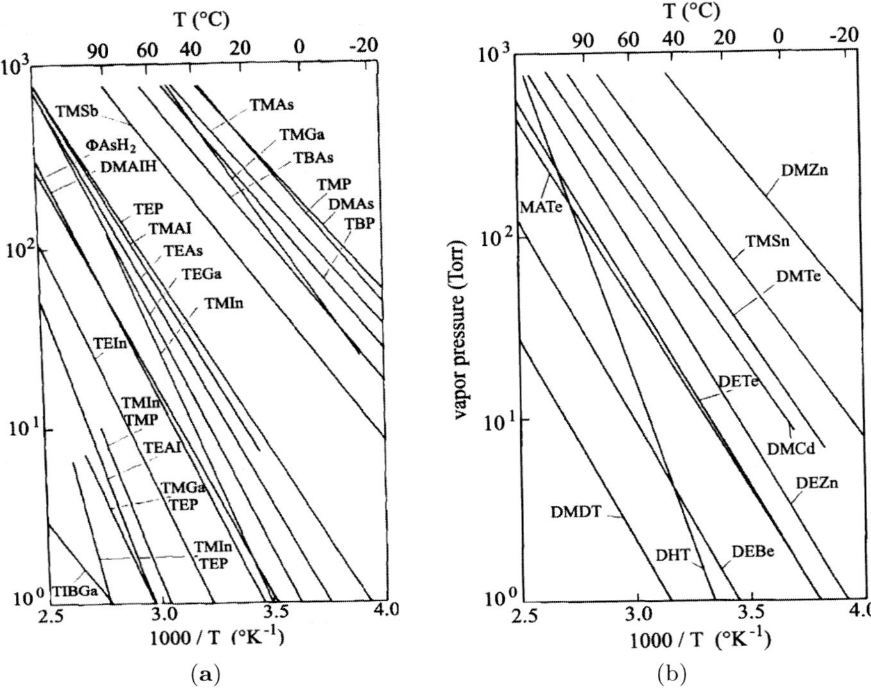


**Fig. 8.23.** (**a**) Temperature dependence of vapor pressures for common group III and group V metalorganic sources and (**b**) temperature dependence of vapor pressures for common group II and group VI metalorganic sources (taken from [8.6])

the lightest molecules. This is clearly seen from the plots of vapor pressure versus temperature shown in Fig. 8.23 for several alkyls useful for MOVPE growth. Of course, the intermolecular interactions in the liquid also strongly affect the vapor pressure, which makes quantitative predictions impossible. Generally, however, higher order, more branched molecules have weaker interactions, which enhances the vapor pressure. For example, the vapor pressure of TIBAl is much larger than for the lighter TEAl.

## 8.4 Precursor Decomposition and Reactions

Decomposition of precursor molecules can be performed by any kind of energy transfer and excitation into unstable electronic states. We will discuss briefly later plasma-MOVPE and photo-MOVPE where either electronic collisions in a plasma discharge or higher energetic photons cause the molecule to dissociate. In the standard MOVPE process, however, it is just the thermal

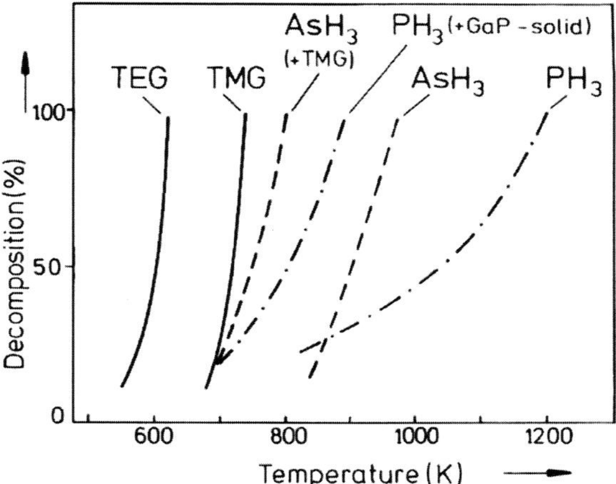


**Fig. 8.24.** Pyrolysis of different standard precursors. Note also the dependence on environment (taken from [8.21])

energy which causes the molecules to break up in smaller parts. This process is usually called pyrolysis. For standard precursor it takes place across a temperature interval of approximately $\Delta T = 50$ to 100 K. Fig. 8.24 gives some examples. The temperature where 50 % of the precursor molecules is decomposed is called $T_{50}$ or generally the decomposition temperature. The decomposition temperature is, besides the vapor pressure, the most important quantity of the organometallic precursor. The substrate temperature has to be set on the appropriate level in order to have sufficient decomposition at the growing surface but also to avoid, at low decomposition temperatures, predepositions on the reactor walls. On the other hand epitaxial growth needs a certain surface mobility of the growing species which sets a lower value for useful decomposition temperatures. In practice since the standard susceptor have an upper temperature limit around 1000 K useful decomposition temperatures are normally between 500 and 900 K.

Decomposition can take place in the gas phase and is then called a homogeneous reaction or can take place at the surface and is then termed a heterogeneous reaction. Both contribute but the latter is quite often, because of the two dimensional geometry (surface catalysis), the dominant one. This can be seen for example from $PH_3$ decomposition curves with and without the GaP surface in Fig. 8.24. Similarly the conditions of the reactor walls (deposits) have a large influence, too. A clean reactor gives completely different results compared to one in which growth has already been taken place. Examples ($AsH_3$) can be found in [8.22].

Thus decomposition temperature data have to be taken with extreme care, a circumstance further enhanced by the fact that reliable measurements are difficult to perform.

In order to understand the MOVPE process it would be desirable to know all the detailed decomposition steps of the hydrocarbons containing precursors because a direct consequence of such reactions is the incorporation of radicals (carbon) into the epitaxial layers or the adsorption and reaction of atomic hydrogen on the surface. However, the reaction and their rates are largely unknown. The problem is that even the homogeneous pyrolysis of single molecular species already needs to consider many reaction to be described properly. The pyrolysis of $Ga(CH_3)_3$, for example, should be governed by reactions like

$$Ga(CH_3)_3 \rightarrow Ga(CH_3)_2 + CH_3 \tag{8.4}$$

$$Ga(CH_3)_2 \rightarrow GaCH_3 + CH_3 \tag{8.5}$$

$$CH_3 + H_2 \rightarrow CH_4 + H \tag{8.6}$$

$$GaCH_3 \rightarrow Ga + CH_3 \tag{8.7}$$

$$Ga + H \rightarrow GaH \tag{8.8}$$

$$GaH + H \rightarrow Ga + H_2 \tag{8.9}$$

$$GaH + CH_3 \rightarrow Ga + CH_4 \tag{8.10}$$

$$GaCH_3 + CH_3 \rightarrow GaCH_2 + CH_4 \tag{8.11}$$

Experimental data from absorption spectroscopy for TMGa pyrolysis are shown in Fig. 8.25. Some of the species appearing on the right-hand sides of reactions (8.4)–(8.11) are clearly detected. However, it is extremely difficult to obtain quantitative data and derive from there values for reaction rates.

The situation becomes more complicated and the number of reactions increases if other reaction partners, as usual in a MOVPE process, are present and have to be considered. The inclusion of surface processes (heterogenous reactions) finally leads to many more reaction equations since adsorption,

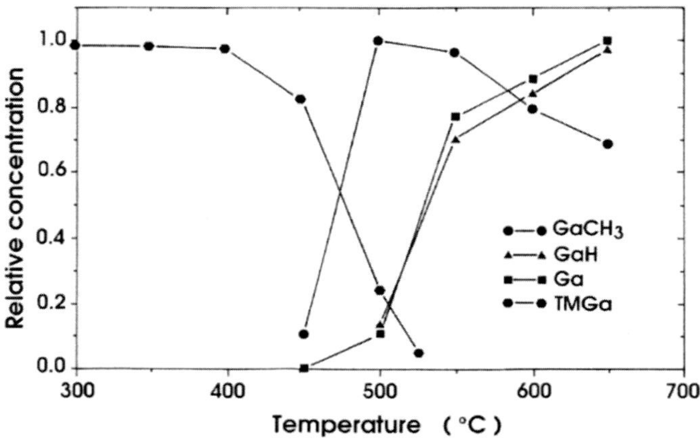


**Fig. 8.25.** Pyrolysis of TMGa and concentration of several pyrolysis products (taken from [8.21])

desorption, occupation of sites and different rates have to be considered. Growth models have taken more than 200 reactions into account and it is clear that with such a large number it is not possible to prove certain mechanisms but at best we might be able to identify certain critical reaction steps [8.23]. A large collection of experimental data concerning the pyrolysis of MOVPE precursors in the presence of other precursors and different surfaces can be found in [8.6].

In such a situation it seems hopeless even to try metalorganic VPE. However, since growth takes place on the surface the important questions are (i) to realize which are the molecules finally arriving from the gas phase at the surface, (ii) to decide whether surface reactions eliminate the last attachments on the atoms desired for epitaxial growth and (iii) finally discover whether the surface allows for sufficient mobility to have the atoms attach to the surface positions necessary for epitaxial growth. For the standard case of tri-methylgallium it has been shown that mono-methylgallium is the molecule present in high concentrations at or near the surface at standard growth temperatures [8.22]. Since the decomposition of arsine releases similar considerable amounts of H the methyl groups react to methane which desorbs easily (see Fig. 9.13):

$$\mathrm{GaCH_3 + H + As \rightarrow CH_4 + GaAs}. \tag{8.12}$$

This simple mechanism explains why growth of GaAs from TMGa and with precursors containing direct As-H bonds is successful but not with tri-methylarsine or just arsenic from an effusion cell. A large concentration of methyl groups will remain on the surface in the latter cases and in contrast reactions like

$$\mathrm{GaCH_3 + CH_3 \rightarrow GaCH_2 + CH_4} \tag{8.13}$$

will take place on the surface. Carbon will be incorporated in the epitaxial layer since carbene ($CH_2$) is not easily removed. This effect will occur also with arsine at low temperatures when a low degree of decomposition provides insufficient hydrogen. This explains why quite often highest purity material is obtained at higher temperatures or vice versa that low temperatures create carbon doping which may be utilized for efficient p-doping in many cases. Also the excellent performance of the alternative precursor TBAs and TBP (Fig. 8.21e) is understood by that. Both contain two hydrogen atoms directly attached to As in contrast to tri-methylarsine with no hydrogen bonded directly. It seems also that atomic hydrogen has the effect of increasing the mobility on the surface (the diffusion lengths turn out to be much larger than in MBE) and thus the epitaxial quality of the MOVPE grown material is high.

Fortunately the surface conditions can be accessed by optical techniques (RAS) during growth and thus it turns out easy to control the growth and optimize the growth conditions even the gas phase chemistry near the surface might be not very well understood. This will be discussed in the next section.

## 8.5 Control of Surfaces Before and During Growth

The drawback of MOVPE compared to MBE until recently was that there was no information on the substrate surface and the grown layers until the growth was finished and the structures were analyzed ex situ by different analytical techniques. In MBE on the other hand substrate deoxidation, recognition of surface reconstruction and growth rate calibration could be applied *in situ* and increased thereby the probability of a successful growth run. The optical techniques developed in the last decade can provide a similar service for MOVPE growth as described in Chap. 10. We give here a few examples for such *in situ* control.

Substrate deoxidation can be followed quite easily as shown in Fig. 8.26 with the example of InP(001) substrates. At low temperatures (a) the reflectance anisotropy on a standard substrate is caused by strains which exist as a consequence of the oxide coverage. At lower temperatures the spectral shapes do not change with annealing time. Around 675 K the spectra change drastically (b) and approach the spectrum of the InP(001)-(2×1) surface (c, d), which above 700 K and at a partial pressure of more than 25 Pa is the stable surface reconstruction of InP(001) [8.24, 25]. Other semiconductors show similar behavior.

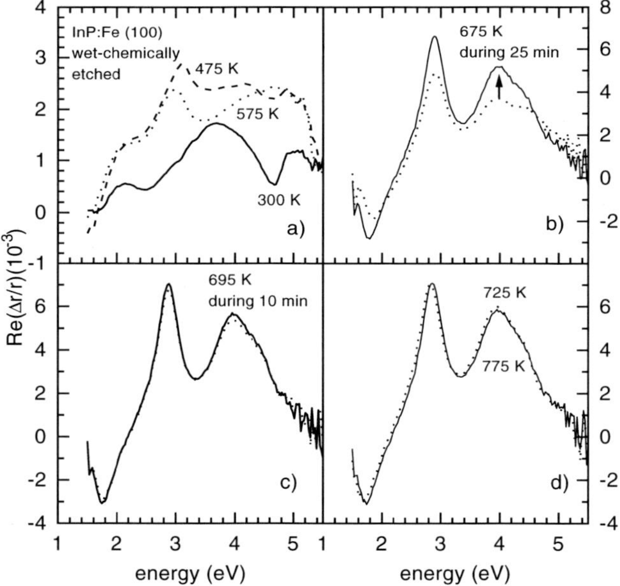


**Fig. 8.26.** RAS spectra during deoxidation of a wet chemical etched InP substrate under phospine with increasing temperature. Above 700 K the spectra correspond to those of InP(001)-(2×1) (taken from [8.17])

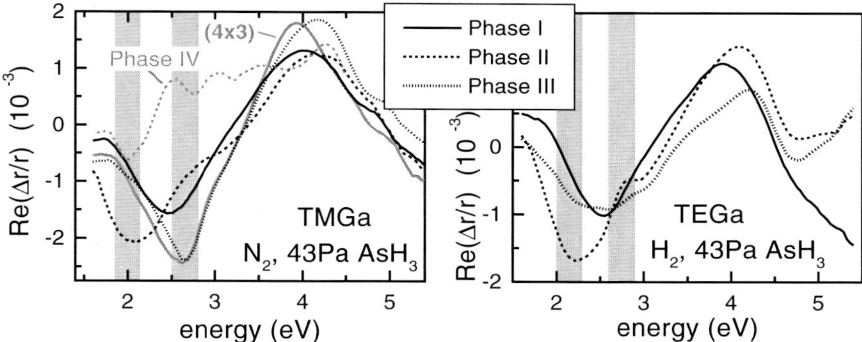


**Fig. 8.27.** Different types of RAS spectra occurring during homoepitaxy of GaAs(001) with TMGa. The partial pressure of TMGa increases from I to IV. The shaded areas mark the typical spectral regions of the RAS minimum for Ga-rich ( 2 eV) and arsenic-rich (2.6 eV) surface reconstructions. In I the surface is essentially identical to the pregrowth c(4×4) surface. In II the surface contains less arsenic and especially at edges of steps or islands Ga-dimers contribute to a minimum at 2 eV. In III and IV the surfaces are covered with hydrocarbons. TEG behaves similar except in phase IV, because of the lower decomposition temperature (taken from [8.26, 27])

The right growth conditions (growing surface) can of course also be well defined via the optical surface response similarly as described in Chap. 10 for the nongrowing stationary surfaces. Figure 8.27 summarizes different types of RAS spectra obtained during MOVPE growth with TMGa and TEGa as precursors [8.26, 27]. The spectra labeled I to IV correspond to different surface phases of GaAs(001). In phase I the surface is still mostly arsenic-rich c(4×4) covered and the spectra represent a large ratio of arsenic to gallium on the surface. With increasing amounts of Ga phase II and III spectra are obtained. In II the minimum is shifted to lower energies thus indicating more Ga-rich surface: ($n$×6). In III and IV finally the optical response is dominated by surfaces covered with hydrocarbon groups. Corresponding to their spectral shape the spectra define then different growing surfaces in a partial pressure vs temperature diagram (Fig. 8.28). In the arsenic-rich phase I growth is by step flow. In II monolayer oscillations can be observed and consequently growth is via island nucleation. It represents also the region relevant for step bunching. In III with its rather low temperatures $CH_3$ and $CH_2$ adsorb on the surface. Since at these temperatures not much arsine is decomposed to provide sufficient atomic hydrogen in order to have the hydrocarbons desorb as methane ($CH_4$) they stick to the surface. The growth rate saturates because the available adsorption sites become limited. This region in the phase diagram however can be used to achieve carbon doping. Phase IV finally represents a surface with probably a monolayer of adsorbed

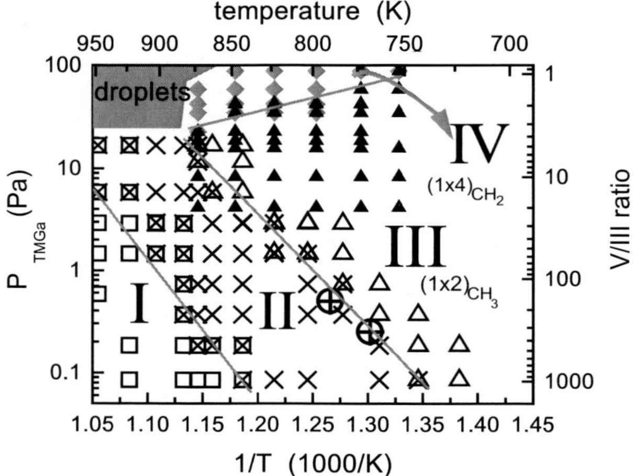


**Fig. 8.28.** Phase diagram for homoepitaxy on GaAs(001) with TMGa. The diagram is derived from spectra such as shown in Fig. 8.27. (taken from [8.17])

methylene ($CH_2$) and is (1×4) reconstructed [8.28]. It is the surface which gives the monolayer self-limitation in ALE.

It turns out that for nearly all cubic III–V semiconductors RAS can classify the epitaxial growth process, a statement which includes also the MBE technique. The optimization procedures become straightforward because they can be performed *in situ* during growth. Especially RAS can also recognize any malfunction of the growth system, such as insufficient supply of precursors, substrate problems, insufficient deoxidation, the necessity to grow buffer layers and similar occurrences. For the hexagonal nitrides and growth on (111) faces RAS is, because of symmetry reasons, not helpful. However, it seems that ellipsometry can provide a similar function for *in situ* control [8.29,30]. For the group IV elemental semiconductors the possible optical control techniques are more difficult to use for growth control. The signals are quite often weak (RAS) or the experiments are more difficult in connection with a growth reactor (SHG).

## 8.6 Nonthermal MOVPE Techniques

In order to lower the deposition temperatures further nonthermal decomposition of the MOVPE precursors has been also tested. The main techniques have made use of photo-decomposition and plasma-decomposition of precursors. The former case is called photo-MOVPE and is easy to perform because only light has to be directed into a standard reactor, which preferably, however, is equipped with a window, and the growth can be followed as a function of light intensity and photon energy. In the latter case the reactor has to be equipped with electrodes which makes larger modifications necessary. Both techniques, however, have not been very successful with respect to crystalline quality as compared to purely thermal MOVPE and thus the practical importance of these techniques for high quality epitaxial growth is not very high.

### 8.6.1 Photo-MOVPE

The influence of light on the epitaxial growth can be two-fold. Firstly by illuminating the substrate with an intense light source the substrate temperature is increased and the growth rate will be influenced and usually increased. This is called the pyrolytic effect and does not depend on the specific photon energies but only on the total radiant power. The physical mechanisms are the same as in standard thermal growth and the deposition temperature is not lowered. However, if the light is tightly focused the temperature increase is more locally defined and so will be the deposition. Light can be used in this way to grow spatially defined structures which due to thermal conductivity of the substrate and the light focusing possibilities are in the micrometer range or larger.

Secondly the photons when chosen with proper energy may directly dissociate the precursor by excitation to higher non stable electronic states or influence the electronic status of the substrate. The process is then called photolytic. Since it involves electronic excitations it should depend on the photon energy and this fact can be utilized to discriminate against pyrolytic growth. Usually UV light sources are required (mercury lamp, excimer laser). The term photo-MOVPE assumes that the photolytic process is the dominant one. Photostimulated MOVPE growth has already been used quite early in order to reduce the growth temperatures of II–VI compounds [8.31–34] because they are plagued by the creation of self-compensating native defects at higher temperatures. Figure 8.29 demonstrates nicely the possibilities of photo-MOVPE in the case of III–V compounds, where especially the high decomposition temperatures of phosphine made photo-MOVPE attractive. In contrast to the thermally driven growth rate (curve: thick solid line) the photostimulated growth rate becomes temperature independent at low temperatures and is just determined by the number of photons. The thermal process is reaction limited and negligible at these temperatures. In principle photo switching of growth is thus also possible. However, the problem turns out to be the epitaxial quality of the material grown at low temperatures which is not very good (polycrystalline at low $T$). This disadvantage together with the additional efforts of introducing a light source and even lasers into the epitaxial laboratory has prevented the photo-methods gaining general importance in MOVPE growth. But practical applications arise from the fact rhat light can be focused and thus selective illumination and selective deposition is possible. This can be utilized for example for conductive metallic connections [8.36, 37].

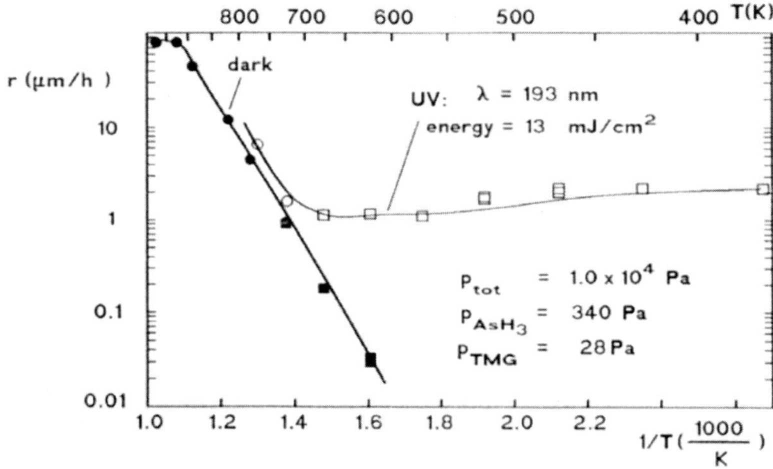


**Fig. 8.29.** Photolytic growth of GaAs with an excimer laser (grey curve) as compared to standard thermal growth (solid curve) (taken from [8.35])

### 8.6.2 Plasma-MOVPE

The motivation for plasma-stimulated MOVPE has similarly been mainly concerned with lowering the growth temperature. The main problem were the high decomposition temperatures of the group-V hydrides. It turned out therefore to be necessary to have just the hydrides flowing through the plasma zone and adding the group III metalorganics afterwards. Two different plasma geometries, a transverse (field perpendicular to the flow direction) and longitudinal (field in flow direction) one, were tested. While the transverse configuration has little effect on the growth rate, the longitudinal plasma (shown in Fig. 8.30), especially when the substrate is biased, gives an increase in growth rate (Fig. 8.31) or correspondingly lowering of the growth temperature and V/III ratio. However, damage induced in the epitaxial layer through the energetic ions had to be considered [8.38].

## 8.7 Safety Aspects of MOVPE

Concluding the presented introduction to the MOVPE growth technique, some important points concerning the safety aspects of this technique should be emphasized. The main reason for this is the use of highly toxic group V hydride source molecules, $AsH_3$ and $PH_3$, in growth of III–V semiconductor compounds. Moreover their vapor pressure is very high at room temperature ($AsH_3$-100%: 12.5 bar, $AsH_3$-10% in $H_2$: 150–200 bar) and therefor experimental modifications and leak tests have to be performed with extreme care.

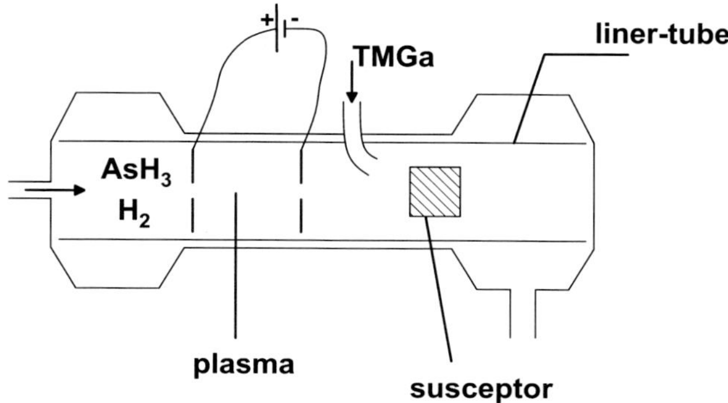


**Fig. 8.30.** Schematic view of a MOVPE reactor with plasma stimulation of growth with longitudinal excitation (field parallel to gas flow) (taken from [8.38])

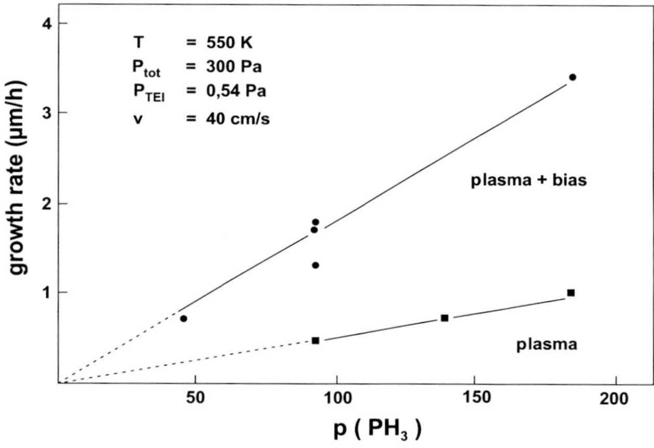


**Fig. 8.31.** Growth rate of InP as a function of the phospine partial pressure in a plasma stimulated MOVPE reactor (Fig. 8.30). Plasma excited by a DC voltage of 1300 V with a power of 6.6 W. Upper curve: substrate with additional bias of minus 300 V, lower curve: without bias. Without plasma their is no InP growth but some In deposition (taken from [8.38])

Toxicity testing experiments are difficult, so comparisons should be based on tests carried out in the same laboratory under similar conditions. An exhaustive summary of the effects on humans to exposure to the As and P hydrides is given by Hess and Riccio [8.39]. The threshold limit values (TLVs) for $AsH_3$ and $PH_3$ are equal to 0.05 and 0.3 ppm (part per million), respectively, however, the lethal doses (determined by experiments concerning the mortality caused by an exposure to hydride atmosphere) are equal to 500 ppm $s^{-1}$, and 2000 ppm $s^{-1}$ respectively, while life-threatening doses are given by (6–15) ppm / (30–60) s, and (400–600) ppm/ (30–60) s.

An attractive solution to these safety issues is to develop much less hazardous group V sources. Fortunately, a number of metal organic group-V sources are indeed very much less toxic than the hydrides. Even more important is the fact that their vapor pressures are below one bar.

Modern practice calls for flow limit valves on each toxic gas cylinder to prevent release of the entire contents of the cylinder in a short time due to an error or failure of some component of the system. Cross-purge-vent assemblies are used to allow thorough purging of toxic gas from the regulator and line before opening the line to change cylinders. Most computer-controlled systems also have an emergency interlock system that shuts the toxic gas valves in the event of loss of air flow or electric power, or exceeding the preset toxic gas alarm level.

In some laboratories the safety precautions have extended to the use of self-contained rooms, remotely observed and operated [8.6]. The room exhaust is purged through a dedicated system with continuous monitoring for toxic gases with the room maintained at a negative pressure.

In all cases, however, the most effective measures involve common sense as well as appropriate safety apparatus.

# Part III

# In-situ Analysis of the Growth Processes

# 9. In-situ Analysis of Species and Transport

The analysis of the growth process is concerned with three questions:

(i) which are the species moving towards the surface and participate in the growth,

(ii) what determines their flux to the surface and

(iii) finally, actually the most important but also most difficult to answer question, what are the processes on the surface?

(i) and (ii) will be discussed within this chapter, (iii) within the next. These questions of course have different importance for the different growth techniques. The answers are most difficult for MOVPE since neither the chemical reactions the precursor molecules undergo in the thermally inhomogeneous reactor are well understood, nor the rather complicated transport of relevant species governed by the hydrodynamical differential equations and the reactions on the growing surface can be very well described. In MOVPE the understanding of growth still derives many conclusions from the comparison with MBE growth. There, species and flux are determined by thermodynamics and the surface processes could always be studied with the full potential of all the surface science tools applicable in UHV (MBE) but not in the gas-phase (MOVPE). For the hybrid growth techniques MOMBE, GSMBE the situation is somewhat in between: more complex chemistry on the source side but again the possibility to involve classical surface science.

## 9.1 Identification of the Growth Relevant Species

In MBE solid sources, in general the elements, are used (Sect. 7.1). Species for the molecular beam are generated by thermal evaporation (Knudsen cells). From vapor pressure data which are tabulated for most solids the, in general, simple species (As. $As_2$, $As_4$, Ga, In, ...) as well as their concentrations are known. In MOMBE, GSMBE, and MOVPE the situation is more complicated. In all cases, however, mass spectrometry can be involved. In MBE and its variants, due to the low pressures, the application of mass spectrometers is straightforward and similar as in MBE. In MOVPE, however, the high pressure prevents a direct mass spectrometric measurement near the sample surface. Sampling of the gas by a capillary and appropriate pressure reduction nevertheless allows also for mass spectrometry. However, the long time needed for the atoms or molecules to reach the spectrometer allows for many gas and wall collisions and thus there is a rather large uncertainty about how much the mass spectrum measured might deviate from that at the sampling point. For that reason optical diagnostics have also been used in MOVPE quite intensively to study the distribution of different species in the gas phase. In this section we will first discuss mass spectroscopy and then optical methods for identification of the epitaxial relevant species.

### 9.1.1 Mass Spectrometry

For mass spectroscopy [9.1] in all cases commercially available quadrupole mass spectrometers (QMS) can be used. In the molecular beam growth techniques (X-MBE = MOMBE or GSMBE) experimental care has to be exercised to reduce the background signal appearing uncorrelated to the molecular beam. This can be achieved with the cryoshroud of the X-MBE equipment, careful mounting and with an extra shroud for the QMS. A schematic set-up is shown in Fig. 9.1 where the QMS has been used in a study of CdZnTe growth. In this example the QMS detects in the reflected beam the number of desorbing atoms (Fig. 9.2). The result of the reflecting flux measurement

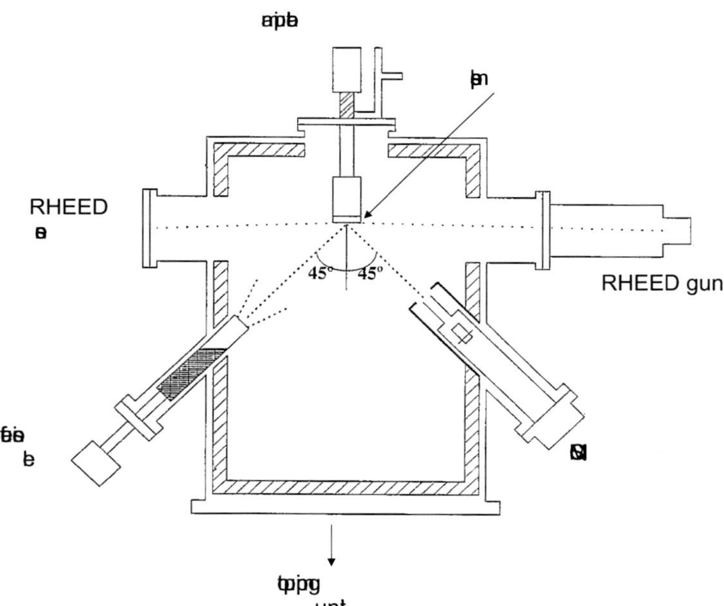


**Fig. 9.1.** MBE equipment with a quadrupole mass spectrometer (QMS) mounted in reflection configuration for growth studies (taken from [9.2])

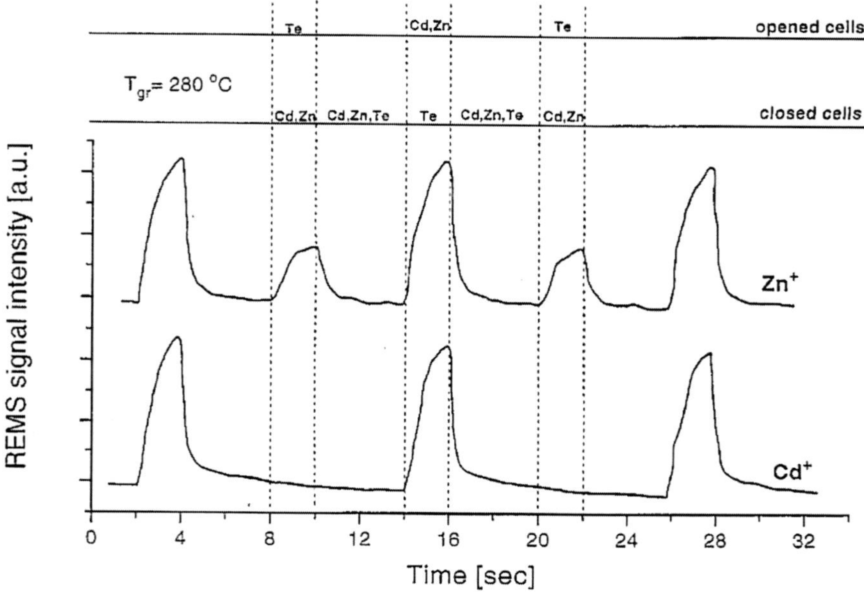


**Fig. 9.2.** Reflectance mass spectrometer (REMS) signal during CdZnTe growth. Extra deposition of Te on the CdZnTe surface leads to an increased desorption of Zn (upper curve) (taken from [9.2])

is that extra deposition of Te on the CdZnTe surface leads to an increase in the desorption of Zn. Similarly of course the incoming flux in the incident molecular beams can be measured.

In the X-MBE techniques the application of MS is similar, however, depending on the molecules used, the requirements for the mass spectrometry are now higher. This is because the, in general, higher masses of the metal-organic molecules are now involved ($M$ up to 300). Secondly the decomposition products appearing in the reactor differ sometimes only by a few hydrogen atoms, resulting in species with small mass differences but a large mass. Thus a large mass but high resolution QMS is required. In MOVPE besides the need for large mass, high resolution mass spectrometer there is the need also to reduce the pressure from 10–100 mbar in the reactor to $10^{-5}$ as required for the QMS. Continuous sampling is done in a two-step pressure reduction: in a first stage the gas is pumped from the reactor via a capillary down to $10^{-1}$ and in a second stage via a nozzle down to the high vacuum needed for the mass spectrometer. This two-step procedure assures that mass discrimination through the pumping process is a small effect, because the inlet and outlet flow types in each of the two stages are the same: viscous in the first and molecular in the second case. A schematic set-up can be seen in Fig. 9.3. The ionization of the large molecules, common in the MO methods,

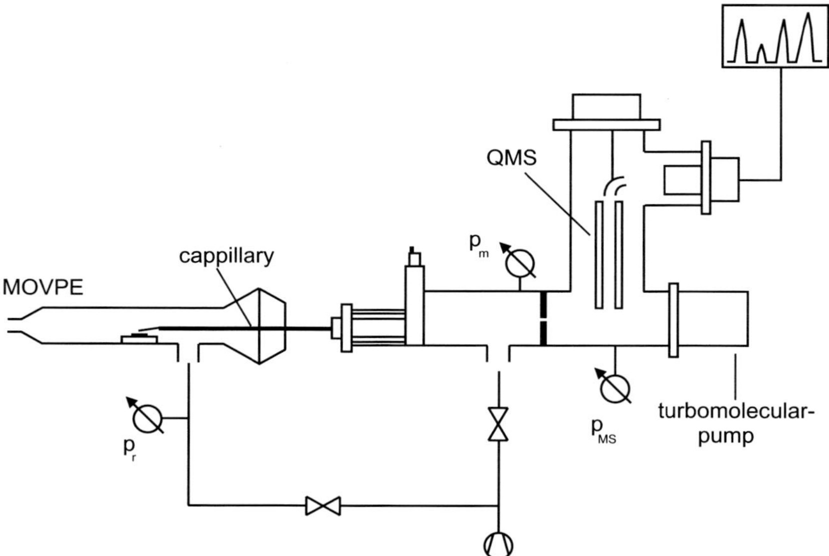


**Fig. 9.3.** Schematic set-up for using a quadrupole mass spectrometer (QMS) in a MOVPE reactor. The gas is sampled from a region inside the reactor through a capillary. The pressure is reduced first by a roughening pump from the reactor pressure ($\approx 20$ mbar) to an intermediate value ($p_m \approx 0.1$ mbar) and in a second stage to the pressure needed for the QMS ($p_{\mathrm{MS}} \approx 10^{-5}$ mbar). This two-step procedure minimizes mass discrimination by the pumping process

leads to their fragmentation and results generally in complicated cracking patterns. Figure 9.4 gives an example for tBAs.

The 300 K spectrum corresponds to a gas mixture containing just tBAs and in addition argon as a reference for calibration. The mass spectrum in a heated reactor (923 K) is clearly different especially in the high mass range due to the thermal decomposition of tBAs. The interpretation of such a pattern in terms of thermal decomposition products however is not always easy and requires reference measurements as well as models about the decomposition products.

Of course in a MS set-up for high pressures like that in Fig. 9.3 the molecules undergo many collisions with each other within the gas phase but even more at the intermediate pressures with the walls of the tubing leading finally to the mass detector. There remains therefore the question of how much the gas composition still reflects the composition above the substrate or in which way it has been modified on the long way to the detector. From that point of view mass spectroscopy cannot be classified as an *in situ* technique. A very special solution which might be termed a true *in situ* mass spectrometer has also been published [9.3]; however, since a hole had to be fabricated into the sample the application is rather limited.

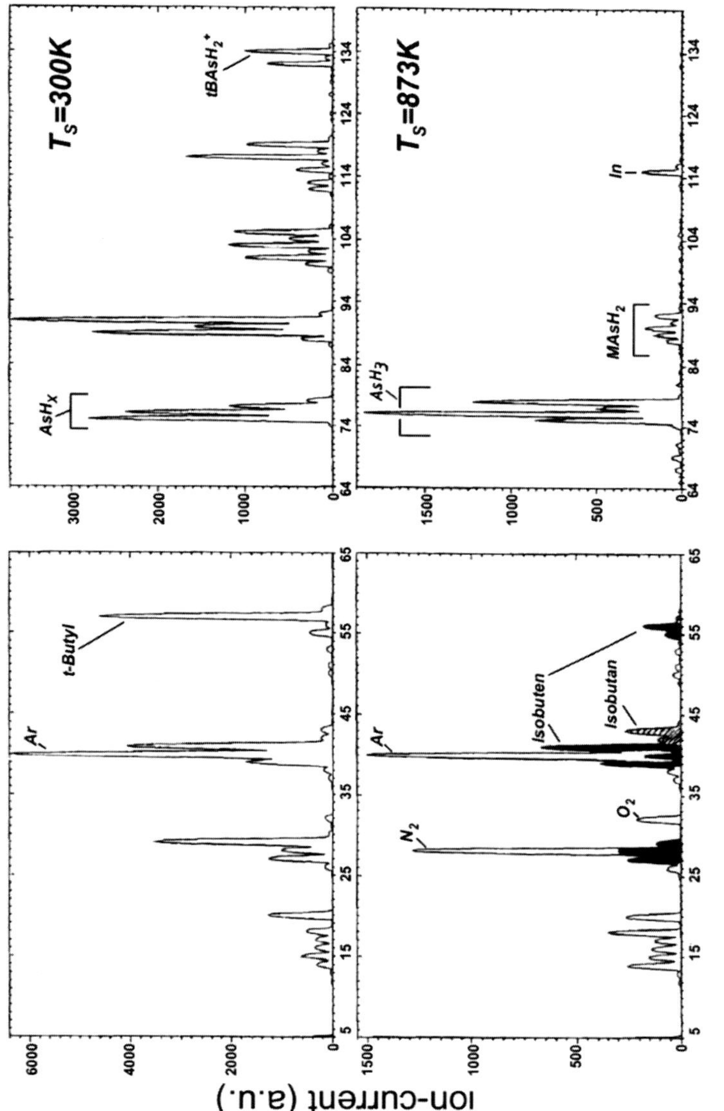


**Fig. 9.4.** In-situ mass spectra from a MOVPE reactor with tBAs and nitrogen as carrier gas sampled with a set-up as shown in Fig.9.3. The plots on the left display the mass range 5–65; those on the right 65–135. The upper plots are taken at a gas temperature of 300 K; the lower ones at 873 K

### 9.1.2 Optical Identification of Species

In order to obtain specific "fingerprint"-like information for each of the different molecular species, interaction of the electromagnetic radiation with either electronic, vibrational or rotational excitations may be utilized. Electronic transitions are conveniently observed in the ultraviolet (UV) and visible (VIS) spectral range by absorption spectroscopy or laser induced fluorescence, while vibrational or rotational excitations may be either observed by Raman type interaction conveniently in the VIS, or by direct interaction in the infrared. Methods which can be performed in the VIS or near UV spectral range are usually preferred since the reactor walls, made of quartz, are transparent in this spectral region. Infrared and deep UV measurements require, in contrast, special window materials and quite often also the removal of the ambient air, which in these spectral regions might absorb light, too. Thus experiments become more complicated. Consequently much less data have been published from those spectral regions. In the following we will describe the optical techniques with respect to their abilities in detecting and determining quantitatively species in the gas phase.

There are many optical techniques which are able to identify the molecular and atomic species present in a gas phase. They differ, however, with respect to their sensitivity, their ease of interpretation and their spatial resolution, a fact which is important under the very inhomogeneous situation in the gas phase environment of an epitaxial reactor. The most important feature required for detection of growth relevant gas phase species is the sensitivity, because their concentration is low, usually several orders of magnitude below the concentration of carrier gas molecules. In laser induced fluorescence, one of the most sensitive techniques, minimum values down to $10^8$ molecules/cm$^3$ have been realized for Si atoms [9.4].

**Absorption Spectroscopy.** Absorption experiments are relatively simple as compared to laser induced fluorescence or Raman type interactions. Different setups have been used for absorption spectroscopy in the gas phase depending on whether electronic transitions with high absorption constants in the UV or vibrational transitions with usually lower absorption constants in the IR are analyzed. For large absorption constants single pass configurations (Fig. 9.5a), for small absorption constants multipass configurations (Fig.9.5b), are in use. As light sources in the UV, xenon high pressure lamps or deuterium lamps are utilized. UV sensitive intensified diode array detectors or photomultiplier tubes with lock-in amplifiers are available for detection. The latter is preferred if long term stability and sensitivity are required. In vibrational IR spectroscopy, tunable diode lasers and multipass configuration allow for very high resolution spectra and, therefore, identification of species is simplified [9.6]. FTIR spectrometers may also be employed; however, in comparison to diode lasers their resolution is smaller, their sensitivity lower and the FTIR instrumentation is more difficult to adapt to a growth reactor [9.7–9]. Absolute number densities may be determined from Beer's law:

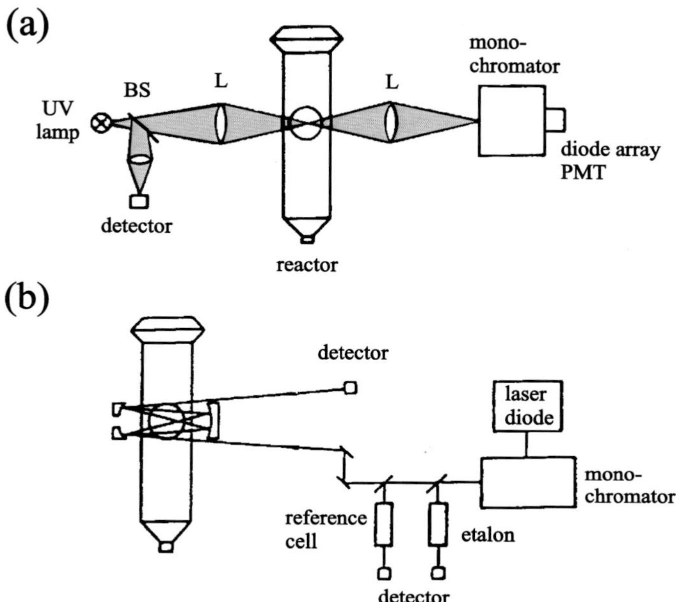


**Fig. 9.5.** Schematic arrangements for absorption measurements in a reactor with single-pass configuration (**a**) and multi-pass configuration (**b**) (taken from [9.5])

$$I = I_0 \exp(-SNL) \tag{9.1}$$

where $S$ is the absorption cross-section, $N$ the number density and $L$ the length of the light path through the reactor. $I$, $I_0$ are the light fluxes with and without absorber. In the unlikely case the absorption cross-section is known, it is of course also possible to determine the number density. Even then only average values are obtained in the nonhomogeneous situation of a cold wall reactor with a heated substrate, because the spatial resolution of an absorption experiment is poor and absorption originates from the whole reactor region traversed by the light beam (Fig. 9.5a). In addition, if the reactor windows (walls) acquire some deposits during the measurement, accurate values for $I_0$ are difficult to obtain. Thus quantitative evaluation of absorption experiments is complicated. However, for monitoring [9.10] and for the detection of atoms and radicals, absorption spectroscopy is well suited [9.11]. At room temperature the electronic transitions, because of the many possible vibrational and rotational combinations in large molecules, generally give only broad and uncharacteristic bands in the UV [9.10, 12, 13]. At higher temperatures when decomposition produces smaller molecules, however, characteristic fingerprints are obtained. Figure 9.6 shows as an example an absorption spectrum of trimethylgallium (TMGa) at room and elevated temperature which displays clearly the decomposition of the precursor into gallium radicals and Gallium atoms. Similar, for example, absorption peaks

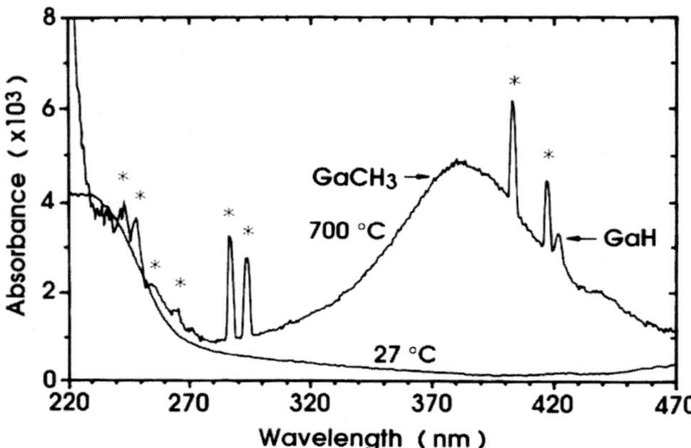


**Fig. 9.6.** Absorbance of trimethygallium (TMGa) at room temperature and 970 K. At the latter temperature structure due to gallium radicals (GaH, GaCH3) and Ga atoms (*) is observed (taken from [9.11])

due to electronic transitions of arsenic subhydrides have been found during the decomposition of arsine at higher temperatures.

**Spontaneous Raman Scattering.** In contrast to absorption spectroscopy spontaneous Raman scattering (RS) has a high spatial resolution, defined by the intersection of the focused laser beam and the aperture collecting the scattered light (Fig. 9.7).

However, the sensitivity is in general lower than in absorption detection since Raman scattering is second order in the electric field dipole interaction but absorption is first order.

Raman scattering was applied quite early to gas phase diagnostics with the goal to determine temperatures by using rotational transitions in the carrier gas molecules (Sect. 9.2.2). This provides no problem experimentally since the concentration of carrier gas molecules is in the order of $10^{19}/\mathrm{cm}^3$ e.g., very high. The situation is, however, different for the detection of reactive

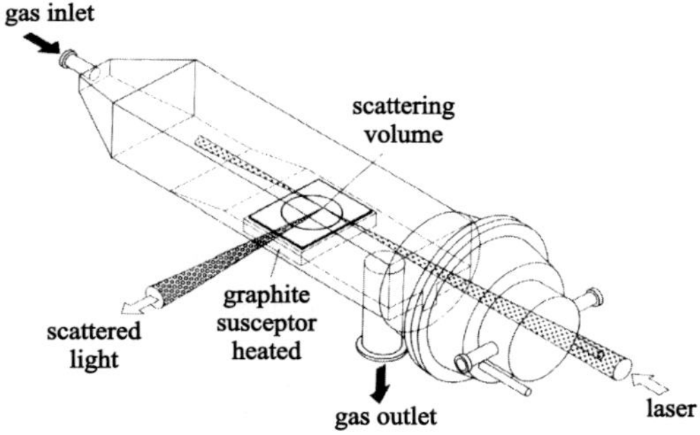


**Fig. 9.7.** Schematic configuration of a Raman scattering experiment in a MOVPE reactor (taken from [9.14])

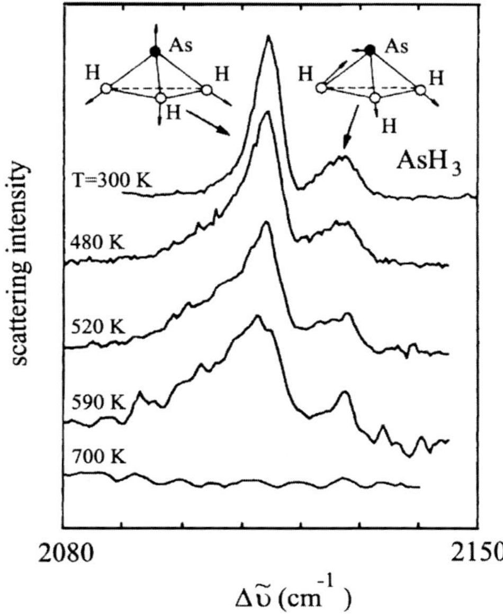


**Fig. 9.8.** Raman spectrum of the $A_1$ vibration of $AsH_3$ at different temperatures. At the highest temperature $AsH_3$ has been decomposed. Input partial pressure to the reactor was around 20 mbar (taken from [9.14])

species with concentrations usually several orders of magnitude lower. Raman scattering from vibrational modes can be utilized as molecule specific information. Only relatively high Raman detection limits for species have been reported (partial pressures between 1 and 10 mbar [9.14, 15]). Figure 9.8 gives an example for Raman spectra of two vibrational modes of arsine for an input partial pressure to the reactor of 20 mbar, a value very large compared to standard growth conditions. The decomposition of arsine at the highest temperature can be recognized by the disappearance of the Raman modes. The usefulness of Raman scattering for the goal of detecting decomposition or reaction products because of the considerable lower concentrations is, however, rather limited. Nevertheless, a few results concerning reaction mechanisms have been reported [9.15–17].

**Laser Induced Fluorescence.** In laser induced fluorescence (LIF) a tunable laser is used to selectively excite electronic transitions of an atom or molecule to be studied and the resulting fluorescence is analyzed spectrally. Corresponding electronic transitions for atoms and molecules are indicated in Fig. 9.9. The experimental setup is equivalent to that for Raman scattering shown schematically in Fig. 9.7. In contrast to Raman scattering, however, a tunable laser is needed for selective excitation. This laser may operate in cw as well as in pulsed mode. In the latter case it provides also temporal resolution. The LIF technique consequently combines the high spatial resolution of Raman scattering with the temporal and spectral resolution of a laser technique. Moreover, whenever fluorescent transitions exist a high sensitivity

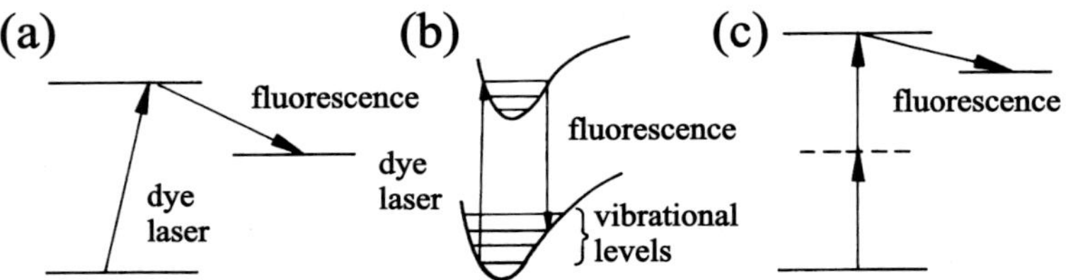


**Fig. 9.9.** Electronic transitions responsible for laser induced fluorescence (LIF) for the case of a one-photon excitation process of an atom (**a**) or a simple molecule (**b**) and for a two-photon excitation process (**c**)

can be obtained. Simple diatomic molecules and atoms quite often produce characteristic, fingerprint-like structures in the fluorescence spectra with a high quantum yield close to one. The sensitivity of LIF in such a case is in the range of $10^8$ molecules/cm$^3$. For larger polyatomic molecules the quantum yield is usually lower because these molecules can alternatively return to the ground state by non radiative processes or possibly photodissociate from the excited state. In such cases the detection sensitivity is substantially lower and other techniques than LIF may be preferred, e.g., spontaneous or coherent Raman scattering.

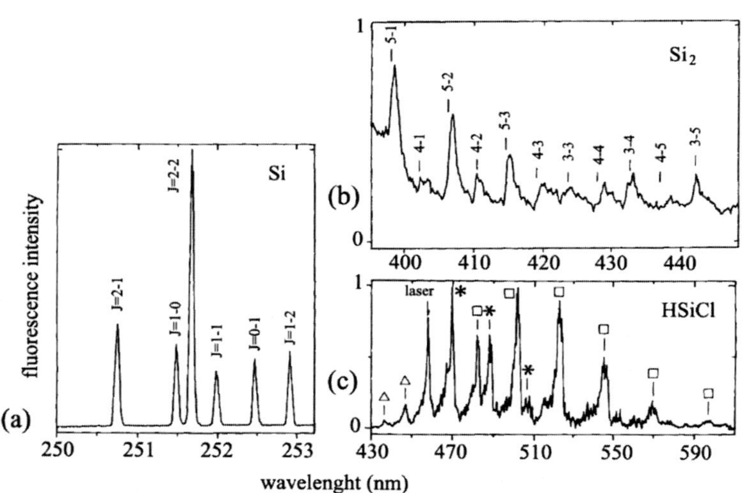


**Fig. 9.10.** LIF spectrum obtained from: (**a**) Si atoms during deposition in a reactor. Excitation was at 250.7 nm. The labels denote the total angular momentum of the states involved in the transition; (**b**) from $Si_2$ molecules excited at 390.83 nm. Labels mark the calculated vibrational origin of the transitions involved; (**c**) of HSiCl (excitation: 457.5 nm) observed in Si deposition from dichlorosilane ($SiCl_2H_2$). Equal labels mark the rotational subband structure of three vibrational bands (taken from [9.18–22])

Absolute determination of number densities in LIF is, however, difficult since usually no simple calibration procedure exists. LIF is especially useful for nonstable species. It has been mostly applied to gas phase diagnostics in silicon chemical vapor phase epitaxy [9.4]. Examples for LIF spectra from Si-atoms, $Si_2$-molecules, and dichloro-silane are presented in Fig. 9.10. Clearly, the decrease in spectral sharpness and signal-to-noise ratio from the simple atom to the more complicated molecule can be noticed.

The results of similar experiments for the hydrides of group V elements have been published, too [9.12]. On the other hand, surprisingly few investigations have been reported for metalorganic precursors. Results have been quoted for example for dimethylzinc [9.23] or trimethylgallium [9.24]. The high spatial resolution in addition allows us to measure the height distribution of the growth relevant species above the substrate and to compare with calculations and growth results. Results for the density of Si atoms density above a Si substrate have been quoted for example in [9.19, 20].

**Coherent Anti-Stokes Raman Scattering.** CARS is a third-order non linear optical technique. It is described as a four wave mixing process (Fig. 9.11) where the frequency difference of two of the input laser beams matches a vibrational–rotational transition frequency of the molecules under study. This creates a coherent set of molecules for which the corresponding vibrational–rotational states are occupied. The CARS process may then be

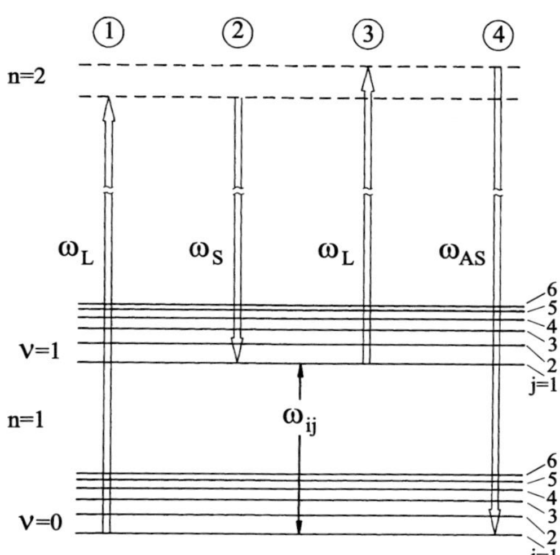


**Fig. 9.11.** Schematic energy diagram of transitions between molecular levels for a CARS process. Rotational ($J$), vibrational ($v$) and electronic ($n$) quantum numbers are indicated. Transitions 1 and 2 can be thought to be responsible for coherent excitation of molecules and steps 3 and 4 may be viewed as an anti-Stokes Raman process (taken from [9.5])

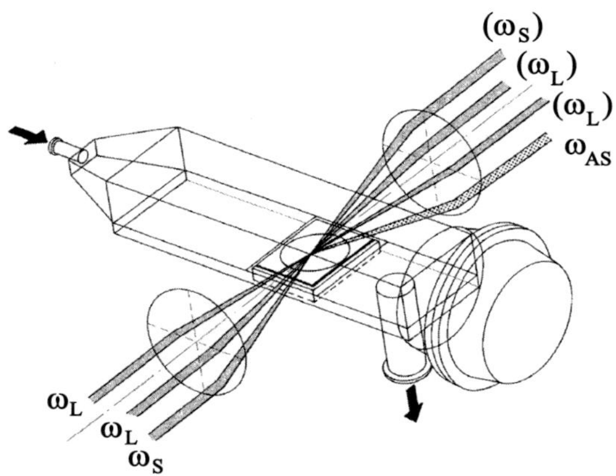


**Fig. 9.12.** Experimental arrangement for CARS in a VPE reactor. Wave vector conservation, (9.3), requires the anti-Stokes beam to appear under a different angle than the other beams in a so-called folded BOXCARS configuration (all three input beams not in one plane). This allows for easy rejection of input laser light (taken from [9.26])

thought of as an anti-Stokes Raman scattering process of the photons in the third input laser beam caused by the coherent molecular excitations.

Not only energy has to be conserved in the CARS process

$$2\omega_L - \omega_S - \omega_{AS} = 0 \tag{9.2}$$

but, since the excitations are coherent, the wave vectors of the light beams, too:

$$2\boldsymbol{k}_L - \boldsymbol{k}_S - \boldsymbol{k}_{AS} = 0. \tag{9.3}$$

The anti-Stokes beam, $\omega_{AS}$, therefore emerges in a well-defined direction (Fig. 9.12) which makes its separation from the other laser beams and the corresponding intensity measurements relatively simple.

The total power in the CARS beam may be written as [9.27]

$$P_{\mathrm{CARS}} \sim \left| N \sum_{i,j} \left[\frac{d\sigma}{d\Omega}\right]_{ij} \frac{\Delta\rho_{ij}}{\omega_{ij} - \omega_L + \omega_S - i\Gamma_{ij}} + \chi_{\mathrm{nr}}^{(3)} \right|^2 P_L^2 P_S \tag{9.4}$$

where the matrix elements in the resonant part of the third-order nonlinear susceptibility $\chi^{(3)}$ have been expressed by the spontaneous Raman cross-section $(\mathrm{d}\sigma/\mathrm{d}\Omega)_{ij}$ and $\chi_{nr}^{(3)}$ is the non-resonant part containing the contributions from all transitions which do not fulfill energy conservation as expressed by equation 9.2. $\Delta\rho_{ij}$ is the difference in occupation between states $i$ and $j$, $\Gamma_{ij}$ the transitional line width and $N$ the number density. By increasing the laser powers ($P_L$, $P_S$) the CARS signal may be enhanced considerably; however, the background ($\chi_{nr}^{(3)}$) is enhanced, too. Background discrimination may be achieved by exploiting the different polarization properties of the non resonant and resonant part of the CARS signal [9.25, 27, 28].

In contrast to the incoherent methods described before, the number density enters the CARS signal squared and the contributions from the different vibrations with frequency $\Omega_{ij}$ may interfere with each other. This effect occurs especially when the eigenfrequencies are close to each other. In such a case a careful modeling of the spectra is required in order extract the number density.

The experimental setup for CARS measurements is quite elaborate and involves a Nd:YAG and a dye laser as well as filter monochromator, and a reference cell (argon, 1 bar) in order to normalize for the shot to shot variation in laser pulse power [9.5]. The detectivity of the CARS method in general is limited by laser noise (especially in the case of multimode lasing) and by the background signal originating from all other molecules present. The main background signal contribution is naturally caused by the carrier gas molecules being present in large number. Detectivity levels as low as $10^{-3}$ mbar may be achieved for diatomic molecules with sharp spectral features, but typical values for larger metalorganic molecules used as reac-

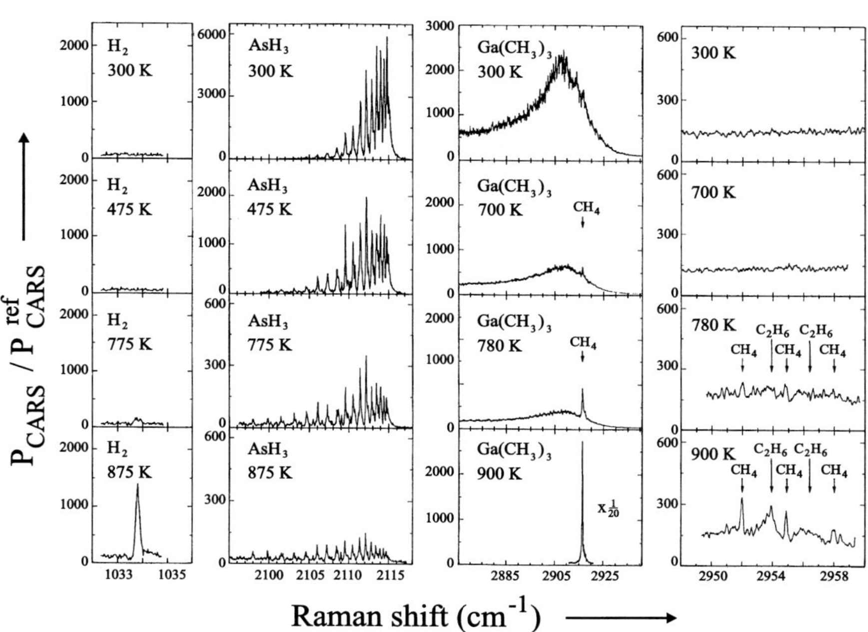


**Fig. 9.13.** CARS spectra taken at different temperatures in a MOVPE reactor for the growth of GaAs from arsine ($AsH_3$) and trimethylgallium ($Ga(CH_3)_3$). The spectra demonstrate with increasing temperature and from left to right: The production of molecular hydrogen, the decomposition of arsine and trimethylgallium and the production of hydrocarbons (taken from [9.25])

tants in a MOVPE environment are around $10^{-1}$ mbar. Spatial resolution for a crossed beam arrangement ("Folded BOXCARS") like that in Fig. 9.12 is of the order of a few mm along the beam direction, but much better (μm) perpendicular to it.

By tuning the dye laser frequency $\omega_S$, the denominator in (9.4) becomes resonant with the different frequencies $\omega_{ij}$, and spectra such as those in Fig. 9.13 are obtained at different temperatures. They show how with increasing temperature (top to bottom) arsine and TMGa are decomposed and reaction products like $CH_4$, $C_2H_6$ and $H_2$ are created.

**Other Techniques.** The previously described methods have been actively employed in gas phase diagnostics of the VPE processes. Other methods, however, which have been mainly used in distinct circumstances (glow discharges, flames) might be useful, too. These include two- or three-photon processes (Fig. 9.9) used to excite certain species to electronic states from which fluorescence may then be observed. An interesting example of this multiphoton fluorescence with respect to MOVPE is the possibility of detecting atomic hydrogen [9.29]. Since atomic hydrogen is supposed to play a major role in the removal of carbon-containing radicals in MOVPE, such a diagnostic technique would turn out to be quite useful. Another interesting technique is opto-galvanic spectroscopy, which is based on the photoionization of molecular species. The appearance of ions is then detected by a current between two electrodes in the photoexcited volume [9.30]. While this method obviously has a high sensitivity, the presence of electrodes in a VPE environment might disturb severely the purity of the epitaxial layers grown.

## 9.2 Mass Transport to the Surface

For molecular beams the mass transport of the growth relevant species to the growing surface is simply given by the flux emitted from the source into the solid angle determined by the growing surface. In MBE the flux from Knudsen cells is easily calculated according to (6.5). It can of course also be measured by placing a mass spectrometer ("flux meter") into the beam. In CBE or MOMBE the metalorganic precursors are decomposed in the source with a hot wire. The flux out of the source cell depends on chemical reactions and cannot be simply calculated. In these cases it is best measured with a mass spectrometer.

As already discussed in Sect. 6.2.2 in a VPE environment the mass transport, i.e. the flux $\boldsymbol{j}_i$ of reactant $i$ to the gas–solid interface, is given by convective flow, by diffusion in a concentration gradient and by thermodiffusion (temperature gradient driven) [9.31]:

$$\boldsymbol{j}_i = \frac{p_i \boldsymbol{v}}{kT} - \frac{D_i}{kT}\left[\nabla p_i + \frac{\alpha_i}{T} p_i \nabla T\right] \tag{9.5}$$

where $p_i$ is the partial pressure of the species $i, D_i$ is the diffusion constant of species $i$ within the carrier gas and $\alpha_i$ is the thermodiffusion coefficient. In writing this equation multicomponent diffusion and gas phase reactions have been neglected. The important parameters which determine the flux under these conditions are the velocity $\boldsymbol{v}$, the temperature $T$ and the partial pressures. While the partial pressures of the growth relevant species will be the result of the chemical gas phase reactions, gas velocity and temperature are essentially a property of the carrier gas only. This is because of the large difference in partial pressures which allow the contribution of the precursor and their reaction products to the hydrodynamical and thermodynamical properties of the gas mixture to be neglected in a good approximation. Thus for the experimental determination of $\boldsymbol{v}$ and $T$ as well as for theoretical calculations it is sufficient just to consider the carrier gas. This simplifies the experimental studies (no precursors are needed) as well as the theoretical work by reducing the number of necessary equations.

The outcome of those calculations performed by finite element methods gives two-dimensional solutions for the four parameters $p_i$, $T$, $v_x$ and $v_y$ mentioned above. Among these, $T$ is certainly the most important parameter with respect to the thermally driven gas phase reactions in VPE. The velocity profile will deviate of course from the standard parabolic in $v_x$ because of the non isothermal situations and because natural convection may create (depending on $T$) a $v_y$ component. As long as no turbulences are created in the gas flow, which might lead to an irregular flow pattern, transport is in general not too critically affected by this, since the flow velocities anyway have to be zero at the wall or the substrate. The pressure finally is to a good approximation a constant within the reactor and for that reason also not a very crucial quantity. Thus, it appears that temperature and the velocity profiles are the most strongly influenced variables and therefore good test candidates for the verification of experimental data.

### 9.2.1 Measurement of Velocities

For velocity determination, Doppler frequency shifts (laser Doppler anemometry (LDA)) from small particles ($TiO_2$, diameter 1 µm) moving with the gas stream can be utilized [9.32]. A different technique utilizes two focused laser beams which produce two light pulses when a particle passes through [9.33].

The more common Doppler experiment is sketched schematically in Fig. 9.14.

The moving particle receives the laser light wavelength $\lambda_0$ (frequency $\nu_0$) with a frequency $\nu_p$

$$\nu_p = \nu_0 \left(1 - \frac{\boldsymbol{v} \cdot \boldsymbol{e}_s}{c}\right) \tag{9.6}$$

where $c$ is the velocity of light. The light scattered from the moving particles (Mie scattering) $\nu_p$ is detected with frequency $\nu_{\mathrm{DE}}$

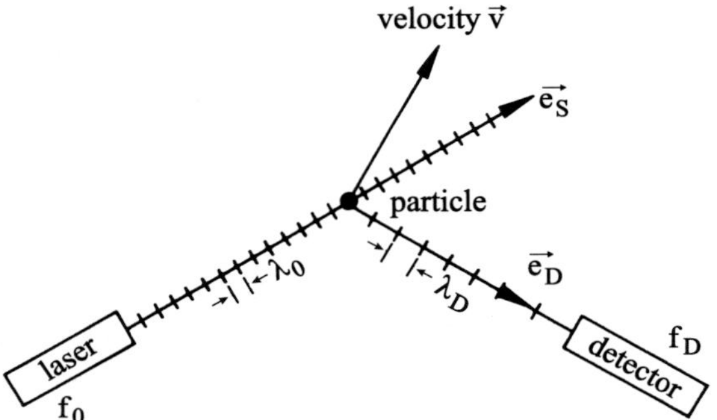


**Fig. 9.14.** Principle of velocity determination by measuring the Doppler shifted light scattered from the moving particle (taken from [9.5])

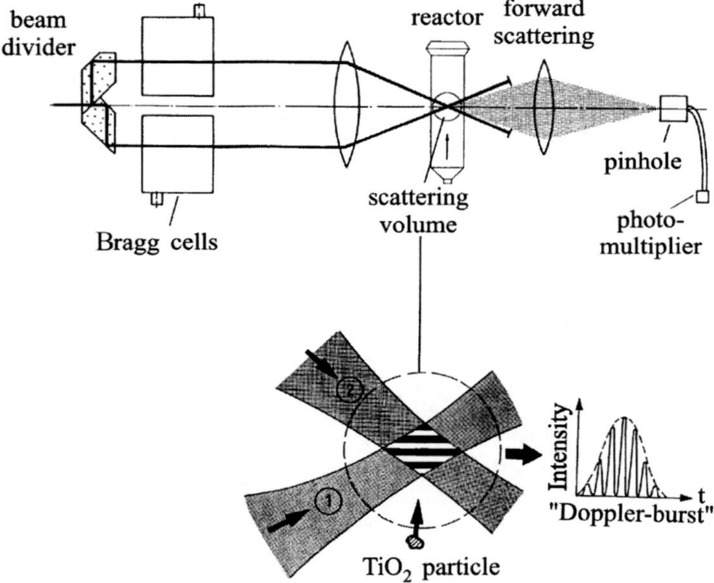


**Fig. 9.15.** Symmetrical experimental arrangement for measuring the velocity by Doppler shifted scattered light: laser Doppler anemometry (LDA) (taken from [9.5])

$$\nu_{\mathrm{DE}} = \nu_p \left(1 - \frac{\boldsymbol{v} \cdot \boldsymbol{e}_{\mathrm{DE}}}{c}\right)^{-1} \tag{9.7}$$

and therefore:

$$\nu_{\mathrm{DE}} = \nu_0 \left(1 - \frac{\boldsymbol{v} \cdot \boldsymbol{e}_s}{c}\right) \left(1 - \frac{\boldsymbol{v} \cdot \boldsymbol{e}_{\mathrm{DE}}}{c}\right)^{-1} \tag{9.8}$$

The Doppler frequency shift:

$$\nu_\mathrm{D} = \nu_\mathrm{DE} - \nu_0 \tag{9.9}$$

is obtained as the difference frequency from mixing with light unshifted in frequency. By introducing the wavelength $\lambda = c/f$ and for small shifts with

$$(\boldsymbol{v} \cdot \boldsymbol{e_s})(\boldsymbol{v} \cdot \boldsymbol{e}_\mathrm{DE})/c^2 \ll 1 \tag{9.10}$$

one obtains

$$\nu_\mathrm{D} = \frac{n}{\lambda_0}\boldsymbol{v}(\boldsymbol{e}_\mathrm{DE} - \boldsymbol{e}_s) \tag{9.11}$$

where $n$ is the refractive index and $\lambda_0$ the vacuum wavelength. Since the evaluation of (9.11) requires an additional measurement of the angle between the directions of the laser beam and the scattered light, a symmetrical setup as shown in Fig. 9.15 is generally preferred today. There, two beams generated by a prism arrangement from a single laser (He-Ne) intersect at an angle $\Theta$, which is easily predetermined, and the Doppler frequency is then given by:

$$\nu_\mathrm{D} = \frac{2\,v\,n}{\lambda_0}\sin\frac{\theta}{2} \tag{9.12}$$

where $v$ is the velocity component perpendicular to the interference fringes.

The experimental arrangement includes a Bragg cell to shift the Doppler frequency by a known amount for the detection of small velocities with higher precision. Particles are introduced into the gas stream by flowing the carrier gas over a powder of $TiO_2$-particles. For the measurement of all velocity components either the reactor or the optical arrangement has to be rotated in order to have interference fringes perpendicular to the component to be determined. In the present problem, where two components are to be measured, this can be achieved simply by rotating the optical arrangement by 90° around its optical axis without moving the reactor. The "Doppler burst" generated by the particle when moving through the interference fringes has a typical frequency in the order of a few Hz. After detection by a photomultiplier it can be directly converted to a velocity component.

In Fig. 9.16 examples of velocity measurements in a cold wall reactor taken with the LDA method are shown. The velocity near the susceptor is increased by increasing the flow rate through the reactor [9.14]. A similar effect is obtained by reducing the pressure considerably below atmospheric (low pressure VPE) [9.31]. In case the natural convection becomes more dominant (high pressure, large height of the reactor and large temperature gradient) than the forced convection, vortex flow (roll, cells) may be induced in addition to the laminar flow. Such examples have been given in many flow visualization studies [9.31, 35] and there discussed critically in terms of thermophoretic forces. More accurate velocity measurements are presented in [9.36, 37].

In general one can state, however, that most reactors in use today are working in well controlled flow regimes which in most cases have been substantiated by finite element calculations, for which programs are also commercially available.

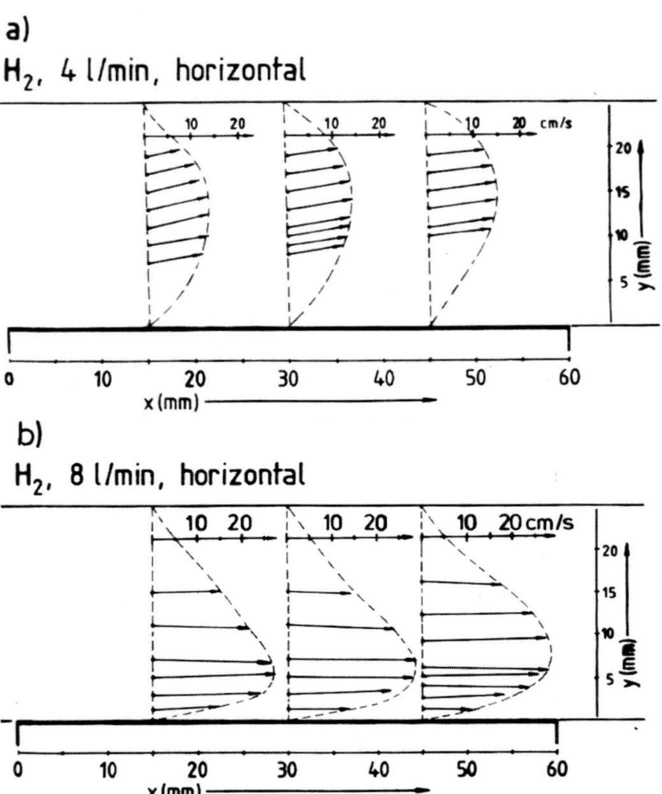


**Fig. 9.16.** Velocity profiles above the susceptor measured in a reactor midplane (for coordinates see Fig. 6.16) by LDA for different flow rates (taken from [9.34])

### 9.2.2 Measurement of Temperature

Temperature constitutes the most important parameter in gas phase epitaxial growth. It determines the constituents in the gas phase as well as on the surface. Remote temperature measurements of matter (gas or solid) can be performed via elementary excitations whose strength of interaction with the optical radiation depends strongly on the thermal distribution functions and thus gives the possibility to determine the temperature. Phonons in condensed matter and rotational or vibrational excitations of molecules in gases have typical energies in the range from 10 to 200 meV. This results in strongly varying values for their occupational probability function in the temperature range of relevance for epitaxial growth. Signals depending on the occupational numbers can be most easily generated by Raman-type light interaction processes. This includes spontaneous Raman scattering, which will be discussed here, as well as nonlinear Raman type interactions (for example coherent anti-Stokes Raman scattering CARS, Sect. 9.1.2). Raman-type interactions have the advantage especially for gaseous media that their signal intensities in most cases can be evaluated directly with respect to temperature, without many significant corrections originating from the influence of electronic states. This is somewhat more complex in solids and thus for the determination of bulk or surface temperatures different approaches for temperature measurement may be advisable.

In the gas phase containing molecular components, however, the large number of rotational or vibrational–rotational transitions may be utilized for a very accurate and convenient temperature determination. In a typical VPE environment with a carrier gas at higher partial pressures, the measurement of the rotational transitions of the carrier gas ($N_2$, $H_2$) is the most appropriate choice. Since the concentration of molecules is high, large scattering intensities can be expected. In addition the molecular structure is simple and quantitative expressions for the temperature dependence of the intensities can be given [9.33]:

$$\begin{aligned} I(J) = {} & A(\omega)\frac{N}{Q}g_j(2J+1)\exp\left(-\frac{\hbar\Omega(J)}{kT}\right) \\ & \times\gamma_0^2 f(J)\frac{3(J+1)(J+2)}{2(2J+1)(2J+3)}\,\frac{45(2\pi)^4}{7}\,\omega^3 I_0, \end{aligned} \tag{9.13}$$

where the factor $\omega^3$ assumes photon counting and $A(\omega)$ is the spectral sensitivity of the spectrometer, which at a known temperature can be easily

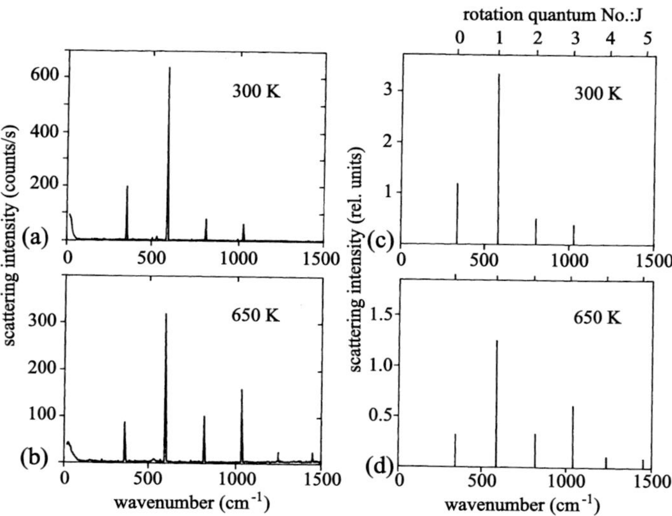


**Fig. 9.17.** Raman scattering intensities from rotational transitions of $H_2$ for two temperatures measured (**a**, **b**) and calculated (**c**, **d**) by using (9.13) (taken from [9.5])

calibrated with the help of (9.13). $N$ is the number of molecules in the scattering volume, $Q$ is the partition function, $g_j$ is the nuclear spin degeneracy, $J$ is the rotational quantum number, $\Omega(J)$ are the frequencies of the rotational transitions, $\gamma_0$ is the anisotropic matrix element of the Raman tensor, $f(J)$ is a correction term accounting for the anharmonicity and $I_0$ is the incident laser intensity.

Raman scattering measurements can be performed for example in a configuration like the one displayed in Fig. 9.7 using a conventional Raman setup [9.38]. The first measurements have therefore been performed shortly after laser excited Raman spectroscopy became a common experimental technique [9.39, 40]. A comparison of measured and calculated rotational scattering intensities is shown in Fig. 9.17. Exactly the same temperature dependence of the rotational intensities can be seen in both plots.

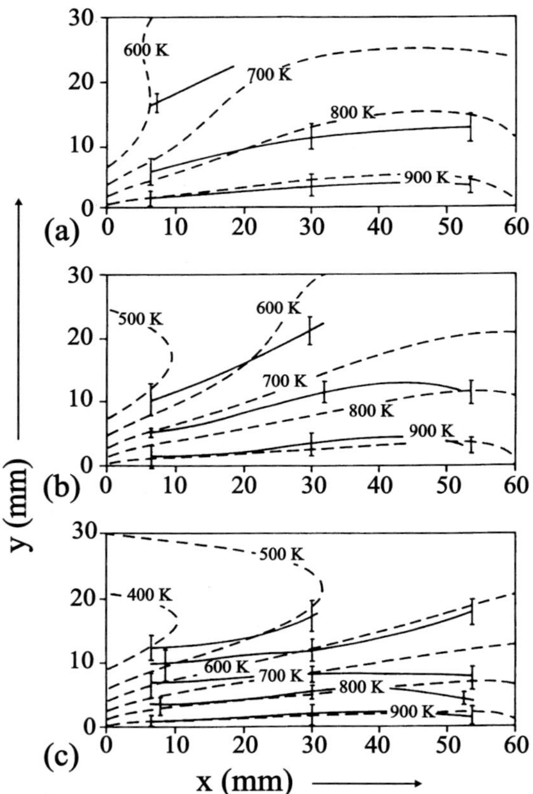


**Fig. 9.18.** Temperature profiles in a reactor midplane for different flow rates: (**a**) 2 sl min$^{-1}$, (**b**) 4 sl min$^{-1}$, (**c**) 8 sl min$^{-1}$. Solid lines: measured by Raman scattering, dashed lines: finite element calculations with equations (11.15)–(11.22) with appropriate boundary conditions (taken from [9.38])

For a fast online analysis of the data, (9.13) may be rewritten in the form

$$I(J) = m(J) \exp\left[-n(J)\frac{1}{T}\right]. \tag{9.14}$$

which clearly reveals the simple dependence of intensity on temperature corresponding to the Boltzmann distribution function which as a result leads to a linear dependence of $\ln(I(J))$ versus $n(J)$. This can be used very efficiently to generate a temperature value from intensity data immediately after measurement by linear regression. Typical errors are in the order of a few percent or less depending on effort.

The result of many of such temperature measurements at different locations within the reactor allows two or three dimensional temperature profiles to be generated. An example is given in Fig. 9.18 in a reactor midplane for different flow rates of the carrier gas. Such measurements may also be used to test hydrodynamical calculations. The theoretical results included in Fig. 9.18 were generated by a finite element calculation including detailed boundary conditions for thermal conductivity and also heat transfer by radiation [9.31].

# 10. In-situ Surface Analysis

The probing tools for epitaxial growth must of course have to be surface science tools. They preferably should work in real time such as to be used not only for analysis of equilibrium situations but also for growth control and, secondly, they should be applicable in vacuum as well as in gas phase environments. This will allow us to compare the different growth techniques like molecular beam epitaxy and metalorganic vapor phase epitaxy.

The vacuum-based growth methods like MBE, CBE and MOMBE could take advantage in the past of the already available vacuum-based surface science tools, especially RHEED. As a consequence, a high degree of analysis and control was available from the beginning and most epitaxial structures and devices have been grown first and for a long time exclusively with these growth techniques. However, with the increasing industrial use of MOVPE, the desire for more control became stronger, for example in VCSEL (vertically surface emitting laser) growth and especially in nanoscale growth situations concerned with interface engineering, nanostructures (quantum dots and wires) and thin interlayers. Probing devices can help us to understand and analyze the basic mechanism of epitaxial growth but also help for technological growth control.

Roughly, three groups of *in situ* probing devices exist: local scanning probes (AFM, STM) [10.1,2], diffraction probes (LEED, RHEED, GIXS) [10.3,4] and optical probes (linear ones like SE, RAS, LS [10.5] and non-linear techniques like SHG [10.6,7]). With the exception of the electron diffraction methods all techniques operate in vacuum as well as at higher pressures. The scanning probes, however, are difficult to operate under the high temperatures of growth and in a hot gaseous environment, because the scanners loose their piezoelectric properties at high temperatures. Moreover, in order to obtain a nearly real time analysis they have to be operated with extremely high scan rates or under unrealistic growth conditions. Thus, not many results have been reported. Nevertheless, very impressive results about growth morphology have been obtained [10.8–10].

## 10.1 Scanning Microscopes

The invention of the scanning tunneling microscope (STM) has been an excellent example of the creation of a new research tool by innovative implementation of scientific and technological knowledge: the quantum mechanical phenomenon of electron tunneling and the know-how of controlling motions on a nanometer scale [10.1, 2]. Figure 10.1 gives the well known set-up, where a piezo-driven tip scans the surface with all three coordinates controlled electronically via a feedback loop. The success of the STM in real space visualization of surfaces on an atomic scale soon gave rise to the development of other scanning probe microscopes (SPM) exploiting other interactions than just electron tunneling with the surface. The most common is the atomic force microscope (AFM) which commonly utilizes the repulsive force between a tip at the end of a cantilever and the sample surface [10.2]. The scanning near-field optical microscope (SNOM) is another example where the higher resolution in the optical near field is used to obtain a microscopic image of the surface [10.12]. Other techniques like the magnetic scanning microscope are at present on their way to becoming standard tools [10.13]. STM and AFM are the most common ones in use. Both are commercially available for table-top or vacuum usage.

All scanning microscopes operate in vacuum as well as in non vacuum environments. In order to image a surface the STM requires a conducting

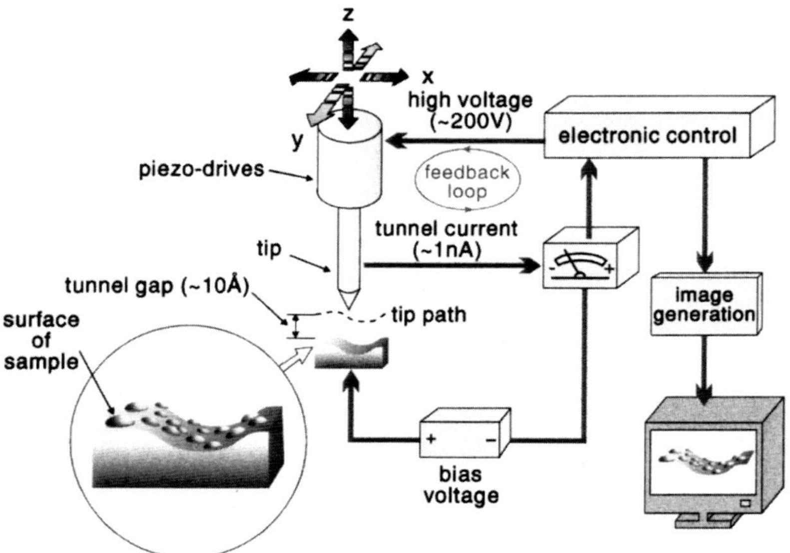


**Fig. 10.1.** Schematic illustration of a scanning tunneling microscope (STM). (taken from [10.11])

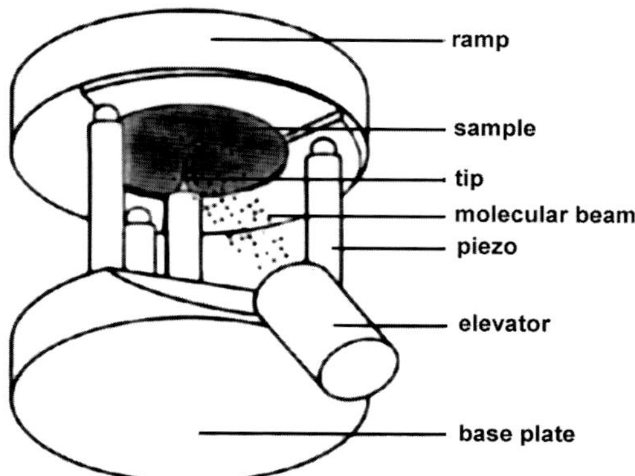


**Fig. 10.2.** Schematic illustration of a scanning head for an *in situ* STM (taken from [10.8])

material (metals, semiconductors) excluding insulators. The AFM in contrast works with all kinds of solids. The resolution on the other hand is in STM generally more local since essentially only electron transfer to or from the electronic atomic orbitals is involved (sub-angstrom resolution) while the AFM exploiting van der Waals forces is more susceptible to structure on the nanoscale (quantum dots, surface roughness). The resolution of the other scanning probes is in general less than that of STM and AFM. For *in situ* studies of epitaxial growth first of all one has to take care that the tip does not represent an obstacle for the flux of species to the surface and thus influences the growth itself. Figure 10.2 showing a STM for *in situ* studies of MBE growth, illustrates the problem.

However, the STM studies performed so far seemed to have the "shadowing" under control. More severe, however, is a general problem that all the scanning probes have in operating under high temperatures. This appears because the scanning technology relies on piezoelectric scanners which lose their piezoelectric properties at moderate temperatures. Therefore the scanners have to be shielded thermally. In the vacuum-based growth methods this requires shielding against radiation and can be accomplished with metallic sheets. In the gas phase techniques, however, not only radiation but also the, in general, much larger heating through the thermal conduction of the carrier gas has to be taken into account. Moreover, in order to analyze the time evolution of growth extremely high scan rates are needed to push the time needed for one image into the growth relevant range of seconds. As a result most SPM studies have been performed not *in situ* but in an extra transfer chamber. There exist only very few true *in situ* studies of epitaxial growth. Figure 10.3 shows an example of a study in MBE growth on Si(111). The triangular island on Si(111) is growing by adding a row with the width of a 7×7 unit cell on the right edge. Only one report seems to have been given about a study in VPE growth [10.10]. Most of the STM reports deal with samples which have been transferred from the high temperature growth

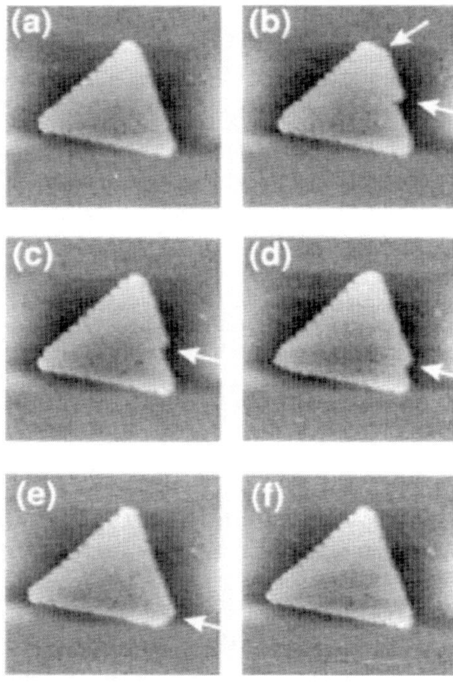


**Fig. 10.3.** Sequence of images showing the lateral growth of a triangular Si(111) island. A row with the width of the 7×7 unit cell is growing along the right edge of the islands (taken from [10.9])

chamber to a room temperature analysis chamber and thus the question of how much this corresponds to the *in situ* growth situation always arises. In addition time resolution as in Fig. 10.3 cannot be reached.

## 10.2 Diffractions Techniques

Electron diffraction techniques like LEED (low energy electron diffraction) and RHEED (reflection high energy electron diffraction) are the standard methods in surface science to monitor the structure and the status of surfaces. Because both methods require a vacuum they can be applied only in MBE and its variants. In LEED surface sensitivity is reached by choosing the electron energy around the minimum of the electron escape depth (50 – 100 eV see Fig. 10.10 below). In RHEED the electrons have high energies (5 – 50 keV) and the small information depth is reached by a very small glancing angle of incidence ($0 < \vartheta < 5°$) as illustrated in Fig. 10.4.

This figure also explains why RHEED in contrast to LEED is used nearly exclusively in MBE. RHEED is mounted far away from the sample (not in front like LEED) and thus does not block the molecular beams. In GIXS (grazing incidence X-ray scattering) the surface sensitivity is obtained by a very small angle ($< 1°$) of incidence (grazing). The penetration depth is then

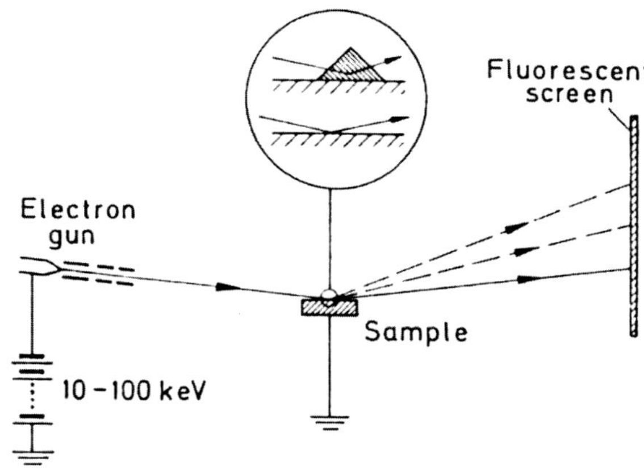


**Fig. 10.4.** Schematic of RHEED set-up showing the glancing angle of the incident high energy electrons. The inset explains the sensitivity of RHEED to morphological surface features: surface roughness, three-dimensional growth (taken from [10.11])

in the order of the atomic distances. Since X-rays do not require a vacuum for propagation, GIXS can also be applied to gas phase epitaxial growth.

### 10.2.1 Diffraction

For the purpose of structural analysis in a first approximation the kinematic theory is sufficient. There the plane monochromatic wave incident on the sample generates, through coherent elastic scattering, secondary waves of low power such that the loss of energy in the incident wave as well as scattering of the secondary waves can be neglected. The total amplitude is then obtained by integration of all secondary waves originating from the scattering volume as a Fourier-like integral

$$A \propto \int \rho(\boldsymbol{r}) \cdot e^{-i\boldsymbol{K}\boldsymbol{r}} \mathrm{d}\boldsymbol{r} \tag{10.1}$$

where $\rho(\boldsymbol{r})$ denotes the scattering density at position $\boldsymbol{r}$ and $\boldsymbol{K} = \boldsymbol{k} - \boldsymbol{k}_i$. In the case of a periodic three-dimensional distribution of scattering centers (crystalline solid) the result may be written in reciprocal space

$$A \propto \sum_{\boldsymbol{G}} \rho \int e^{i(\boldsymbol{G}-\boldsymbol{K})\boldsymbol{r}} \mathrm{d}\boldsymbol{r} \tag{10.2}$$

where $\boldsymbol{G}$ is a reciprocal lattice vector defined in terms of the unit vectors of the reciprocal lattice vectors $\boldsymbol{g}_1$, $\boldsymbol{g}_2$, $\boldsymbol{g}_3$:

$$G = h\boldsymbol{g}_1 + k\boldsymbol{g}_2 + l\boldsymbol{g}_3. \tag{10.3}$$

From (10.2) immediately the Laue conditions for diffraction

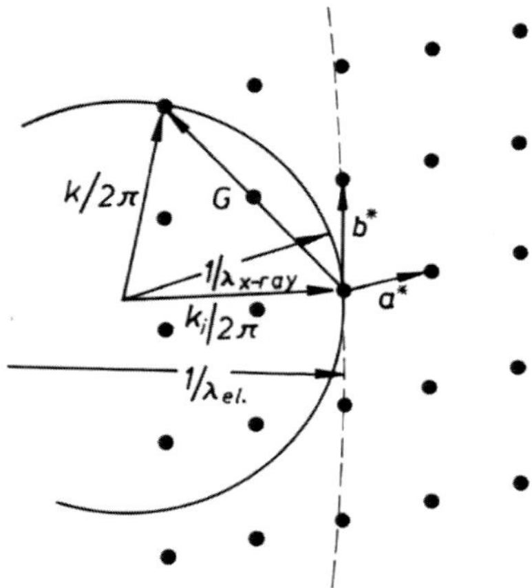


**Fig. 10.5.** Reciprocal lattice and Ewald sphere construction for X-rays (small $k$-vector) and electrons (large $k$-vector) (taken from [10.11])

$$\boldsymbol{G} = \boldsymbol{K} = \boldsymbol{k} - \boldsymbol{k}_i \tag{10.4}$$

and the Ewald construction are derived. For the three dimensional case and a monochromatic wave (10.4) in general will have no solution and the Ewald sphere might not cut through any reciprocal lattice point. In X-ray diffraction (the large curvature case in Fig. 10.5 therefore only a few reflections will be observable. In electron diffraction because of the larger $k$-vector, the Ewald sphere is rather flat, and many reflection spots will occur simultaneously when the scattering vectors match the reciprocal lattice vectors. In two dimensions appropriate for a surface the remaining two equations always have solutions (two variables: magnitude and direction of $\boldsymbol{k}$) and in the Ewald construction the reciprocal lattice points degenerate in one direction (normal to the surface), not specified, to rods which intersect with the Ewald sphere. The large $\boldsymbol{K}$ range of intersection (Fig. 10.6) explains the appearance of streaky features in RHEED and similarly in GIXS.

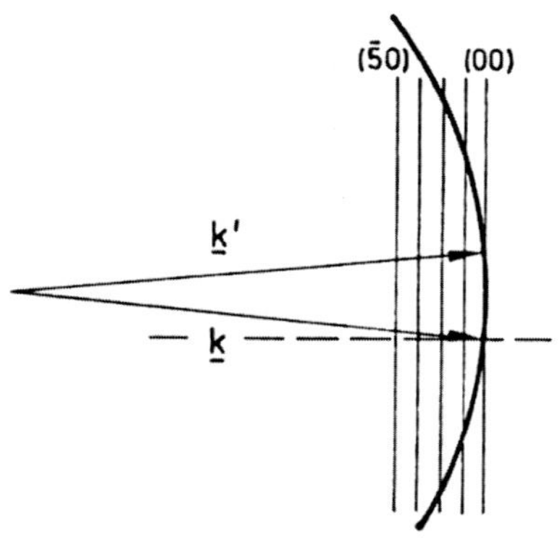


**Fig. 10.6.** Ewald sphere construction for a two-dimensional periodic structure explaining the formation of streaks in RHEED or GIXS. The rods extend in the direction not defined in reciprocal space (taken from [10.11])

### 10.2.2 RHEED

A schematic setup for RHEED is shown in Fig.10.4 which clearly displays also the advantages (full access to the sample) of RHEED for epitaxial growth. The distance between gun and screen may be of the order of 50 cm.

The differences as compared to LEED are slightly more complex electron guns because of the higher electron velocities and also the higher voltages requiring special power supplies and vacuum feedthroughs. But because of the high velocities no extra voltage is required for the screen. The surface sensitivity (a few atomic layers) and the coherence length (the area in which atoms can be considered to be illuminated by a plane wave) are similar to LEED. The main difference occurs in the large $k$-vector, which causes the diameter of the Ewald sphere to be much much larger than the reciprocal lattice constant. The very small curvature of the sphere lets it touch the two-dimensional rods over a large distance and gives rise to the streaks observed in general in RHEED. Figure 10.7 gives an example the GaAs(001)$-\beta2(2\times4)$ reconstruction.

The assurance of the right starting surface for epitaxial growth is one of the main applications of RHEED in MBE. The appearance of the surface periodicity gives evidence for the complete deoxidation of the wafer. Moreover, from measurements of the intensity oscillations the growth rate can be determined (Fig. 1.6). This analysis is possible in the island growth mode (not in the step flow mode) where the morphology oscillates between a two-dimensional nucleated surface and a completed monolayer (Fig. 1.7). RHEED detects these changes since in the highly nucleated surface electrons are scattered out of the regular beam direction and thus cause a loss of intensity in the diffracted spot. Usually this kind of analysis is performed before actual growth takes place but not during growth since the hot filaments in the electron gun might lead to contaminations (carbon) in the epitaxial layer. This of course can be avoided by using instead of RHEED reflectance anisotropy spectroscopy through an optical port of the MBE chamber (Sect. 10.3.2).

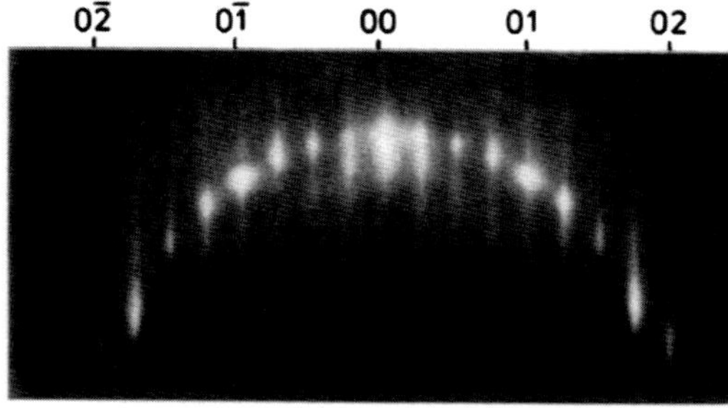


**Fig. 10.7.** RHEED image of a $\beta$2-GaAs(001)(2×4) surface (taken from [10.11])

### 10.2.3 GIXS

Grazing incidence X-ray scattering is actually the method of choice for structural analysis of surfaces in MOVPE. Crucial to the use of this surface sensitive X-ray technique is the availability of high brightness X-ray sources, preferably synchrotrons. Wavelengths in the order of the lattice constant corresponding to photon energies of 10 keV are used. The high brightness of the synchrotron sources, allows not only structural studies but also time resolved analysis in a range of milliseconds. Apparative MBE arrangements for X-ray scattering can be found for example in [10.14]. The more difficult design for a MOVPE reactor for such studies is shown in Fig. 10.8. Besides verifying for the first time the ordered surface structure also in MOVPE growth (Fig. 10.9) morphological features like step separations similar as in RHEED or LEED are observed [10.4]. Monolayer oscillations in the scattering intensity for island growth mode on GaAs(001) have been observed as well, similar to RHEED and RAS [10.16–18]. These oscillations were explained for GaAs(001) to originate from the out-of-phase scattering of successive Ga and As bilayers. As a consequence the bulk contribution besides that of the toplayer cancels in general. The maximum contribution thus corresponds to about one completed monolayer. At half monolayer coverage, however, there is additional compensation between the bilayers in the islands and those of the surrounding terraces and thus the scattering signal is reduced. One of the most interesting possibilities of GIXS is concerned with diffusive scattering

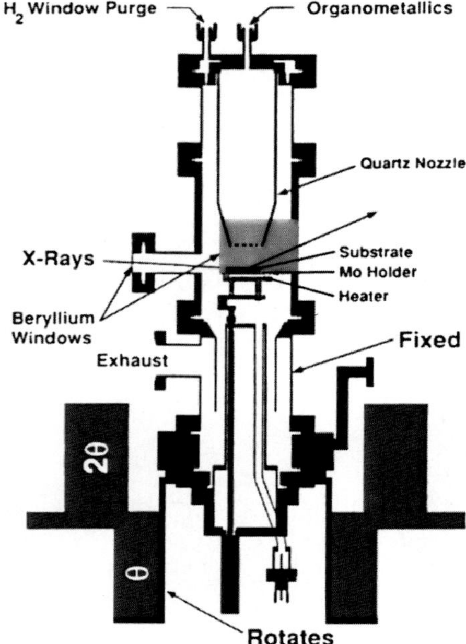


**Fig. 10.8.** MOVPE reactor used for grazing incidence X-ray scattering studies at a synchrotron (taken from [10.15])

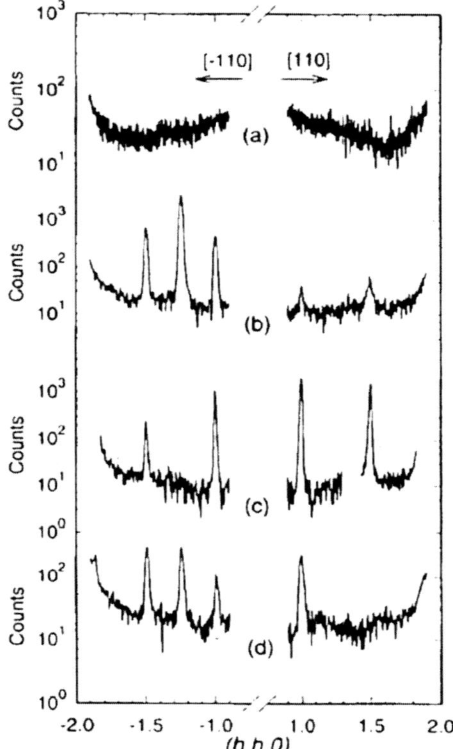


**Fig. 10.9.** Development of surface reconstructions during the initial phases of MOVPE: (**a**) initially there are no reflections observable along either of the [hh0] azimuths, (**b**) after heating in hydrogen (10 min, 850 K) the surface deoxidizes and shows a Ga-rich (4×2) reconstruction, (**c**) annealing in TBAs transforms the surface into a disordered form of the c(4×4) structure, (**d**) dosing with TMGa results in recovery of the (4×2) (taken from [10.4])

[10.19] appearing at wavevectors $\boldsymbol{k} = 0$. This diffusive scattering will have a width inversely proportional to the size of the islands and may be influenced by correlation between the islands. Results have shown larger islands in MOVPE than in MBE indicating a large diffusion length, a fact which is also compatible with the lower transition temperature from island to step flow growth in MOVPE. This has also been confirmed in RAS measurements. While the more basic work in the last decade has been mainly concerned with GaAs surfaces recently also the application to the technologically important GaN(0001) surface in the MOVPE environment has revealed surface reconstruction in the gas phase environment [10.20].

It is clear that time resolved, surface sensitive X-ray scattering techniques can yield a great deal of information of fundamental importance especially to vapor phase crystal growth. Nevertheless it remains a technique which is rarely used. Besides the complexity of epitaxial growth apparatus in combination with a high brightness synchrotron facility the great difficulty of introducing the highly reactive chemicals of VPE into the general user environment of a large scale facility puts a large barrier to VPE experiments at synchrotrons.

## 10.3 Reflectance Based Optical Techniques

Optical probes, especially the linear ones, have been in use as growth monitor for quite some time. For the UHV techniques they deliver additional information with respect to the electron based techniques for gas phase growth they are nearly the only surface analysis tools applicable. The optical techniques moreover possess a number of advantages compared to other surface science tools with respect to growth monitoring: they interact only weakly with the surface and in general do not perturb the growth process. The spectral analysis of optical probes, in addition, gives chemical information (stoichiometry, doping) about the surface. Moreover, optical methods can be made fast and therefore time resolution is in general not difficult to achieve. With the present modulation type optical methods time resolution is limited by the modulation frequencies and is around 1 ms. This allows for studies of desorption/adsorption kinetics or growth dynamics (quantum dots). Since the gas phase generally absorbs radiation in the UV (molecular electronic transitions) and in the IR (molecular vibrations) the visible spectral range is the standard region for optical growth monitoring. The optical techniques are quite simple to install externally as long as there are windows available in the growth equipment. Windows are standard in the stainless steel UHV growth chambers and they are presently about to become standard in VPE equipment too. One should stress also the spectral information available in optical experiments. This is directly related to the material under study (elec-

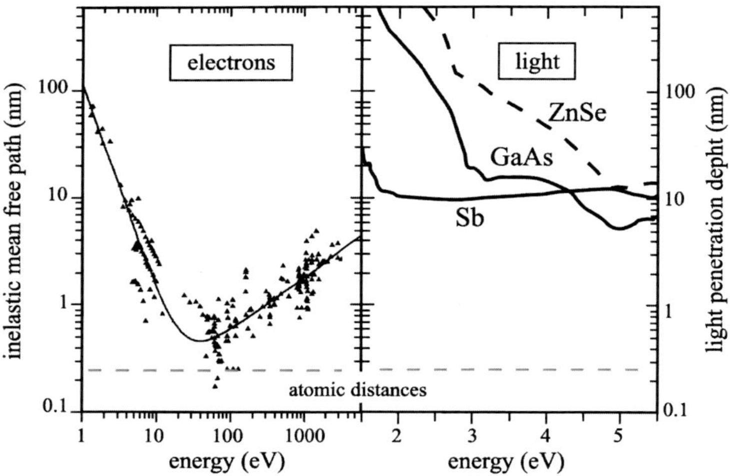


**Fig. 10.10.** Mean free path of electrons (left) and typical optical penetration depth of different solids (right) compared to typical interatomic distances (dashed line at the bottom) (taken from [10.5])

tronic states, phonons) and thus gives chemical information (not available for example in STM) and identifies the material.

Of course there are drawbacks. One of the most severe is the large penetration depth of the light. This is clearly demonstrated by a plot of the penetration depth $d_p$ (inverse of the optical absorption coefficient) of light with photon energies from 2 to 6 eV (Fig. 10.10). For the three representative materials shown the optical penetration depth $d_p$ at its lowest value is around 10 nm (50 monolayers) and is still much larger than the mean free path of low energy electrons (two monolayers). In any ordinary optical measurement, for instance in reflectance, the contribution of the bulk is therefore dominant. The surface contribution to the signal is in general negligible except for the near-ultraviolet range where it may reach percentage levels. Clearly, by penetration depth optics is not a surface sensitive technique.

Nevertheless, there are possibilities, as specified in Fig. 10.11, to obtain surface sensitivity without the need of a small penetration depth. Surface sensitivity has been achieved for example with standard optical methods (Fig. 10.11a) like ellipsometry [10.22], infrared spectroscopy [10.23] and Raman scattering [10.24, 25] (Fig. 10.11a). These methods have been developed today to such a high level of sensitivity and accuracy that small surface signals can be recovered from large signals dominated by contributions from the bulk. Usually this is assisted by differential procedures of changing the surface, which, however, prevent in general real time analysis [10.24, 26]. Exper-

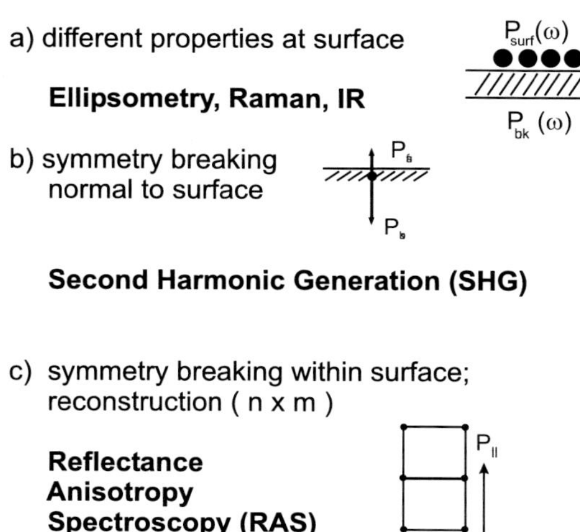


**Fig. 10.11.** Possibilities of obtaining surface sensitivity with optical methods: (**a**) through different physical properties of the surface, (**b**) through breaking of inversion symmetry at the surface and (**c**) through breaking of symmetry within the surface (**c**) (taken from [10.21])

imental methods falling into this category are surface differential reflectivity (SDR) [10.26] or surface photo absorption (SPA) [10.27], a technique named also p-polarized reflectance (PRS) [10.25]

The reduction of symmetry at surfaces, however, can be exploited (Fig. 10.11b,c) to concentrate the signal generation to the surface. This is done by either non-linear techniques using two-frequency mixing (commonly second harmonic generation: SHG) or by the linear reflectance anisotropy spectroscopy (RAS). If these methods are applied in high symmetry (cubic, isotropic) materials the bulk does not or only marginally contributes to the total signal. Therefore these methods may be called true surface methods in the sense of classical surface science. Fortunately, the high symmetry materials constitute a large part of the materials being of scientific or technological interest and therefore these methods can be applied in many studies of growth.

The application of nonlinear techniques in growth is still rare [10.6, 7] and, in addition, spectral capabilities are not available everywhere. Therefore, we will concentrate here on the linear techniques SE and RAS exploiting the polarization status of light. The focus will be on RAS which has been applied in the last decade in many growth processes under very realistic conditions.

Besides these surface sensitive techniques often simple reflectance measurements are useful. They can be obtained sometimes more or less as a by-product of these methods, but also in separate measurements (reflectometry). Thereby, additional information like film thickness (through Fabry–Perot interferences [10.28]), morphology (via elastic light scattering [10.29, 30]), temperature (via relative values of reflectance [10.31]) or doping [10.32] is obtained.

### 10.3.1 Reflectance of Polarized Light

The general scenario for the reflection of polarized light is sketched in Fig. 10.12 The dielectric function $\varepsilon_s$ of the surface (of thickness $d$ in the order of a few lattice constants and much smaller than the light wavelength

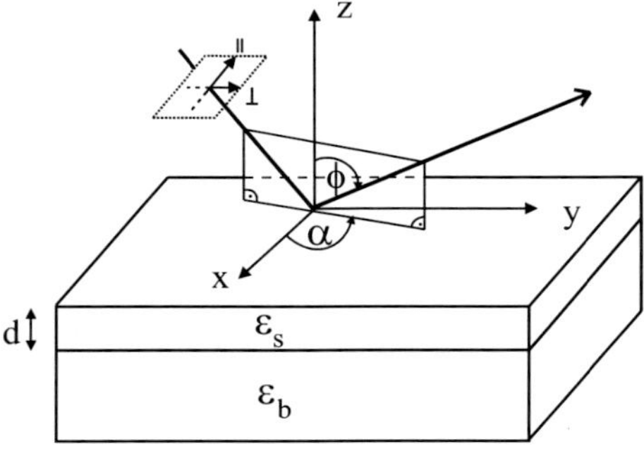


**Fig. 10.12.** Schematic sketch of a surface reflectance experiment with polarized light (taken from [10.33])

$\lambda$) is assumed to be anisotropic while the bulk is described by an isotropic dielectric function $\varepsilon_b$:

$$\varepsilon_s = \begin{pmatrix} \varepsilon_{xx} & 0 & 0 \\ 0 & \varepsilon_{yy} & 0 \\ 0 & 0 & \varepsilon_{zz} \end{pmatrix} \qquad \varepsilon_b = \begin{pmatrix} \varepsilon_b & 0 & 0 \\ 0 & \varepsilon_b & 0 \\ 0 & 0 & \varepsilon_b \end{pmatrix} \tag{10.5}$$

where $x$, $y$ are the anisotropic eigenvectors of the surface and $z$ is the surface normal.

Different experimental techniques are practiced in order to become sensitive to the surface dielectric function. These are sketched in Fig. 10.13. Reflectance anisotropy spectroscopy measures under normal incidence the reflectance for two polarizations of incident light (Fig. 10.13a), thus becoming sensitive to the anisotropic dielectric surface contribution. Surface photoabsorption (SPA or PRS) obtains sensitivity to small changes in the surface dielectric properties by measuring reflectance in p-polarization near the reflectance minimum at the Brewster angle (Fig. 10.13b). The general case of ellipsometry measures the ratio in reflectance of s- and p-polarized light also near the reflectance minimum at the Brewster angle (Fig. 10.13c). By rotation of the sample the anisotropic surface part can be determined too.

For normal incidence ($\phi = 0$) one obtains the result for the complex reflectance anisotropy with $\Delta r = r_{xx} - r_{yy}$ [10.34]:

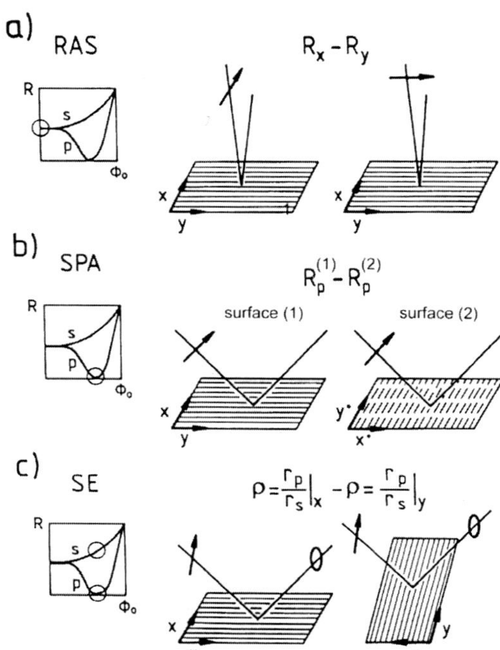


**Fig. 10.13.** Schematic illustration of surface sensitive linear optical methods utilizing polarized light (taken from [10.21])

$$\frac{\Delta\tilde{r}}{\tilde{r}}|_{RAS} = \frac{\Delta r}{r} + i\Delta\Theta = \frac{4\pi i d n_a(\varepsilon_{xx} - \varepsilon_{yy})}{\lambda(\varepsilon_b - \varepsilon_a)} \tag{10.6}$$

where $\varepsilon_a$, $n_a$ denote the dielectric function and the refractive index of the ambient.

For interpretation and comparison to theory it is especially useful (but very seldom performed) to extract the surface dielectric anisotropy $(\varepsilon_{xx}-\varepsilon_{yy})$ from this expression because this is the material property of the surface layer. The reflectance on the other hand is geometry dependent and secondly contains the bulk dielectric function $\varepsilon_b$ which introduces an additional spectral dependence. In order to extract the desired surface dielectric anisotropy (SDA) $\Delta\varepsilon = \varepsilon_{xx} - \varepsilon_{yy}$ one needs of course the dielectric function of the bulk $\varepsilon_b$, which can be obtained from an ellipsometric measurement at oblique incidence. The measured quantity in ellipsometry is the complex ratio $\rho$ of the reflection coefficients $r_p$, $r_s$ defined by

$$\rho_{SE} = \frac{r_p}{r_s} = \tan\Psi e^{i\Delta}. \tag{10.7}$$

It is related to the dielectric functions and the other experimental parameter as given in the following equation [10.34]

$$\begin{aligned} \rho_{SE} = \rho^0\{1 + &\frac{4\pi i d n_a \cos\phi}{\lambda(\varepsilon_b - \varepsilon_a)(\varepsilon_b\cos^2\phi - \varepsilon_a\sin^2\phi)} \\ &\cdot\ [(\varepsilon_b - \bar{\varepsilon})(\varepsilon_b + \varepsilon_a)\cdot\sin^2\phi - (\frac{\varepsilon_b^2}{\varepsilon_{zz}} - \bar{\varepsilon})\varepsilon_a\sin^2\phi \\ &+ (\varepsilon_{xx} - \varepsilon_{yy})((\varepsilon_b(1+\cos^2\phi) - 2\varepsilon_a\sin^2\phi)\cos(2\alpha) \\ &+ 2\frac{(\varepsilon_b - \sin^2\phi)^{1/2}}{n_a}((\varepsilon_b - \sin^2\phi)^{1/2} \\ &\cdot\ n_a\cos\phi csc(2P) - \varepsilon_a\sin^2\phi\cot(2P))\sin(2\alpha))]\}\,, \end{aligned} \tag{10.8}$$

where $\rho^0$ is the ratio of the reflection coefficients for the bare substrate, $\bar{\varepsilon} = (\varepsilon_{xx}+\varepsilon_{yy})/2$ the isotropic part of the surface dielectric function and $P$ is the polarizer angle. The other quantities are defined in Fig. 10.12 and (10.5). This lengthy equation is first of all quoted here to indicate the more complex experimental nature (exact knowledge of the angles) of an ellipsometric measurement compared to the simpler normal incidence ($\phi = 0$) RAS technique. Nevertheless, (10.8) also contains $(\varepsilon_{xx} - \varepsilon_{yy})$ and thus can be utilized to measure the surface anisotropy. This has been exploited for the observation of monolayer growth oscillations [10.35, 36] or for the determination of the isotropic surface dielectric function [10.37, 38]. As can be seen from (10.8) the coefficient of $(\varepsilon_{xx} - \varepsilon_{yy})$ varies with experimental geometry ($\alpha$) and can be also made zero.

The change in p-polarized reflectance induced by a change of the surface (reconstruction or adsorbate) is given similarly by

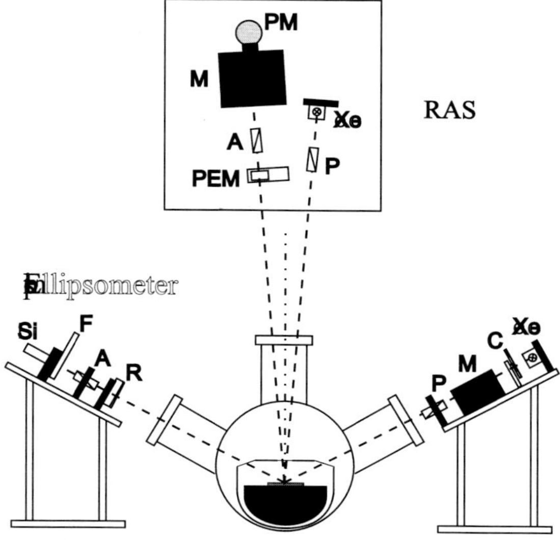


**Fig. 10.14.** Cross-section (in flow direction) through a MOVPE reactor with SE and RAS attached (taken from [10.21])

$$\frac{r_p^{(2)} - r_p^{(1)}}{r_{pp}}|_{SPA} = \frac{4\pi i d n_a}{\lambda(\varepsilon_b - \varepsilon_a)} \cdot \frac{(\varepsilon_{yy}^{(2)} - \varepsilon_{yy}^{(1)})(1 - \frac{\varepsilon_a}{\varepsilon_b}\sin^2\phi) + (\frac{1}{\varepsilon_{zz}^{(2)}} - \frac{1}{\varepsilon_{zz}^{(1)}})\varepsilon_a\varepsilon_b\sin^2\phi}{(\cos^2\phi - \frac{\varepsilon_a}{\varepsilon_b}\sin^2\phi)}, \tag{10.9}$$

where the indices (1) and (2) indicate the different surfaces. $\varepsilon_{zz}$ appears in the expressions for non normal incidence (see also (10.8)) since the normal component of the light electric field is reduced by $1/\varepsilon_{zz}$ by the induced surface charges. For practical purposes these terms are therefore in general neglected. For a more general description of the methods the reader is referred to [10.5, 34, 39, 40] where also experimental details are discussed.

Typical set-ups combining the growth apparatus and the optical equipment are shown in Fig. 10.14 for a MOVPE reactor and in Fig. 10.15 for a MBE apparatus. The epitaxial reactors have to be fitted with three windows: one for normal incidence reflectance and two allowing for reflectance at angle $\phi$ around 70 °. They have to be of low strain in order not to influence the polarization status of the light, a fact which often complicates the analysis considerably or even makes it impossible. Experimentally one has to deal also with rotating substrates. A simple rotation usually simplifies the analysis because background polarization, e.g., by strained windows can be easily eliminated because the rotation modulates the electronic signal [10.41]. However, the rotation also introduces new problems through wobbling of the rotational axis. This has to be taken care of either mechanically in the rotation construction or with additional optics (anti-wobble mirror). With multi wafer planetary motion, found especially in commercial reactors, however,

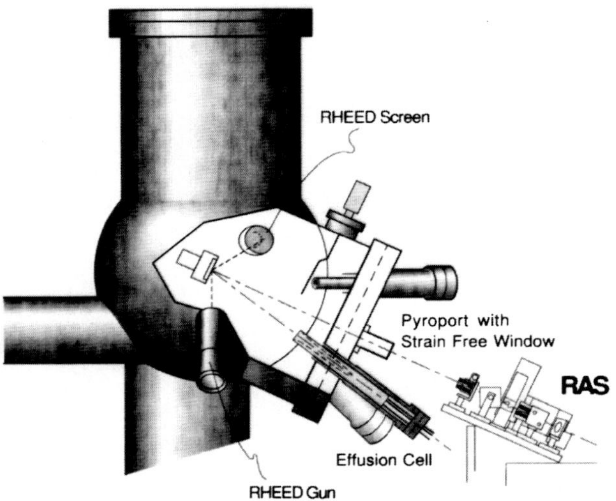


**Fig. 10.15.** MBE equipment with RAS apparatus attached (taken from [10.21])

the signal extraction becomes considerably more complicated since in addition the substrates and their orientations have to be recognized. However, this technical problem has been solved [10.42, 43].

### 10.3.2 Reflectance Anisotropy Spectroscopy (RAS)

Reflectance anisotropies at semiconductor surfaces were already measured in 1966 by Cardona et al. by rotating a sample (Si(110)) around the surface normal, a technique which they called rotoreflectance [10.44]. The RAS measurement is basically an ellipsometric measurement performed just at near normal incidence (Fig. 10.13). Therefore, names referring to the experimental arrangement like perpendicular incidence ellipsometry (PIE) or Normal Incidence Ellipsometry (NIE) have also been used for this technique [10.45, 46]. The principal experimental setup described here (Fig. 10.16) and which today is nearly used everywhere, was basically developed by Aspnes and coworkers [10.47, 48]. They were also the first to recognize the potential of anisotropic reflectance with respect to surface science and epitaxial growth. The success of their setup is essentially founded in the use of the photoelastic modulator, which avoids mechanical rotations (as in ellipsometry with the rotating analyzer) and allows a high modulation frequency (50 kHz).

The incident light is polarized midway between the two eigenvectors of the surface. In the absence of any anisotropy the linear polarization is then restored in the reflected beam, which is then linearly polarized along the principal axis of the photoelastic modulator. As a result the output is not modulated by the PEM and there is no modulated RAS signal, just a DC signal corresponding to $<r>$. In case the reflectance is different along the principal axes it will cause a change in amplitude and/or phase and a phase modulation will result behind the PEM. This is converted to an intensity modulation by the analyzer. The setup thus works as an optical null bridge. By using the Jones matrix formalism one obtains in lowest order for the ratio between the AC and DC signal at the detector [10.47]

$$\frac{\Delta I(t)}{I} = 2\left[\Im m\left(\frac{\Delta r}{r}\right)\right] \cdot J_1(\delta_c) \cdot \sin\omega t + 2\left[\Re e\left(\frac{\Delta r}{r}\right)\right] \cdot J_2(\delta_c) \cdot \cos 2\omega t \quad (10.10)$$

where $\omega$ is the modulation frequency of the PEM, $\delta_c$ is the retardation due to the modulator and $J_1$ and $J_2$ are Bessel functions. The correction terms due to non ideal polarization and due to window strain have been neglected here. The latter influence most seriously the phase and thus appear dominantly in the imaginary part. By measuring in two sample positions (rotation) it can be eliminated.

The schematic set-up in Fig. 10.16 allows the real as well as the imaginary part of the RAS signal to be evaluated by detecting according to (10.10) the signal at $2\omega$ and $\omega$ respectively. This is a very useful feature since it allows us to do a Kramers-Kronig consistency check on the data. This setup or very similar versions are now in use in several research laboratories around the world and are also commercially available. A detailed discussion and a comparison of this configuration with those using rotating samples or analyzers is given in [10.39, 47].

A typical RAS spectrum can be recorded in a few minutes limited by the time to scan the monochromator and additionally optimizing the phase shift of the PEM at every wavelength. Thus, at present real-time monitoring of

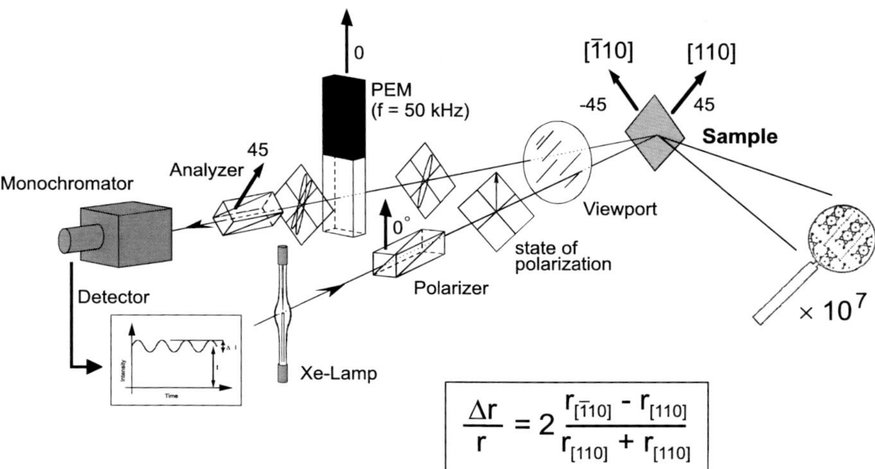


**Fig. 10.16.** Diagram of a RAS setup utilizing a photo-elastic modulator (PEM)

growth is hardly feasible utilizing the full spectral range. This problem of course can be overcome by the use of optical multichannel analyzers (CCD). Presently, the detection of changes in the RAS signal at a fixed photon energy can be performed within 100 ms or less. From such transients taken at different photon energies RAS spectra can be assembled and photon energies identified at which maximum changes occur during growth and which are suited for growth control.

The technique can be further improved [10.50] by using different sample alignments (i.e., angles unequal to 45°) in order to eliminate the offsets. If possible in the experimental apparatus, an elegant way of doing this is by introducing an additional modulation through a slow rotation of the sample at approximately 0.1 Hz. These approaches allow systematic errors in the determination of the RAS signal to be eliminated. Accurate values of $\Delta r/r$ may then be converted into the surface dielectric anisotropy $\Delta(\epsilon \cdot d)$ with $d$ being the thickness of the surface layer using [10.50]

A more complicated situation arises in the analysis of RAS if layer systems are under consideration (Fig. 10.17). In this case multiple reflections have to be taken into account. However, in the case of RAS they have to be calculated separately for the two polarizations along the surface eigenvectors. Anisotropic properties may occur in such a multilayer system not only at the surface but also at all interfaces:

$$\frac{\Delta r}{r} = C_1 \cdot \frac{\Delta r_{\mathrm{SF}}}{r_{\mathrm{SF}}} + C_2 \cdot \frac{\Delta r_{\mathrm{IF}}}{r_{\mathrm{IF}}} + C_2 \cdot \Delta n_o \frac{4\pi i d_o}{\lambda} \qquad (10.11)$$

with the constants

$$C_1 = \frac{r_{\mathrm{SF}}(1 - Z r_{\mathrm{IF}})(1 + Z r_{\mathrm{IF}})}{(1 + Z r_{\mathrm{IF}} r_{\mathrm{SF}})(Z r_{\mathrm{IF}} + r_{\mathrm{SF}})},$$

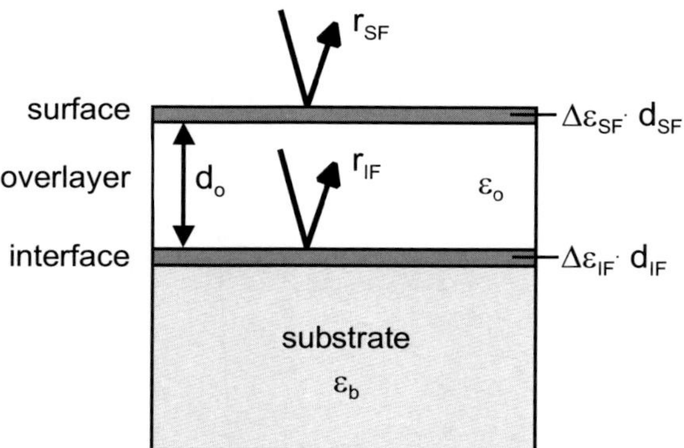


**Fig. 10.17.** Schematic sketch of an anisotropic overlayer on a substrate with an anisotropic interface. The index -"o" marks the properties of the overlayer (taken from [10.52])

$$C_2 = \frac{Zr_{\mathrm{IF}}(1 - r_{\mathrm{SF}})(1 + r_{\mathrm{SF}})}{(1 + Zr_{\mathrm{if}}r_{\mathrm{SF}})(Zr_{\mathrm{IF}} + r_{\mathrm{SF}})} \tag{10.12}$$

and

$$Z = e^{\frac{4\pi i}{\lambda} n_o d_o} . \tag{10.13}$$

III–V-semiconductor systems consisting of two layers (InGaAs/InP [10.51] and AlAs/GaAs [10.52]) have been studied. Anisotropies at the interface as well as on the surface were found. In the case of a Si/$SiO_2$ layer system in contrast Yasuda et al. found that only anisotropy at the Si/$SiO_2$ interface has to be considered [10.53]. The analysis requires in any case RAS spectra taken at several layer thicknesses. This of course can be done easily during real-time studies. In ex-situ studies a number of samples with different layer thicknesses should be available.

**Experimental Results.** Most experiments and as well theory up to now have been performed in III–V-semiconductor growth. Examples from this area will be discussed. However, studies on II–VI compounds [10.54–56], group IV semiconductors [10.57] and metals [10.58] have been performed as well.

The main success of RAS originates from the fact that typical spectra correlate with certain surface reconstructions and secondly that under standard growth conditions the spectra appear identically in MBE and MOVPE. Thus the diffraction techniques (RHEED) in MBE can be utilized to define the RAS spectra in terms of surface reconstructions. Figure 10.18 gives an examples of GaAs growth where spectra taken in MOVPE and MBE growth equipment under group V element stabilization at different temperatures are shown [10.5]. At the higher temperatures minor differences between MBE and MOVPE are of the same order as the possible wafer to wafer variations. At lower temperatures in MOVPE differences might occur since the decomposition of precursors is no longer complete and in addition to the III and V elements other molecular groups may adsorb on the surface. Surface reconstructions not even possible in MBE may be obtained. These special features can be exploited for example in intrinsic carbon doping at low V/III ratios with TMGa [10.32].

This equivalence at standard conditions between RAS spectra in MOVPE and UHV-based growth techniques allows us then to assign the surface reconstruction found with RHEED or LEED in the latter also to the corresponding MOVPE surface with the same RAS spectrum. This was first done for GaAs [10.59]. That this procedure seems to be correct has been verified also by the application of GIXS in MOVPE in a few cases [10.4, 18, 60]. From such studies correlations between surfaces and RAS spectra have been established like the one for GaAs shown in Fig 10.19.

Thus RAS can fulfill in MOVPE and MBE the same function as RHEED has, namely to identify and define the starting surface. This includes deoxidation of the substrate, and setting temperature and group V-partial pressure.

For practical growth purposes we show here the very important task of wafer deoxidation with the example of InP(001).

There are some advantages using RAS as a RHEED replacement in UHV growth system. The RAS can be added externally without breaking the vacuum and moreover since no electron gun is involved there is not the problem of contamination, which is the reason for switching off the RHEED in serious growth experiments.

There are other advantages. One is related to the fact that the surface area needed to create a typical RAS spectrum is much smaller (a few unit cells) [10.61] than the one needed for a diffraction pattern (100 unit cells). This appears because optics is more sensitive to near range order than the long range measured in a diffraction experiment. This manifests itself in the speed in which RAS and RHEED react to changes in surface structure, RAS being much faster than RHEED which needs the time to establish the long range order. In simultaneous RAS and RHEED experiments like the ones displayed in Fig. 10.18 RHEED often gives a diffuse pattern since the size of the corresponding domains becomes too small for generating a clear diffraction pattern. This is the case for example in the range between the c(4×4) and the $\beta2(2\times4)$ while RAS shows well defined spectra. In fact these spectra

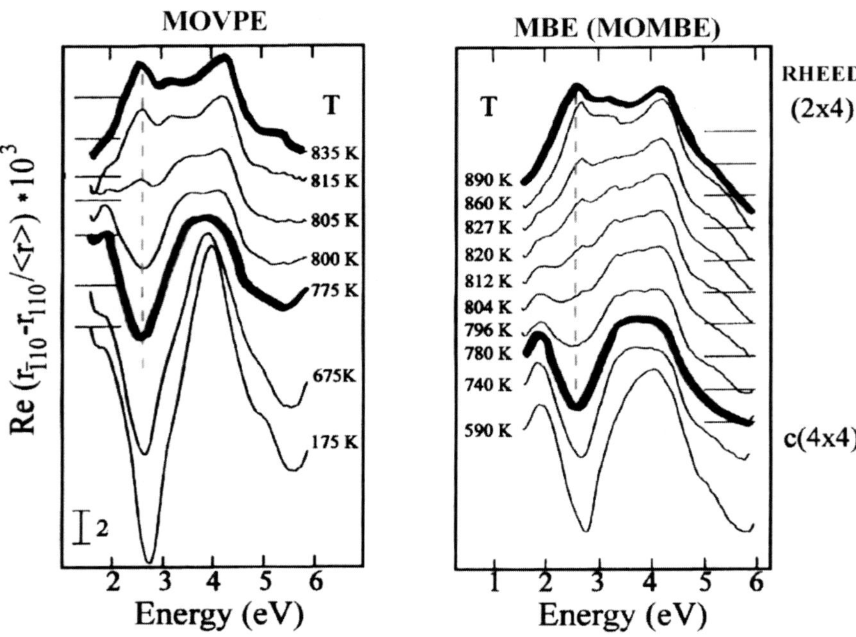


**Fig. 10.18.** RAS spectra from GaAs(001) surfaces at different temperatures under group V element stabilization in MOVPE and MBE. The marked curves represent RAS spectra for the (2×4) and c(4×4) reconstruction (taken from [10.33])

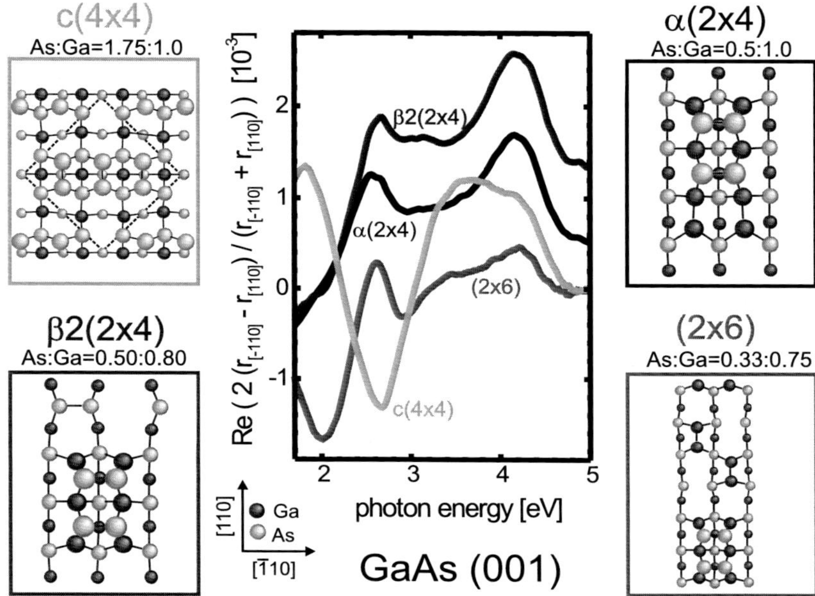


**Fig. 10.19.** RAS spectra of different GaAs(001) surface reconstructions (taken from [10.21])

can be described by a linear combination of the two marked spectra ((2×4), c(4×4)), and thus serve to define the intermediate surfaces quantitatively in terms of the two reconstructions present. Figure 10.20 shows two such fits which give 40 % and 80 % c(4×4) contribution, respectively to the surface coverage [10.62], a change occurring with a 5 K substrate temperature variation! Although the GaAs(001) surface has been discussed here mostly as the best studied example, other semiconductors and metals have been under investigation as well. RAS result, have been reported for example for InP(001), GaP(001), GaSb(001), InAs(001) and many non-(001)-GaAs surfaces (see [10.21]). Similarly but not so intensively II–VI semiconductors have been studied [10.55]. All surfaces give characteristic features and also similarity between surfaces in MBE and MOVPE seem to hold. (111)-surfaces in contrast by their threefold symmetry should not give any anisotropic signal. Si(001), on the other hand, as one of the most important semiconductor surfaces, cannot be well characterized by RAS since it usually forms monosteps and thus the contribution of $(n \times m)$ reconstructions on one half of the terraces is always canceled or at least strongly compensated by $(m \times n)$ reconstructions on the other terraces [10.63]. While the application of RAS was first in semiconductors, in the last few years also metals have been studied (see [10.58]).

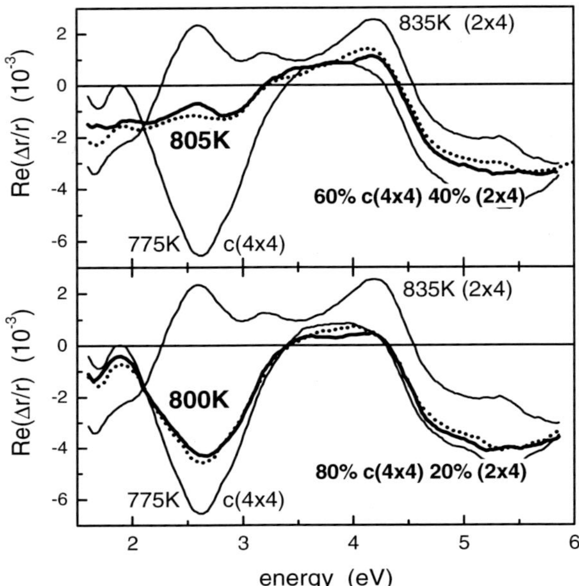


**Fig. 10.20.** Spectra from MOVPE surfaces between the (2×4) and c(4×4) reconstructions of Fig. 10.19 at 800 K and 805 K. They can be represented by linear combinations of (2×4) and c(4×4) with percentages as given in the top of the figures (taken from [10.21])

Besides having established, by empirical correlation, RAS as a surface science tool for cubic solids, the next question is of course: do we understand the spectra? The answer is to a certain extent yes; however, many features are still not understood. The theoretical problem is that simple models just based on surface dimers do not work. Tight binding calculations have given a first satisfactory description of RAS spectra for GaAs [10.64]. Similarly, the first use of *ab initio* theory in the GW approach has shown promising results for the GaAs(110) surface [10.65]. Very recent application of an approximated GW method (GWA) [10.66] has been quite successful in describing the low temperature RAS spectra of InP(001)-(2×4) [10.67]. However, the more complex GW *ab initio* approach together with the large number of atoms in the surface supercells is presently at the computational limits. Thus, not many results have been produced up to now. Thus one is far away from a situation were one could predict the response or model spectra. The application of RAS in epitaxial growth analysis, therefore, still relies on the experimentally established relations between reconstructions and spectral features in the RAS spectra.

The large amount of fundamental scientific knowledge available already for the III–V semiconductors has pushed, on the other hand, the introduction

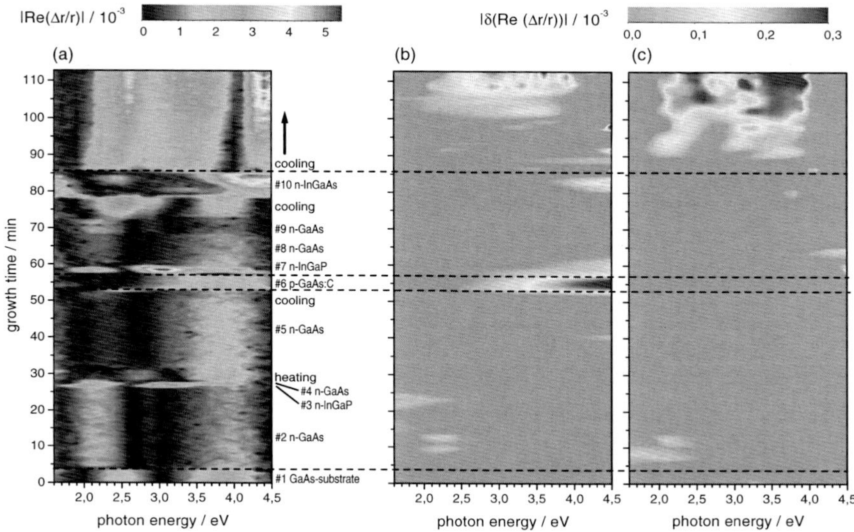


**Fig. 10.21.** RAS spectra (a) taken in real time during growth of a layer structure for a heterobipolar transistor (HBT) (b) and (c) are difference RAS spectra to (a) with a different doping (b) and differences in cooling (b, c) (taken from [10.70])

of RAS into technological applications. HBTs (hetero bipolar transistors) and VCSELs (vertically surface emitting laser) have been successfully grown under optical control [10.68, 69]. We show as an example the growth of the layer structure for a hetero bipolar transistor (HBT) in Fig. 10.21. Clearly the different layers can be recognized (also different doping) and can be utilized for example to control deviations from a reference HBT in series production. Moreover, the first commercial RAS equipment interfaced with a commercial epitaxial MOVPE reactor is now on the market. The goal there is of course not just monitoring but also to have in the future automatic control at least for some steps of growth. For that purpose the optical signals will be fed back to the flow controllers for the precursor.

### 10.3.3 Ellipsometry

In ellipsometry we determine the average optical response (Fig. 10.22) of a material over the penetration depth of the light. This general response is described by the ellipsometric angles (see (10.7)). For a semi infinite, homogeneous material one can extract the dielectric function from those ellipsometric angles in terms of a two-phase model (bulk material with dielectric function $\varepsilon$ and vacuum) as

$$\varepsilon = \sin^2\phi\left[1+\left(\frac{1-\rho}{1+\rho}\right)^2\tan^2\phi\right]. \tag{10.14}$$

However, since surfaces have different optical properties a two-phase model is never correct. Nevertheless, the evaluation with (10.14) is very simple and the description with a dielectric function of a homogeneous bulk is quite often a very good approximation that (10.14) is used. One designates in this case the result as the pseudo- or effective dielectric constant $<\varepsilon>$ and writes then for a non homogeneous bulk

$$<\varepsilon> = \sin^2\phi\left[1+\left(\frac{1-\rho}{1+\rho}\right)^2\tan^2\phi\right]. \tag{10.15}$$

This is often quite useful since the characteristic spectral features from electronic interband transitions can still be recognized and used for a first interpretation (direct inspection method [10.71]). In case the deviation from a homogeneous bulk is larger than just a thin surface layer (multilayer structures) of a non-semi-infinite bulk these effective dielectric functions may also contain geometrical contributions, e.g., interference from the coherent superposition of partially reflected waves. The description with an effective dielectric function is then less useful. If the geometrical structure is known (layer thicknesses in a multilayer structure) as well as first approximations of the dielectric function, optical layer models (multi phase models) can be utilized to simulate in a forward calculation the optical response of the structure. Such models are based on the superposition of electromagnetic coherent plane waves [10.72] and are included in the software packages of today standard commercial ellipsometers. They also include the effective medium (Maxwell-Garnet [10.73], Brüggeman [10.74], Looyenga [10.75] or Bergman [10.76]) modeling of materials (layers) with an inhomogeneous two-phase mixture for situations like porous silicon (vacuum and crystalline silicon) or interface / surface roughness. These models are also applied then in general

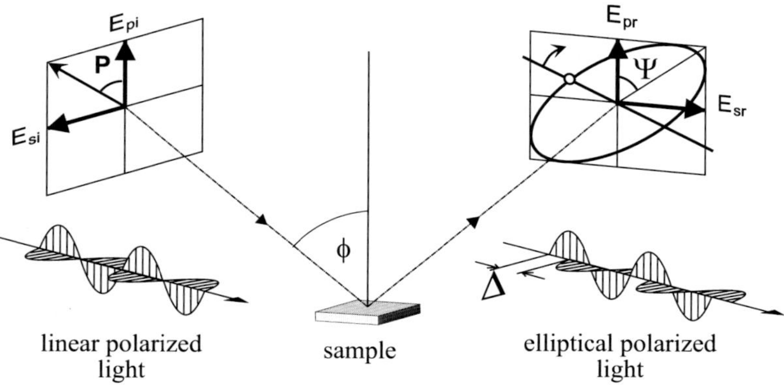


**Fig. 10.22.** Principle function of an ellipsometer

to simulate surface or interface roughness. A schematic setup for ellipsometry is shown in Fig. 10.23

The importance of ellipsometry for growth monitoring is especially related to situations where RAS cannot be applied, that is optical isotropically reconstructed surfaces like often the (111) surfaces of cubic materials or equivalently (0001) surfaces of hexagonal crystals. Moreover, for the deposition of non crystalline layers SE is the monitoring device of choice. The apparative realization requires low aperture beams in order to have well defined angles of incidence ($\phi$ in Fig. 10.22) and therefore the illumination as well as the detection part may be rather long or in case the long beams are folded, rather

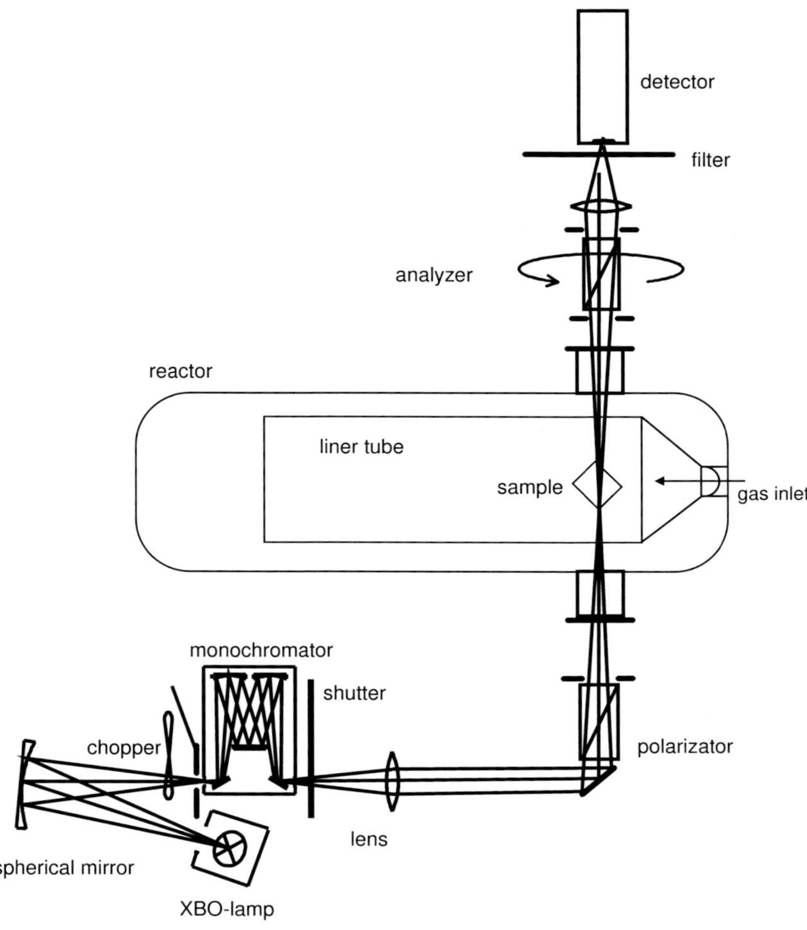


**Fig. 10.23.** Optical layout and set-up of a spectroscopic ellipsometer with the example of a MOVPE reactor (taken from [10.77])

bulky. The polarization status is measured either with phase modulation like in RAS or with a rotating analyzer. The former one is somewhat faster than the latter; however, a higher accuracy is reached more easily in the latter. The rotating analyzer version is realized in the majority of ellipsometers. Depending on the optical constants one has to measure it can also be useful to add a phase shifter (retarder) in order to increase the accuracy of measurement. A detailed discussion can be found in [10.39].

Since ellipsometry is a technique more than 100 years old there exist countless reports on surface and adsorption studies. With the advent of small computers ellipsometry also became capable of growth monitoring and one of the first reports was concerned with growth of AlGaAs heterostructures [10.78, 79]. A single wavelength helium neon laser was used for faster monitoring. The by now classical result is shown in Fig. 10.24. Monitoring over a wider spectral range is desirable for interpretation but requires at least the time needed to scan the monochromator similar as in standard RAS equipment. However, also fast multichannel ellipsometers are available today and spectral real time studies should be in principle no problem.

The dependence of the dielectric function on stoichiometry allows for monitoring and control in ternary or quaternary compounds. For that purpose the ellipsometric signal is utilized to control the mass flow controllers and thus exercise closed loop control. Examples are the growth of a parabolic quantum well [10.80] and the growth of lattice matched InGaP/InP heterostructures [10.81].

The growth of (0001)-GaN (and other nitrides) represents a very important case for ellipsometry where RAS cannot be applied because to the

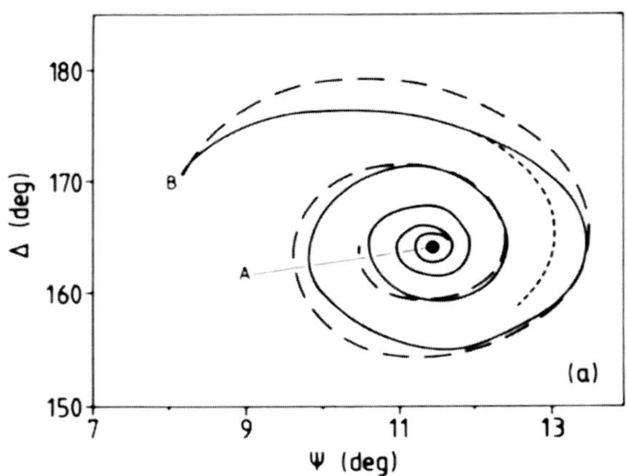


**Fig. 10.24.** Real-time examination of AlGaAs –GaAs heteroepitaxy (solid line). The long dashed line is the $(\Delta, \Psi)$ locus calculated for an abrupt interface. The short dashed line assumes a 29 nm linear transition region (taken from [10.78])

present knowledge the surface reconstructs optically isotropic. Recent work with ellipsometers extending also into the VUV because of the high bandgap of GaN have shown successfully that growth and especially the important nucleation layer growth process can be analyzed [10.82, 83]. Similarly ellipsometry helped to establish growth procedures for MOVPE of InN [10.84].

One important application of ellipsometry concerns morphology related effects. Roughness effects on the ellipsometric spectra are well known and effective medium approximations are used in many applications to quantify the surface or interface roughness. For most epitaxial growth situations today directed to the growth of atomically flat layers roughness is not a major topic. However, for the growth of nanostructures morphological effects are desired and it has become important to create substrates with large steps by step bunching or to utilize the Stranski–Krastanow growth modus to generate quantum dots. In such situations ellipsometry can give complementary information to RAS which mainly detects the material specific surface reconstructions. Figure 10.25 shows SE and RAS spectra taken during the growth of InAs quantum dots on GaAs. In the submonolayer deposition regime RAS exhibits an immediate strong response (the surface reconstruction from GaAs(001)-c(4×4) towards InAs(001)-(2×4) [10.86, 87]. SE in contrast hardly changes (see insert): the effective dielectric function is not much affected by a monolayer (ML) of a material with similar dielectric contribution. However, after the formation of quantum dots around 2 ML a 20 % decrease of the signal at 4.5 eV is observed. Now the gas phase with

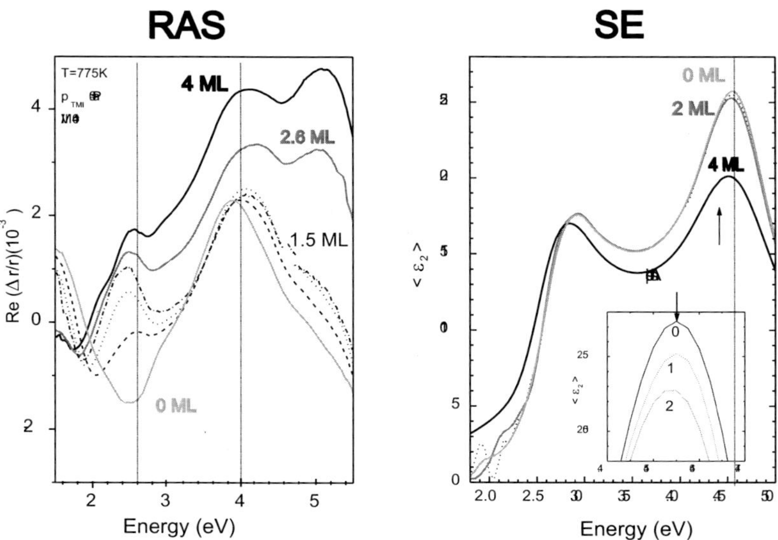


**Fig. 10.25.** RAS and SE spectra for deposition of up to 4 InAs monolayers on GaAs(001)-c(4×4) (taken from [10.85])

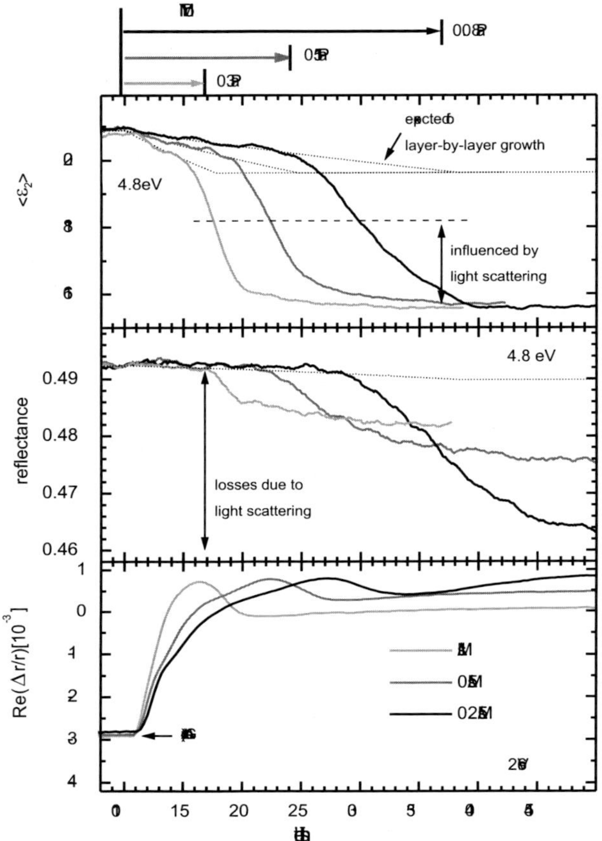


**Fig. 10.26.** Time dependence of effective dielectric function (top), reflectance (middle) and RAS (bottom) during deposition of 4 ML InAs on GaAs(001) with three different growth rates (taken from [10.85])

dielectric constant of approximately 1 contributes a major decrease to the effective polarization. The transition point from 2d to 3d growth determined by ellipsometry from the transients in Fig. 10.26 agrees excellently with that determined by RHEED as was shown in simultaneous measurements [10.88]. The losses due to light scattering indicated in that figure are a measure of the QD size: the larger the losses the larger are the QD caused in this example by the low growth rate.

It should be mentioned that in case of anisotropic surface reconstructions the anisotropic part of the surface dielectric function of course can also be seen by ellipsometry (10.8) and indeed monolayer oscillations have been measured [10.35, 36].

### 10.3.4 P-polarized Reflectance Spectroscopy (PRS) Surface Photoabsorption (SPA)

SPA or PRS are essentially the same techniques [10.27, 89]. In these methods the p-polarized reflectance is measured at or near the Brewster angle. The sensitivity originates from a modulation of the surface structure similarly as in differential reflectivity spectroscopy (DRS) [10.90] but under oblique angles near the Brewster case. This method was developed by Kobayashi and Horikoshi who coined the more common name surface photoabsorption (SPA) [10.91]. Its main application lies in non stationary growth techniques like atomic layer epitaxy.

In the SPA experiment the light hits the sample surface at a very shallow angle of typically 70° with respect to the surface normal. For most semiconductors this is close to the Brewster angle at which the bulk contribution to the reflected intensity is minimal (a few percent) for light polarized parallel to the plane of incidence. One may therefore expect that any modification of the surface which alters the surface reflectivity will reveal itself most markedly in a change of reflectance for p-polarized light.

The experimental setup is sketched in Fig. 10.27. The SPA technique has been applied to growth methods like atomic layer epitaxy (ALE) which alternately supply the group III and V elements in quantities which allow one atomic layer to be formed upon deposition. Such an approach is also feasible in MBE, e.g., by supplying elemental Ga and $As_4$ molecules alternately as well as in MOVPE when the constituents are supplied in their gaseous form as, e.g., TMGa and $AsH_3$. The former method is also called migration-enhanced epitaxy (MEE) while the latter in a more general sense has also been named flow-rate modulation epitaxy (FME). According to this growth

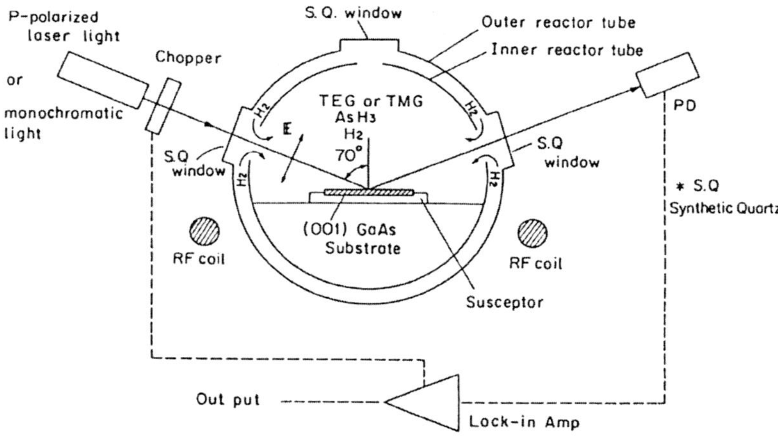


**Fig. 10.27.** Schematic diagram of a possible SPA experimental setup (taken from [10.92])

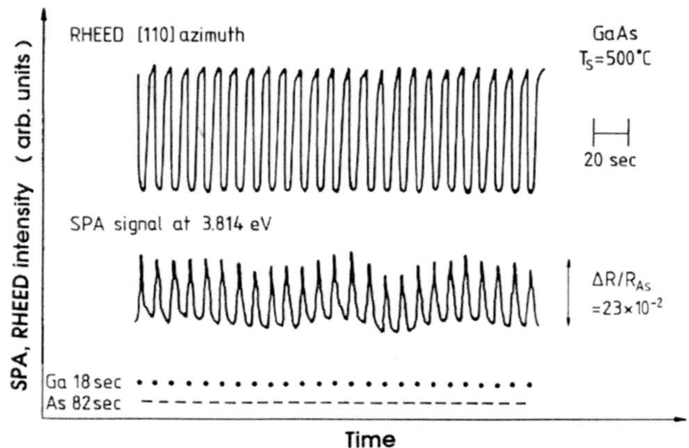


**Fig. 10.28.** RHEED and SPA intensities for MEE growth of GaAs. Ga and As MBE cell shutters are opened alternately at 1.8 s and 8.2 s, respectively. 1.8 s corresponds to a number of Ga atoms equal to the number of surface sites. A He-Cd laser ($h \cdot \nu = 3.814\,\mathrm{eV}$) was used. The azimuth of incidence is [110] for both techniques (taken from [10.91])

procedure the surface is alternately terminated by either Ga or As, respectively. Consequently, the SPA signal may be defined as:

$$SPA = \frac{R_{\mathrm{Ga}} - R_{\mathrm{As}}}{<R>} \qquad (10.16)$$

where $R_{\mathrm{Ga}}$ and $R_{\mathrm{As}}$ are the absolute reflectance intensities of the Ga- and As-terminated surfaces, respectively, and $<R>$ is the average reflectance.

Figure 10.28 shows as a typical example the kind of signal obtained in SPA/PRS. The oscillating signal structure results from the alternate pulsed fluxes of Ga and As. They generate different surface reconstructions with different translational periodicities (RHEED signal) and surface polarizibilities (SPA signal). This oscillatory behavior appearing because of the pulsed flux structure should not be confused with the oscillatory behavior related to the monolayer growth oscillations observed during stationary growth by RHEED or RAS as described before.

Measurements of the spectral dependence of the SPA signal have been performed mainly on different GaAs surfaces but also for the growth of heterostructures [10.93].

In spite of its high sensitivity a drawback of SPA compared to RAS is of course that SPA measures only the difference between different surfaces but is not capable of characterizing a certain surface. Its main application lies therefore in epitaxial growth techniques like ALE, FME or MEE where the precursor does not flow continuously through the reactor but instead a gas switching procedure is employed.

### 10.3.5 Reflectometry

Reflectometry uses the simplest normal incidence arrangement to measure the reflectance. This can be achieved for example with the DC channel of a standard RAS setup. But a dedicated setup containing just light source, monochromator and detector may be designed to be rather small. Reflectance measurements are extremely useful for layers with optical thicknesses in the order of the wavelength of the light. The resulting Fabry-Perot interference structure allows for very accurate thickness or growth rate determination [10.94]. Figure 10.29 shows the importance of such *in situ* measurements for optimizing VCSEL laser structures [10.68]. As a consequence of this it was found afterwards that VCSEL #1 showed no cw operation while VCSEL #2 produced 0.16 mW. The temperature dependence of the dielectric function can moreover be exploited to serve for in situ measurements of surface temperature, the true temperature for growth [10.95]. A recent report shows that the temperatures given by the indicators of epitaxial equipment, usually derived from thermocouples in the substrate holders, might differ by up to 50°C around 600°C from the true temperature [10.31]. This is certainly not acceptable by today's epitaxial technology.

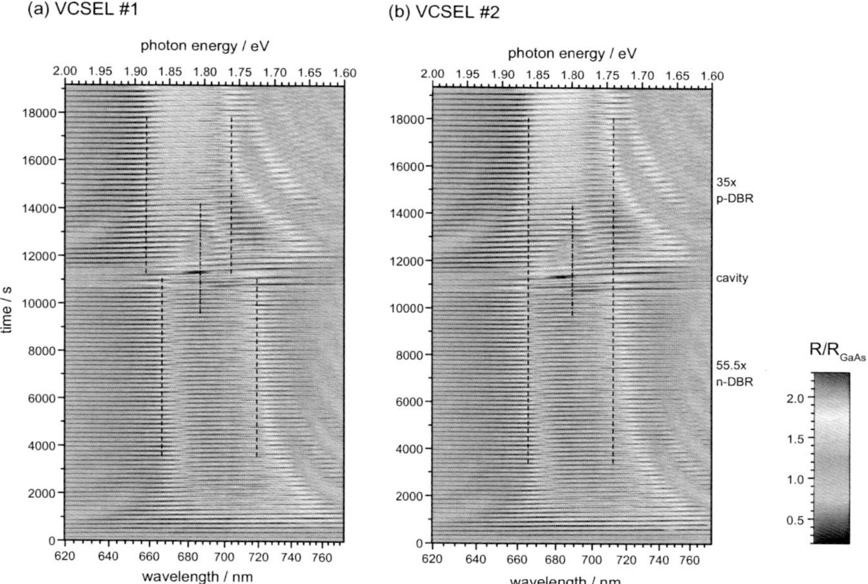


**Fig. 10.29.** In situ reflectometry data taken during growth of a VCSEL structure. (**a**) Center wavelength of the Bragg reflectors slightly misaligned with respect to each other (dashed vertical lines), (**b**) well-aligned Bragg reflectors (taken from [10.68])

## 10.4 Other Optical Techniques

### 10.4.1 Laser Light Scattering (LLS)

Laser light scattering (LLS) is a mostly qualitative but very simple and useful technique to monitor the surface morphology. The experimental arrangement consists of a laser directed at the growing surface and a light detector (photodiode) which samples the light intensity outside the direction given by specular reflection. Irregularities at the surface are responsible for this scattered light.

The surface roughness can be described by a height function $z(x, y)$ where $x, y$ are the coordinates within the surface and $z$ gives the height of the irregularities. For a flat surface one has $z(x, y) = 0$. The goal of the experiment is to determine $z(x, y)$ from the angular distribution of the scattered light. A convenient description for this purpose is given by the Fourier transform of $z(x, y)$

$$\phi(k) = F(z(x, y)). \tag{10.17}$$

This defines the power spectral density function (PSD)

$$\mathrm{PSD}(k) = \phi(k) \cdot \phi^*(k) = |\phi(k)|^2 \tag{10.18}$$

which gives the spatial roughness frequencies (the phase information, however, is lost) and is identical to the Fourier transform of the autocovariance function (nonnormalized autocorrelation function). $\mathrm{PSD}(k)$ describes the scattered power $p$ in terms of the incident power $P$ per solid angle [10.96]

$$\frac{\mathrm{d}p}{P\mathrm{d}\Omega} = OF \cdot \mathrm{PSD}(k). \tag{10.19}$$

The factor $OF$ describes the optical response of the material. It can be expressed analytically in certain cases. For example for a situation where the surface irregularities are small against the wavelength of the light

$$OF = \cos^2\theta \frac{16\pi|1-\varepsilon|^2}{\lambda^4|1+\sqrt{\varepsilon}|^2}\left[\frac{\cos^2\phi|q'|^2}{|q'+q\varepsilon|^2} + \frac{\sin^2\phi\left(\frac{2\pi}{\lambda}\right)^2}{|q'+q|^2}\right] \tag{10.20}$$

with

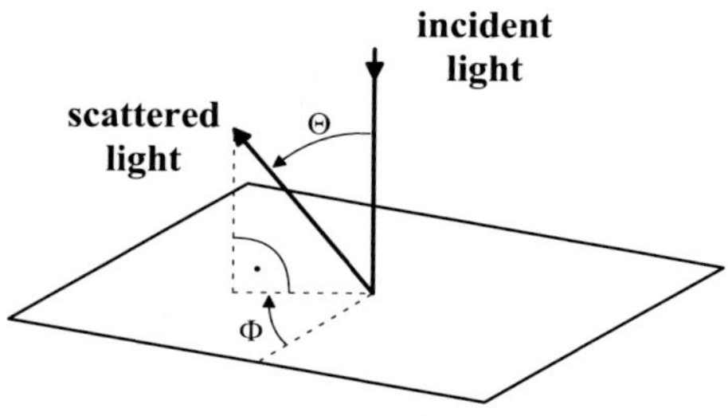


**Fig. 10.30.** Schematic sketch of a light scattering experiment for definition of the quantities in (10.20) (taken from [10.])

$$q' = \sqrt{\varepsilon \left(\frac{2\pi}{\lambda}\right)^2 - k^2} \tag{10.21}$$

and

$$q = \frac{2\pi}{\lambda} \cos\phi \qquad \text{and} \qquad k = \frac{2\pi}{\lambda} \sin\phi \tag{10.22}$$

with the angles as defined in Fig. 10.30.

Such an analysis requires, however, the measurement of the angular dependence of the scattered light, a task difficult to perform in the epitaxial growth environment. Therefore, in general only monitoring at a fixed angle as a function of time is performed and the result is interpreted more or less within the philosophy: the larger the scattering intensities the larger the surface roughness. Nevertheless this gives valuable information on the time development of the morphology and thus many reports also utilizing LLS data among others have been published [10.93, 97–103]. LLS may also be strongly anisotropic and so indicates anisotropic morphologies [10.105]. Such contributions have also been observed superimposed on RAS signals (see Fig. 10.25). As far as the sensitivity to morphological features is concerned it has been found that LLS can detect correlation lengths between 0.5 and 10 wavelengths and can have a height resolution of a few atomic steps (nanometers) [10.101]. An experimental example is shown in Fig. 10.31 [10.30, 104]. Such data could be interpreted in terms of the formation of dislocations (from 4 min growth time on in Fig. 10.31). A study performed during GaAs growth has shown also that oscillations periodic with the growth of a

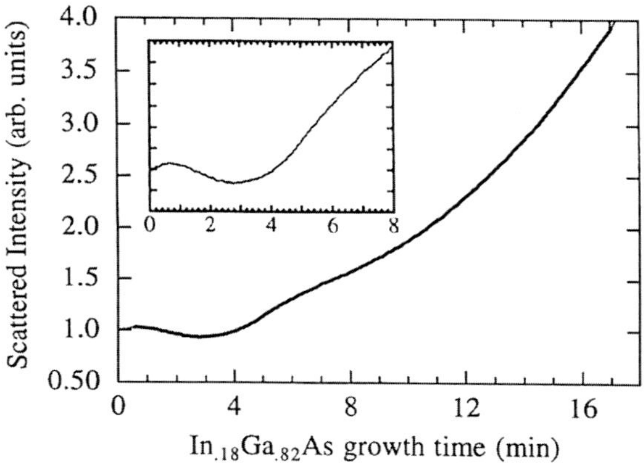


**Fig. 10.31.** Evolution of light scattering intensity along the [110] direction during growth of $In_{0.18}Ga_{0.82}As$ on GaAs at 490°C (spatial frequency 5.4 μm$^{-1}$). The inset shows intensity oscillations due to interference of light scattered at the interface and at the surface. The increase of the LS signal indicates the formation of misfit dislocations from 50 nm (approximately 4 min) on (taken from [10.104])

GaAs monolayer can be observed [10.106]. The existence of such oscillations opens, similar as in RAS, new possibilities for growth control.

The LLS technique which is simple in its setup certainly requires additional information as for instance supplied by STM/AFM pictures in order to explore the origin of the scattering intensity changes. It is basically limited to the investigation of surface roughness. This, however, affects the signals provided by RHEED and RAS. It therefore seems that this easy to use technique is a very useful complementary diagnostic tool for *in situ* growth studies. It would certainly help to obtain a better understanding of the damping observed in RHEED oscillations and the roughness induced changes in RAS spectroscopy.

### 10.4.2 Second Harmonic Generation (SHG)

When two strong electromagnetic fields of frequency $\omega$ interact with matter they can combine to produce a field of frequency $2\omega$. This non linear process is called second harmonic generation (SHG). It is coherent and the radiating fields have well-defined directions. For example, SHG in the usual reflection geometry emerges along the path of the primary reflected beam and it can easily be separated from the light with frequency $\omega$ by the use of appropriate filters. Since SHG is of higher order, however, the cross-section for such three-wave mixing is low, with typically one signal photon per $10^3$–$10^6$ incident photons. In a macroscopic picture SHG can be described by

$$\boldsymbol{P}(2\omega) = \epsilon_0 \chi^{(2)} \boldsymbol{E}(\omega)\boldsymbol{E}(\omega) \tag{10.23}$$

where $\chi^{(2)}$ is a second rank non linear susceptibility tensor. There are bulk as well as surface contributions to this tensor. In the electric dipole approximation this susceptibility vanishes for the bulk of materials such as Si or Ge because of their inversion symmetry. There may, however, be bulk contributions of higher order like magnetic dipole and electric quadrupole terms but these are typically some orders of magnitude smaller than the electric dipole contribution [10.107]. The reduced symmetry at a surface or interface can then lead to a considerable surface (interface) SHG signal which is often comparable to the higher-order bulk contribution [10.108]. Information on the surface structure can be deduced by an appropriate choice of the sample geometry and the polarization vectors [10.109–113].

SHG has been widely used to study surfaces and interfaces of materials with inversion symmetry such as centrosymmetric materials. The potential of the technique for surface analysis and growth has been reviewed in [10.7, 114].

Compound semiconductors like GaAs, on the other hand, do not have inversion symmetry in the bulk. As a result the bulk contribution to the SHG signal is usually dominant and exceeds the surface contribution by several orders of magnitude [10.115]. For that reason experimental results for compound semiconductors are very limited. Growth related SHG studies

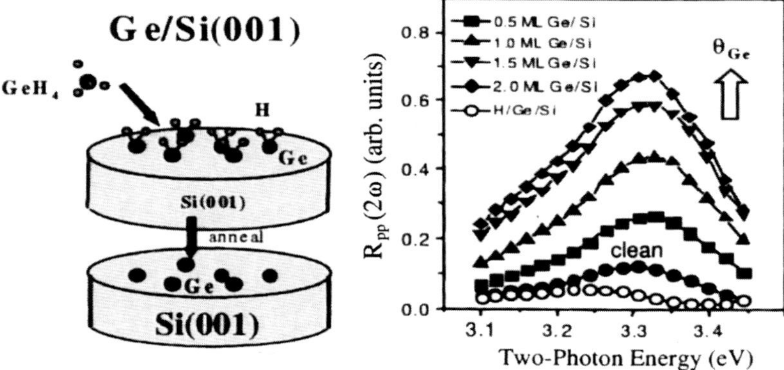


**Fig. 10.32.** Room temperature SHG signal ($R_{pp}(2\omega)$) as a function of two-photon energy for Ge deposition on Si(001) in UHV. The coverage is increasing from 0 to 2 ML (top five spectra). The $E_1$-gap intensity increases by the replacement of Si–Si dimers by Si–Ge and Ge–Ge. (taken from [10.114])

on Si(001) can be found in [10.114, 116] and for GaAs growth by MOVPE [10.117].

The example in Fig. 10.32 implies that SHG could be a useful tool for the *in situ* investigation of growth. However, as pointed out above the SHG signal intensity is fairly weak and the high laser powers involved may influence the growth. Thus it seems questionable whether it is a suitable method for real-time growth monitoring at elevated temperatures. Nevertheless, the development of SHG in this area is continuing. In particular, tunable light sources may lead to a better understanding of SHG by spectroscopy and may give further enhancement of the surface SHG signal when the photon energy is tuned into resonance with surface electronic states.

### 10.4.3 Raman Spectroscopy

Raman spectroscopy as a higher-order optical process is, by sensitivity, certainly not so well suited to determine vibrational frequencies of adsorbed molecules but is sufficiently sensitive to determine material properties through measurement of phonon frequencies, linewidth or intensities. Bulk as well as surface phonons may be detected with a very high (0.1 meV) spectral resolution.

In Raman spectroscopy, inelastic light scattering processes are analyzed, i.e., scattering processes in which energy is transferred between an incident photon with energy $\hbar\omega_{\mathrm{i}}$ (incident) and the sample, resulting in a scattered photon of a different energy $\hbar\omega_{\mathrm{s}}$ (scattered). The amount of transferred energy corresponds to the eigenenergy $\hbar\Omega_j$ of an elementary excitation labelled "$j$" in the sample, e.g., a phonon, a polariton, a plasmon, a coupled plasmon–phonon mode or a single electron or hole excitation. A Raman spectroscopy experiment thus yields the eigenfrequencies of the elementary excitations through the analysis of the peak frequencies $\omega_\mathrm{s}$ in the scattered light, since the frequency of the incident light $\omega_\mathrm{i}$ is well defined by the use of a laser-light source. Energy conservation yields:

$$\hbar\omega_\mathrm{s} = \hbar\omega_\mathrm{i} \pm \hbar\Omega_j . \tag{10.24}$$

Here the "−" sign stands for those Raman processes in which an elementary excitation is generated. These processes are called Stokes processes. Those which imply the annihilation of an elementary excitation correspond to the "+" sign. They are referred to as anti-Stokes processes. In most experimental investigations only Stokes processes are studied.

In analogy to energy conservation, the quasi-momentum conservation law gives the correlation between the wave vector $\boldsymbol{k}_\mathrm{i}$ of the incident light, $\boldsymbol{k}_\mathrm{s}$ of the scattered light and the excitation wave vector $\boldsymbol{q}_j$:

$$\boldsymbol{k}_\mathrm{s} = \boldsymbol{k}_\mathrm{i} \pm \boldsymbol{q}_j . \tag{10.25}$$

Since the light wave vectors are small compared to the size of the Brillouin Zone (or equivalently $\lambda_\mathrm{light} \gg$ lattice constant) only phonons in the center of the Brillouin Zone are observed ($k = 0$). The conservation of momentum is an idealized picture which may be limited in real experiments. This can occur through attenuation of the wave or inhomogeneities (impurities, dislocations) of the sample. Such limitations reflect themselves in broadening of the Raman line.

In the dominant Raman-scattering process in solids the interaction between the photons and the elementary excitations is indirect. It is mediated by electronic interband transitions, because these transitions define the dielectric susceptibility $\chi$ in the visible spectral range, where Raman experiments are usually performed. Raman scattering occurs when the interband transitions are influenced by a phonon excitation for example. The generation of scattered light with frequency $\omega_\mathrm{s}$ by incident light with frequency $\omega_\mathrm{i}$ can be described by a generalized dielectric susceptibility tensor $\tilde{\chi}(\omega_\mathrm{i}, \omega_\mathrm{s})$:

$$\boldsymbol{P}(\omega_\mathrm{s}) = \epsilon_0 \tilde{\chi}(\omega_\mathrm{i}, \omega_\mathrm{s}) \boldsymbol{E}(\omega_\mathrm{i}) . \tag{10.26}$$

Here $\boldsymbol{P}(\omega_\mathrm{s})$ is the oscillating polarization which gives rise to the scattered light wave and $\boldsymbol{E}(\omega_\mathrm{i})$ is the oscillating electric field of the incident light wave. The Raman scattering intensity can thus be expressed by the dipole radiation intensity using the generalized dielectric susceptibility $\tilde{\chi}(\omega_\mathrm{i}, \omega_\mathrm{s})$:

$$I_\mathrm{s} = I_\mathrm{i} \frac{\omega_\mathrm{s}^4 V}{(4\pi\epsilon\epsilon_0)^2 c^4} \left| e_\mathrm{s} \tilde{\chi}(\omega_\mathrm{i}, \omega_\mathrm{s}) e_\mathrm{i} \right|^2 . \tag{10.27}$$

Here $I_\mathrm{i,s}$ and $e_\mathrm{i,s}$ denote the intensity and polarization unit vector of incident and scattered light, and $V$ is the scattering volume. The Raman scattering efficiency can be defined from (10.27) by normalizing to the incident power $I_\mathrm{i}$.

Similarly to the susceptibility, the Raman scattering tensor, $\tilde{\chi}(\omega_\mathrm{i}, \omega_\mathrm{s})$, may become resonant when the photon energies are close to electronic transition

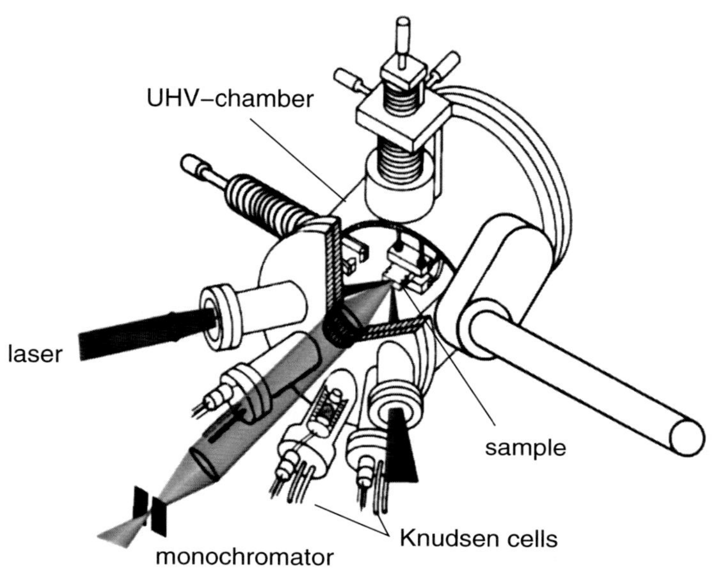


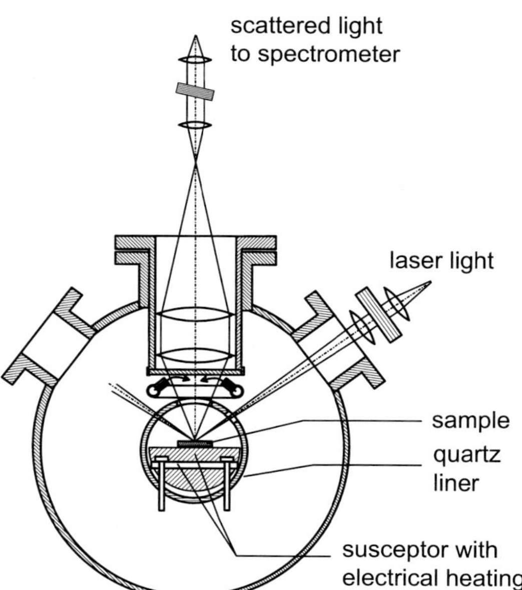


**Fig. 10.33.** Schematic view of experimental arrangements for *in situ* Raman spectroscopy during MBE (top) and MOVPE (bottom) (taken from [10.5] and [10.123])

energies (gaps) [10.118–120]. While resonance Raman scattering (RRS) is an interesting topic in itself it is of the highest importance for setting the experimental conditions, e.g., the choice of laser line. The resonances quite often change the scattering intensities over several orders of magnitude and thus the choice of laser line quite often decides success or failure of the experiment.

Raman scattering equipment is quite bulky because the high spectral resolution monochromators (double or triple versions) have large focal lengths (1 m) and suitable laser sources have a corresponding size [10.120]. Therefore not too many reports in connection with epitaxial growth have been published.

Raman setups for MBE and MOVPE are shown schematically in Fig. 10.33. Raman spectra taken during growth of ZnSe on GaAs are seen in Fig. 10.34. One clearly observes how the Raman spectrum of ZnSe develops during growth. The scattering intensities increase since the ZnSe scattering volume increases through the deposition. From the intensity variation with time it is possible to determine the growth rate [10.121]. The details of the spectral shape allow for detailed material characterization like stoichiometry, structural quality and in doped samples also determination if free carrier concentration. The growth of GaN has also been studied [10.122].

The time resolution of RS is moderate. The sensitivity of modern Raman equipment derives from the data accumulation in CCD detectors. Acquisition times of a couple of minutes for spectra such as those in Fig. 10.34 are standard. Measurements in a MOVPE reactor have also recently been pub-

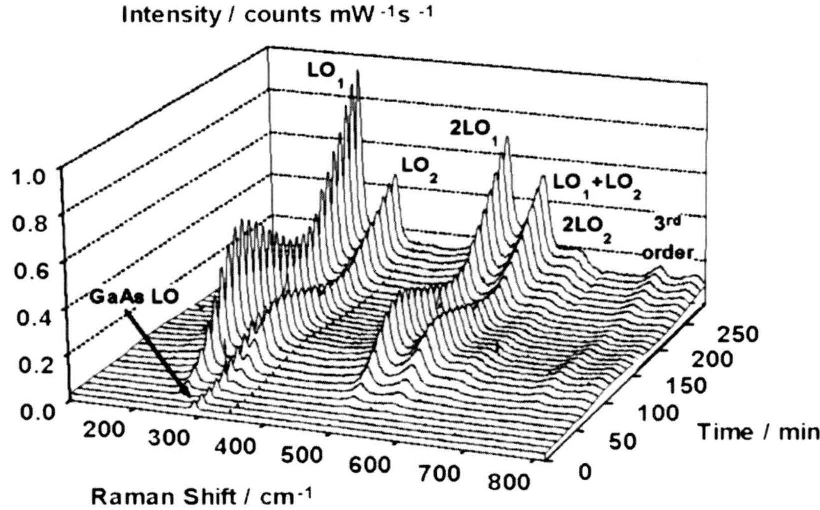


**Fig. 10.34.** Raman spectra taken during growth of ZnSe on GaAs(110) by MBE in UHV. The intensity oscillations in each of the modes are ascribed to Fabry–Perot interference in the overlayer with increasing thickness (taken from [10.121])

lished for III–V semiconductor growth with a laser source coupled to the growth reactor by optical fibers [10.123]. The scattering signals turn out to be surprisingly large at temperatures up to 1200 K. This gives hope for the feasibility of *in situ* monitoring of surface features in epitaxial growth with Raman scattering. The results show, moreover, that the high temperature Raman spectra published so far are not representative at all since the surface had not been actively stabilized in an epitaxial growth environment.

### 10.4.4 Infrared Reflection Absorption Spectroscopy (IRRAS)

Considering any growth method using metalorganic compounds as precursors it is of particular interest to identify exactly how the metalorganic species such as TMGa or TEGa adsorb and decompose on the growing surface. In principle that can be done by looking at the vibrational modes of the molecules on the surface in combination with symmetry considerations in order to determine the adsorption site (see, e.g. [10.124]). The frequencies or energies of the various vibrational modes of TMGa, for instance, lie in the range from approximately 500 to $3000\,\mathrm{cm}^{-1}$ or 60 to 370 meV (see, e.g. [10.125]). The conventional method for detecting vibrational modes of adsorbates on surfaces is high resolution electron energy loss spectroscopy (HREELS) which necessarily depends on UHV conditions. The adsorption behavior of several metalorganic compounds on different Si surfaces, for example, was studied using HREELS [10.125].

In a MOVPE environment which prevents the application of HREELS, IR optical techniques are candidates for studying the vibrational properties of the adsorbed species.

However, when IR measurements are performed to study adsorbates on surfaces in the monolayer coverage regime, the signals are usually very weak and often difficult to detect against the background. Furthermore, the only experimental geometries which seem to be compatible with *in situ* growth studies are those which utilize external reflection from the front surface for two reasons. Firstly conventional growth reactors usually do not allow access to both substrate surfaces. This would be required for both attenuated total (internal) reflection or single pass transmission measurements. Secondly, these measurements are anyway hampered by the increasing free carrier absorption in semiconductor substrates at the elevated growth temperatures. In MOVPE one has in addition to pay attention to the gas phase absorption.

External reflection has already been successfully applied to study adsorbates on metallic surfaces (see, e.g. [10.127–129]). It was found that in particular for p-polarized light close to grazing incidence, typically 85° with respect to the surface normal can provide sensitivities which allow 1/1000th of a monolayer of adsorbed molecules to be detected in favorable cases. The molecules absorb some of the light at their vibrational frequencies and absorption peaks occur in the spectrum of the reflected light. This technique is

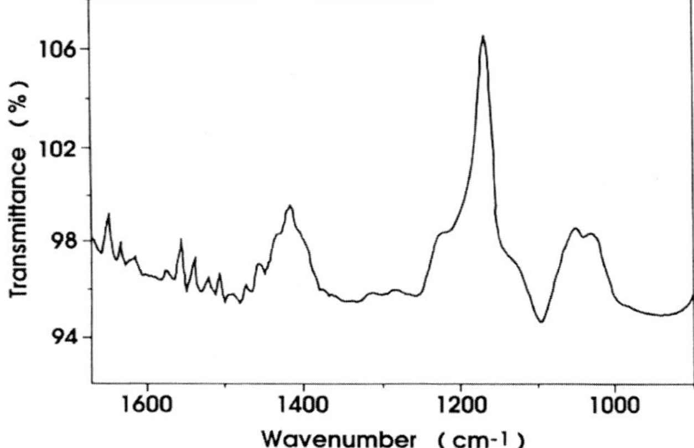


**Fig. 10.35.** IRRAS spectrum recorded from a GaAs(001) surface exposed to TMGa. The transmittance is defined as $[R_{\text{covered}} - R_{\text{clean}}]/R_{\text{clean}}$. Spectra are obtained with 1000 scans of a FTIR spectrometer fitted with a MCT detector [10.126]. The peaks can be assigned to C–H deformation modes [10.126] (taken from [10.5])

then called infrared reflection-absorption spectroscopy (IRRAS) or reflection-absorption infrared spectroscopy (RAIRS). The spectral resolution of IRRAS (typically around 1 meV) is superior to that of conventional HREELS (typically 2 – 10 meV) [10.124]. The general advantage of vibrational spectroscopy is that it can provide direct information on the chemical state of the adsorbed molecules from the analysis of the vibrational frequencies observed.

In comparison to semiconductors (Si) a similar optimum angle connected, however, with a very small increase (approximately 20 times smaller than that on metals) was found. The corresponding changes in reflectance amounted to 0.01 – 0.1, values which are in the order of the inherent noise [10.130]. This low sensitivity may impose a limit to the applicability of IRRAS as a growth monitoring technique; however, because of its molecular fingerprint character it is still a very useful technique for research purposes.

Figure 10.35 gives an example of the usefulness of such measurements. However, the surface was exposed to $10^5$ Langmuir TMGa at 300 K which very probably created physisorbed multilayers on the sample as was noted also in the report. This fact is also responsible for the very large signals seen in the figure [10.126]. Thus, not the properties of TMGa molecules on the GaAs(001) surface were measured.

Such data show that IRRAS is in principle applicable in growth-like situations. However, further development is needed in order to explore the application of IRRAS during real growth conditions. For the moment it is not clear whether monolayer or even submonolayer sensitivity will be achievable at elevated temperatures.

# Part IV

# Physics of Epitaxy

# 11. Thermodynamic Aspects

There are, in general, four basic conceptual tools required for a formal description of epitaxy. These are:

(i) thermodynamics of phase transitions and interface formation [11.1–3],

(ii) fluid dynamics (hydro-, and gasodynamics) of mass transport [11.4–6],

(iii) statistical mechanics of crystal growth processes [11.7], including the kinetics of surface migration/diffusion and ordering processes [11.8], and

(iv) quantum mechanics of chemical bond formation (chemisorption and lattice incorporation) [11.9]

(the problem of heat transport in the solid or liquid phases, important for understanding bulk crystal growth, is omitted, because of the fairly low growth rates characteristic of epitaxial growth). The extent of applicability of these formal tools for analyzing the definite epitaxial growth process depends on the physical approach which is chosen for describing this process.

In general, three approaches are most frequently used when studying epitaxy [11.10]. The first is the phenomenological or macroscopic approach, which provides a boundary between what is possible and what is not (this results from phase diagram analysis). This approach is suitable for analyzing phase transitions, phenomenological aspects of interface formation, as well as mass transport phenomena. The second approach is based on the atomistic description of epitaxy, involving statistical thermodynamics (the one-dimensional two-body interaction potentials are usually sufficient for a description of adsorption processes essential for epilayer formation). In the framework of this approach tangible previsions may be advanced which concern the topology of epitaxial growth, discrete island formation (2D and 3D nucleation), growth in continuous layers (SF-, FM-, and SK-modes) and influence of external parameters (photo- or plasma-assisted growth) modifying the growth process.

However, the most fundamental processes of the growth, i.e., incorporation into the crystal lattice, and the anisotropic effects, resulting from crystallographic orientation and reconstruction of the surface on which the epitaxial growth proceeds, require quantum mechanical treatment (the wave functions of the valence electrons of the relevant atoms here play a dominant role).

Thus, the quantum mechanical wave function approach to epitaxy is the third, most fundamental, and simultaneously most sophisticated approach.

In this chapter, these approaches will be discussed, taking as the most general model of the epitaxial growth system the one illustrated schematically in Fig. 1.1 (Sect. 1.1).

## 11.1 The Driving Force for Epitaxy

Epitaxial crystal growth is an example of a dynamical phase transition. A stable phase, the epilayer, grows out from a metastable phase, that is from an amorphous solid, melt (or liquid solution) or vapor (including the special case of growth from the rarified gas phase of intersecting atomic or molecular beams). The driving force for the growth is the chemical potential difference of the stable and the metastable phases. The simple assumption of the linear response, which means that the growth velocity is proportional to the driving force, gives ideal linear growth laws. When the epilayer is finite in its size (e.g., the growth occurs on a patterned substrate surface), or when the crystallization interface deforms, the surface stress or surface tension has to be included in the thermodynamics of the epitaxial growth. Knowledge of the surface stress/tension plays an important role in the determination of the shape of a finite sized epilayer, or the deformation of the flat interface [11.1].

### 11.1.1 Basic Concepts and Terminology of Thermodynamics

Let us begin this subsection with a brief review of the basic concepts and terminology of thermodynamics [11.11]. Phase is the first important phenomenological concept used when considering thermodynamics of crystal growth. This is a region (usually homogeneous) which is physically distinguishable and distinct from other phases. Thus, we have the vapor and the liquid phases as well as various solid phases in most cases when growing different material systems by epitaxy.

The basic goal of thermodynamics as applied to epitaxy is to define the compositions of the various phases in an equilibrium system at constant temperature and pressure. Equilibrium is defined as the state where the Gibbs free energy per mole, $G$, is a minimum [11.1]. The Gibbs free energy is defined in terms of the enthalpy, $H$, and entropy, $S$:

$$G = H - TS \tag{11.1}$$

where

$$H = U + pV \tag{11.2}$$

and $U$ is the internal energy, $V$ is the volume, and $p$ is the pressure of the system. $G, H, S, U$ and $V$ are all extensive quantities, i.e., they depend on the size of the system. For convenience, they are expressed on a per mole basis. For a two-phase system, the total free energy is $G_t = G_t^\alpha + G_t^\beta$. Since $G_t$ is a minimum at equilibrium, the change in $G_t$ by moving an infinitesimally small number of moles of component $i$, $\mathrm{d}n_i$, between two phases causes no change in $G_t$. This may be expressed mathematically in the form

$$\left(\frac{\partial G_t}{\partial n_i}\right)^\alpha_{T,p,n} - \left(\frac{\partial G_t}{\partial n_i}\right)^\beta_{T,p,n} = 0 \tag{11.3}$$

where the superscripts $\alpha$ and $\beta$ represent the two phases. The partial derivative of $G_t$ with respect to $n_i$ is so important a quantity for thermodynamic calculations that it is given a name, the chemical potential, represented as $\mu_i$ Thus, the equilibrium condition may be expressed in the simple form

$$\mu_i^\alpha = \mu_i^\alpha \tag{11.4}$$

for each component in the system. For a reversible perturbation of the system, it can be shown from equations (11.1) and (11.2) and the relationship $\mathrm{d}U_t = T\,\mathrm{d}S_t - p\,\mathrm{d}V_t$ that

$$\mathrm{d}G_t = V_t\,\mathrm{d}p - S_t\,\mathrm{d}T. \tag{11.5}$$

This is one of Maxwell's equations of thermodynamics [11.1]. For an ideal system (e.g., the ideal gas) $pV_t = nRT$, thus, at constant $T$, (11.5) yields for a change in pressure

$$\mathrm{d}G_t = nRT\mathrm{d}\ln(p). \tag{11.6}$$

Hence, for an ideal single gas

$$\mu = RT\ln p \tag{11.7}$$

and

$$\mu = \mu^0 + RT\ln(p/p^0) \tag{11.8}$$

where $\mu^0$ and $p^0$ represent the chemical potential and pressure of an arbitrary standard state. For an ideal gas mixture

$$\mu_i = \mu_i^0 + RT\ln(p_i/p_i^0) \tag{11.9}$$

where $p_i$ is the partial pressure, equal to the mole fraction $x_i$ multiplied by $p$, and the standard state is usually a pure component $i$.

For an ideal liquid or solid solution, the same expression holds with $p_i/p_i^0$ replaced by $x_i/x_i^0$. However, the standard state is pure $i$, so $x_i^0 = 1$. The form of (11.9) is so useful that it is retained even for a non-ideal solution with $x_i$ replaced by the activity $a_i$, which may also be considered as a product of $x_i$ and the non ideality factor $\gamma_i$, the activity coefficient

$$\mu_i = \mu_i^0 + RT\ln a_i = \mu_i^0 + RT\ln(x_i\gamma_i). \tag{11.10}$$

### 11.1.2 The Interphase Exchange Processes

Let us now consider a simple exchange process in which the particles of the metastable phase $A$ and particles of the solid crystalline phase $B$ (substrate crystal or the already grown epilayer) pass from one phase to the other, and back, as shown in the scheme (Fig. 1.1)

$$A \rightleftharpoons B \tag{11.11}$$

with relevant rate constants $k_1$ (for $A \to B$) and $k_{-1}$ (for $B \to A$). For these transitions the equilibrium condition is given by

$$\mu_A^0 + RT\ln(a_A^e) = \mu_B^0 + RT\ln(a_B^e) \tag{11.12}$$

where $a^e$ is the activity at equilibrium. Thus,

$$\frac{a_B^e}{a_A^e} = \exp\left(-\frac{\mu_B^0 - \mu_A^0}{RT}\right) = K_1 = \frac{k_1}{k_{-1}} \tag{11.13}$$

which is the basic law of mass action.

When the system is not at equilibrium, the thermodynamic force to restore equilibrium is $\Delta\mu = \mu_B - \mu_A$

$$\Delta\mu = \mu_B^0 + RT\ln(a_B) - \mu_A^0 - RT\ln(a_A) = RT\ln\left(\frac{a_A^e a_B}{a_A a_B^e}\right). \tag{11.14}$$

The difference in chemical potentials of the phases $A$ and $B$, which is quantitatively given by (11.14), is the driving force for epitaxy. A non equilibrium situation is intentionally created which drives the system to produce the desired epilayer. The maximum quantity of the epitaxially grown solid which can be produced depends on the intentionally created supersaturation of the metastable phase $A$, i.e., it is simply the amount which would establish equilibrium. This amount is thus fundamentally limited by thermodynamics and the total size of the system, i.e., the melt (solution) volume in the case of LPE, or the total volume of gas passing through the reactor in VPE.

Ordinarily, the growth rate is considerably slower than calculated from thermodynamics. Kinetics, the rate at which the reactions occur, is not fast enough to allow for reaching equilibrium throughout the system at all times. Often equilibrium is established at the growing solid surface (the crystallization interface), but not in the bulk of the whole metastable phase. In this case, the reaction rate (or growth rate) is limited by mass transport, which occurs by some combination of convection and diffusion in both of the mentioned cases of epitaxial growth techniques. The other common limitation to the growth rate is the surface reaction rate. In what follows, we will discuss the mass transport phenomena in epitaxial growth systems.

## 11.2 Mass Transport Phenomena

Mass transport in epitaxial growth systems is an extremely complex subject when treated correctly and in detail [11.11]. Two processes are involved, namely, fluid flow and diffusion. Fluid flow in a simple constant temperature growth system is fairly complex, but temperature and solutal concentration gradients, leading to diffusion, further complicate the problem. Furthermore, thermal gradients may affect diffusion what has the consequence that thermal and mass transport processes are coupled. Techniques for treating mass transport problems in epitaxial systems usually belong to one of the two extreme categories. The first is physically and mathematically rigorous. It involves solving the mass continuity equation, the Navier–Stokes equation (momentum conservation), and the energy transport equation simultaneously. This often leads to accurate results, however, after a great deal of mathematical analysis [11.12]. The second, and very popular, category is to use such simple approaches that they cannot give physically reasonable results. They give approximate information, leading to some understanding of the important physical mass transport processes occurring during the growth, which is of great value for crystal growers using epitaxy.

### 11.2.1 Basic Equations Describing Mass Transport in VPE Systems

The mathematical framework for describing continuous mass transport phenomena in VPE growth systems consists of nonlinear, coupled partial differential equations that represent the conservation of momentum, energy, total mass, and individual species [11.5]. The general derivation and form of these equations is given in standard references in transport phenomena textbooks [11.6]. Numerous modifications of these equations have been used in models ranging from simple boundary-layer-type description to complete three-dimensional models [11.13, 14]. The general equations applicable to most crystal growth systems, and thus also to VPE operating in the continuum regime, can be summarized as follows [11.5]:
Total mass:

$$\frac{\partial \rho}{\partial t} + \nabla(\rho \nu) = 0 \tag{11.15}$$

Momentum:

$$\rho\left(\frac{\partial \nu}{\partial t} + \nu \cdot \nabla \nu\right) = -\nabla p + \nabla\cdot \mu\left(\nabla \nu + (\nabla \nu)^T - \frac{2}{3} I \nabla\cdot \nu\right) + \rho g e_z \tag{11.16}$$

Energy:

$$\rho C_p \frac{\partial T}{\partial t} + (\nu\cdot\nabla T) = \nabla \cdot (k \nabla T) - \sum_{i=1}^{s}\left(C_{pi} J_i \cdot\nabla T + \sum_{j=1}^{n^g} \Delta H_{ji} \nu_{ij}^g R_j^g\right) \tag{11.17}$$

Chemical species:

$$\frac{(\partial\rho\,\omega_i)}{\partial t} + \nu\cdot\nabla(\rho\,\omega_i) = -\nabla\cdot j_i + \sum_{j=1}^{n^g} \nu_{ij}^g M_i R_j^g\,, \quad i = 1,\ldots,S-1\,. \tag{11.18}$$

In these equations $\rho$ is the density, $\nu$ the velocity, $p$ the pressure $\mu$ the viscosity, $g$ the gravitational constant and $\omega_i$ the mass fraction of chemical species $i$. Since time scales for diffusion and convection in VPE reactors are short (on the order of a few seconds) relative to the film growth time, pseudo-steady-state flow conditions may usually be assumed; i.e., the time derivatives in (11.15) – (11.18) may be set equal to zero. Dynamic effects need only be included in the simulation of switching between gas compositions during the growth of layered structures and in prediction of rapid thermal processing [11.5].

The solution of the conservation equations for total mass (continuity equation (11.15)) and momentum (11.16) gives the pressure and velocity distribution throughout the reactor enclosure. The temperature distribution is given by the solution to the energy balance (11.17). Contributions from viscous energy dissipation [11.6] and pressure effects are negligible and have been omitted. With the possible exception of very high deposition rate systems (growth rate $\gg 1\,\mu\mathrm{m\,h}^{-1}$) the last term in (11.17), representing the contributions from heats of reaction, will be insignificant.

The balance over the $i$-th chemical species (equation. (11.18)) consists of contributions from diffusion, convection, and the loss or production of the species in gas phase reactions. Only $S-1$ equations are needed, since the mass fraction of the last component is given by the constraint

$$\sum_{i=1}^{s} \omega_i = 1\,. \tag{11.19}$$

An equation of state for the density $\rho$ and constitutive relations for the diffusion flux $j_i$ are needed to complete the modeling equations. The temperature variation of the density is a critical element in the model, since gas expansion effects caused by density changes when heating the gas phase play a major role in the flow behavior. The diffusion flux $j_i$ combines ordinary diffusion driven by concentration gradients ($j_i^C$) and thermodiffusion, or Soret diffusion, driven by thermal gradients ($j_i^T$), thus, $j_i = j_i^C + j_i^T$.

The diffusion fluxes stemming from concentration gradients are given by the Stefan–Maxwell equation [11.6], which can be expressed in terms of mass fractions and species fluxes in the form [11.5]:

$$\nabla\omega_i + \omega_i\nabla(\ln\overline{M}) = \sum_{j\neq 1}^{S} \frac{1}{\rho D_{ij}}\left(\frac{\overline{M}\omega_i}{M_j} j_j^C - \frac{\overline{M}\omega_j}{M_i} j_i^C\right), \tag{11.20}$$

where $D_{ij}$ is the binary diffusion coefficient. This equation may be solved iteratively subject to the constraint

$$\sum_{i=1}^{S} j_i^C = 0 \tag{11.21}$$

to yield diffusion fluxes as functions of mass fractions.

The thermodiffusion component, $j_i^T$, which drives molecular species away from hot regions towards cold regions, is given by the expressions [11.5]:

$$j_i^T = D_i^T \nabla(\ln T), \quad D_i^T = \sum_{j\neq 1}^{S} \frac{c^2}{\rho} M_i M_j D_{ij} k_{ij}^T, \tag{11.22}$$

where $D_i^T$ is the thermodiffusion coefficient and $k_{ij}^T$ is the thermodiffusion ratio. Because of the large temperature gradients present in crystal growth from the vapor, thermodiffusion is often significant, reducing the growth rate by 10 – 20 % relative to the growth rate achievable in the absence of thermodiffusion. Finally, to achieve accurate predictions, it is necessary to include also the temperature variations of the physical parameters [11.5].

In order to simulate real reactor flow patterns, boundary conditions must be specified for the above given equations. These depend, however, on the particular geometry of the reactor. An example of how the mass transport equations and their boundary conditions may be formulated in a particular reactor case is given in [11.14] (see also Sect 6.2.2 for a description of experimental data given in this reference).

### 11.2.2 The Boundary Layer at the Substrate Surface

The physical situation that occurs during the growth of epitaxial films in VPE growth reactors with atmospheric or near atmospheric pressure gaseous flow is illustrated in Fig. 11.1. A characteristic feature of these reactors is the formation of a boundary layer adjacent to the stationary substrate surface. The decreased velocity of the gaseous species in this region leads to a laminar, near stagnant flow [11.15] that controls the transport of the source molecules toward the substrate. Because this transport is mainly diffusive in its nature, the stagnant gaseous layer near the substrate surface is frequently called the diffusion boundary layer [11.16].

To obtain more detailed information about the flow of the gaseous species in VPE reactors in the neighborhood of the substrate surfaces, the boundary- layer approach [11.6, 11, 15–17] will be described, on the example of boundary-layer development as a function of position in steady flow [11.6].

To this end, the incompressible flow pattern near the leading edge of a flat plate immersed in a fluid stream, as shown in Fig. 11.2, will be considered in detail. The differential equations that describe the flow field in this case are: (i) the two-dimensional, steady-state equation of continuity (compare with (11.15))

$$\frac{\partial v_x}{\partial x} + \frac{\partial v_y}{\partial y} = 0 \tag{11.23}$$

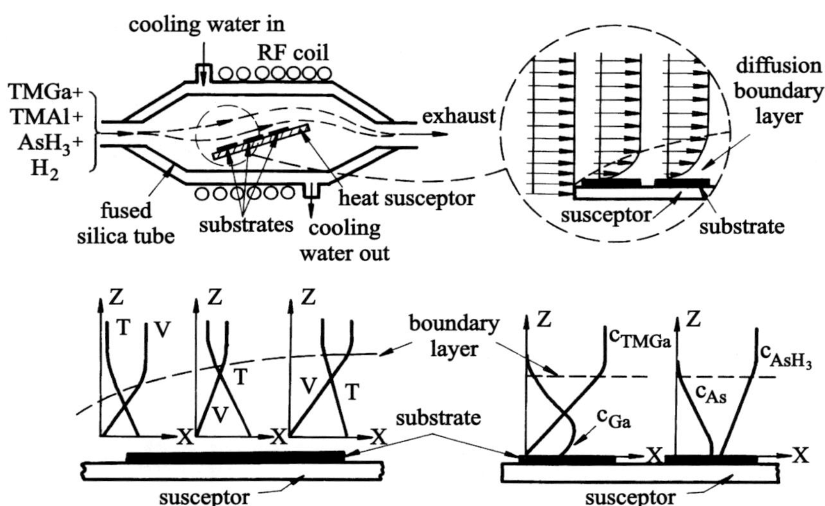


**Fig. 11.1.** Schematic drawing of the fused silica horizontal tube crystallization reactor for the MOVPE cold wall technique and the temperature, flow velocity, and concentration gradients in the diffusion boundary layer (taken from [11.16])

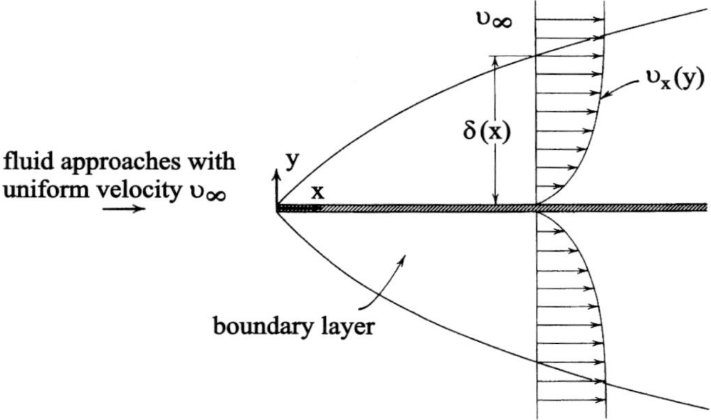


**Fig. 11.2.** Boundary-layer development near a flat plate of negligible thickness (taken from [11.6])

and (ii) the $x$-component of the equation of motion (compare with (11.16))

$$v_x \frac{\partial v_x}{\partial x} + v_y \frac{\partial v_y}{\partial y} = \frac{\mu}{\rho} \left( \frac{\partial^2 v_x}{\partial x^2} + \frac{\partial^2 v_y}{\partial y^2} \right), \tag{11.24}$$

where $\mu$ and $\rho$ are the viscosity and density of the fluid. The $y$-component of the equation of motion is ignored here because the flow is not great in that direction. Moreover, one can neglect the term $\partial^2 v_x/\partial x^2$ on the ground that it is small in comparison with $v_x \partial v_x/\partial x$ [11.17]. Solving (11.23) for $v_y$, with the boundary condition that $v_y = 0$ at $y = 0$, and substituting the result into (11.24) one gets

$$v_x \frac{\partial v_x}{\partial x} - \left( \int_0^y \frac{\partial v_x}{\partial x} \right) \frac{\partial v_x}{\partial y} = \frac{\mu}{\rho} \left( \frac{\partial^2 v_x}{\partial y^2} \right) \tag{11.25}$$

This is the starting equation to determine $v_x(x, y)$; it is clearly nonlinear. It is to be solved with the boundary conditions:

$$v_x = 0 \quad \text{at} \quad y = 0, v_x = v_\infty \quad \text{at} \quad y = \infty \tag{11.26}$$

$v_\infty$ is the value of $v_x$ in the main stream of the fluid, far from the plate surface, and

$$v_x = v_\infty \quad \text{at} \quad x = 0 \quad \text{for all} \quad y. \tag{11.27}$$

Assuming now that the velocity profiles at various values of $x$ have the same shape one gets

$$\frac{v_x}{v_\infty} = \phi(\eta), \text{ where } \quad \eta = \frac{y}{\delta(x)} \tag{11.28}$$

in which $\delta(x)$ is the boundary-layer thickness a distance down the plate. Assuming further that $v_x = v_\infty$ outside the boundary layer, i.e., for $y > \delta$, one may calculate the derivatives in (11.25)

$$\frac{\partial v_x}{\partial x} = v_\infty \phi' \left( -\frac{\eta}{\delta} \right) \frac{\mathrm{d}\delta}{\mathrm{d}x} \tag{11.29}$$

$$\frac{\partial v_x}{\partial y} = v_\infty \phi' \frac{1}{\delta} \tag{11.30}$$

$$\frac{\partial^2 v_x}{\partial y^2} = v_\infty \phi'' \frac{1}{\delta^2}, \tag{11.31}$$

in which primes denote differentiation with respect to $\eta$. Substituting these expressions into (11.25) and integrating this equation with respect to $\eta$, one gets

$$(B - A)\delta \frac{\mathrm{d}\delta}{\mathrm{d}x} = \frac{\mu}{\rho v_\infty} C \tag{11.32}$$

where

$$A = \left( \int_0^1 \phi \phi' \eta \, \mathrm{d}\eta \right) \tag{11.33}$$

$$B = \left( \int_0^1 \phi' \left( \int_0^{\bar{\eta}} \phi' \eta \, \mathrm{d}\eta \right) \mathrm{d}\bar{\eta} \right) = -A + \left( \int_0^1 \phi' \eta \, \mathrm{d}\eta \right) \tag{11.34}$$

$$C = \left( \int_0^1 \phi'' \mathrm{d}\eta \right) = \phi' \Big|_0^1 . \tag{11.35}$$

Integration of (11.32) then gives

$$\delta(x) = \sqrt{2\left(\frac{C}{B-A}\right)\frac{\mu}{\rho}\frac{x}{v_\infty}}\,, \tag{11.36}$$

where the boundary condition that $\delta = 0$ at $x = 0$ has been used. The conclusion from the presented calculations is that the boundary-layer thickness is proportional to the square root of the distance down the plate.

When the velocity profile is selected to be

$$\phi(\eta) = (3/2)\eta - (1/2)\eta^3 \tag{11.37}$$

one finds that $A = 9/35$, $B = 33/280$, and $C = -(3/2)$ [11.6]. Hence the boundary-layer thickness is given by the simple expression

$$\delta(x) = \sqrt{\frac{280\mu x}{13\rho v_\infty}} = 4.64\sqrt{\frac{\mu x}{\rho v_\infty}} \tag{11.38}$$

and the velocity distribution is

$$\frac{v_x}{v_\infty} = \frac{3}{2}\left(\frac{y}{\delta(x)}\right) - \frac{1}{2}\left(\frac{y}{\delta(x)}\right)^3 ; \quad 0 < y < \delta(x). \tag{11.39}$$

### 11.2.3 Effusion from Solid Sources in MBE

When analyzing mass transport by effusion from solid sources one can proceed according to at least two approaches. The first consists in considering an ideal effusion cell (also called a Knudsen-like cell). Such a cell contains the condensed phase and the vapor of the charge material being in thermodynamic equilibrium with each other [11.18]. The effusion aperture of this cell is a small orifice in an infinitesimally thin cell lid (the diameter of the orifice is much smaller than the mean free path of the gas particles in the cell), and there is no orifice wall. The gas reservoirs outside and inside the cell are large enough that the molecules encounter each other more frequently than the walls. The gas pressure inside the cell is much higher than the pressure outside, i.e., in the MBE growth chamber. The second approach consists in introducing experimentally determined correction factors to the transport equations characteristic of ideal cells, in order to get for the real effusion cells reasonable predictions concerning the mass transport toward the substrate surface.

Let us consider the ideal case first. As shown in Fig. 11.3, the molecules or atoms of the charge vapor escape from the cell through the orifice into the UHV environment in all directions encompassed by a hemisphere with its center coincident with the orifice center [11.19].

An individual particle of mass $m$ moving along a straight line in the direction $\vartheta$ toward the cell orifice center passes it and escapes into the vacuum in the same direction, because the thickness of the orifice is assumed to be infinitely thin ($L_0 \approx 0$). The differential angular effusion rate $\mathrm{d}\Gamma_\vartheta$ from the

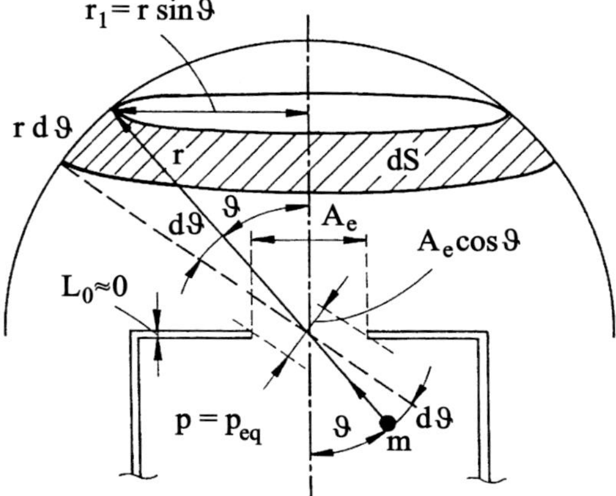


**Fig. 11.3.** Schematic illustration of the upper part of an ideal Knudsen-like effusion cell, indicating geometrical quantities relevant to flux distribution calculations (taken from [11.19])

orifice area $A_e$ into the vacuum in the directions contained between $\vartheta$ and $\vartheta + \mathrm{d}\vartheta$ is proportional to:

1. the surface of the orifice seen by the particle

$$\mathrm{d}\Gamma_\vartheta \sim A_e \cos(\vartheta)$$

2. the number of particles entering unit area of the orifice in unit time

$$\mathrm{d}\Gamma_\vartheta \sim \frac{\Gamma_e}{A_e} = 3.51 \times 10^{22} \frac{p}{\sqrt{MT}}$$

(see (6.5) in Chap. 6)

3. the probability $P$ that a particle $m$ enters the orifice in a direction between $\vartheta$ and $\vartheta + \mathrm{d}\vartheta$.

$P$ is proportional to the solid angle $\mathrm{d}\omega$ at $\mathrm{d}S$ subtended by these directions. The solid angle element $\mathrm{d}\omega$ is measured by the area on a sphere of radius $r$ (Fig. 11.3) of the circular zone between the angle $\vartheta$ and $\vartheta + \mathrm{d}\vartheta$, and is equal to $2\pi \sin\vartheta \mathrm{d}\vartheta$. The total solid angle into which the particles escape from the cell orifice is equal to $2\pi$, so the probability $P = \mathrm{d}\omega/2\pi = \sin\vartheta \mathrm{d}\vartheta$, and $\mathrm{d}\Gamma_\vartheta \sim \sin\vartheta \mathrm{d}\vartheta$.

The differential angular effusion rate $\mathrm{d}\Gamma_\vartheta$ of particles escaping from the cell orifice may be expressed by the formula

$$\mathrm{d}\Gamma_\vartheta = (C_0 A_e \cos\vartheta)\, \frac{\Gamma_e}{A_e}\, (\sin\vartheta \, \mathrm{d}\vartheta), \tag{11.40}$$

where $C_0$ is a proportionality constant. This constant can be estimated, taking into consideration the obvious fact that $0 \leq \vartheta \leq \pi/2$ and that the total effusion rate of particles escaping from the orifice must be equal to $\Gamma_e$ given by $3.51 \times 10^{22} p A_e (MT)^{-1/2}$ (see (6.5)). Thus

$$\Gamma_e = \int_0^{\pi/2} \mathrm{d}\Gamma_\vartheta = \int_0^{\pi/2} C_0 \Gamma_e \cos\vartheta \sin\vartheta \, \mathrm{d}\vartheta = \frac{C_0 \Gamma_e}{2} \tag{11.41}$$

which means that the proportionality constant is $C_0 = 2$. The formula (11.40) may now be written in the form

$$\mathrm{d}\Gamma_\vartheta = \frac{\Gamma_e}{\pi} \cos\vartheta \, \mathrm{d}\vartheta. \tag{11.42}$$

Equation (11.42) represents the so-called cosine law of effusion, which is equivalent to Lambert's law known in optics (see [11.20]; Sect. 1.3). Using the Knudsen equation in the form (6.5) and the cosine law (11.42), one may easily calculate the impingement rate $I_A$ (also called in the literature the "beam flux") of the particles beam (molecular or atomic) at the central point A of the substrate axially mounted in front of the effusion cell (Fig. 11.4a). This impingement rate is given by the angular effusion rate for $\vartheta = 0$ per unit substrate area $\mathrm{d}S = \mathrm{d}\omega r_A^2$ around point $A$:

$$I_A \equiv \left. \frac{\mathrm{d}\Gamma_\vartheta}{\mathrm{d}S} \right|_{\vartheta=0} = \frac{\Gamma_e}{\pi r_A^2} = 1.118 \times 10^{22} \frac{p A_e}{r_A^2 \sqrt{MT}} \quad \left[ \frac{\text{molecules}}{\text{cm}^2\,\text{s}} \right] \tag{11.43}$$

where $p$ is expressed in torrs, and the other quantities in cgs units. If SI units are used, the numerical factor must be changed to $2.653 \times 10^{22}$ [11.19].

The flux at the edge point $B$ is defined by $I_B = \mathrm{d}\Gamma_\vartheta / \mathrm{d}S(\vartheta)$, where $\mathrm{d}S(\vartheta)$ is the surface element around point $B$ given by $\mathrm{d}S(\vartheta) = ((\mathrm{d}\omega r_A^2)/\cos\vartheta)(r_B^2/r_A^2)$. Thus $I_B$ can be written in the form

$$I_B = \frac{\Gamma_e \cos^2\vartheta}{\pi r_A^2} \left( \frac{r_A^2}{r_B^2} \right) = I_A \cos^4\vartheta. \tag{11.44}$$

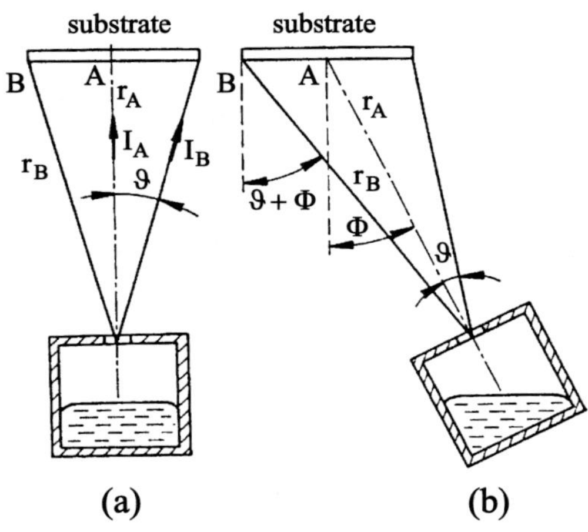


**Fig. 11.4.** Particle flux distribution across a substrate mounted in the MBE growth chamber axially (**a**) and non-axially (**b**) with respect to the effusion cell orifice (taken from [11.19])

Proceeding in an analogous way the impingement rate from a source which is tilted by an angle $\phi$ from the perpendicular substrate axis can be calculated (Fig. 11.4b):

$$I'_A = I_A \cos\phi, \tag{11.45}$$

$$I'_B = I_A \frac{r_A^2}{r_B^2} \cos\vartheta \cos(\vartheta + \phi). \tag{11.46}$$

In the second approach to effusion transport phenomena, real effusion cells are treated by modifying the ideal effusion laws, most frequently by introducing to these laws relevant correction factors. The effusion cells used in MBE systems are either cylindrical crucibles with a single circular collimating orifice in the crucible lid of a definite thickness, or single-channel-type crucibles cylindrical, and more frequently, conical in shape without a lid (see Fig. 6.2), however, often with a conical insert (see Fig. 7.6).

The first kind of effusion cells are also called near-ideal cylindrical effusion cells [11.18, 19], because the vapor pressure exhibits a near-equilibrium value at the exit orifice. Such cells are used in MBE systems as dopant or as calibration cells. The theoretical analysis of such cells was given by Clausing [11.21, 22], who assumed a random return of particles to the gas reservoir from the orifice walls according to a cosine law of return. According to Clausing's theory the Knudsen effusion equation (see (6.4)), and consequently also (11.43), should be multiplied by a correction factor $W_a$, i.e., the orifice transmission factor. This factor has the meaning of the probability that a particle which enters the orifice in the effusion cell lid at the side of the vapor volume, over the solid charge of the cell, goes directly to the vacuum environment ($p \approx 0$) without having been back scattered to the vapor by the orifice walls. The Clausing's orifice transmission factor $W_a$ depends only on the geometry of the orifice. As an example, in Table 11.1 some values are given of $W_a$ for a straight tubular orifice of length $L_0$ and diameter $d_0$. Further modifications to the effusion equation introduced later, after Clausing's work, may be found in [11.19]. Here, we will give in Table 11.2 the values of another correction factor, relevant when analyzing the effusion from a near-ideal cylindrical effusion cell with symmetrical orifice ($L_0 = d_0$) into various directions, and show in Fig. 11.5 a plot in polar coordinates of the angular distribution of

**Table 11.1.** Values of Clausing's orifice transmission factor $W_a$ for straight tubular orifices with various orifice dimensions [11.22].

| $L_0/d_0$ | $W_a$ | $L_0/d_0$ | $W_a$ |
|---|---|---|---|
| 0 | 1.0000 | 0.75 | 0.5810 |
| 0.1 | 0.9092 | 1.00 | 0.5136 |
| 0.2 | 0.8341 | 1.50 | 0.4205 |
| 0.4 | 0.7177 | 2.00 | 0.3589 |
| 0.5 | 0.6720 | 4.00 | 0.2316 |

**Table 11.2.** Values of the correction factor $C_0$ for the case of a symmetrical orifice for various effusion angles $\vartheta$ [11.21].

| $\vartheta$ | $C_0$ | $\vartheta$ | $C_0$ |
|---|---|---|---|
| 0° | 1.0000 | 50° | 0.4259 |
| 10° | 0.8882 | 60° | 0.3687 |
| 20° | 0.7721 | 70° | 0.3221 |
| 30° | 0.6463 | 80° | 0.2811 |
| 40° | 0.5183 | 90° | 0.2426 |

gas particles effused from this kind of cell.

It is evident that a significant collimation of the particle beam generated in an effusion cell can be reached by enlarging the value of the ratio $L_0/d_0$ for the orifice of the cell. This problem has been analyzed by Dayton [11.23], who proposed more general modification of Clausing's theory [11.18]. The angular term $C_0 \cos\vartheta$ of the flux angular distribution for several positions of the charge surface in a single-channel effusion cell of diameter $d_0$ according to Dayton's theory is shown in Fig. 11.6.

Conical effusion cells are at present frequently used as sources of the main molecular or atomic beams in MBE systems, i.e., for beams of the constituent elements of the film to be grown [11.19]. They are nonequilibrium cells, which means that the vapor and the condensed phase of the contents are not in ther-

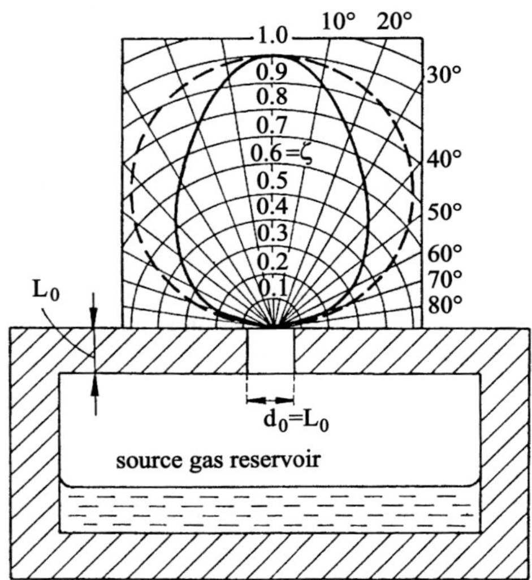


**Fig. 11.5.** A plot in polar coordinates of the angular distribution of gas particles effused from a near-ideal cylindrical cell with a symmetrical orifice ($L_0 = d_0$). The dashed line gives the relevant distribution for an ideal effusion cell ($L_0 = 0$), i.e., according to the cosine law (11.42) (taken from [11.21])

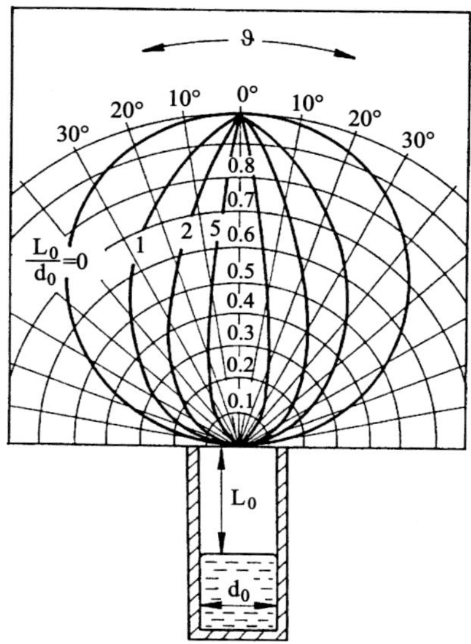


**Fig. 11.6.** The angular term $C_0 \cos \vartheta$ of the flux angular distribution for several positions of the charge surface in a single-channel effusion cell of diameter $d_0$ (taken from [11.23])

modynamic equilibrium with one another. Therefore, the effusion equation (6.4) as well as the cosine law (11.42) cannot be directly applied in analyzing the effusion patterns of conical cells. Both of these equations should be corrected for this case. One way of doing this is based on the procedure proposed by Shen[11.24] of treating the effusing surface of the cell as a combination of many small point sources, and subsequently superimposing their fluxes.

As an example of such a procedure, the analysis given by Curless [11.25] can be considered. In this analysis the following model of the effusion process has been used. The source material is assumed to be at a single temperature (the evaporation rate for the material is independent of position). It is assumed that the effusion cell consists of a truncated cone shown in Fig. 11.7, or its limiting case of a cylinder, and is positioned with respect to the substrate as shown in Fig. 11.8. The rate of evaporation at a point on the wall of the crucible is taken to be equal to the flux incident on the wall at that point, unless that point is defined as being wetted by the source material, in which case the rate of evaporation is given by the rate of evaporation for the source material. The bottom of the crucible is assumed to be covered by the evaporating source material with its surface normal to the axis of symmetry of the crucible. All fluxes leaving the surfaces under consideration obey a cosine distribution. The effect of surface diffusion is ignored, as is any interaction between particles (molecules or atoms).

Using the procedure of Curless [11.25], Yamashita et al. [11.26] have analyzed the beam fluxes from conical crucibles arranged slantingly with respect to the substrate and containing liquid Ga or Al. They have analyzed the

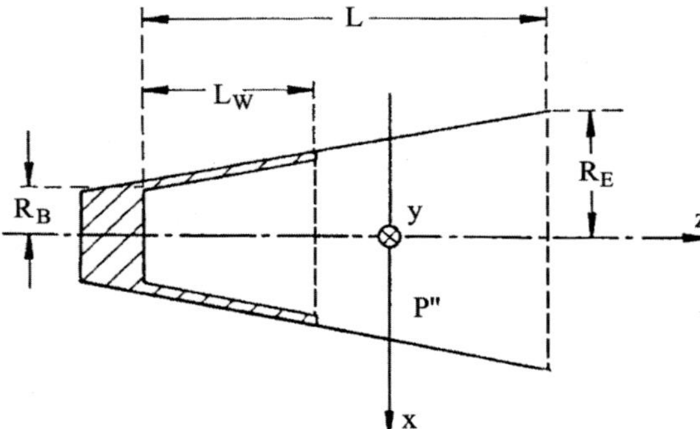


**Fig. 11.7.** Cross-sectional view of an idealized conical effusion cell (taken from [11.25])

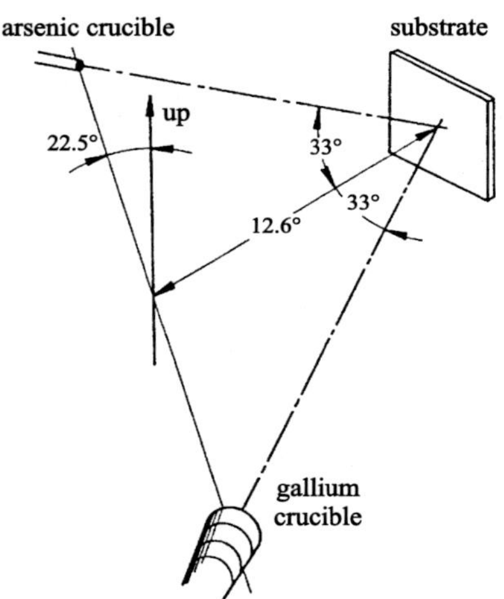


**Fig. 11.8.** Geometry of the "source-substrate" system considered by Curless for simulation of MBE growth of GaAs (taken from [11.25])

influence of the shape of the conical crucible on the distribution of the beam flux upon the substrate surface, as well as the influence of the arrangement angle $\alpha$ of the beam source on the flux distribution. The configuration parameters of the beam source are shown in Fig. 11.9, while the results of the calculations are shown in Fig. 11.10. It is evident from these calculations that in the case of a liquid source, a cylindrical crucible is more effective in improving the uniformity of the flux distribution than a conical crucible with a large tapering angle $\zeta$, since the effusion cell is slantingly arranged. Since the direct flux from the source becomes dominant as the tapering angle of a conical crucible increases, the distribution of the beam flux is then mostly affected by the source surface determined by the arrangement angle.

For more information on the effusion characteristics of conical effusion cells the reader is referred to [11.27–30].

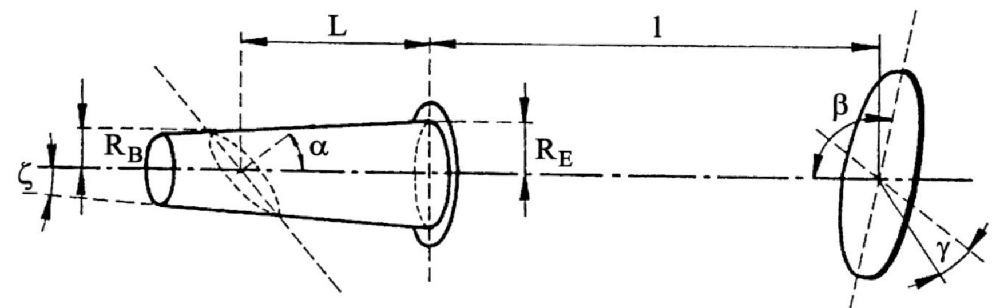


**Fig. 11.9.** Configuration parameters of the conical effusion cells analyzed by Yamashita et al. [11.26]. The following numerical data were used for the Ga (Al) cell: $L = 6.5(5.5)\,\text{cm}$, $R_B = 0.78(0.61)\,\text{cm}$, $R_E = 0.96(0.66)\,\text{cm}$, $\alpha = 58(58)°$, $l = 12(12)\,\text{cm}$, $\beta = 95(95)°$ and $\gamma = 11(-11)°$ (taken from [11.19])

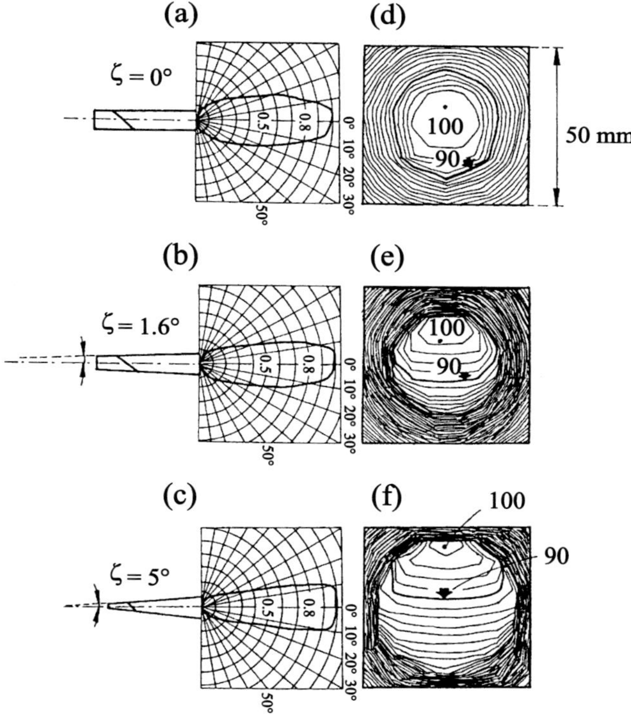


**Fig. 11.10.** Distributions of the calculated beam fluxes (on the 5 cm × 5 cm substrate square) generated by conical liquid charge effusion cells with different tapering angles $\zeta$. **(a,b,c)** show angular distributions, while **(d,e,f)** shows the distributions on the substrate surface (taken from [11.26])

## 11.3 Phase Equilibria and Phase Transitions

The condition of thermodynamic equilibrium is expressed in (11.4) and (11.10). Thermodynamics can, however, be used not only to describe the driving force and maximum growth rate in epitaxy, but also to calculate the composition of multicomponent solids grown by epitaxy. The equilibrium between solid and liquid or vapor phases is of concern for understanding epitaxy. The vapor phase is commonly considered to be ideal (compare the ideal gas model), i.e., $x_i = p_i/p_i^0$. Calculations involving the liquid and/or solid phases must deal with their non ideality. Fortunately, for most of the material systems grown, so far, by epitaxy the liquid and solid phases can be described using the solution thermodynamic models, namely the concepts of an ideal and a regular solution.

### 11.3.1 Ideal and Regular Solutions

Let us consider first a binary solid solution $AB$ (a binary compound), i.e., a solution consisting of two constituent species $A$ and $B$. The Gibbs free energy of this solution can be calculated from the free energies of pure $A$ and pure $B$ species as follows. Assume that $A$ and $B$ have the same crystal structures in their pure states and that they can be mixed in any proportion to make a homogeneous solid solution with the same crystal structure. Mixing together $x_A$ moles of $A$ and $x_B$ moles of $B$ to form one mole of solid solution gives the obvious relation $x_A + x_B = 1$, where $x_A$ and $x_B$ are the mole fractions of $A$ and $B$ in the solution (alloy). To calculate the free energy of the alloy, the mixing can be envisioned to occur in two steps: (**a**) bring together $x_A$ moles of pure $A$ and $x_B$ moles of pure $B$, and (**b**) allow the $A$ and $B$ atoms to mix together to make a homogeneous solid solution, the binary alloy $AB$. After step (**a**), the free energy of the system is given by

$$G_1 = x_A G_A + x_B G_B, \tag{11.47}$$

where $G_A$ and $G_B$ are the molar free energies (energies taken per mole) of pure $A$ and pure $B$ at atmospheric pressure and the temperature of interest. The quantity $G_1$ can be conveniently represented on a molar free energy diagram, where this energy is plotted as a function of the mole fraction $x_B$, as shown in Fig. 11.11. In this case, $G_1$ lies on a straight line between $G_A$ and $G_B$ for all alloy compositions.

When the $A$ and $B$ atoms are allowed to mix in step (**b**) mentioned above, the free energy of the solid solution changes by $\Delta G_M$ so that the resulting Gibbs free energy after mixing $G_2$ is given as $G_2 = G_1 + \Delta G_M$. Using the (11.1) and the notation $\Delta H_M = H_2 - H_1$ and $\Delta S_M = S_2 - S_1$ the change in Gibbs free energy caused by mixing can be expressed by

$$\Delta G_M = \Delta H_M - T\Delta S_M, \tag{11.48}$$

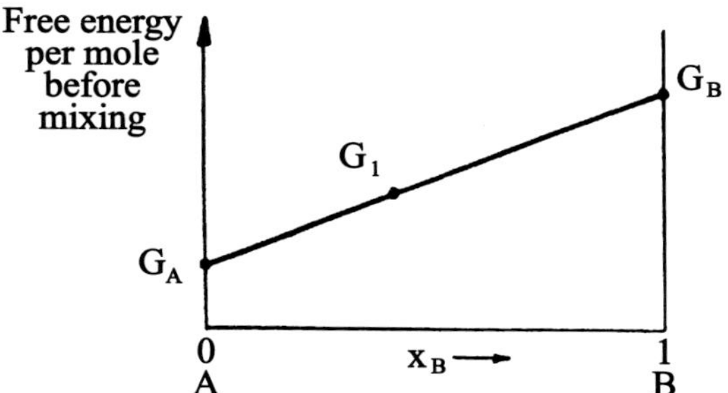


**Fig. 11.11.** Variation of $G_1$ (the free energy before mixing) with alloy composition $x_B$ (or $x_A$) (taken from [11.31])

when the volume change of the system during mixing is ignored. $\Delta H_M$ (the change in enthalpy) is the heat absorbed or evolved during step (**b**), while $\Delta S_M$ is the difference in entropy between the mixed and unmixed states.

The simplest type of mixing occurs for a so-called ideal solution where $\Delta H_M = 0$ and the free energy change on mixing is due only to the change in entropy $\Delta G_M = -T\Delta S_M$. There are two contributions to the entropy, a thermal contribution and a configurational contribution. If there is no volume or heat change during mixing then the only contribution to $\Delta S_M$ is the change in configurational entropy, which is quantitatively related to the randomness of the solid solution by the Boltzmann equation

$$S = k_B \ln Z, \tag{11.49}$$

where $k_B$ is Boltzmann's constant and $Z$ is the number of distinguishable ways of arranging the atoms in the solid solution. Assuming that $A$ and $B$ atoms mix to form a substitutional solid solution where all configurations of $A$ and $B$ atoms are equally probable gives an expression for $\Delta S_M$ (see [11.1,31,32])

$$\Delta S_M = -R(x_A \ln x_A + x_B \ln x_B). \tag{11.50}$$

Note that since $x_A$ and $x_B$ are less than unity, $\Delta S_M$ is positive and there is an increase in entropy on mixing as expected physically. The free energy on mixing is then obtained as

$$\Delta G_M = RT(x_A \ln x_A + x_B \ln x_B). \tag{11.51}$$

The actual free energy of the solution depends on $G_A$ and $G_B$ and combining (11.47) and (11.51) yields

$$G_2 = G_1 + \Delta G_M = x_A G_A + x_B G_B + RT(x_A \ln x_A + x_B \ln x_B), \tag{11.52}$$

which is illustrated graphically as a function of temperature and composition in Fig. 11.12.

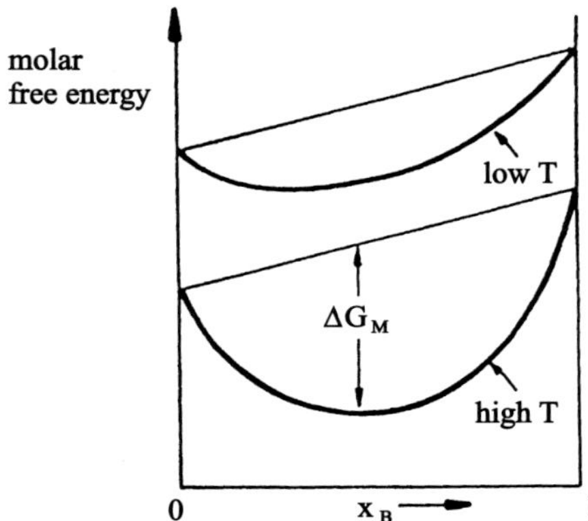


**Fig. 11.12.** Molar free energy of mixing for an ideal solid solution where $\Delta H_M = 0$ (taken from [11.31])

When growing alloys, often the relation between the free energy of the alloy and the changes in the number of atoms belonging to each phase, creating the alloy, should be known. This relation can be expressed by using the chemical potentials of the relevant phases. At constant temperature and pressure the change of free energy of the alloy, resulting from adding or removing of atoms from the phases $A$ and $B$, is given by

$$\mathrm{d}G_t = \mu_A \mathrm{d}n_A + \mu_B \mathrm{d}n_B . \tag{11.53}$$

If $A$ and $B$ atoms are added to a solution in the same proportion as the original composition of the solution (i.e., such that the ratio $\mathrm{d}n_A/\mathrm{d}n_B = x_A/x_B$), then the free energy of the solution increases by the molar free energy $G$. Thus, from (11.53)

$$G = \mu_A x_A + \mu_B x_B . \tag{11.54}$$

When $G$ is known as a function of $x_A$ and $x_B$ as shown in Fig. 11.12 for example, $\mu_A$ and $\mu_B$ can be obtained by extrapolating the tangent to the $G$ curve to the sides of the molar free energy diagram as shown in Fig. 11.13. It then becomes evident that $\mu_A$ and $\mu_B$ vary systematically with the composition of the phase $x_B$. Comparison of (11.52) and (11.54) gives $\mu_A$ and $\mu_B$ for an ideal solution as

$$\mu_A = G_A + RT \ln x_A \quad \text{and} \quad \mu_B = G_B + RT \ln x_B \tag{11.55}$$

which is a simple way of representing (11.52). These relationships are also shown on the free-energy composition diagram in Fig. 11.13, where the distances $ac$ and $bd$ are $-RT \ln x_A$ and $-RT \ln x_B$, respectively.

So far, it has been assumed that $\Delta H_M = 0$. This occurs rarely because mixing is usually endothermic (heat absorbed) or exothermic (heat evolved). However, we can readily extend the simple model developed for an ideal solution to include $\Delta H_M$ using the so-called regular solution model, which is basically a nearest-neighbor bond-counting approach to alloying [11.33].

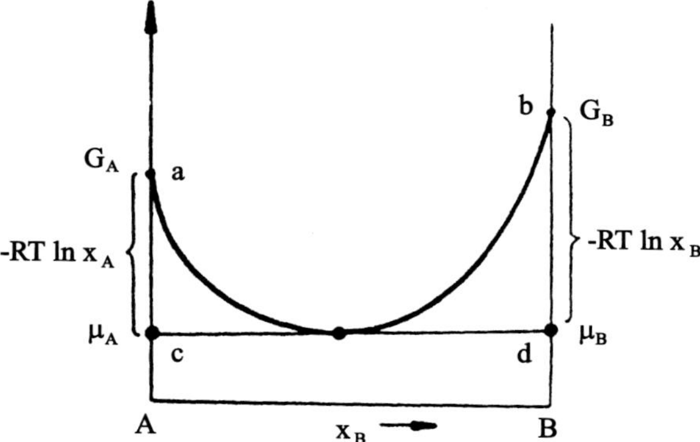


**Fig. 11.13.** Illustration of the relationship between the free energy curve and the chemical potentials for an ideal solution (taken from [11.31])

Three simple assumptions form the basis of the regular solution model:

(i) all constituents are located at points on a lattice,
(ii) all interactions in the solution occur only between nearest-neighbor pairs,
(iii) the configuration of the constituents is random.

In this model it is also assumed that the heat of mixing $\Delta H_M$ is only due to the bond energies of adjacent atoms. For this latter assumption to be valid, it is necessary that the volumes of pure $A$ and $B$ are equal and do not change during mixing so that the inter atomic distances and bond energies are independent of composition. The structure of a simple, binary solid solution is shown in Fig. 11.14. Three types of bonds are present

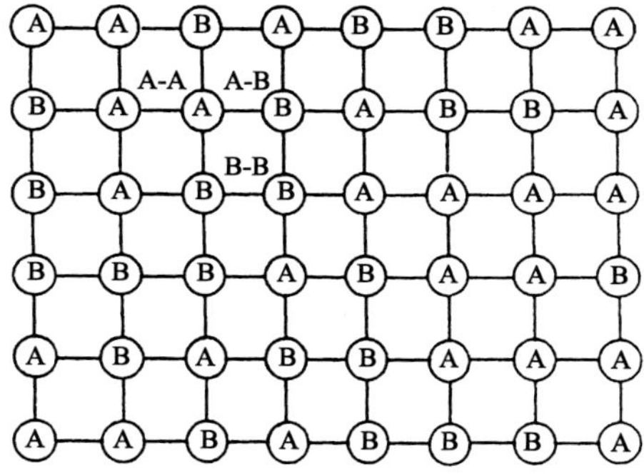


**Fig. 11.14.** Different types of interatomic bonds in an AB solid solution (taken from [11.31])

(i) A–A bonds, each with an energy $H_{AA}$,

(ii) B–B bonds, each with an energy $H_{BB}$, and

(iii) A–B bonds, each with an energy $H_{AB}$.

in the structure:

In the frames of the regular-solution model it is possible to express the entropy and enthalpy of mixing of a binary solution $A + B$ in simple terms. The entropy of mixing is simply the ideal configurational entropy of mixing, given by (11.50). The enthalpy of mixing is obtained by summing the nearest-neighbor bond energies

$$\Delta H_M = x_A x_B \Omega = x_A(1 - x_A)\Omega \tag{11.56}$$

where the interaction parameter is $\Omega = zN_A(H_{AB} - (1/2)(H_{AA} - H_{BB}))$. $H_{AB}, H_{AA}$ and $H_{BB}$ are the nearest neighbor bond energies for the combinations of $A$ and $B$ atoms, $z$ is the number of nearest-neighbors and $N_A$ is Avogadro's number. The activity coefficient in the solution $A+B$ is obtained from the expression

$$RT \ln \gamma_i = \frac{\partial \Delta H_M}{\partial n_i} \tag{11.57}$$

$$\ln \gamma_i = (1 - x_i)^2 \frac{\Omega}{RT}. \tag{11.58}$$

### 11.3.2 The Liquid–Solid Phase Diagram

Consider first the simplest system of an $AB$ solid in equilibrium with the liquid solution $A + B$. The most important manifestation of solution thermodynamics of liquid and solid solutions is the liquid–solid phase diagram for multicomponent systems such as, e.g., the semiconductor two- and three-component (binary and ternary) III–V compound systems. Considering this case as an example of thermodynamic treatment of the phase equilibria in epitaxy, we will follow the review of Stringfellow [11.11]. In the considered case, (11.4) takes the form

$$\mu^s_{AB} = \mu^l_A + \mu^l_B \tag{11.59}$$

where $\mu_i$ is given by (11.10). The solid AB is the pure standard state, so $a_{AB} = 1$ and (11.59) can be rewritten as

$$RT \ln(a^l_A a^l_B) + (\mu^{0l}_A + \mu^{0l}_B - \mu^{0s}_{AB}) = 0. \tag{11.60}$$

The quantity in brackets expresses the Gibbs free energy of fusion $\Delta G_F$ of $AB$ minus the free energy of mixing $\Delta G_M$ of the stoichiometric liquid. It represents the free energy change upon melting a mole of $AB$ and then separating the liquid with $x_A = x_B = \frac{1}{2}$ into one gram-atomic weight, each of pure components $A$ and $B$. In addition, $\Delta G_{F(AB)}$ may be written in terms of the entropy of fusion, $\Delta S_{F(AB)}$

$$\Delta G_{F(AB)} = \Delta S_{F(AB)}(T_{F(AB)} - T) \tag{11.61}$$

by assuming $\Delta S_{F(AB)}$ and $\Delta H_{F(AB)}$ to be temperature-independent. Assuming the liquid to be a regular solution, for which $\Delta G_M$ may be obtained from (11.50) and (11.56) and using (11.58) to obtain the activity coefficient in solution, and (11.60) may be rewritten in the form [11.34]

$$\begin{aligned}\ln(4x_B(1-x_B)) + 2\left(x_B - \tfrac{1}{2}\right)^2 \frac{\Omega^l_{AB}}{RT} \\ = -\left(\frac{\Delta S_{F(AB)}}{R}\right)\left(\frac{T_{F(AB)}}{T} - 1\right).\end{aligned} \tag{11.62}$$

This expresses the composition of the liquid in equilibrium with the pure solid $AB$ as a function of temperature. Values of $\Delta S_F$ and $T_F$ may be found in [11.35], while $\Omega^l$ may be determined by fitting (11.62) to experimental data using $\Omega^l$ as an adjustable parameter. The interaction parameter generally varies between 0 and - 6000 cal mol$^{-1}$ for liquid solutions of the III–V compounds group [11.11]. These values are indicative of fairly weak interactions, characteristic of the metallic bonding in these liquids. By adding a term due to the screened electronegativity difference between the group III and the group V elements in the liquid one may extend the theory of metal solutions [11.36] to describe the thermodynamic properties of the liquid III–V compound solutions. The result is the following equation [11.37] for the enthalpy of mixing for the considered system $AB$

$$\begin{aligned}\Delta H_M = (x_A V_A + x_B V_B) \\ \cdot \left[(\delta_A - \delta_B)^2 - 3 \cdot 10^4 \frac{(\chi_A - \chi_B)^2}{\sqrt{V_A V_B}}\right] \varphi_A \varphi_B\end{aligned} \tag{11.63}$$

where $V_i, \chi_i, \varphi_i$ and $\delta_i$ are the molar volume, electronegativity, volume fraction and solubility parameter of each element.

The thermodynamics of mixing in the solid phase is illustrated by considering ternary compounds of the type $A_x B_{1-x} C$, a solution of compounds $AC$ and $BC$, where $A$ and $B$ are both bonded to $C$ and thus share a common sublattice in the crystal [11.38].

The equilibrium conditions for the ternary system may be obtained in exactly the same way as described for binary systems, by equating the chemical potentials of the two solid components in the two phases, i.e., $\mu^l_A + \mu^l_C = \mu^s_{AC}$ and $\mu^l_B + \mu^l_C = \mu^s_{BC}$. In addition the activities in the solid must be taken into account using the regular solution model. The result is a set of two equations

$$\begin{aligned}\ln(x^s_{AC}\gamma^s_{AC}) &= \ln(4x^l_A x^l_C \gamma^l_A \gamma^l_C) \\ &\quad + \Delta S_{F(AC)} \frac{T_{F(AC)} - T}{RT} - \frac{\Omega^l_{AC}}{2RT} \\ \ln(x^s_{BC}\gamma^s_{BC}) &= \ln(4x^l_B x^l_C \gamma^l_B \gamma^l_C) \\ &\quad + \Delta S_{F(BC)} \frac{T_{F(BC)} - T}{RT} - \frac{\Omega^l_{BC}}{2RT}.\end{aligned} \tag{11.64}$$

The solid activity coefficients are calculated using (11.58), with $\Omega$ replaced by $\Omega^s$. The activity coefficients in the regular ternary liquid may be written in the form [11.11]

$$\begin{aligned} RT\ln\gamma_A^l &= \Omega_{AC}^l(x_C^l)^2 + \Omega_{AB}^l(x_B^l)^2 + (\Omega_{AC}^l + \Omega_{AB}^l - \Omega_{BC}^l)x_B^l x_C^l \\ RT\ln\gamma_B^l &= \Omega_{BC}^l(x_C^l)^2 + \Omega_{AB}^l(x_A^l)^2 + (\Omega_{BC}^l + \Omega_{AB}^l - \Omega_{AC}^l)x_A^l x_C^l \\ RT\ln\gamma_C^l &= \Omega_{AC}^l(x_A^l)^2 + \Omega_{BC}^l(x_B^l)^2 + (\Omega_{AC}^l + \Omega_{BC}^l - \Omega_{AB}^l)x_A^l x_B^l . \end{aligned} \tag{11.65}$$

There are two characteristic curves in the liquid–solid phase diagram. These are the solidus and liquidus curves (in a ternary compound phase diagram these curves become surfaces). The first, the solidus, is the border line on the "temperature–concentration/composition" diagram between the pure solid region and the region where solid and liquid can coexist, while the second, the liquidus, is the border line on this diagram between the pure liquid region and the solid–liquid coexistence region [11.1] (see Fig. 11.16a, where the upper line is the liquidus, while the lower line is the solidus).

The calculation of the ternary liquidus and solidus can be carried out using the values of the three liquid interaction parameters discussed above, and $\Omega^s$ obtained from the pseudobinary solidus data [11.35]. A schematic III–V ternary compound phase diagram is shown in Fig. 11.15. It is remarkably simple for a ternary phase diagram [11.11]. A single liquidus surface covers nearly the entire phase diagram and the solidus phase, the III–V alloy, is represented by a very thin plane running between $AC$ and $BC$. Two representations of the actual Al-Ga-As phase diagram are shown in Fig. 11.16; the pseudobinary section (**a**) is viewed from perpendicular to the $T$ axis and section (**b**) is viewed parallel to the $T$ axis.

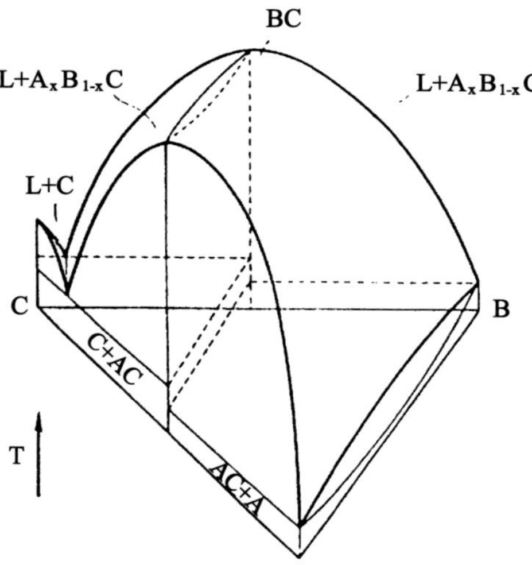


**Fig. 11.15.** Schematic ternary III–V compound phase diagram (taken from [11.11])

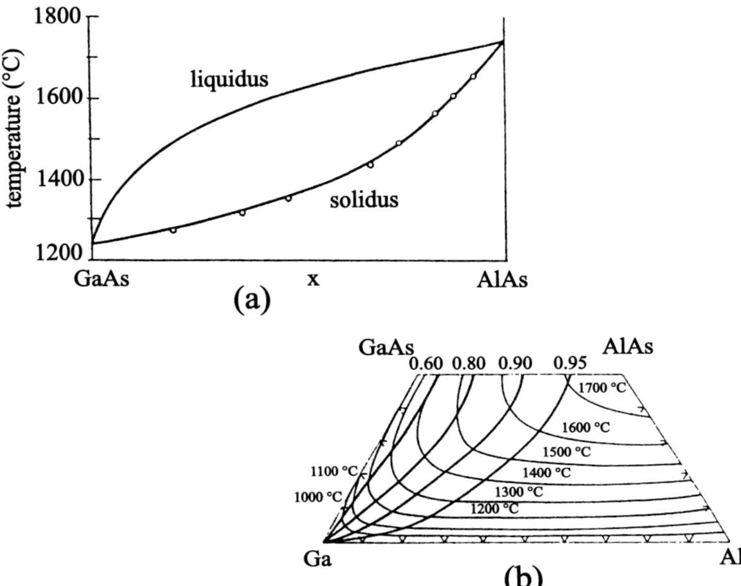


**Fig. 11.16.** Al-Ga-As phase diagram: (**a**) pseudobinary section and (**b**) isothermal sections and isosolidus concentration curves for the metal-rich portion of the diagram (taken from [11.11])

The fundamental significance of the liquid-phase interaction parameter was discussed above. For the solid phase the same valence electrons which determine the optical and electrical properties of the semiconductor also determine the bonding [11.9]. This has allowed simple models (developed to interpret the dielectric function, band gap and other optical properties) to be used to treat the bonding in solid alloys, which determines $\Delta H_M$ and hence $\Omega^s$. Using the Phillips–van Vechten dielectric theory of electronegativity [11.39], Stringfellow [11.38] developed a model which allows for an accurate calculation of $\Omega^s$ in terms of the difference in lattice constants between $AB$ and $AC$:

$$\Omega^s_{AB-BC} = 5 \cdot 10^7 (a^0_{AC} - a^0_{BC})^2 [(a^0_{AC} + a^0_{BC})/2]^{-4.5}. \tag{11.66}$$

This model for the solid phase can also be used to describe the thermodynamics of VPE, where the vapor is treated as an ideal gas. In this way, it is possible to predict solid alloy composition versus temperature and vapor composition, as it is required for growing by VPE high quality epilayers of well defined chemical composition [11.11].

Concluding the consideration on the regular solution model, it is important to note that, although we have presented its application to problems of semiconductor compound phase equilibria, the regular solution model has more general applicability and parallel treatments exist, for example, for metallic alloys [11.33] or for polymer solutions [11.40].

### 11.3.3 Phase Transitions in Epitaxy

Let us now proceed to problems of phase transitions, that is to thermodynamics of crystal growth, as related to epitaxy. We consider in the following a liquid–crystal phase transition as an example. Isobaric variation of the Gibbs free energies of liquid and crystal, $G^l$ and $G^s$, respectively, are shown in Fig. 11.17 as a function of temperature $T$. At a low temperature $G^l$ lies higher than $G^s$ showing that the crystal is more stable than the liquid, but with increasing $T$, $G^l$ decreases faster than $G^s$ due to the large entropy $S^l$ of liquid, compared to $S^s$ of the crystal. At the melting point $T_m(p), G^l$ and $G^s$ cross each other, thus $G^s(T_m, p) = G^l(T_m, p)$, and the liquid becomes more stable than the solid for $T > T_m$.

On heating the crystal under constant pressure, its temperature first increases, as shown in Fig. 11.18. At the melting point $T_m$ the temperature stops increasing and the applied heat is consumed to change the state of the matter from crystal to liquid. Since the absorbed heat does not appear explicitly as a temperature rise, it is called the latent heat $L$. Accord-

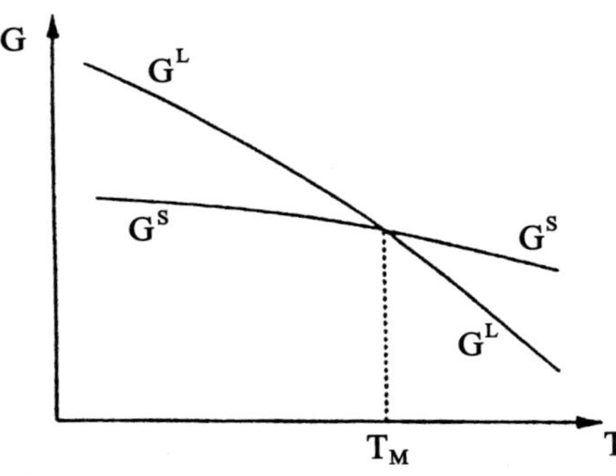


**Fig. 11.17.** Isobaric variation of the Gibbs free energies of the liquid, $G^l$ and of the crystal $G^s$ as functions of temperature (taken from [11.7])

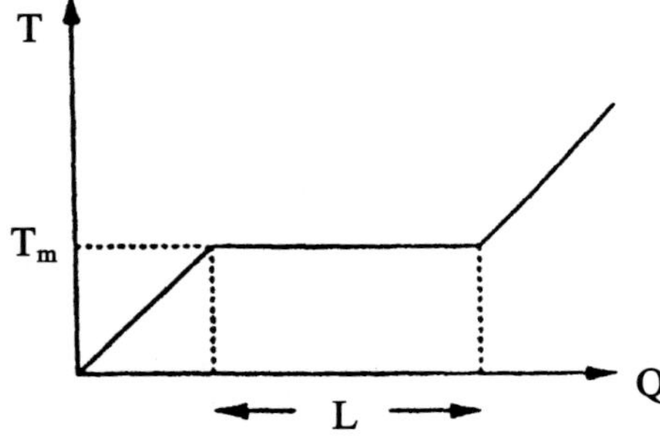


**Fig. 11.18.** Isobaric variation of the temperature due to the heating. At the melting temperature $T_m$, the crystal starts to melt, and until the completion of melting, the system absorbs the latent heat $L$. (taken from [11.7])

ing to the first law of thermodynamics the applied heat changes into work $p\mathrm{d}V$ done to the environment and the increment of the internal energy $\mathrm{d}U$. Since the pressure is kept constant, the heat changes into the enthalpy (see (11.2)), thus, $\mathrm{d}H = \mathrm{d}U + p\mathrm{d}V$. Consequently, the latent heat observed at the melting point corresponds to the enthalpy difference of the two phases as $L = H^l(T_m, p) - H^s(T_m, p)$. Since Gibbs free energies of both phases are equal at the melting point, $G^s = H^s - T_m S^s = G^l = H^l - T_m S^l$, the latent heat is proportional to the entropy difference $\Delta S = S^l - S^s$:

$$L = H^l - H^s = T_m \Delta S. \tag{11.67}$$

According to (11.5), entropy is the temperature derivative of the Gibbs free energy as $S = -(\partial G/\partial T)_p$. Thus the phase transition with latent heat is associated with a discontinuity in the slope of the Gibbs free energy.

$$\left(\frac{\partial G^l}{\partial T}\right)_p - \left(\frac{\partial G^s}{\partial T}\right)_p = -\frac{L}{T_m}. \tag{11.68}$$

This type of phase transition with a discontinuity in the first derivative of $G$ is known in thermodynamics as a first-order phase transition.

Let us now cool the liquid to a temperature $T$ below the melting point $T_m$. The Gibbs free energy of a crystal $G^s(T, p)$ is lower than that of the liquid $G^l(T, p)$ as shown in Fig. 11.17, and the true equilibrium state is a crystalline phase. However, the whole liquid cannot instantaneously turn into the crystal. We often experience that the liquid is supercooled for a long while near the melting point. Eventually, a small crystalline nucleus is formed in the liquid, and then it grows. The crystal grows by the advancement of the crystallization interface (see Fig. 1.1). The driving force in this case is the difference of the Gibbs free energies of the liquid and the solid $\Delta G = G^l - G^s$. Since this difference vanishes at the melting point, one can expand $\Delta G$ up to the first order of the undercooling $\Delta T = T_m - T$, as

$$\Delta G \approx \left(\frac{\partial G^l}{\partial T}\right)_p (T - T_m) - \left(\frac{\partial G^s}{\partial T}\right)_p (T - T_m) = L\frac{\Delta T}{T_m}. \tag{11.69}$$

Under this chemical driving, the phase transition proceeds, i.e., the crystal grows. For a liquid molecule to be incorporated into the crystalline order, it has to change the configuration. But around the liquid molecule there is a high density of other molecules and they hinder the free motion of the molecule. It can mainly vibrate around its average position with a frequency $\nu$, which is of the order of that of the lattice vibration [11.7]. In order for a liquid molecule to change the configuration drastically, it has to overcome the energy barrier $E_d$ of molecular diffusion, as shown in Fig. 11.19. At a temperature $T$ a molecule acquires the energy fluctuation $E_d$ with a probability proportional to the Boltzmann factor (the statistical weight) which has the form $\exp(-E_d/k_BT)$. Therefore the crystallization rate per unit time is given as $\nu \exp(-E_d/k_BT)$.

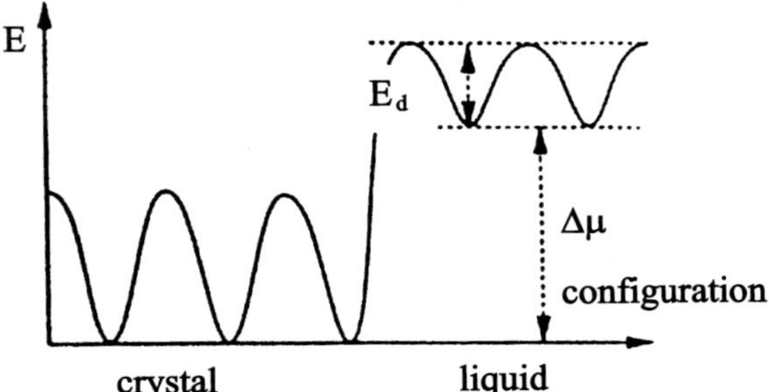


**Fig. 11.19.** Schematics of the potential surface in configuration space. The crystal phase corresponds to a stable phase, the liquid to a metastable phase, and in between is the diffusion activation energy (taken from [11.7])

There is, however, a counter-effect, namely the melting, which means a passage of a crystal molecule into the liquid state. Since the Gibbs free energy per molecule, i.e., the chemical potential, is higher in the liquid than in the solid, $\mu^l > \mu^s$, the rate of melting is smaller than that of crystallization by a factor of $\exp(-\Delta\mu/k_BT)$. By the crystallization of one molecular layer the crystallization interface (the growth front) moves inside the metastable liquid phase by a distance equal to the molecular layer thickness $a_{\mathrm{mol}}$. Thus the growth rate is given by

$$V_g = a_{\mathrm{mol}}\nu \exp(-E_d/k_BT)(1 - \exp(-\Delta\mu/k_BT)). \tag{11.70}$$

In epitaxial growth from the liquid phase (the case of LPE) the phase transition between the liquid and the solid phases concerns liquid solutions and the substrate (or epilayer) crystal. In this case solute atoms should occur in front of the crystallization interface to start the liquid–solid phase transition of epitaxial growth. The probability to find a solute atom at a certain crystallization point with a unit volume $a_{\mathrm{sol}}^3$ is equal to the product of $c$, the concentration of the solute species in the solution, and the volume $a_{\mathrm{sol}}^3$. The solute atom is oscillating around the average position with a frequency $\nu$, and tries to pass to the solid phase (to crystallize). In the solution, however, the solvent molecules make some chemical bonding with the solute molecules. Thus, for the solute molecules to be incorporated into the crystalline structure, the solvent molecules have to be separated from the solute molecules. For this separation (desolvation) process there is an energy barrier $E_{\mathrm{des}}$. Among $\nu$ trials of solidification, the rate which overcomes this energy barrier is given by the Boltzmann factor $\exp(-E_{\mathrm{des}}/k_BT)$. Then the velocity of crystallization (of liquid–solid phase transition) is given by

$$V_g(\mathbf{c}) = a_{\mathrm{sol}}\nu(ca_{\mathrm{sol}}^3) \exp(-E_{\mathrm{des}}/k_BT). \tag{11.71}$$

There is here also an inverse process, i.e., melting of the crystalline phase. Its rate is determined by the temperature, and should balance the crystallization from the solution with the equilibrium concentration $c_{\mathrm{eq}}$ of the solute species ($V_{\mathrm{melt}} = V_g(c_{\mathrm{eq}})$). The net rate of crystallization is given by the relation:

$$V_g = V_g(\mathbf{c}) - V_{\mathrm{melt}} = \nu a_{\mathrm{sol}}^4 \exp(-E_{\mathrm{des}}/k_B T)(c - c_{\mathrm{eq}}). \tag{11.72}$$

This growth rate may be expressed in terms of the chemical potential difference $\Delta\mu = \mu_{\mathrm{sol}}(T, c) - \mu^s(T) = \mu_{\mathrm{sol}}(T, c) - \mu_{\mathrm{sol}}(T, c_{\mathrm{eq}})$ (note that the chemical potential of the solid is equal to the chemical potential of the solution with equilibrium concentration of the solute species). The relevant expression has the form

$$V_g = K(\exp(\Delta\mu/k_B T) - 1) \approx K(\Delta\mu/k_B T). \tag{11.73}$$

With the kinetic coefficient $K = \nu a_{\mathrm{sol}}^4 c_{\mathrm{eq}} \exp(-E_{\mathrm{des}}/k_B T)$. It is important to note that by the crystallization process the number of solute molecules in front of the crystallization interface decreases in time. In order to compensate this solute deficiency, molecules of the solute have to be transported from the far bulk of the solution toward the crystallization interface. By including this process of material transport one comes to a different growth law. This subject has been treated in detail in Chap. 5, devoted to implementation of the LPE growth technique.

Epilayers can grow not only from a liquid phase but also from a gas phase, as shown in Figs. 8.1 and 7.4. It is, therefore, useful to consider here also the gas–solid phase transition.

For a gas phase an ideal gas (consisting of monatomic species) is a good approximation due to its low density [11.7]. At a temperature $T$ and a pressure $p$ the average number density of gas atoms $n = p/k_B T$, and the velocity distribution follows the Maxwell distribution [11.41]

$$P(\boldsymbol{v})\mathrm{d}\boldsymbol{v} = (m/(2\pi k_B T))^{3/2} \exp(-(mv^2)/(2k_B T))\mathrm{d}\boldsymbol{v}, \tag{11.74}$$

where $\boldsymbol{v} = (v_x, v_y, v_z)$ is the velocity of an atom and $m$ is its mass. Multiplying expression (11.74) by the number density $n$ and integrating then the product over all possible velocity values of the gas atoms, one gets the number of atoms impinging on a unit substrate surface and trying to crystallize there:

$$f = \frac{p}{\sqrt{2\pi m k_B T}}. \tag{11.75}$$

Inversely, there are some atoms desorbing from the crystal surface at a finite temperature $T$. The desorption flux is independent of the deposition flux from the gas phase, but is a function of temperature. If the substrate crystal is facing the gas phase which is under the equilibrium pressure, $p_{\mathrm{eq}}(T)$, the deposition rate balances with the desorption rate. The desorption flux from the crystal $f_{\mathrm{des}}$ is equal to the deposition flux $f_{\mathrm{eq}}$ of the gas under the pressure $p_{\mathrm{eq}}(T)$. Assuming that an atom is cubic with a linear dimension $a_{\mathrm{gas}}$, the net atomic flux in an atomic area $a_{\mathrm{gas}}^2$ contributes to the crystal growth, and the crystallization interface advances toward the gas phase by the atomic height $a_{\mathrm{gas}}$. The growth rate is thus given by the formula

$$V_g = a_{\text{gas}}^3 (f - f_{\text{eq}}) = \frac{\Omega(p - p_{\text{eq}})}{\sqrt{2\pi m k_B T}}. \tag{11.76}$$

Here $\Omega$ is the specific volume of a single gas phase species, equal to $a_{\text{gas}}^3$. The linear growth law, given in the form of (11.76), is known as the Hertz–Knudsen formula [11.42, 43] (see also (6.2)).

Since the chemical potential of an ideal gas is given by (11.8), then the chemical potential difference of the gas and the solid (the driving force of the phase transition considered here) is written as

$$\Delta\mu = \mu^g(T,p) - \mu^s(T) = \mu^g(T,p) - \mu^g(T,p_{\text{eq}}) = k_B T \ln(p/p_{\text{eq}}). \tag{11.77}$$

This relation gives $p = p_{\text{eq}} \exp(\Delta\mu/k_B T)$, which may be substituted into the Hertz–Knudsen formula. Consequently, the growth rate may be expressed as

$$V_g = \Omega f_{\text{eq}}(\exp(\Delta\mu/k_B T) - 1) \approx K(\Delta\mu/k_B T) \tag{11.78}$$

with the kinetic coefficient $K = \Omega f_{\text{eq}}$. The growth rate $V_g$ is thus proportional to the driving force, when $\Delta\mu$ is small enough.

## 11.4 Interface Formation in Epitaxy

Under the term "interface formation" we understand here the first stages of epitaxy [11.44], i.e., formation of the first nuclei and subsequently the first deposited epilayers. The most important elements of interface formation are related to its topology and structural characteristics. To be more precise in analyzing the phenomena occurring in the early stages of epitaxial layer growth, let us begin by defining some of the basic thermodynamic quantities required for discussion of the relevant surface phenomena.

Atoms at a surface or interface are in an environment that is markedly different from the environment of atoms in the bulk of the solid. They may be surrounded by fewer neighbors and these neighbors may be more anisotropic than in the bulk. Thus, we need to be able to quantify the excess properties of surfaces and interfaces [11.33] (see also Chap. 13 for more information).

### 11.4.1 The Interface Energy

One often encounters three terms in the literature relating to surfaces:

(i) the surface tension,
(ii) the surface energy, and
(iii) the surface stress.

All three quantities have units of energy per unit area (J/m$^2$) or force per unit length (N/m). The term surface tension is appropriate when referring to liquids, because liquids cannot support shear stresses and atoms in the liquid can diffuse fast enough to accommodate any changes in the surface area. This is not the case for solid surfaces and solid–solid interfaces, which usually possess elastic stresses up to the melting temperature. Consequently, the surface energy and surface stress will be used in what follows to define the state of all surfaces [11.45–47].

Solid surfaces can change their energy in two ways: first, by increasing the physical area of the surface, for example, by cleaving a surface or adding atoms to the surface with the arrangement of the atoms being identical to those in the bulk; second, by changing the position of atoms at the surface through elastic deformation, for example, by phenomena such as surface relaxation or reconstruction.

The first case involves simply creating more or less surface area and is independent of the nature of the surface, whereas the second case involves the detailed arrangements of atoms within a solid surface and may be thought of as the work involved in straining a unit area of surface.

Consider a homogeneous crystalline body that contains $n$ atoms at temperature $T$ and pressure $p$, which is surrounded by plane surfaces (e.g., a convex polyhedron). By creating a surface $S_j$ in the $j$ direction, a specific surface free energy is defined by $\sigma_j = (\partial G/\partial S_j)_{T,p,n,S}$ in which the subscript $S$ indicates that all other surfaces $S_i$, if they exist, are kept constant. The surface free energy $\sigma_j$ is supposed to be independent of the size of the face $S_j$, but anisotropic in the direction $j$ [11.44]. The considered convex polyhedron is supposed to be limited by faces $S_j$. From an arbitrary point inside the polyhedron the central distances $h_j$ to the different faces are defined as shown in Fig. 11.20a. If from the bulk of the metastable phase A (vapor phase for example), being under the partial saturated pressure $p_\infty$ , a molecule $A$ is brought to the crystal $A$ where the vapor pressure is $p$, the chemical potential variation is $\Delta\mu = k_B T \ln(p/p_\infty)$. If $n$ particles are included in the crystal which then expands its surface, the balance of work performed at bulk and surface is expressed by the formula

$$\Delta G(n) = -n\Delta\mu + \sum_1^j \sigma_j S_j \tag{11.79}$$

the summation being carried out over all the faces of the polyhedron. The volume of the crystalline polyhedron may be split into pyramids with summit at the point 0 and bases at $S_j$; this may be expressed as $V = (1/3)\sum_1^j h_j S_j$. Because the volume is a third-order homogeneous function, any surface variation leads to the corresponding variation in volume $\mathrm{d}V = (1/2)\sum_1^j h_j \mathrm{d}S_j$. Taking now the volume of one particle as being equal to $V_1 = V/n$ one gets for the differential of (11.79)

$$\mathrm{d}\Delta G = -(\Delta\mu/2V_1)\left(\sum_1^j h_j \mathrm{d}S_j\right) + \sum_1^j \sigma_j \mathrm{d}S_j. \tag{11.80}$$

Thermodynamic equilibrium is reached when all partial derivatives $(\partial\Delta G/\partial S_j)_{T,S,\Delta\mu}$ are zero, that is according to (11.80), when

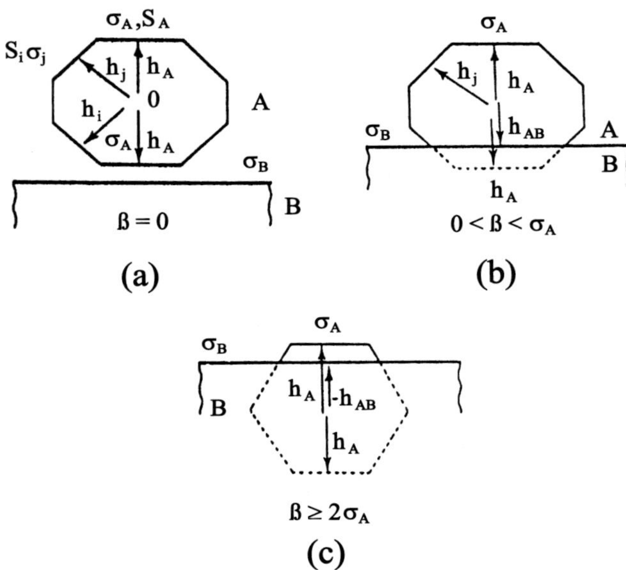


**Fig. 11.20.** Equilibrium forms of a deposit $A$ on a substrate $B$: (**a**) without contact with the substrate ($\beta = 0$); (**b**) and (**c**) when in contact if $\beta \neq 0$ (taken from [11.44])

$$\sigma_j/h_j = \sigma_i/h_i = \cdots = \Delta\mu/2V_1 . \tag{11.81}$$

This expression – the so-called Wulff theorem – permits one to build concentric polyhedrons, the faces of which are perpendicular to the $h_j$'s, when one knows $\sigma_j$ in every direction. The innermost polyhedron of such a crystalline form corresponds to the equilibrium shape [11.44]. This is thus the homogeneous critical nucleus.

Let us now suppose that the crystal of the species considered is in contact with the upper face of a substrate crystal $B$ as shown in Fig. 11.20b, and that the epitaxial orientation is imposed by the contacting face of the substrate as well as by the mutual azimuth in the interface between $A$ and $B$. The faces which are in contact are defined by $\sigma_A$ and $\sigma_B$, respectively. Let the area of the interface created by the two contacting faces be equal to $S_{AB}$, and let a specific interfacial free energy $\sigma^*$ be defined by the transformation shown in Fig. 11.21.

When creating two surfaces of equal sizes, one in the crystal $A$ and the second in the crystal $B$ by a reversible and isothermal separation, the work per surface unit area is given by $\sigma_A$ and $\sigma_B$ (the definitions of the $\sigma$'s are taken from Born and Stern [11.48]). When putting then both of the crystals $A$ and $B$ into contact in epitaxial orientation in such a way that the interface reaches its equilibrium, the work gained per surface unit area is equal to $-\beta$. The balance of the transformation reveals the excess quantity $\sigma^*$ given by

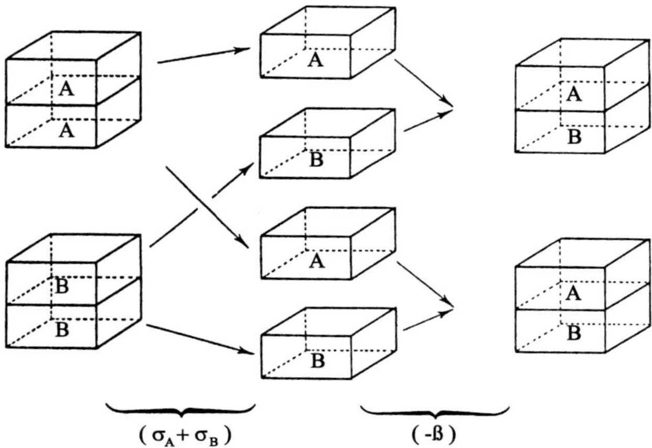


**Fig. 11.21.** Cleavage and adhesion balance leading to Dupre's relation (taken from [11.44])

$$\sigma^* = \sigma_A + \sigma_B - \beta\,, \tag{11.82}$$

which characterizes the interface ((11.82) is known as Dupre's relation [11.44]). The separation work $\beta$ of the crystals $A$ and $B$ is the specific adhesion free energy which we will call in brief the adhesion energy. Like $\sigma_A$ and $\sigma_B, \beta$ is an anisotropic quantity depending moreover on the azimuth of the contact area.

### 11.4.2 Initial Stages of Epitaxial Growth

In the following we will assume that the substrate $B$ is inert in the sense that its vapor pressure is negligible compared with the vapor pressure of the deposit. We will also assume that $A$ and $B$ do not mix in the solid state even if alloying is thermodynamically possible.

When a crystal $A$ grows on $B$, the balance of formation may be written in a manner similar to (11.79):

$$\Delta G_{3\mathrm{D}}(n) = -n\Delta\mu + \sum_1^j \sigma_j \mathrm{d}S_j + (\sigma^* - \sigma_B)S_{AB}\,. \tag{11.83}$$

The summation is now carried out only on the free faces of $A$ (see Fig. 11.20b); the third term represents the energy of changing a free surface $S_B$ to an interface $S_{AB}$ when $S_{AB} = S_B$. In the considered case the variation of volume may be expressed in the form

$$\mathrm{d}V = \frac{1}{2}\Big(\sum_1^j h_j \mathrm{d}S_j + h_{AB}\mathrm{d}S_{AB}\Big) \tag{11.84}$$

where $h_{AB}$ is the distance from point 0 to the interface. The derivative of (11.83) thus yields

$$\mathrm{d}\Delta G_{3\mathrm{D}}(n) = -(\Delta\mu/2V_1) \left(\sum_1^j h_j \mathrm{d}S_j + h_{AB}\mathrm{d}S_{AB}\right) + \sum_1^j \sigma_j \mathrm{d}S_j + (\sigma^* - \sigma_B)\mathrm{d}S_{AB} \qquad (11.85)$$

Upon taking this derivative the quantities $\sigma$ were again supposed to be independent of $S$, which is the case occurring in practice, except for crystallites built with only a few atoms. Under this condition the partial equilibria, for the given mutual orientation of $A$ and $B$, are determined by the following two equations

$$\left(\frac{\partial \Delta G}{\partial S_{AB}}\right)_{S,T,\Delta\mu} = 0; \quad \left(\frac{\partial \Delta G}{\partial S_i}\right)_{S,T,\Delta\mu} = 0. \qquad (11.86)$$

The first condition represents the interfacial equilibrium, the others represent the equilibrium of the free faces of $A$. When expanded, (11.86) leads to the generalized Wulff theorem [11.44]:

$$\frac{\Delta\mu}{2V_1} = \frac{\sigma_j}{h_j} = \frac{\sigma_i}{h_i} = \cdots = \frac{(\sigma^* - \sigma_B)}{h_{AB}} = \frac{(\sigma_A - \beta)}{h_{AB}}. \qquad (11.87)$$

When $\beta = 0$ (11.87) yields (11.81) of the homogeneous case. When both surfaces, $A$ and $B$, are just in contact, $h_{AB} = h_A$ (which corresponds to the equilibrium shape without a substrate (Fig. 11.20a)), they have no affinity, thus, according to (11.82) the excess quantity $\sigma^* = \sigma_A + \sigma_B$ is maximum. It is also noted that a finite value of the adhesion energy leads to $h_{AB} < h_A$, that is a polyhedron truncated by the substrate (Fig. 11.20b). The case $\beta = \sigma_A(\sigma^* = \sigma_B)$ is trivial, $h_{AB} = 0$. The real polyhedron is reduced by one half compared with the homogeneous case. When $\sigma_A < \beta < 2\sigma_A$ one has $h_{AB} < 0$; the polyhedron grows flat (Fig. 11.20c) and the limiting case $\beta \to 2\sigma_A(\sigma^* \to \sigma_A - \sigma_B)$, i.e., $-h_{AB} = h_A$ means that the lateral faces of the crystal $A$ are evanescent. In this important case one may no longer speak of a 3D phase of $A$ but of a phase of molecular thickness, that is a 2D phase lying on the substrate $B$. The case $\beta > 2\sigma_A(\sigma^* < \sigma_A - \sigma_B)$, although physically conceivable and important, cannot be discussed with the relation (11.87).

Two very important conclusions may be drawn from the presented phenomenological thermodynamic considerations of epitaxial crystallization. The first concerns 3D nucleation on the substrate, and states that:

> *epitaxial 3D nucleation can take place on a substrate only at an existing supersaturation in the metastable phase, which means that* $\Delta\mu > 0$.

The second conclusion concerns 2D nucleation, and states that:

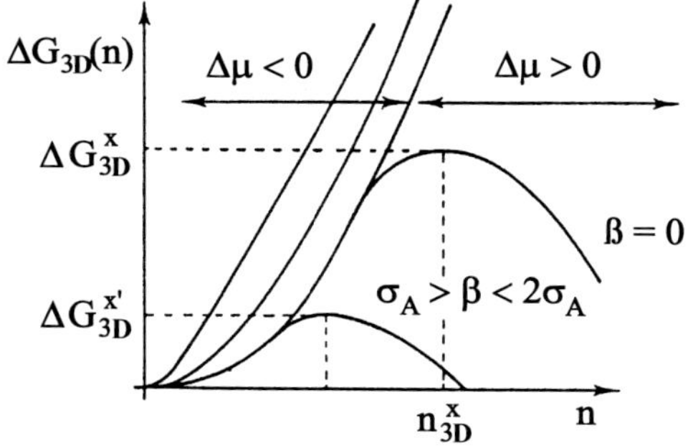


**Fig. 11.22.** Free energy of 3D crystal formation at constant supersaturation $\Delta\mu$, as a function of the number of particles $n$. If $\Delta\mu > 0$ the curve shows a maximum. The $\Delta G_{3\mathrm{D}}$ barrier is lowered when the adhesion energy is positive (taken from [11.44])

*in contrast to the 3D nucleation case, epitaxial 2D nucleation can occur on the substrate only at an undersaturation in the metastable phase, which means that* $\Delta\mu \leq 0$.

The important point, resulting from the second conclusion, is that if $\beta > 2\sigma_A$, $A$ may be deposited on $B$ even when the temperature $T_A$ of the source of $A$ is lower than the temperature $T_B$ of the substrate.

The characteristic surface quantities defined above permit one to differentiate between the two epitaxial growth modes which are initiated by nucleation, i.e., 3D growth and 2D growth:

(i) when $\beta < 2\sigma_A(\sigma_A > \sigma_B - \sigma^*) \Rightarrow$ 3D growth mode occurs, while
(ii) when $\beta > 2\sigma_A(\sigma_A < \sigma_B - \sigma^*) \Rightarrow$ 2D growth mode occurs.

The physical meaning of these inequalities is quite obvious. When the adhesion energy $\beta$ is smaller than the doubled value of the surface energy of the crystalline nucleus $A$ ($2\sigma_A$ presents here the so-called cohesion energy of the crystal $A$) then 3D growth occurs. In the opposite case when adhesion is larger than cohesion, 2D growth occurs.

The detailed calculations leading to these conclusions are given in [11.44], but the resulting dependencies of the free energies of 3D and 2D crystal formation at constant $\Delta\mu$ as functions of the number of particles $n$ inside the growing crystal $A$ (compare this with (11.83)) are given after [11.44] in Figs. 11.22 and 11.23.

After the nucleation stages, the nuclei will sooner or later start to agglomerate. This agglomeration process will intensify until islands are formed. At a later stage, island chains join up until only channels are left. Eventually, even these will disappear and a continuous film will be obtained; the surface irregularities will be smoothed out by further growth.

Thin film growth experiments carried out inside an electron microscope have shown that the islands drift on the substrate during their growth, that they rotate and that they rapidly "flow together" after their contact (they undergo coalescence). This movement of the nuclei leads to a picture of liquid-like behavior [11.49]. However, it could be shown that the nuclei are small

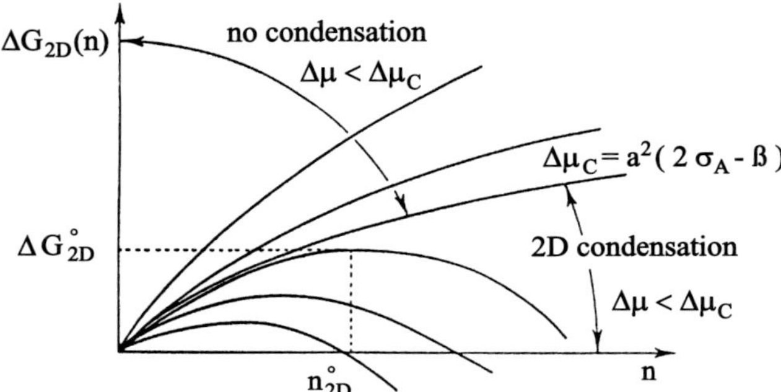


**Fig. 11.23.** Free energy of 2D crystal formation at constant $\Delta\mu$ as a function of the number of particles. The $\Delta G_{2D}$ barrier exists only if $\Delta\mu > \Delta\mu_C (\Delta\mu_C < 0)$ (taken from [11.44])

single crystals during all stages of the growing process, and that the agglomeration mass transport mechanism is one of surface diffusion. An important feature of the coalescence phenomena is that the time taken for the coalescence of two islands is a function of island size, the smaller the island the more rapid is the diffusion process. Given this, one may conclude that there is a critical stage of growth at which the coalescence time of an island is greater than the time required for the island to continue its growth. In this case the island coalesce only with the nearest neighbors.

Examination of growing films in the electron microscope has also shown that recrystallization by grain boundary migration occurs during growth. Grain boundaries are formed in the necks between islands of different orientations and stay in the islands until the neck is eliminated. This recrystallization is a rapid process in the early stages of growth when the islands are small, but after the critical growth stage is reached, no further recrystallization can occur by this process.

We have already mentioned (see Chap. 2) that the ordering process in epitaxy is not only influenced by nucleation, but also by the subsequent growth. The effect of growth parameters on the coalescence process has been analyzed theoretically by Stowell [11.50]. He has shown that cases can exist in which growth is more important than nucleation in determining epitaxy. This result is in experimental agreement with results of other authors [11.51].

## 11.5 Self-Organization Processes

The ordering processes occurring in the early stages of epitaxy, when the structure of the first epilayers is created, can lead in special growth conditions to a new growth phenomenon known under the names self organization effects

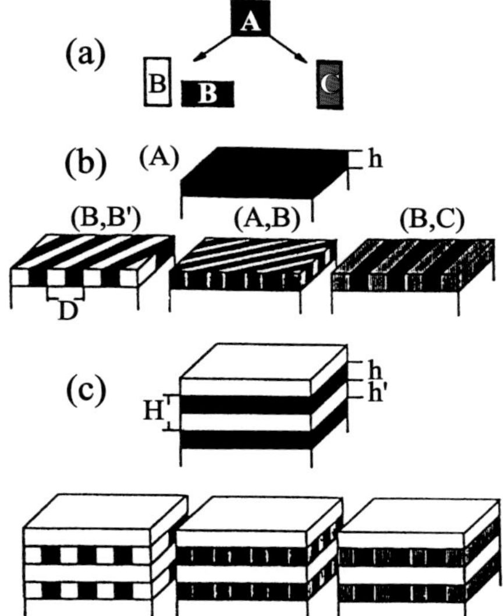


**Fig. 11.24.** Schematic illustration of polydomain heterostructures. (**a**) phase states of the active layer; phase $A$ can transform to two variants of phase $B$ or phase $C$, (**b**) formation of the polydomain layer, (**c**) formation of polydomain multilayer structure (taken from [11.54])

[11.52], or self-ordering (self-assembling) effects [11.53]. This phenomenon, which will be called hereafter the self organization effect (SO effect), consists in spontaneous structural ordering/reordering of the epitaxial deposit during the growth process.

So far, the SO-effect is observed in step-mediated epitaxy (see Sect. 1.2 and Fig. 1.7) and in strained layer heteroepitaxy (growth of epilayers highly mismatched against the substrate). This effect is gaining increasing interest because of the possibility of growing sophisticated low dimensional heterostructures by using it in epitaxial growth (see Sect. 3.2 and Fig. 3.6). Even a new class of materials, so-called polydomain heterostructures, has been proposed, which can be grown by using the SO effect in epitaxy [11.54]. Due to elastic interaction between the layers of a heterostructure, these layers transform into a set of periodically alternating lamellae or elastic domains. A polydomain heterolayer can consist of either differently oriented domains of the same phase (e.g., twins) or domains of different phases (e.g., short period superlattices [11.55]). A schematic illustration of polydomain heterostructures is shown in Fig. 11.24.

Because the SO effect in step-mediated epitaxy is determined more by surface kinetic phenomena and bonding nature to the steps, than by thermodynamics [11.56], we will postpone the consideration of this case until Sect. 12.3. Consequently, let us then begin with the strain-induced self organization phenomena [11.57–70].

### 11.5.1 Strain-Induced Self-Ordering; Quantum Dots

In principle, there are two physical mechanisms which lead to SO effects in strained layer heteroepitaxy. The first is based on the Stranski–Krastanov mode of epitaxial growth (see Sect. 1.2 and Fig. 1.2 for definition), while the second, called strain-induced lateral ordering (SILO), is based on spinodal decomposition in elastically anisotropic epilayers, followed by lateral modulation of the epilayer composition at the heterointerface area [11.55, 63–65]. Both mechanisms are currently investigated by different research groups because of their importance for fundamental research as well as for technological applications.

Let us begin with the SK-mode related SO effect. The SK-growth morphology is intimately related to the accommodation of elastic strain associated with epitaxial lattice misfit. Over the years, this point has been made many times by many authors using theoretical models of varying degrees of sophistication [11.70, 71]. In all these treatments, the first few monolayers of the film material are strongly bound yet strained to match the substrate lattice constant. Subsequent deposited material collects into islands that are regarded as essentially bulk-like due to the presumed existence of interfacial misfit dislocations (MD) to relieve the strain.

The forgoing model prediction definitely occurs in many heteroepitaxial systems. However, another scenario apparently can occur as well; i.e., the SK morphology with dislocation free but strained islands. To date, this phenomenon has been documented most thoroughly for the Ge-on-Si(100) system [11.72], and for the $In_xGa_{1-x}As$-on-GaAs(100) system [11.73].

The limiting case of the latter system, namely the InAs-on-GaAs case, is most interesting because of the quantum dot formation processes. The energetics of an array of 3D coherent, strained islands (quantum dots) on a lattice mismatched substrate has been theoretically studied in [11.57, 70, 74, 75] and compared with experimental data gained with MBE-growth of pyramidal InAs/GaAs islands. It has been shown that:

(i) coherent, strained, essentially 3D equal-sized islands form stable, periodic, 2D arrays under certain growth conditions, on lattice mismatched substrates,

(ii) for islands on the (100) surface of a cubic crystal, the arrangement of the array in a 2D periodic square lattice with primitive lattice vectors oriented along the lowest-stiffness directions [100] and [010] of the substrate lattice is energetically preferred.

An exhaustive investigation of the strain distribution in and around self-organized InAs/GaAs islands of pyramidal shape has been performed by Grundmann et al. [11.58]. They considered an InAs pyramidally shaped island on a thin wetting layer (typical geometry for the SK-mode) within GaAs, as is observed for self-organized quantum dots grown by MBE (Fig. 11.25). In order to obtain the 3D strain distribution for this system, the authors

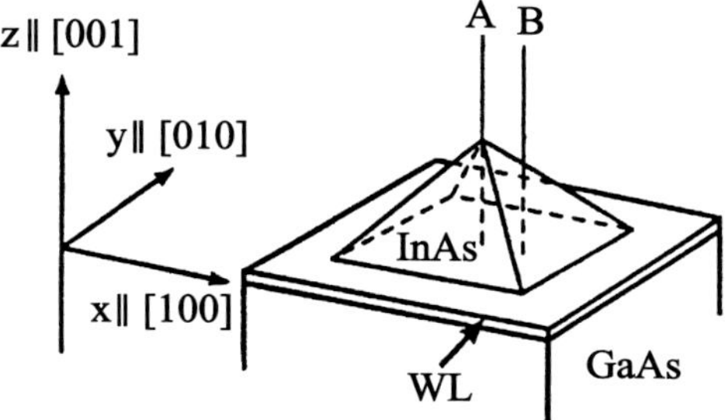


**Fig. 11.25.** Schematic drawing of the quantum dot pyramid geometry. Lines A and B denote line scans in the [001] direction. The origin of the $z$-axis is at the lower interface of the wetting layer (WL) (taken from [11.58])

performed a numerical simulation, in which the total strain energy of the structure was minimized in the elastic continuum theory. The approach of the elastic continuum theory, used in the calculations, has been shown to be valid down to one monolayer thin epitaxial films [11.76].

The typical considered pyramid base length was equal to 12 nm, with a distance of 30 nm between neighboring pyramids. This work showed that the strain distribution in the pyramid does not depend on the actual size of the island but on its shape. The main effects can be discussed with line scans along the $z$-direction at different positions of the $xy$ plane (Fig. 11.25). The solid line shown in Fig. 11.26 denotes $\varepsilon_{zz}$, the dashed line $\varepsilon_{xx}$, and the dotted-dashed line $\varepsilon_{yy}$. The line scan intersecting the wetting layer far from the pyramid is shown in Fig. 11.26a. Figure 11.26b shows the case when the intersection goes through the summit of the pyramid (line A shown in Fig. 11.25). In both cases symmetry imposes that $\varepsilon_{xx} = \varepsilon_{yy}$. The case when the line scan intersects the pyramid at one half of the base length in the [100] direction from the center (line B in Fig. 11.25) is shown in Fig. 11.26c. In the wetting layer (Fig. 11.26a) the strain is biaxial and entirely confined to InAs. Comprehensive (negative) interfacial strain causes an expansion along the $z$-direction ($\varepsilon_{zz}$ is positive), known as tetragonal distortion (see Fig. 2.17a). The wetting layer is affected by the pyramid only in its vicinity, within a distance of about half of the pyramid's base length.

Along the line scan through the summit of the pyramid (Fig. 11.26b), a very different situation is revealed. Close to the lower interface, $\varepsilon_{zz}$ is still positive but much smaller ($\approx 3\,\%$) than in the wetting layer because the substrate can no longer force the interface lattice constant to be that of GaAs. With increasing height within the pyramid, $\varepsilon_{zz}$ changes its sign and becomes negative at the top of the pyramid. This happens because at the pyramid summit only small forces act on it in the $xy$ plane, but the GaAs barrier environment of the pyramid compresses the pyramid mainly from the sides along the $z$-direction, imposing tensile strain components in the $xy$ plane ($\varepsilon_{xx} = \varepsilon_{yy}$ at the summit become positive). Generally, however, the strain is still compressive even at the summit of the pyramid. Around the

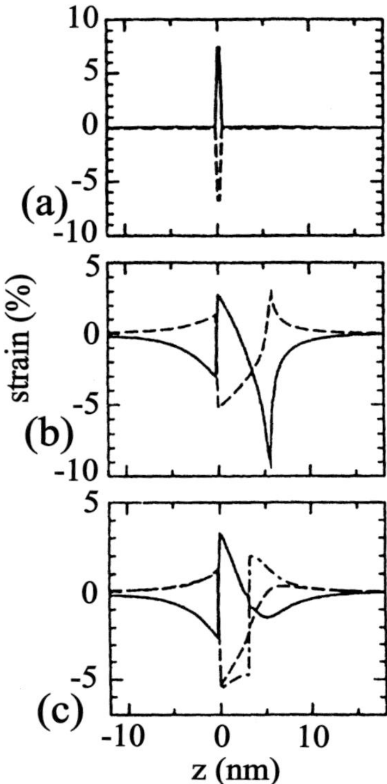


**Fig. 11.26.** Strain distribution in and around a pyramidal quantum dot for line scans in the [001] direction: (**a**) through the wetting layer far away from the pyramid, (**b**) along the line A, and (**c**) along the line B shown in Fig. 11.25. The solid lines denotes $\varepsilon_{zz}$, the dashed line $\varepsilon_{xx}$, and the dotted-dashed line $\varepsilon_{yy}$ (taken from [11.58])

pyramid the barrier environment (GaAs in-between the pyramids of InAs) also becomes significantly strained ($\approx 3\,\%$ close to the interfaces).

The character of the strain is not determined by the separate components of the strain tensor, but by decomposing the strain tensor into the isotropic (hydrostatic) and anisotropic parts. Additionally shear strains $\varepsilon_{ij} (i \neq j)$ also exist, which turn out to be significant close to the pyramid edges (intersections of the $\{011\}$ side faces). In Fig. 11.27 the lines cans through the pyramid center (line A in Fig. 11.25) of the hydrostatic part $I$ and the biaxial part $B$ of the strain tensor are compared. As expected, the inner part of the pyramid contains nearly homogeneous hydrostatic strain, while the barrier exhibits almost no hydrostatic strain. The biaxial strain is transferred to a significant amount into the barrier around the pyramid and has a distinct minimum in the pyramid.

The theoretical results concerning the strain distribution in and around the InAs/GaAs pyramidal quantum dots (QDs) have been proved experimentally by measuring some optical parameters of the QDs which depend on the strain inside the dot. The optical phonon energies in the dots were estimated and perfect agreement with experimental data was found. From the variation of the strain tensor the local band-gap modification was calculated,

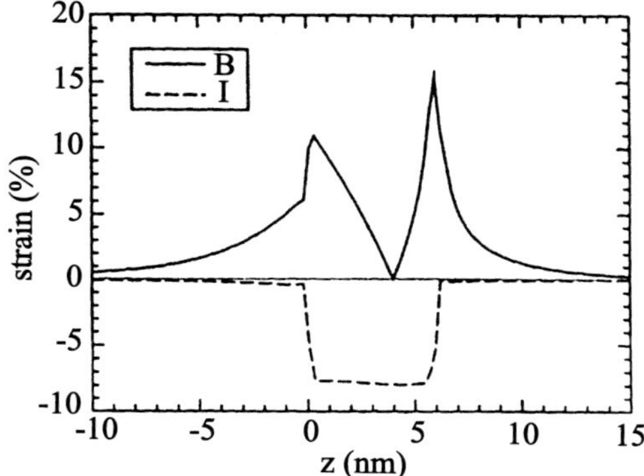


**Fig. 11.27.** Hydrostatic (dashed line) and biaxial (solid line) strains in the pyramidal quantum dot along the line A shown in Fig. 11.25 (taken from [11.58])

including in addition piezoelectric effects, and the thus determined electronic structure of the dots was compared with luminescence data characteristic of the dots. An agreement between the theoretical data and the luminescence experiments has been found. Further experimental evidence of the validity of the results gained by dot strain simulation has been presented in [11.60].

An interesting feature of the QDs grown due to the SK-mode has been presented in [11.61, 70]. The energetics of multisheet arrays of 2D QD islands was studied there theoretically. The structure of the surface sheet is determined by thermodynamic equilibrium under the constraint of a fixed structure of sheets of buried islands. For the arrangement of islands in a single surface sheet both a 1D structure of stripes and a 2D structure of square-shaped islands were examined (Fig. 11.28). The buried islands were considered as planar elastic defects characterized by a uniaxially anisotropic straining double force density, and the surface islands were considered as 2D islands characterized by an isotropic intrinsic surface stress tensor. It was shown that in cubic crystals with a negative parameter of elastic anisotropy the elastic interaction between successive sheets of islands parallel to the (001) crystallographic plane exhibits an oscillatory decay with the separation between the sheets. By varying the distance between successive sheets of islands, a transition occurs from vertical correlation between islands (where islands of the upper sheet are formed above the buried islands of the lower sheet) to anticorrelation between islands (where islands of the upper sheet are formed above the spacing in the lower sheet). The separation between successive sheets of islands corresponding to this transition depends sensitively on the anisotropy of the double force density of buried islands. These results are in agreement with experimental data concerning anticorrelation in multisheet arrays of CdSe islands in the ZnSe matrix [11.70].

The physical mechanism of spontaneous formation of ordered semiconductor nanostructures usually involves two possible processes. According to the

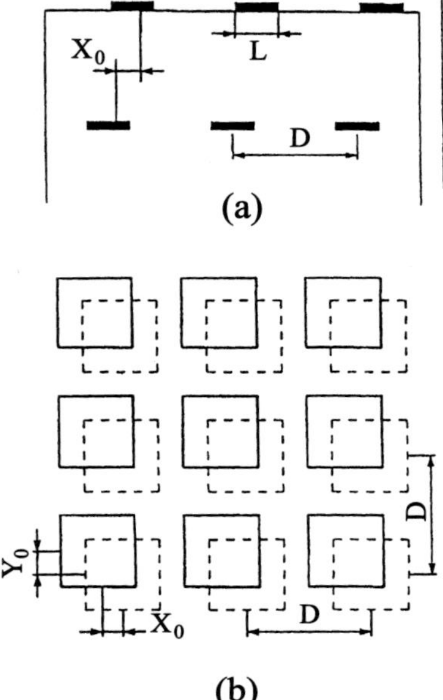


**Fig. 11.28.** Geometry of double-sheet arrays of 2D islands. The array of surface islands has the same structure as the array of buried islands but is shifted as a whole. (**a**) Each sheet of islands forms a 1D array of stripes. The cross-section of the double-sheet structure is shown. (**b**) Each sheet of islands forms a 2D array of square-shaped islands. The plan view of the double-sheet structure is plotted. Buried islands are depicted by dashed lines, and solid lines are used for surface islands (taken from [11.61, 70])

first, equilibrium domain structures can be formed in closed systems. Such formation is realized by long-time growth interruption or by post- growth annealing. Thermodynamics can be applied to describe the equilibrium structures that meet the conditions of the Gibbs free energy minimum. Second, nonequilibrium structures can be formed in open systems. Such structures are formed in growth processes and observed *in situ* or ex situ in as-grown samples. These structures are additionally governed by growth kinetics.

Multisheet arrays of islands are distinct from other types of nanostructures for the following reasons:

(i) Formation of multisheet arrays of 2D or 3D islands is a process that is dominated by both equilibrium ordering and kinetic-controlled ordering. If the deposition of the first sheet of islands of material 2 on a material 1 is followed by a growth interruption or just the growth rate is sufficiently low, islands of the equilibrium structure are formed. If then islands are regrown by material 1, and the second cycle of the deposition of material 2 is introduced, a new growth mode occurs. For typical growth temperatures and growth rates, the structure of the buried islands of the first sheet does not change during the deposition of the second sheet. The second sheet of islands grows, however, in the strain field created by the buried islands of the first sheet. And the structure of the second sheet of islands reaches the equilibrium under the constraint of the fixed structure of buried islands of the first sheet.

(ii) A variation of the separation between successive sheets gives an additional possibility (as compared to single-sheet arrays) to tune geometrical and electronic characteristics of nanostructures.

(iii) In multisheet arrays of 3D islands the buried islands in successive sheets are spatially correlated, which means that at the surface new islands are observed to be formed directly above the buried islands. This spatial correlation can be explained theoretically by accounting the strain created by buried islands. In [11.77] and [11.78] the strain-induced migration of adatoms of the growing layer was shown to drive adatoms to positions above buried islands. In [11.79], energetically preferred sites for nucleation of islands of the second sheet were shown to occur above buried islands. In these papers, buried islands were approximated as elastic point defects, and the crystal was treated as an elastically isotropic medium.

In seeming contradiction to the above experimental and theoretical results other experiments on multisheet arrays of 2D islands of CdSe in the ZnSe matrix [11.80] surprisingly revealed vertical anticorrelation between islands in successive sheets. Surface islands in this material system are formed above the spacings in the sheet of buried islands. This effect is explained as the consequence of the influence which separation between successive sheets of islands has on the growth process of the upper sheet [11.70].

According to the considerations presented above, QD formation requires significant lateral mass transport during growth. This kinetic effect (the lateral mass transport) seems to be a very surprising phenomenon, given that the melting temperature $T_m$ at atmospheric pressure of InAs is equal to 1215 K. An attempt to identify the physical origin of the significant mass transport on the top of the wetting layer of critical thickness (see Sect. 2.3, Fig. 2.17 and Sect. 14.1.4 for definition) has been presented by Bottomley [11.62], who based his consideration on the experimentally well established fact that hydrostatic pressure induces lowering of the melting point in zincblende structure compounds by several hundred K. The model proposed for explaining the reduction of the melting temperature of an epitaxial layer due to heteroepitaxial stress is as follows.

In the limit of equilibrium thermodynamics, at zero pressure and at a temperature $T < T_m$, the molar Gibbs free energy of the liquid exceeds that of the solid by an amount $\Delta G(T)$ given by

$$\begin{aligned}\Delta G(T) &= H_l(T_m) - (T_m - T)\, C_{p,l} \\ &\quad - T\left(S_l(T_m) - C_{p,l}\ln\left(\frac{T_m}{T}\right)\right) - G_s(T)\end{aligned} \tag{11.88}$$

where $H, C_p, S$, and $G$ are the molar enthalpy, specific heat at constant pressure, entropy and Gibbs free energy. The subscripts "$l$" and "$s$" refer to the liquid and solid phases, respectively. For InAs and GaAs all the quantities in (11.88) are available in standard tables. Consistent with its tabulated data

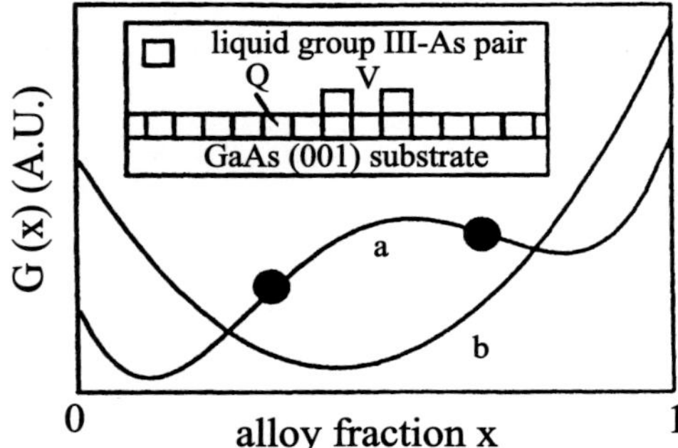


**Fig. 11.29.** Schematic diagram of the liquid Gibbs free energy $G$ as a function of alloy fraction $x$. Curve (**a**) has two spinodal (inflexion) points denoted by the black circles, as for $In_xGa_{1-x}As$ at 770 K. Curve (**b**) has no spinodal points as for $Si_xGe_{1-x}$ at 970 K. Inset shows a schematic diagram of liquid $In_{0.8}Ga_{0.2}As$ layers on GaAs (001). The squares represent liquid group III–As pairs. Q is a pair in the second bilayer, and V is a vacancy in the third bilayer. The one monolayer thick wetting layer has been omitted for simplicity (taken from [11.62])

for $T > T_m, C_{p,l}$ is assumed to be constant for $T < T_m$ [11.81]. Heteroepitaxial misfit causes the molar Gibbs free energy of the solid film to increase by an amount $\int V\mathrm{d}p$, where $V$ is the solid molar volume (compare with (11.5)). The result of the integral can be understood if one recalls that the stress tensor for hydrostatic pressure for a thin film of cubic symmetry on the (001) face is

$$\sigma_{xx} = \sigma_{yy} = \left(\frac{\varepsilon_x}{c_{11}}\right)(c_{11} - c_{12})(c_{11} + 2c_{12})) \tag{11.89}$$

where the $c_{ij}$ are the elastic stiffness constants of the film, $\varepsilon_x$ is the epitaxial strain, and the other elements of $\sigma_{ij}$ are zero [11.62]. Assuming the heteroepitaxial liquid phase is not stressed, the critical strain required to melt a thin solid film on the (001) face at a temperature $T$ is

$$\varepsilon_x(T) = -(3c_{11}/(2(c_{11} - c_{12})(c_{11} + 2c_{12})))(\Delta G(T)/V) \tag{11.90}$$

where $\Delta G(T)$ is given by (11.88). For InAs and GaAs at a temperature of 770 K, one obtains critical strains of −1.7 % and −1.9 %, respectively. When InAs is grown on GaAs, the strain is −6.7 %. Therefore, for $In_xGa_{1-x}As$ heteroepitaxy on GaAs (001) at 770 K, melting is predicted for compositions $x > 0.3$. The form of $G(x)$ for $In_xGa_{1-x}As$ at 770 K is indicated schematically in curve (**a**) in Fig. 11.29. The points marked by circles are spinodal (inflexion) points: bulk compositions intermediate to the spinodal points exhibit so-called spinodal decomposition, or compositional instability in the limit of equilibrium thermodynamics [11.1, 82].

The prediction of melting for $x > 0.3$ is consistent with the 3D island-shaped interfacial morphology reported in [11.83] for $In_xGa_{1-x}As$ growth at compositions $x \geq 0.35$. For InAs grown on GaAs (001) at 770 K, the theory predicts that the InAs is in liquid phase, and equilibrium thermodynamics [11.82] can be applied in the limit that the liquid phase's viscosity is negligible. In general, dot or alloy formation may be hindered or prevented if the liquid phase viscosity is sufficiently large.

Concluding this consideration on the physical origin of InAs QDs on GaAs(001) one has to emphasize that the large heteroepitaxial stress causes InAs to melt when deposited on GaAs(001) at 770 K. This leads to mixing with the substrate in order to realize a local minimum in the Gibbs free energy of the liquid phase, producing an approximate liquid composition of $In_{0.8}Ga_{0.2}As$. QD formation occurs after about two monolayers of liquid material accumulate in order to minimize the surface tension but without reducing the net coordination of the liquid phase atoms. It is worth pointing out that in the case of melting induced by heteroepitaxial stress, in the limit of equilibrium thermodynamics, interdiffusion with the substrate is expected to be more substantial than the interdiffusion which would occur for the epitaxial growth of a solid film on a solid substrate.

### 11.5.2 Strain-Induced Lateral Ordering; Quantum Wires

The second SO effect, i.e., the SILO effect, is related to self-organized quantum wire (QWR) structures. So far, this has been observed when $(GaP)_n/(InP)_n$ and $(GaAs)_n/(InAs)_n$ short-period ($n = 1, 2$) superlattices (SPS) are grown on GaAs (001) and InP (001), respectively [11.55].

The SILO effect leads to formation of lateral quantum wells in vertical SPS of $(GaP)_n/(InP)_n$ and $(GaAs)_n/(InAs)_n$. The strain induced from the deviation of superlattice periodicity $d_{2e}$ from $na_s$ is the major driving force of the lateral modulation of the composition along the [110] direction, where the integer $n$ is the number of monolayers of each binary compound within a period of SPS structure, $d_{2e}$ is the thickness of the two binary compound layers creating the thickness period of the SPS, and $a_s$ is the lattice constant of the substrate material. The percentage thickness deviation of the superlattice periodicity from $na_s$, i.e., $(d_{2e} - na_s)/na_s$, is denoted by $\Delta T$. When $\Delta T$ is larger than $\approx 4\,\%$, both $(GaP)_n/(InP)_n$ and $(GaAs)_n/(InAs)_n$ ordered vertical SPS layers were found to have a lateral periodic modulation of composition with periodicities as small as $\approx 200$ Å. This effect is enhanced when $n$ was increased from 1 to 2.

A schematic illustration of the $(InP)_2/(GaP)_2$ bilayer superlattice and the type-I to type-II superlattice [11.84] modulation for electron–valence band transitions is shown in Fig. 11.30, and Fig. 11.31 shows the cross-sectional TEM images from a $(GaAs)_1/(InAs)_1$ SPS structure with $\Delta T = 11.5\,\%$. A uniform image appears in the [110] cross-section (the plane shown in the TEM image is perpendicular to the [110] direction) and a modulated image composed of moderate dark and light fringes, each approximately 100 Å wide, parallel to the growth direction appears in the $[\bar{1}10]$ cross-section. Similar

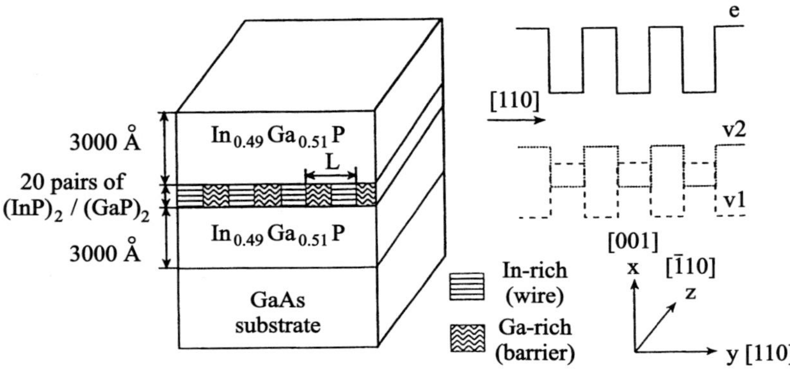


**Fig. 11.30.** A schematic diagram of the $(InP)_2/(GaP)_2$ bilayer SPS and the type-I to type-II superlattice modulation for electron (**e**)– valence band (v1) and electron (**e**)– valence band (v2) transitions (taken from [11.55])

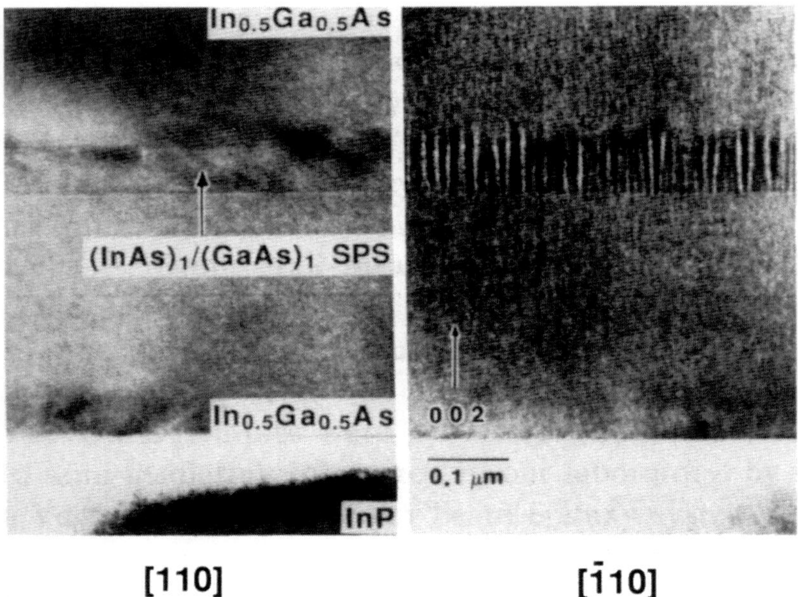


**Fig. 11.31.** Dark field cross-sectional transmission electron micrographs of the $(GaAs)_1/(InAs)_1$ SPS grown on the $Ga_{0.5}In_{0.5}As$ buffer layer along the [100] direction perpendicular to the on-axis InP (100) substrate. The TEM micrographs show: (left) the (110) and (right) the $(\bar{1}10)$ planes. $\Delta T = 11.5\,\%$ (taken from [11.65])

cross-sectional TEM images of a $(InP)_2/(GaP)_2$ SPS sample with a thickness deviation of $\Delta T = 6.2\,\%$ have been reported in [11.67].

The experimental results of [11.65] and [11.67] indicate that a lateral modulation of the composition exists only in SPS structures with a moderate to large deviation $\Delta T$ of the superlattice periodicity from $na_s$. Therefore, the lateral composition modulation process is strongly related to the magnitude of the induced strain in the SPS structure. The following thermodynamic arguments support this statement.

The occurrence of both vertical long-range ordering and lateral compositional modulation (phase separation or lateral layer ordering) in the binary SPS structures is unfavorable. This is because the thermodynamic driving forces for their formation are opposite to each other. However, the experimentally evidenced existence of both of these ordering features of the heterostructure with SPS suggests that the thermodynamic requirements are relaxed by factors other than the formation enthalpies and entropies. One of these factors is the induced strain associated with the deviation of superlattice periodicity in the SPS structure from $na_s$. Taking into account elastic strain relaxation, Glas [11.63] demonstrated that when the composition was modulated in an epitaxial layer in directions parallel to the substrate surface, i.e., when the SILO effect occurs, the layer was more thermodynamically stable than its unmodulated counterpart. The implication of this model is that a strained epitaxial layer, if kinetically feasible, will undergo a lateral compositional modulation first, rather than a less favorable reaction of generating misfit dislocations to relieve the misfit strain.

Thus, the driving force for the lateral ordering is attributed to the surface strain generated during the initial stages of heteroepitaxial growth of the SPS and attendant spinodal decomposition of these structures (decomposition due to a miscibility gap in solid solutions). In addition, once a lateral modulation of the composition has started, it will continue with the same periodicity. This latter property leads to the creation by the SILO effect of a lateral QWR array in the SPS heterostructure in the $(GaP)_n/(InP)_n$ and $(GaAs)_n/(InAs)_n$ material systems.

A general analysis of the SILO effect is presented in [11.54], part II, in the framework of the thermodynamic theory of polydomain heterostructure formation. The QWR arrays shown in Figs. 11.30 and 11.31 present a special case of polydomain heterostructures (compare it with Fig. 11.24).

The crucial point related to generating a QWR-like polydomain heterostructure is the thickness of the epilayer (in the above considered cases the thickness of the binary compound layer in the SPS). Let the lateral period of the polydomain structure (the QWR array) be denoted by $D$, and the thickness of the epilayer by $h$ (see Fig. 11.24). Then, for structures of $D \ll h$, microstresses occurring in the heterostructure contribute only to the effective energy of the interfaces. They do not cause a real interaction between the in-

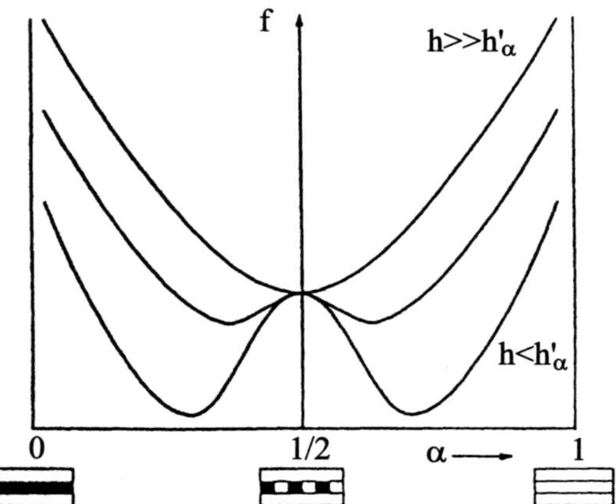


**Fig. 11.32.** The dependence of the free energy on domain fraction $\alpha$, for a two phase polydomain heterostructure of constant domain period. The thickness of the active layer $h$ is taken as the parameter here (taken from [11.54])

terfaces, which is needed for generating the lateral ordering. In the limit of $D/h \to 0$ no QWRs will exist.

The theory presented in [11.54] is quantitatively correct only for heterostructures with $D \ll h$; however, in the case of QWRs shown in Fig. 11.31 $D \approx 200\,\text{Å}$ while $h \approx 6\,\text{Å}$. Nevertheless the theory gives useful qualitative estimates for $D/h \approx 1$. Extension of the theory to cases of $D/h > 1$ has to be based on the involvement of the factors closely related to the interaction between interfaces through their microstresses. The first results indicating considerable progress in this direction have already been published in [11.85–87]. For interacting interfaces the energy of microstresses depends on mutual arrangement of the stress source at neighboring interfaces and therefore has different fractional dependence for different polydomain geometry. We will not go further into the theoretical details, but present a few of the most interesting results. These results are shown in Figs 11.32 and 11.33 after [11.54], part II. The first figure concerns the dependence of the free energy of the heterostructure on domain fraction, with the thickness of the active layer treated as a parameter (the domain fraction $\alpha$ gives the partial amount of the phase $A$ in a multiphase heterostructure $A + B + C + \ldots$).

For $h > h_{\mathrm{cr}}$, where $h_{\mathrm{cr}}$ is the critical thickness below which the polydomain heterostructure becomes unstable, the equilibrium domain structure is a symmetrical equidomain structure with equal domain fractions $\alpha_{\mathrm{equ}} = \frac{1}{2}$ (in the structure shown in Fig. 11.32 there are two phases, the black one and the white one; $\Delta\alpha = \alpha - \frac{1}{2}$, and the Gibbs free energy is denoted as $f$). If the active layer thickness $h$ becomes less than $h_{\mathrm{cr}}$, a break of symmetry occurs in the heterostructure, which causes the energy minimum at $\Delta\alpha = 0 (\alpha = \frac{1}{2})$ to

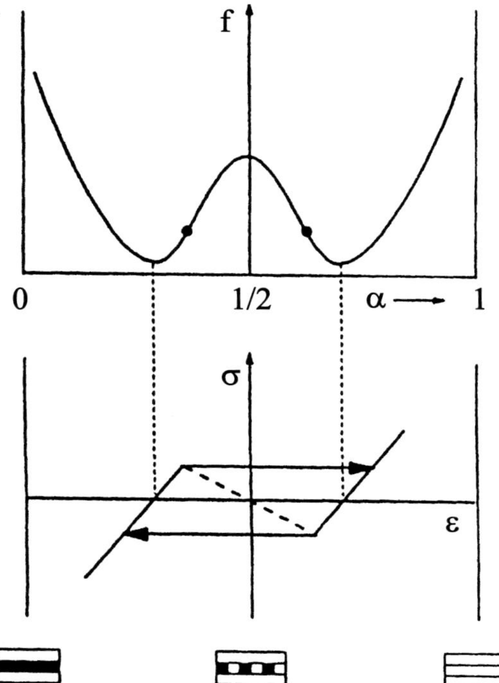


**Fig. 11.33.** The free energy and stress–strain relation of the two phase polydomain heterolayer with a subcritical thickness. $\sigma = ((1/\varepsilon)\mathrm{d}f)/(\mathrm{d}\Delta\alpha)$, where $\varepsilon$ is the strain; the spinodal inflection points on $f(\alpha)$ correspond to extrema on the $\sigma(\varepsilon)$ curve (taken from [11.54])

disappear and two minima corresponding to the equivalent nonequilibrium states with $\Delta\alpha = \pm\frac{1}{2}\sqrt{h - h_{\mathrm{cr}}/h}$ to appear.

Figure 11.33 shows the free energy and the stress–strain relation of the polydomain heterostructure of subcritical thickness ($h < h_{\mathrm{cr}}$). It is important to note that the elasticity of heterostructures with active layers of subcritical thickness is a strongly nonlinear function of stress. Thus, the presence of two minima separated by a nonconvex section of $f(\alpha)$ makes it possible to switch the equilibrium domain structures from one to the other. The compliance (a quantity which characterizes the dependence of the equilibrium strain on stress) becomes infinity at some stresses when the domain fraction reaches the spinodal points, where $\mathrm{d}^2 f/\mathrm{d}\alpha^2 = 0$. At these stresses the domain structure jumps from one equilibrium state to the other. The reverse switching requires reaching the other spinodal point. Thus there is a hysteresis in strain response of a heterostructure to reversible stress [11.54].

## 11.6 Morphological Stability in Epitaxy

In many technically important epitaxial multilayer structures nonplanar interfaces between different heteroepitaxial layers appear. Surface corrugations for grating waveguide couplers or distributed-feedback lasers, buried ridge waveguides, V-grooved substrates for stripe geometry or quantum wire lasers, may serve as examples [11.88]. Most frequently the nonplanar interfaces are covered by epitaxial layers which at the counter-side are bounded by planar surfaces. Consequently, the crystallization of a uniform, planar coverage on the top of a nonplanar substrate surface, regardless of the shape of this surface, can be believed to be one of the most important goals in the fabrication process of these devices.

An epitaxial layer crystallized over a profiled substrate will, at the counter-side, exhibit a surface whose shape depends on the definite growth conditions chosen. For studying the evolution in time of the profiled interface during the crystallization of an epitaxial overgrowth, one may use the linear morphological stability theory of Mullins and Sekerka (hereafter the MS theory). The theory has been extensively developed [11.89–94], following the now-classic papers by these authors [11.95–97].

### 11.6.1 The Mullins–Sekerka Theory

The MS theory deals with the crystal–melt interface during solidification of a dilute binary alloy. However, its main ideas and mathematical formalism can be used for studying also the crystal–vapor or crystal–solid interfaces when crystallizing from the vapor phase [11.98–100], or from the solid phase [11.90]. The principal physical assumptions on which the MS theory is based are those of isotropy of bulk and surface parameters in the metastable fluid phase and on the growing interface, as well as local equilibrium at all points of this interface. The critical mathematical simplification is the use of steady-state values for thermal and concentration (diffusion) fields. Moreover, dissolution or thermal decomposition effects on the profiled substrate surface are excluded [11.7, 95–97]. This means that the MS theory, when applied to the problems of morphological stability in epitaxy, delivers quantitative results which should be treated merely as an approximation to reality. Nevertheless, even approximate information obtained in a relatively simple way may be of considerable use for crystal growers involved in the epitaxial technology of semiconductor devices with nonplanar interfaces. A morphological stability theory which would take into consideration all specific features of the epitaxial growth processes should be quite sophisticated from the mathematical point of view [11.89, 101, 102], which exceeds the scope of this book.

Let us consider a profiled interface separating the substrate crystal from a fluid (liquid or gaseous) phase which is called the metastable phase (see Sect. 1.1, Fig. 1.1). Suppose this interface has a shape which can be represented in a Cartesian coordinate system $(x, y, z)$ by a periodic function $z(x)$,

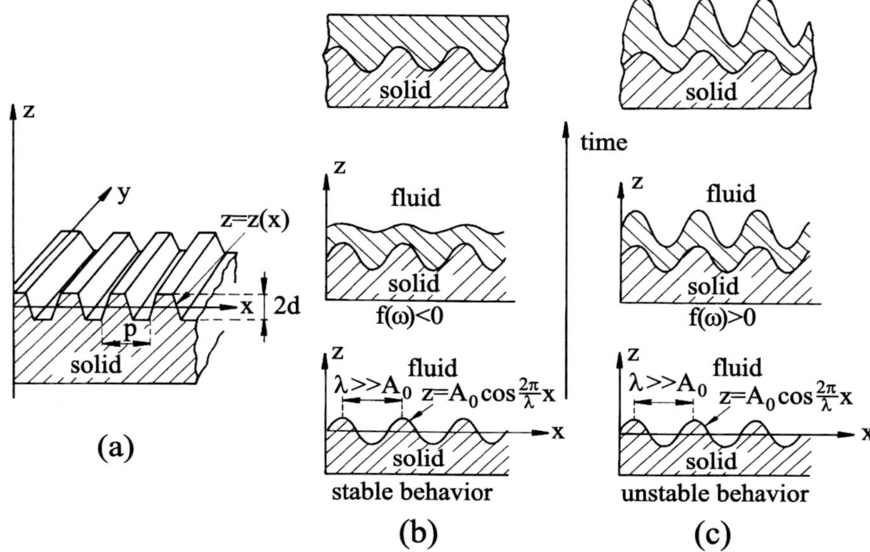


**Fig. 11.34.** Schematic illustration of periodically profiled substrate surface (**a**) and examples of stable (**b**) and unstable (**c**) behavior of this interface shape during crystallization of a covering epitaxial layer (taken from [11.88])

with the spatial period $p$ along the $x$-coordinate (Fig. 11.34a). For the sake of simplicity we consider here only interfaces for which $z(y)$ = const. Such interfaces (periodically profiled along one direction) are met in practice when a planar substrate becomes patterned by etching.

Following the MS theory we consider the periodically profiled substrate surface $z(x)$ to be a small perturbation to the initially planar (flat) interface $z(x, y) = 0$, between the growing crystal and the crystallizing metastable phase. This means that the amplitude $d$ of the function $z(x)$ (Fig. 11.34a) has to be small, compared to the spatial period $p$ of this function. For the surface shapes most frequently met in optoelectronic devices, this is more or less true [11.88]. Consequently, $z(x)$ can be Fourier-analyzed [11.100], yielding

$$z(x) = b_0/2 + \sum_{n=1}^{\infty} (a_n \sin n(2\pi/\lambda)x + b_n \cos n(2\pi/\lambda)x), \tag{11.91}$$

where the Fourier coefficients $a_n$ and $b_n$ are given by

$$a_n = (2/\lambda) \int_0^{\lambda} z(x) \sin n(2\pi/\lambda)x \mathrm{d}z, \tag{11.92}$$

and

$$bn = (2/\lambda) \int_0^\lambda z(x) \cos n(2\pi/\lambda) x \mathrm{d}z. \tag{11.93}$$

The evolution in time of the profiled interface during the crystallization of an epitaxial overgrowth can be found by considering the development of its general Fourier component [11.92]

$$z_F(x, t) = A(t) \exp(i\omega x) = A_0 \exp(i\omega x + f(\omega)t). \tag{11.94}$$

$A_0$ is here the initial amplitude of the Fourier component of the substrate surface disturbance with periodicity $\lambda = 2\pi/\omega$. This disturbance will be considered to be small, compared to its lateral extent, if $A_0 \ll \lambda$. The function $f(\omega)$ appearing in the exponent is called the stability function [11.98]. The development in time of the amplitude $A(t)$ depends on the sign and the absolute value of this function. If $f(\omega)$ is negative the initial perturbation will decay (Fig. 11.34b), while for positive value of this function the perturbation will grow (Fig. 11.34c).

### 11.6.2 Morphological Stability in LPE

The stability function depends on the crystallization parameters describing the epitaxial growth process. When these parameters are known, the function $f(\omega)$ can be calculated, and the evolution in time of the nonplanar interface can be predicted. Proving whether the periodic perturbation grows or decays for specified crystallization conditions, one has to evaluate the time derivative $\mathrm{d}A(t)/\mathrm{d}t$. To this end the growth rate at each position of the interface should be calculated in terms of the suitable thermodynamic driving force of the crystallization process considered. These calculations differ in detail for each epitaxial growth technique we are interested in.

For the sake of brevity the relevant considerations will be restricted only to one epitaxial growth technique. This will be liquid phase epitaxy (LPE), the technique which is most suitable for being analyzed by using the MS theory. The LPE growth process will be considered in its confined solution slider variant applied to the GaAs/AlGaAs material system [11.103]. This variant of LPE enables precise control of thickness, morphology, and uniformity of the crystallized layers [11.11].

The idea of the LPE slider technique with confined solution volume is illustrated in Fig. 11.35. A GaAs slice is placed on top of the liquid solution and constrained to remain parallel to the substrate, separated by a very thin solution layer of only a few millimeters thickness. The growth process is entirely controlled by the diffusion of nutrients (As atoms in the diluted Ga solution of GaAs or AlGaAs) through the solution layer toward the substrate. It is slow (typical growth rate is in the range of $10\,\text{Å}\,\mathrm{s}^{-1}$ [11.103]) and the heat transport can be ignored. For such crystallization conditions the assumptions of the MS theory are well justified, and the morphological stability function $f(\omega)$ defined by (11.94) as

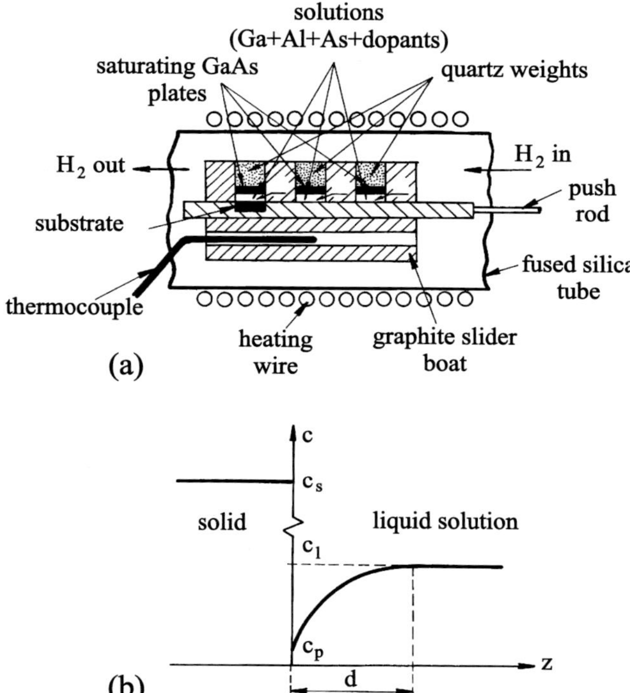


**Fig. 11.35.** The central part of a fused silica tube LPE reactor (**a**) The cross-section of the graphite slider boat typical for the confined solution volume variant of LPE is shown. The Ga solutions 1, 2, 3 are confined to very small volumes by the substrate platelet or the slider rod from the bottom side and by the saturating GaAs plates from the top side. (**b**) The spatial distribution of the solute concentration near the liquid–solid interface (taken from [11.88])

$$f(\omega) = \frac{\mathrm{d}A(t)/\mathrm{d}t}{A(t)} \tag{11.95}$$

has, for a definite point $x = x_0$ of the interface, the form

$$f(\omega)_{\mathrm{LPE}} = \frac{(D\omega)/(c_s - c_p)}{G - c_L \Gamma \omega^2}. \tag{11.96}$$

In this formula $D$ stands for the diffusion coefficient of the solute in the solution (As atoms in Ga solution), and $G$ is the concentration gradient of the solute atoms (As) at the solid–liquid interface ($\mathrm{d}c/\mathrm{d}z\|_{z=0}$). The coefficients $c_s, c_p$, and $c_L$ (Fig. 11.35b) are the concentrations of As atoms in the epitaxial layer of GaAs (or AlGaAs), in the liquid at the planar interface, and in the bulk of the solution, respectively. $\Gamma = \gamma/L$ is the capillary constant, with $\gamma$ and $L$ standing for the solid–liquid interface free energy per unit area (surface tension) and for the latent heat of solidification. $\omega = 2\pi/\lambda$ is the spatial frequency of the Fourier component of the profiled interface (see (11.94)). The first term in the parentheses of this equation may be interpreted physically as being due to an increase in the concentration gradients of the solute onto the hills on the perturbed interface. This term always causes instability. The second term is due to the concentration gradients along the surface (in its lateral extent) which cause solute transport and so tend to smooth out the cosinusoidal disturbance (see Fig. 11.34b and c). As one can see, the sign of the stability function is determined by the difference of the terms in the parentheses. This means that the competition between the diffusion toward the growing interface and the migration over the surface (of the solute in the liquid phase) defines the sign of the function $f(\omega)$.

Morphological stability functions may be calculated for other epitaxial growth techniques, too [11.88]. For example, in the case of MOVPE the relevant function has the form similar to the LPE function, i.e.,

$$f(\omega)_{\mathrm{MOVPE}} = \frac{D\omega}{(c_s - c_g)}(G_g - c_V \Gamma \omega^2) \tag{11.97}$$

and for MBE

$$f(\omega)_{\mathrm{MBE}} = \frac{D\omega}{(c_s - c_g)}(G_g - c_g \Gamma \omega^2 - c_g \Gamma (D_s \delta / D)\omega^3) \tag{11.98}$$

where $D$ is the diffusion coefficient of the reactive component in the diffusion boundary layer shown in Figs. 8.1 and 11.2 or in the near surface transition layer shown in Fig. 1.1b-d, $G_g$ is the relevant concentration gradient at the crystallization interface, $c_g$ stands for the equilibrium concentration of the reactive component at the planar interface in the boundary layer (near surface transition layer), $c_V$ is the respective concentration in the main gas stream in the MOVPE reactor (this is replaced by $c_g$ in the case of the ultrahigh vacuum of the MBE reactor), and $c_s, \Gamma$ and $\omega$ have the same meanings as in (11.96). In the case of MBE, $D_s$ means the surface diffusion constant of the reactive component over the crystallization interface, while $\delta$ is the thickness of the near surface transition layer (see Fig. 1.1). The third term in parentheses in (11.98) represents the surface diffusion smoothing effect [11.104], which in MBE is much more pronounced than in LPE or MOVPE.

# 12. Atomistic Aspects

The physics of epitaxial crystal growth is usually based on two kinds of theoretical approximations, i.e., a macroscopic approach using phenomenological thermodynamics [12.1], and a microscopic, or atomistic approach, using stochastic models [12.2]. So far the principles of epitaxy have been discussed in a macroscopic, thermodynamic approach. Here, the microscopic aspects of epitaxial crystallization will be considered, mainly in the frames of statistical physics [12.3] and quantum mechanics of chemical bonds in crystals [12.4]. Since the epitaxial growth process takes place at the surface of a single-crystalline substrate, one has to be aware of the influence which the surface structure exerts on this process (for more information see Chap. 13)

## 12.1 Incorporating of Adatoms into a Crystal Lattice

Before considering the crystallization in the atomic scale let us identify when, or in what situations, an atom impinging onto the surface is said to be crystallized. In the gas phase an atom is free and makes no bonds with other atoms. In the bulk of a crystal it makes $z_b$ bonds with neighboring atoms to lower the energy ($z_b$ is here the coordination number characteristic of the lattice of the crystal). The energy gain for each connected bond is set equal to $-\Phi$. Thus, for $N$ crystallized atoms there are in total $z_bN/2$ bonds, and the total cohesive energy of the crystal is $-z_bN\Phi/2$, or $-z_b\Phi/2$ per crystal atom. When the crystal melts, the atoms become free without any bond connections, and the cohesive energy is zero.

### 12.1.1 Kossel's Model of Crystallization

Let us consider an atom freely moving in the gas phase, which impinges on the substrate crystal and makes there bonds to $z_s$ nearest neighbors of the substrate. If $z_s$ is the maximal possible number of neighbors which an atom can have in the first atomic monolayer of the surface then this atom can be regarded to be incorporated into the substrate crystal (is crystallized). However, as a result of the fact that the crystal lattice of the substrate is truncated at the surface, $z_s$ is always less than $z_b$. Now, an important problem

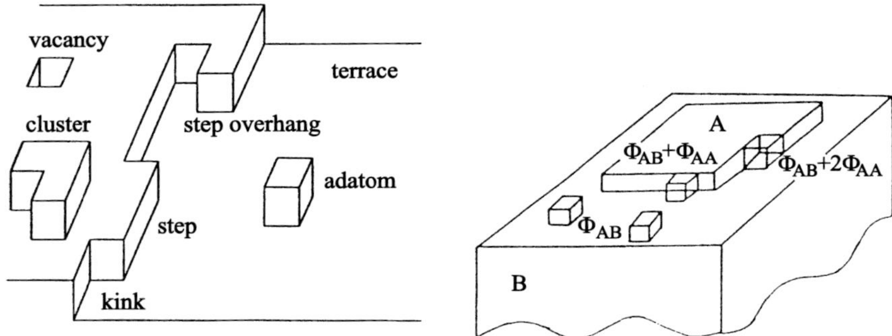


**Fig. 12.1.** (**a**) Surface configuration with terraces bounded by a step with kinks (taken from [12.3]). (**b**) Kossel's model of a crystal $A$ deposited on a substrate $B$. Bond energies: (i) of an isolated adsorbed atom $\Phi_{AB}$ and of two atoms adsorbed (ii) at a repeatable step, $\Phi_{AB} + \Phi_{AA}$, and (iii) at the kink of the step, $\Phi_{AB} + 2\Phi_{AA}$, are indicated (taken from [12.5])

arises, i.e., when will an atom be able to acquire $z_s$ bonds on the substrate surface?

On a completely flat crystal surface, the adsorbed atom cannot make so many bonds. But usually on a real crystal surface, there are various defects as shown in Fig. 12.1 a. A flat portion of the surface is called a terrace. When the heights of two consecutive terraces differ, there is a step between them. A step can be straight as it runs in a closed packed direction. It can also bend and change the orientation at a kink position. When an atom impinges on a flat terrace, the atom makes only a few bonds with the underlying atoms. The bond connection is so weak that the atom can migrate on the surface. This atom is called for short an adatom (adsorbed atom). During migration, the adatom may come in contact with an uprising step. Such a step provides some additional bonds parallel to the terrace, but the number of bonds is still less than $z_s$. Therefore, the adatom can slide along the step edge by diffusion, and may reach the kink position where there are $z_s$ nearest neighbors. At this kink site, the adatom becomes incorporated into the substrate crystal (becomes crystallized). The striking feature of the kink site is that it never disappears by crystallization or melting; it only slides along the step.

Much of the theoretical work on epitaxial growth has been based on atomistic models of two crystals, the substrate and the growing epilayer. Such models including only first-neighbor interactions are, at first, sufficient to draw the main points of the growth phenomena [12.5]. No fundamental restriction is involved if one assumes that both crystals have a simple cubic structure. This assumption is the basis of Kossel's model of a crystal $A$ deposited on a substrate $B$ (see Fig. 12.1b) [12.5]. This model, when considering both the kinetics and mechanisms of epitaxial growth, takes into account the presence of steps, kinks, ledges, etc., on the substrate surface otherwise free of defects such as screw dislocations or impurities. The consequence of Kossel's model is the treatment of the substrate surface as being divided into crystal sites (see Sect. 7.1.1, Fig. 7.4).

Very often in studies on epitaxy, based on atomistic models, the so-called solid-on-solid (SOS) approximation is used [12.6]. In this approximation one assumes that:

(i) an atom can become bound to the solid surface only on top of another, thus, preventing any overhangs (and vacancies) in the solid,

(ii) crystal sites are fixed positions on a crystal surface, and

(iii) the kinetic processes occurring during epitaxy that cause transitions of the adatoms among these sites are described by Arrhenius-type rates

$$K(T) = K_0 \exp\left(-\frac{E_a}{k_B T}\right) \tag{12.1}$$

where $K_0$ is the rate constant of the kinetic process (the attempt rate), $E_a$ is the activation barrier (the activation energy), $k_B$ is Boltzmann's constant and $T$ is the absolute temperature. The essential point is that the rates are determined by the transition-state barrier $E_a$, which is not necessarily associated with either the initial state or the final state of the definite process.

The epitaxial growth process is initiated by adding atoms randomly to the atomic columns at each site at an average rate of $t^{-1}$, where $t$ is the layer completion time. The surface migration of adatoms is taken as a nearest neighbor hopping process, whose rate is described by a relevant rate equation of the form of (12.1) (the temperature across the substrate is assumed to be constant, while the activation barrier is taken as depending only on the initial environment of the migrating adatom). The lateral interactions of the migrating adatoms with other adatoms, or crystal sites of steps and kinks on the substrate surface, as well as the perpendicular interactions with substrate surface sites, are described in this microscopic approach by one-dimensional two-body interaction potentials (see Sect. 7.1.1, Fig. 7.5).

With these conditions established, let us consider the substrate $B$ and the deposit $A$ realizing parallel contact between two (100) faces having the same unit mesh parameter (the same lattice constant). The interaction energies between two nearest neighbor atoms of the same solid phase are denoted $\Phi_{AA}$ and $\Phi_{BB}$, while $\Phi_{AB}$ stands for the interaction energy of atoms of two different phases in contact with each other. These quantities are positive, however, with a minus sign they are equivalent to the relevant potential energies. The surface energies defined in Sect. 11.4.1 are related to these interaction energies by

$$\sigma_A = \Phi_{AA}/2a^2, \quad \sigma_B = \Phi_{BB}/2a^2, \quad \beta = \Phi_{AB}/a^2 \tag{12.2}$$

where $a^2$ is the area occupied by an atom in the surface atomic plane. Dupre's relation (11.82) leads with (12.2) to the definition of a parameter $\Phi^* = \sigma_B^* 2a^2$, such as

$$\Phi^* = \Phi_{AA} + \Phi_{BB} - 2\Phi_{AB}, \tag{12.3}$$

which characterizes the interface. If no adhesion to the substrate surface occurs ($\beta = 0$), then $\Phi_{AB} = 0$, and $\Phi^* = \Phi_{AA} + \Phi_{BB}$. On the contrary, if $\beta = 2\sigma_A$, then $\Phi_{AB} = \Phi_{AA}$ which means that the adhesion between $A$ and $B$ is the same as the adhesion of $A$ on $A$ which would correspond to a situation of homoepitaxy. One may conclude that the case $\beta > 2\sigma_A$ mentioned in Sect. 11.4.1 has physical significance because nothing forbids the occurrence of the case in which $\Phi_{AB} > \Phi_{AA}$. The two inequalities given in Sect. 11.4.1, differentiating between 3D and 2D growth modes in epitaxy ($\beta < 2\sigma_A$ and $\beta > 2\sigma_A$, respectively), now become:

$$\begin{array}{llll} (i) & \Phi_{AB} < \Phi_{AA} & \Rightarrow & \text{3D growth} \\ (ii) & \Phi_{AB} > \Phi_{AA} & \Rightarrow & \text{2D growth.} \end{array} \tag{12.4}$$

The following conclusion is now obvious: *the 3D (2D) growth mode occurs when the interaction energy of an atom of the deposit with the substrate is less than (greater than) to the energy that this atom would have on its own substrate* [12.5].

### 12.1.2 Lattice Gas Models

A quantitative study of the phase transitions in epitaxial crystal growth (in the atomistic approach) requires definitions of microscopic models for the crystal–melt (solution) and crystal–vapor interfaces [12.7]. Among the variety of models for the melting–freezing or solidification–evaporation transitions, the lattice gas models (LG models) seem to be most suitable [12.8–14]. These models [12.10] assume on both sides of the interface about the same density and short-range order (SRO) of the relevant phases, but a different long-range order (LRO) or/and different symmetry.

The simplest LG model is the one-component lattice gas with nearest-neighbor interaction [12.15]. It represents fluid (disordered, metastable phase) and solid (ordered phase exhibiting both SRO and LRO) phases of the same species $A$ in contact at the interface. This model introduces a lattice system, in which the space is divided into discrete lattice points. Each lattice point can be occupied by an atom of the solid, by an atom of the fluid, or can be empty (Fig. 12.2). Double occupancy of a lattice point is forbidden. Stochastic evolution of the system consists of the following steps. Every gas atom hops freely from a lattice site to one of its nearest neighbors, unless it is occupied. When a gas atom diffuses to come into contact with solid atoms, it crystallizes with a definite probability $W$. Inverse to this crystallization process is evaporation, where an atom of the solid phase at an interface tries to turn into a the gas phase atom. These crystallization and evaporation processes at the interface should satisfy the thermodynamic detailed balance condition [12.16]. Evaporation in the bulk crystal and solidification in the bulk of the gas phase are neglected in this approximation. A mobile atom

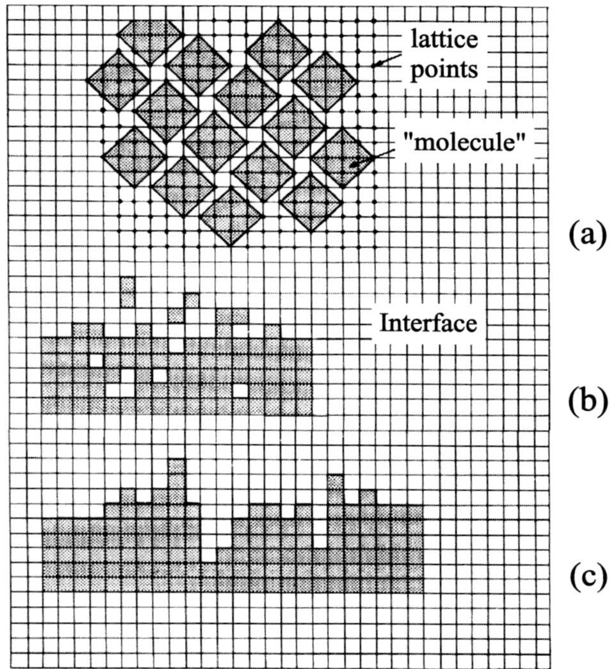


**Fig. 12.2.** (**a**) LG model for a solid–liquid type transition. The molecules (shaded areas) cover more than one lattice point, have a strong short-ranged repulsion (hard-core repulsion) and their close-packed structure do not match the lattice symmetry. (**b**) Ising model. Every atom/molecule occupies one lattice point. There is a hard-core repulsion between nearest neighbors and a short-ranged attractive interaction. An interface may be introduced by appropriate boundary conditions. There exists a critical temperature $T_c$, above which the two phases (formerly separated by the interface) become equivalent (the interface then disappears). (**c**) Solid-on-solid model. The interface between solid and vapor exists at all temperatures, since overhanging structures (admissible in the Ising model) are excluded by definition (taken from [12.7])

of the fluid gives an entropy contribution to the chemical potential of the system (see Sect. 11.1.1), whereas an atom of the solid is immobile but has a contribution to the chemical potential from its energy [12.13].

A simple generalization of the one-component lattice gas is the binary lattice gas [12.14]. In this model each site of the lattice either is occupied by an atom of species $A$, an atom of species $B$, or remains vacant. This is a special case of a ternary alloy, thus, the binary lattice gas model is suitable to be applied in studies on epitaxial growth of alloys [12.17, 18].

In the frames of the LG models, atoms are allowed to move only stepwise between the lattice points. There occur two overlapping interactions between the nearest neighbors in the lattice, i.e., infinite hard-core repulsion, and short-range attraction. The question which naturally arises in relation to the LG models concerns the relationship between the lattice gas and:

(i) the real gas in which the atoms are not restricted to move on lattice points only,

(ii) the real liquid which may consist of atoms or molecules.

The admissibility of the LG models to be used for analyzing crystallization phase transitions is based on statistical-thermodynamic arguments [12.8]. If one replaces the configurational integral in the partition function of the real gas by a summation over lattice points, then one will obtain the partition function of the lattice gas. Subsequently, by making the lattice constant (the distance between the nearest neighbors in the lattice) smaller and smaller, successively better approximations to the partition function of the real gas can be obtained. In the case of liquids one has to take into consideration the possibility of covering more than only one lattice point by the liquid molecule (see Fig. 12.2a).

An important advantage of the LG models is their mathematical equivalence to the ferromagnetic Ising model in an external magnetic field [12.8], which is the simplest lattice model exhibiting first-, and second-order phase transitions [12.19, 20]. For example, the problem of order–disorder transitions in two dimensions was solved with this model in 1944 by Onsager [12.21]. Since that time, a number of studies concerning higher dimensional Ising models have been published [12.22–25]. Let us compare these models in brief.

In the Ising model one considers a lattice of interacting spins, each of which can assume two possible positions, i.e., an upward position symbolized by $\uparrow$, and a downward position $\downarrow$. In the LG models one considers a corresponding lattice with each lattice point either vacant or occupied by an atom. To each configuration of the lattice of spins there corresponds a configuration of the lattice gas in which a lattice point is vacant or occupied according as whether the corresponding spin is $\uparrow$ or $\downarrow$. Using this geometrical correspondence, the mathematical equivalence of the two models could be established [12.8]. In doing this the following identification of corresponding quantities has been used:

| *In the Ising model:* | | *In the LG models:* |
|---|---|---|
| number of spins in the system | $\leftrightarrow$ | volume of the system |
| number of downward spins | $\leftrightarrow$ | number of atoms in the lattice points |
| number of upward spins | $\leftrightarrow$ | number of empty lattice points |
| $2/(1-I)$ | $\leftrightarrow$ | specific volume per atom |

where $I$ stands for the intensity of magnetization of the spin lattice system

| | | |
|---|---|---|
| $-(F+H_e)$ | $\leftrightarrow$ | pressure in the lattice gas |

where $F$ stands for the free energy per spin, while $H_e$ is the intensity of the external magnetic field.

The formal similarity of the two Hamiltonians, representing the total energies of the spin system and of the lattice gas, may also be shown. To this end let us consider two atoms of the lattice gas, namely an atom in the solid phase and an atom in the gas phase of the same species. With each atom in the solid an energy gain $\mu_s$ equal to $-z_b\Phi/2$ is associated, while a gas atom gains entropy by changing its position with a neighboring empty site. This exchange process also mimics the diffusion in the gas phase. The effect of the surface tension can, for example, be incorporated by an energy cost of $2J = \Phi_{AA} > 0$, in breaking a solid–solid nearest-neighbor bond. These energetic parameters are summarized in the Hamiltonian of the form

$$H = 2J\sum_{i,j}\left(C_i(1-C_j)+(1-C_i)C_j\right)+\mu_s\sum_i C_i \tag{12.5}$$

where the first summation runs over all the nearest neighbor pairs, and the crystallization order parameter $C_i$ on the $i$-th lattice site is unity when it is occupied by a solid atom and vanishes otherwise. In terms of an Ising spin variable $S_i \equiv 2C_i - 1 = \pm 1$, the Hamiltonian reduces to that of a ferromagnetic Ising model in an external magnetic field,

$$H = -J\sum_{i,j} S_iS_j + \frac{\mu_s}{2}\sum_i S_i + \frac{V}{2}(zJ-\mu_s). \tag{12.6}$$

Here $J$ is the quantum mechanical exchange energy, $\mu_s$ is the magnetostatic interaction energy of the nearest-neighbor spins, the volume $V$ is the total number of lattice sites and $z$ is the coordination number of the spin lattice.

In view of the presented considerations, one may recognize that the solid-on-solid (SOS) lattice gas approximation (see Fig. 12.1b) corresponds to a restricted version of the Ising model, because no overhangs or vacancies in the lattice are allowed in this approximation (see Fig. 12.2b and c). It is worth noticing that an adatom adsorbed at a kink site in a lattice of the SOS approximation corresponds in the Ising model to a spin with three of its nearest neighbors having downward spins (on sites where the SOS lattice gas atom is bonded to the three kink sites), and the other three nearest neighbors having upward spins (on the sites where no bonds occur to the adatom (represented in the SOS approximation by a cube)). Reversing the spin in the Ising model corresponds to evaporation or to impingement with subsequent adsorption in all of the LG models, and thus, also in the SOS approximation [12.26].

### 12.1.3 Stochastic Model of Epitaxy

In conclusion to this section, let us consider the stochastic model of epitaxy given by Nakayama et al. in [12.27]. We will treat here a zincblende

(ZB) crystal system of a semiconductor epilayer of $(A_{\mathrm{x}}^{\mathrm{III}}B_{1-\mathrm{x}}^{\mathrm{III}})C^{\mathrm{V}}$ formed on the binary compound $A^{\mathrm{III}}C^{\mathrm{V}}$ substrate (001) surface. The growing (001) plane is a face-centered cubic (fcc) sublattice plane of the zincblende lattice. Therefore, in order to describe the ordering behavior of the adparticles in the epitaxial growth of the $(A,B)$ column–III sublattice, the growth of $(A,B)$ (001) two-dimensional square lattice will be considered at first. Interatomic interactions which are key factors in description of ordering phenomena in epitaxial growth are taken into account for a binary system within the framework of the Ising Hamiltonian of surface atoms. In the presented study, the Ising parameters were given a priori as calculating parameters. The master equation for the site-occupation probability, describing the binary growth system, is given by

$$\begin{aligned}\frac{\partial p_\mu(x,t)}{\partial t} &= -\sum_r p_\mu(x,t)p_v(x-r,t)w_\mu(x\to x-r)\\ &\quad+\sum_r p_\mu(x-r,t)p_v(x,t)w_\mu(x-r\to x)\\ &\quad+J_\mu p_{ad,\mu}(x,t)p_v(x,t),\end{aligned} \tag{12.7}$$

where $p_\mu(x,t)$ is the occupation probability of atom $\mu=(A$ or $B)$ at site $x$ and time $t$, while $w_\mu(i\to j)$ is the atomic-jump probability from lattice site $i$ to a vacant lattice site $j$. $r$ and $J_\mu$ stands here for the site displacement vector of the jumping atom, and the molecular beam flux of atom $\mu$, respectively. The first two terms in the right-hand side of the master equation show the site-exchange (jumping) terms and the last term comes from the adsorption of impinging atoms. $p_v(x,t)=1-\sum_\mu p_\mu(x,t)$ is the probability of finding a vacant site at $(x,t)$. The master equation (12.7) was used in the presented study, so as to take the local atomic configuration into account in the formulation of adsorption and atomic jump probabilities on the basis of the square lattice gas model.

The relevant Ising Hamiltonian of the growing surface, which is characterized by the atomic configuration $\xi(\sigma)$ on a 2D square lattice, is given in the form

$$H(\xi(\sigma)) = -J_x\sum_i \sigma_{i,j}^{\xi}\sigma_{i+1,j}^{\xi} - J_y\sum_j \sigma_{i,j}^{\xi}\sigma\xi_{i,j+1}. \tag{12.8}$$

Ising variables in this Hamiltonian are defined as $\sigma=+1$ for atom $A$, $\sigma=-1$ for atom $B$ and $\sigma=0$ for vacant site. The coupling of the Ising Hamiltonian and the master equation can be simply formulated by describing adsorption and atomic jump probabilities in terms of the Ising Hamiltonian. Thus, the adsorption probability of the $\mu$ atom ($A$ or $B$) at the adsorption side $(i_0,j_0)$ is approximately given by the occupation probability of the $\mu$ atom at the adsorption site, neglecting the thermal evaporation of adatoms. Namely,

$$p_{\mathrm{ad}}(i_0,j_0) = \frac{\exp\left(-\beta H\left[\xi_{\mathrm{loc}}(\sigma),\sigma(i_0,j_0)=\sigma_\mu\right]\right)}{\sum_{\xi_{\mathrm{loc}}(\sigma)}\exp\left(-\beta H\left[\xi_{\mathrm{loc}}(\sigma),\sigma(i_0,j_0)=\sigma_\mu\right]\right)}. \tag{12.9}$$

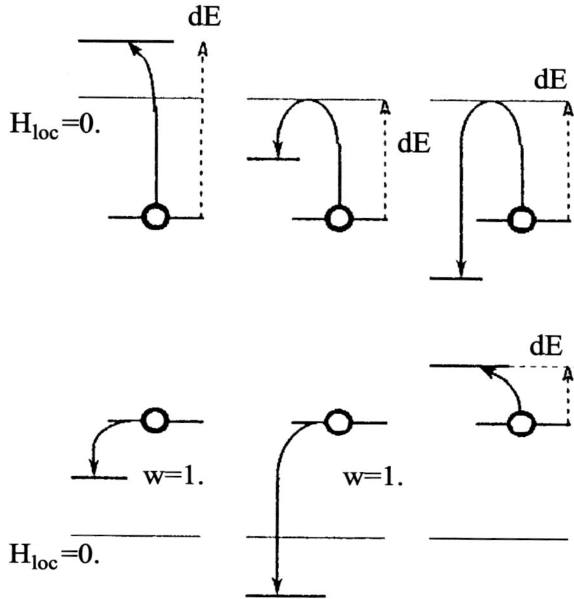


**Fig. 12.3.** Definition of the activation energies related to various types of diffusion paths. The definition of the local energies concerned with the atomic jumps is given in the text by (12.10) (taken from [12.27])

At the adsorption site $(i_0, j_0)$, local equilibrium has been assumed. $\beta$, as usually, stands here for $1/k_BT$. The local energy, for the case of nearest-neighbor interaction, can be estimated as

$$H\left[\xi(\sigma)\right] \cong -J_x \sum_{i=i_0-1}^{i_0} \sigma^{\xi}_{i,j_0}\sigma^{\xi}_{i+1,j_0} - J_y \sum_{j=j_0-1}^{j_0} \sigma^{\xi}_{i_0,j}\sigma^{\xi}_{i_0,j+1} . \tag{12.10}$$

Atomic jump probability from site $(i_0, j_0)$ to the nearest-neighbor vacant site $(i'_0, j'_0)$ can be defined by using the local activation energy as

$$W_\mu((i_0, j_0) \to (i'_0, j'_0)) = \nu_0 \exp\left(-\beta\Delta E_\mu((i_0, j_0) \to (i'_0, j'_0))\right) \tag{12.11}$$

with the pre-exponential factor, $\nu_0$, assumed to be unity. The activation energy for an atomic jump was defined here by using the diagram shown in Fig. 12.3. For example, in the case of an atomic jump from site $(i_0, j_0)$ to site $(i_0 + 1, j_0)$, the initial state energy $H(i_0, j_0)$ is given by

$$H(i_0, j_0) = H\left[\xi(\sigma), \sigma(i_0, j_0) = \sigma_\mu, \sigma(i'_0, j'_0) = 0\right] . \tag{12.12}$$

The key point in the definition of (12.11) is that the activation energy for an atomic jump between nearest-neighbor bound states corresponds to the activation energy to zero-level energy from the initial bound state instead of the energy difference between initial and final states. Therefore, it corresponds to the binding energy of the initial state itself. The detailed balance is realized among the transitions between local bound states.

Structural evolution during epitaxy of the first monolayer can be found by Monte Carlo (MC) simulation based on the master equation (10.7). In the case of an MBE growth process of binary ($A$, $B$) alloy, the algorithm of MC simulation is as follows [12.27]:

(i) choice of the impinging atom, $A$ or $B$, depending on the molecular beam flux ratio,

(ii) choice of a vacant adsorption site,

(iii) calculation of the local site energy of the adsorption site on the basis of the Ising Hamiltonian,

(iv) calculation of adsorption probability,

(v) calculation of atomic jump probabilities from nearest-neighbor occupied sites into the selected vacant site,

(vi) choice of events among adsorption, atomic jump and scattering (desorption without adsorption nor atomic jump),

(vii) determination of a new atomic configuration, $\xi(\sigma)$,

(viii) continue to the next event.

MC calculations have been performed in the presented example for three types of surfaces with $(j_x, j_y) = (F,\ F)$, $(AF,\ F)$ and $(AF,\ AF)$ interatomic interactions. According to the "lattice gas $\leftrightarrow$ ferromagnetic" formal analogy, "$F$" here means the ferromagnetic-type interaction ($J > 0$), where $AA$ and $BB$ nearest-neighbor pairs are more preferable than an $AB$ pair. "$AF$" means the antiferromagnetic-type interatomic interaction ($J < 0$), where an $AB$ nearest-neighbor pair is more preferable than $AA$ or $BB$ pairs. The calculated atomic arrangements of an ($A$, $B$) alloy of a chalcopyrite-type surface with four different surface coverages in time sequence is shown in Fig. 12.4. The interaction parameters are: $(\beta J_x, \beta J_y) = (AF,\ AF) = (-0.5, -0.5)$. The supercell size, in the performed MC procedure, was equal to $50 \times 50 = 2500$ atomic sites. Figure 12.4 demonstrates clearly that the adsorbed atoms gradually form an ordered arrangement with repeating adsorption and diffusion processes. In this case adsorption and diffusion (atomic jumps) processes cooperatively act to form the preferential atomic ordering in the ($A$, $B$) monolayer. The highly ordered atomic arrangement with full coverage, as shown in Fig. 12.4, corresponds to the surface of a chalcopyrite type (001) long-range-ordered surface [12.27].

Many other important problems of crystal growth physics have also been solved by using the LG models. This concerns especially studies performed by using Monte Carlo simulation procedures [12.28] to obtain crystal shapes [12.3, 13]. Let us mention as examples for the application of the LG models in the theory of crystal growth the following, arbitrarily chosen results, which are closely related to epitaxial crystallization:

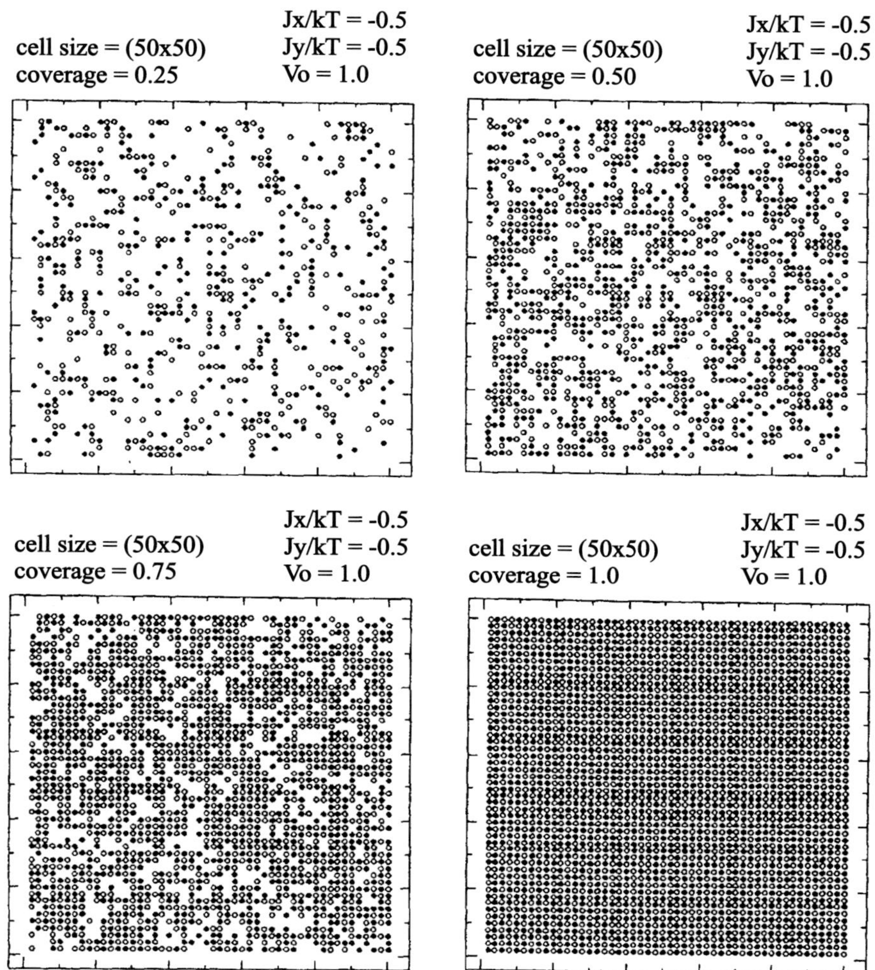


**Fig. 12.4.** Atomic structure evolution of chalcopyrite-type surface during growth of the first monolayer on a zincblende substrate (taken from [12.27])

(i) The role of an attracting substrate in the interplay effects between epitaxy (ordered growth) and layering/wetting in thick film deposition processes has been evaluated; it was shown that strong substrate attraction produces high-density compressed layers and quenches epitaxial growth [12.11].

(ii) The occurrence of a transition layer in the solid–melt interface during the crystallization process has been theoretically predicted; a so-called two-stage interface has been found for the growth on the (001) crystal surface, while a single-stage interface occurs for growth on the (110) surface [12.12].

(iii) Kinetics of ordering (crystallization) and disordering (melting/evaporation) for a two-dimensional binary lattice gas have been studied as a function of substrate coverage and layer stoichiometry; for the case of ordering, the SRO typically reaches a quasi-equilibrium state very rapidly, followed by exponential growth of LRO and final relaxation of both SRO and LRO to equilibrium; however, in disordering transition SRO typically achieves its equilibrium value quickly, but the LRO decreases very rapidly at short times, followed by an exponential decay to the final state [12.14].

## 12.2 Adsorption–Desorption Kinetics

The kinetics of adsorption and thermal desorption processes play a crucial role in many epitaxial growth processes in which vapor–solid interfaces appear, e.g., in VPE and MBE (see 6.2.1 and 7.1.1). In the growth environment in which these processes are performed, usually a dynamic equilibrium occurs between the substrate surface, kept at a definite temperature and covered by adsorbed species, and the vapor molecules or atoms, impinging onto the surface. At this equilibrium the coverage of the substrate surface, in general described by the formula

$$\Theta = n(c) + \sum_{i=n(c)+1}^{m} \frac{N_i}{N_{it}}, \tag{12.13}$$

depends on the vapor pressure characterizing the nearest vicinity to the surface. The curves, expressing graphically the "coverage versus pressure" dependence at constant temperature, are called adsorption isotherms [12.29]. In the definition given by (12.13), $n(c)$ is the number (integer) of completed monolayers, $m$ is the number of the outermost monolayer with sticking adsorbate species, $N_i$ is the number of adsorbed species in the $i$-th monolayer, and $N_{it}$ is the number of adsorption sites available for this species.

### 12.2.1 Adsorption Isotherms; Phenomenological Treatment

The simplest case for which an isotherm may be derived is based on the assumption that on a clean solid surface all adsorption sites are equivalent and that the adsorption probability does not depend on the occupancy of the neighboring adsorption site on this surface [12.30]. The rate of adsorption is then proportional to the pressure $p$ and the number $N(1-\delta)$ of vacant adsorption sites on the surface [12.31]

$$\frac{\mathrm{d}\delta}{\mathrm{d}t} = k_a p N(1-\delta). \tag{12.14}$$

Here, $\delta$ is the fractional coverage, $k_a$ is the rate constant for adsorption and $N$ is the total number of available adsorption sites on the surface. On the other hand, desorption from the surface causes a change in the surface coverage which is proportional to the number $N\delta$ of adsorbed species. Thus, the desorption rate is given by

$$-\frac{\mathrm{d}\delta}{\mathrm{d}t} = k_d N \delta, \tag{12.15}$$

where $k_d$ is the desorption rate constant. When dynamic equilibrium occurs, the two rates should be equal, which leads to the following expression for the "coverage versus pressure" dependence

$$\delta = \frac{Kp}{(1+Kp)} \quad \text{with } K = \frac{k_a}{k_d}. \tag{12.16}$$

This equation defines the Langmuir isotherm for direct adsorption, which may be treated as the zero-order approximation of the adsorption–desorption kinetics problem [12.5].

In the slightly more complicated case of dissociative adsorption, the rate of adsorption is proportional not only to the pressure but also to the probability that both of the adsorbed atoms of the dissociating molecule will find vacant adsorption sites. Thus, $\mathrm{d}\delta/\mathrm{d}t \sim p(N(1-\delta))^2$. On the other hand, the desorption rate is in this case proportional to the probability of the mutual impact of the adsorbed atoms, i.e., $-\mathrm{d}\delta/\mathrm{d}t \sim (N\delta)^2$. Following this, the Langmuir isotherm is now given by

$$\delta = \frac{\sqrt{Kp}}{1+\sqrt{Kp}}. \tag{12.17}$$

One may easily recognize that the coverage at dissociative adsorption depends "weakly" on the pressure. The Langmuir isotherms are plotted in Fig 12.5 for both of the considered cases. It has to be noticed that the simple expres-

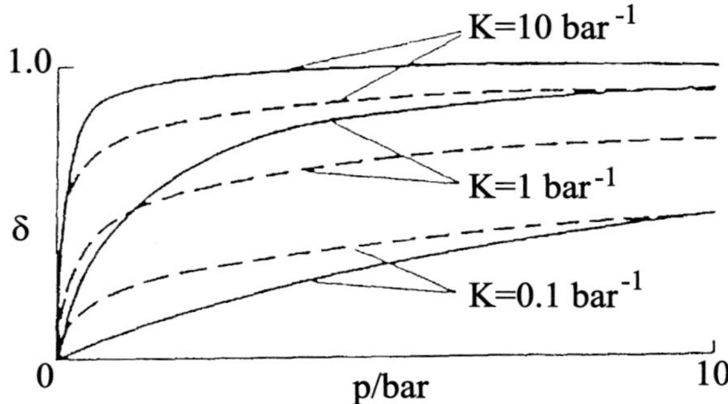


**Fig. 12.5.** The Langmuir isotherms for the direct (solid curves) and the dissociative (broken curves) adsorption. These curves are plots of (12.16) and (12.17) with $K = k_a/k_d$ as parameter (taken from [12.31])

sions presented here restrict the adsorption laws to cases of so called "low" adsorption (submonolayer case) [12.32], which are frequently met in MBE growth. When analyzing multilayer adsorption the Brunauer–Emmett–Teller (BET) adsorption isotherm [12.29] is of importance. In deriving the BET isotherm, a nonvarying value of $k_d$ is assumed for the gas molecules adsorbed on the clean surface, but different $k_a$ values are allowed for the gas molecules adsorbed on the top of the first monolayer of the adsorbate.

The most important conclusion which results from these considerations states that in the case of "low" adsorption and fairly low pressure at the surface of the substrate, or the already grown epilayer, thermodynamic equilibrium of the surface is reached when the adsorbate coverage is less than one monolayer. This means that even the first, bonded by chemisorption, monolayer of atoms on the surface of the substrate (grown epilayer) may desorb partly, if the vapor pressure in the nearest vicinity to the surface is sufficiently low, as it is in the ultrahigh vacuum of the MBE growth system.

### 12.2.2 Adsorption Isotherms; Statistical Treatment

So far, the phenomenological rate constants $k_a$ and $k_d$ have been used in the analysis of adsorption–desorption kinetics. In the atomistic approach to epitaxy, however, microscopic, statistical parameters should appear.

Following the considerations on adsorption isotherms and 2D condensation, given in [12.5], we will apply further Kossel's model of a crystal $A$ deposited on a substrate $B$ (see Fig. 12.1b). The vapor pressure of a bulk crystal of species $A$ is given by

$$p_{\mathrm{3D}} = \frac{f_v}{f_c} \exp\left(-\frac{3\Phi_{AA}}{kT}\right), \tag{12.18}$$

where $-3\Phi_{AA}$ is the potential energy of an atom in a step site, $f_v$ and $f_c$ being the partition functions of the one-component perfect gas and of the crystal expressed by

$$f_v = 2\pi(mkT)^{3/2}\frac{kT}{h^3} \quad \text{and} \quad f_c = \left(\frac{kT}{h\nu}\right)^3, \tag{12.19}$$

respectively, with $m$ the atomic mass, $\nu$ the lattice vibrational frequency, $k_B$ Boltzmann's constant and $h$ Planck's constant.

The (100) atomic plane of the substrate exposed to a vapor pressure $p$ of the deposit $A$, adsorbs this species in well-defined sites, so that the superficial concentration is $n = n_s\Theta$, with $n_s$ being the maximum concentration of adsorption sites ($n_s = 1/a^2$, where $a$ is the area of one particle site in Kossel's model). $\Theta$ is the mean coverage ratio per site; it also represents the occupancy probability of a given site $(1 - \Theta)$ being then the probability for this site to be empty).

An atom $A$ adsorbed on the substrate as an isolated atom has an interaction energy $\Phi_{AB}$. Its partition function $f_{\mathrm{ad}}$ when related to the partition function of this atom in the vapor phase yields the following temperature dependent parameter

$$\varepsilon(T) = \left(\frac{f_{\mathrm{ad}}}{f_v}\right) p \exp\left(\frac{\Phi_{AB}}{kT}\right). \tag{12.20}$$

Let us now consider a certain site of the substrate occupied by an atom of species $A$, which is surrounded by nearest neighbors having a certain probability of being present. These neighbors themselves are surrounded by other atoms. Evaluation of the partition function of such a non-isolated atom is a serious problem, yet more and more refined approximations may be formulated [12.5].

In the mean field approximation of Bragg and Williams [12.15], it is assumed that all other sites have a mean occupancy $\Theta$, especially the sites closest to the one considered. If the given site is occupied, the atom $A$ is located in an average environment and its potential energy is lowered by the value of $-4\Phi_{AA}\Theta/kT$, with $\Phi_{AA}$ as the interaction energy of two, effectively present, nearest neighbors. The relative partition function of the occupied site is then $\varepsilon(T)\exp(4\Phi_{AA}\Theta/kT)$; on the contrary if the given site is empty, the partition function is unity. The ratio of these two partition functions is then equal to the ratio of the probability of finding this site occupied over the probability of finding it empty, that is

$$\frac{\Theta}{1-\Theta} = \varepsilon(T)\exp\left(\frac{4\Phi_{AA}\Theta}{kT}\right). \tag{12.21}$$

This expression, which gives the coverage ratio as a function of the pressure of the vapor phase at a given temperature, is called the Frumkin–Fowler's isotherm [12.33].

An approximation better than that of the mean field consists in including the detailed first neighbor environment. In this, so-called quasi-chemical approximation, a vacant substrate site has a partition function which is split into a sum of terms corresponding to the partition functions when 0, 1, 2, 3, 4 first neighbors are present. When no first neighbor site is occupied the partition function is unity. If only one site is occupied then $\varepsilon$ is related to the vapor phase (in fact it is $4\varepsilon$, as there are four such possible sites). In order to take into account all the atoms on the substrate surface, which interact with the first sites, a factor $\mu$ is introduced, which is due to the mean field of the layer acting on the first neighbor sites of the central site. By continuing similarly for two sites occupied, etc., where the central site always remains empty, one obtains

$$1 + 4\varepsilon\mu + 6(\varepsilon\mu)^2 + 4(\varepsilon\mu)^3 + (\varepsilon\mu)^4 \equiv (1+\varepsilon\mu)^4. \tag{12.22}$$

On the contrary, if the central site is occupied, one has to consider, according to the occupancy of the first neighbor sites, the interaction potential energies $0, -\Phi_{AA}, -2\Phi_{AA}, \ldots$ which result from the geometry of the adatom localization on the surface. If all the closest sites are vacant, the partition function of

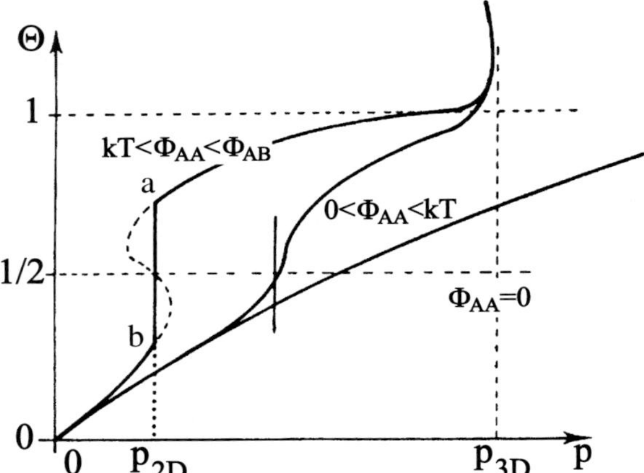


**Fig. 12.6.** Frumkin–Fowler isotherms for different interaction parameters $\Phi_{AA}$. $\Theta$ is the coverage ratio of $A$ at a given partial pressure $p$ of the species $A$; $p_{3\mathrm{D}}$ is the equilibrium pressure of the 3D solid $A$; $p_{2\mathrm{D}}$ is the 2D condensation pressure (taken from [12.5])

the central atom is $\varepsilon$. If only the first neighbor site is occupied, the partition function of the central atom is then $4\,\varepsilon(\varepsilon\mu)\exp(\Phi_{AA}/kT)$, $\mu$ accounting again for the mean field acting on this first neighbor. One thus obtains the complete partition function $\varepsilon(1+\varepsilon\mu\exp(\Phi_{AA}/kT))^4$. Combining this with the partition function of an empty central site gives the ratio $\Theta/(1-\Theta)$ of the occupancy probabilities, i.e., the relevant adsorption isotherm

$$\frac{\Theta}{1-\Theta} = \varepsilon(T)\left(\frac{1+\mu\varepsilon\exp(\Phi_{AA}/kT)}{1+\mu\varepsilon}\right)^4 . \tag{12.23}$$

This isotherm differs from the preceding one (12.21) only by the lateral interaction factor, which remains close to $\varepsilon$. If $\Phi_{AA} \to 0$, both of these isotherms lead to the Langmuir isotherm, which may be expressed as $\Theta/(1-\Theta) = \varepsilon(T)$.

Qualitatively the isotherms (12.21) and (12.23) are not distinguishable and the same phenomena may be foreseen. Consequently, we will focus our attention only on the Frumkin–Fowler isotherm. Figure 12.6 gives the form of this isotherm schematically when the interaction parameter $\Phi_{AA}$ is changed. For any $\Phi_{AA}$, for a very low partial pressure of species A, the isotherm is a single straight line $\Theta = \varepsilon(T)$, which means that at these low pressures the coverage of the surface does not change with $p$. With increasing pressure $p$ in the vapor phase, the behavior is different. If $\Phi_{AA} = 0, \Theta$ increases nearly linearly with $p$, and the layer is continuously filled. As long as $0 < \Phi_{AA} < kT$, an inflection point appears at $\Theta = \frac{1}{2}$ where the slope is positive; the filling of the substrate surface is still continuous, yet the layer no longer behaves like a 2D perfect gas. As soon as $kT < \Phi_{AA} < \Phi_{AB}$, at the inflection point, still located at $\Theta = \frac{1}{2}$, the slope is infinite then negative. A "Van der Waals' loop" appears (shown as the dashed line in Fig. 12.6); it characterizes an instability. This dashed part of the isotherm has no physical significance, but the solid straight line segment a – b does. Along this step a – b two distinct phases coexist in equilibrium: a diluted (2D gas) phase of $A$ and a condensed (2D solid or liquid) one. The ratio of the two phases varies along the step which occurs at a constant partial pressure $p_{2\mathrm{D}}$. This is the 2D condensation phenomenon. The partial pressure of this transition is obtained with the aid of the relations (12.21) and (12.20) setting $\Theta = \frac{1}{2}$

$$p_{2\mathrm{D}} = \frac{f_v}{f_{\mathrm{ad}}} \exp\left(-\frac{\Phi_{AB} + 2\Phi_{AA}}{kT}\right). \tag{12.24}$$

One may recognize (see Fig. 12.1b) that the term $-(\Phi_{AB} + 2\Phi_{AA})$ represents in this relation the potential energy of an atom $A$ located in a step position of a 2D condensed phase of $A$ on the substrate $B$ (relation (12.24) expresses the vapor pressure of this two-dimensional phase). It is worthwhile comparing $p_{2\mathrm{D}}$ with the vapor pressure of the 3D crystal of phase $A$ (see (12.18)) at the same temperature. This leads to the relation

$$\frac{p_{2\mathrm{D}}}{p_{3\mathrm{D}}} = \frac{f_c}{f_{ad}} \exp\left(\frac{\Phi_{AA} - \Phi_{AB}}{kT}\right), \tag{12.25}$$

which again indicates that the 2D phase is more stable than the 3D phase of $A$ ($p_{2\mathrm{D}} < p_{3\mathrm{D}}$) if

$$kT < \Phi_{AA} \leq \Phi_{AB} + kT. \tag{12.26}$$

This condition is close to relation (12.4); the difference lies in the term $kT$ due to the fact that certain temperature dependent terms have not been neglected here. One will notice that the 2D phase grows at undersaturation, as $p_{2\mathrm{D}} < p_{3\mathrm{D}}$.

Upon increasing the vapor pressure beyond $p_{2\mathrm{D}}$, the coverage reaches a monolayer, yet it is obvious that as soon as islands of the 2D phase are formed one has to envisage the adsorption of $A$ on top of these islands. The exact formulation of the phenomenon is complicated, yet a good approximation consists in considering that adsorption at level 2 is correctly described by substituting in the isotherms (12.21) or (12.23) $\Theta_2$, the coverage ratio in the second layer, instead of $\Theta_1$, and $\Phi_{AA}$ instead of $\Phi_{AB}$, as according to the model there is no interaction between second nearest neighbors (hence $\Phi_{AB} \equiv \Phi_{AA}$). Condition (12.26) when applied in the second layer gives $p^{(2)}{}_{2\mathrm{D}} = p_{3\mathrm{D}}$ and besides for any layer (i), $p^{(i)}{}_{2\mathrm{D}} = p_{3\mathrm{D}}$. Thus, the isotherm of Fig. 12.6 has a vertical asymptote at $p = p_{3\mathrm{D}}$ where a layer by layer growth develops at the saturation, for an infinite number of layers [12.5].

### 12.2.3 Thermal Desorption Kinetics

The kinetics of thermal desorption is usually different from the adsorption kinetics. This results from the fact that in desorption the rate constant is most frequently dependent on the surface coverage [12.34, 35], because of the lateral interactions occurring between the adsorbate particles. In studies of the desorption processes the Polanyi–Wigner equation [12.36], having the mathematical structure of the Arrhenius rate equation, is usually used. This equation represents the material balance on the surface and gives the rate of desorption by

$$k_d(\Theta) = -\frac{\mathrm{d}\Theta}{\mathrm{d}t} = \nu(\Theta)\Theta^n \exp\left(-\frac{E_a(\Theta)}{RT}\right), \tag{12.27}$$

where $\nu$ is the pre-exponential factor, $E_a$ is the activation energy of desorption, $n$ defines the order of the desorption process, $R$ is the gas constant and $T$ is the absolute temperature of the desorbing system.

If the dependence on coverage can be ignored in desorption processes, as is usually the case in MBE [12.37], then thermal desorption can be described by a simple Arrhenius equation (see (12.1)) of the form

$$k_d = -\frac{\mathrm{d}\Theta}{\mathrm{d}t} = \nu_0 \exp\left(-\frac{E_a}{RT}\right). \tag{12.28}$$

The inverse value of the rate coefficient $k_d^{-1}$ can be identified as the mean lifetime $\tau$ of the adsorbate particle on the adsorber surface

$$\tau = \tau_0 \exp\left(\frac{E_a}{RT}\right), \tag{12.29}$$

where $\tau_0 = 1/\nu_0$. Often it is assumed that the pre-exponential factor $\nu_0$ is of the order of $10^{13}\,\mathrm{s}^{-1}$, i.e., of the order of the vibrational frequency of a particle bound to the surface (usually a physisorbed molecule). Then for an activation energy characteristic for physisorption, $E_a = 25\,\mathrm{kJ\,mol^{-1}}$ ($\approx 0.25$ eV/atom) is predicted at room temperature [12.31], while for chemisorption the relevant energy is considerable larger, reaching the values of $10^2 - 10^3$ kJ/mol ($1-10$ eV/atom).

The experimental technique most frequently employed for investigation of thermal desorption processes is thermal desorption spectroscopy (TDS), called also temperature programmed desorption (TPD) [12.34]. After an appropriate calibration this technique can be used to determine surface coverages of adsorbates as well as to evaluate the activation energy of desorption. The TDS technique is likely to furnish also quantitative information on the binding energy between deposit and substrate. In a TDS experiment, the temperature of a substrate on which the deposit has been condensed is gradually raised (generally linearly, as a function of time) and one follows with the aid of a mass spectrometer placed in direct view of the substrate the number of atoms desorbed as a function of time, $\mathrm{d}n/\mathrm{d}t = Z_d$ ($\mathrm{cm^{-2}s^{-1}}$). The area under the spectrum represents the quantity initially deposited, that is $\alpha_0$ when expressed by the number of atoms of the deposit per number of surface sites. If the atoms of the deposit are bound in rather different states to the substrate, the desorption will stagger in time, in the form of different desorption peaks.

The Ag/Si(111) system, with growth according to the Stranski–Krastanov (SK) mode at deposition of Ag on Si, clearly illustrates the situation [12.5]. When increasing the desorption temperature of the Ag layer (see Fig. 12.7), a first peak situated at lower temperature appears, which corresponds to desorption of 3D crystallites formed after saturation of the 2D phase, the maximum coverage of which is $\alpha_0 = 2/3$. The second peak, corresponding to desorption of the 2D phase, occurs subsequently, when the temperature is further increased. It has to be acknowledged that the quantitative exploitation of the results gained by TDS has to be handled with caution. Despite this, the determination of important physical quantities related to bonding of the epilayer constituents to the surface of the substrate is possible.

Isothermal desorption spectroscopy (ITDS) is a frequently used alternative to the conventional TDS method. It is based on a study of the surface coverage at different constant temperatures. Let us present as an example of its application, the desorption experiments performed with Te/Cd deposits desorbed from CdTe(111) and GaAs(100) substrates [12.38, 39].

At the beginning of these thermal desorption experiments, a fairly thick noncrystalline film of the element (Cd or Te) is deposited in high vacuum at room temperature on the CdTe substrate surface. At this moment no clearly defined transition layer occurs between the substrate and the deposited film (see Fig. 12.8a). During the thermal desorption process the covered substrate

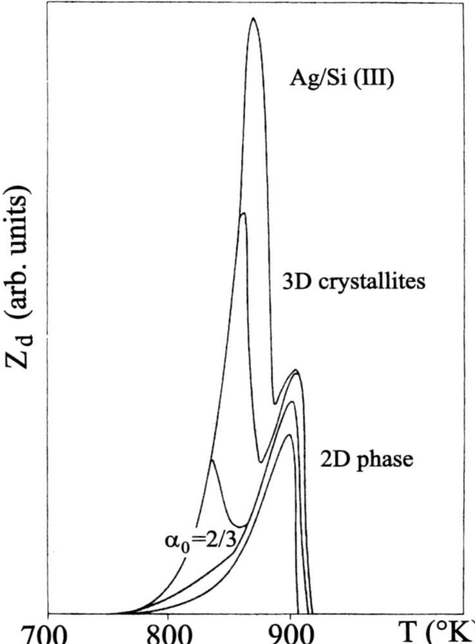


**Fig. 12.7.** Thermal desorption spectra of the Ag/Si(111) material system, measured for three different samples. Two peaks are visible for each case: one (at higher temperatures) which corresponds to the first monolayer of the Ag deposit, and another (at lower temperatures) due to 3D crystallites (taken from [12.5])

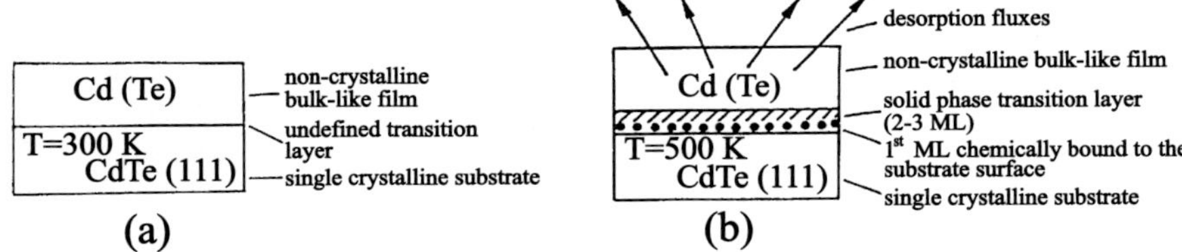


**Fig. 12.8.** Schematic illustration of the geometry of the system "substrate-deposit" occurring: (**a**) before the thermal desorption starts, (**b**) during the desorption, when a crystalline structure is created in the interface zone (taken from [12.38])

is kept at a definite, elevated temperature. Therefore, a solid-phase transition layer may appear at the interface between the single-crystalline substrate and the non-crystalline deposit (as is typical for solid phase epitaxy (see Sect. 4.1)). Taking this into consideration (see Fig. 12.8b), one may assume that when the deposited noncrystalline film, owing to re-evaporation, becomes very thin (3 – 4 ML), a single-crystalline structure, strongly bound to the substrate crystal, remains on the surface of this crystal. Consequently, one may distinguish three different zones in the system "deposited film – single-crystalline substrate", i.e., the bulk-like noncrystalline film of the element (Cd or Te), the single-crystalline substrate of the compound CdTe, and the crystalline transition layer in between, which occurs near to the substrate surface. If we assume now a closely packed structure (coordination number equal to 12) for the non-crystalline solid phase of the bulk film, and a zincblende structure (coordination number equal to 4) for the single-crystalline substrate, then by measuring the activation energies of thermal desorption of the constituent elements (Cd or Te) for both of these solid phases, we will be able to determine the strength of the individual atom–atom bonds in these phases.

The relevant experiments performed in a high vacuum system ($p \approx 10^{-4}$ Pa) have been described in [12.38]. Figure 12.9 shows the arrangement and dimensions of the experimental setup (upper panel). The substrate wafers covered with deposits were mounted on an oven below a quadrupole mass spectrometer (QMS) inlet to study the desorption processes of Cd or Te. This oven was heated to temperatures where desorption of the deposits could be detected. The mass analyzer of the QMS was fixed to the masses 114 and 256, which represents the most prominent mass-spectra peaks of Cd and $Te_2$, respectively. The lowest detectable desorption rates were about 0.005 ML $s^{-1}$ for Cd and 0.03 ML $s^{-1}$ for Te. To measure the activation energies of the evaporation process the experiments were performed for free Langmuir-type evaporation (see Sect. 6.1.2) of the deposits.

The samples were heated stepwise. Figure 12.9 (the lower panel) shows the variation of the sample temperature together with the intensity of the

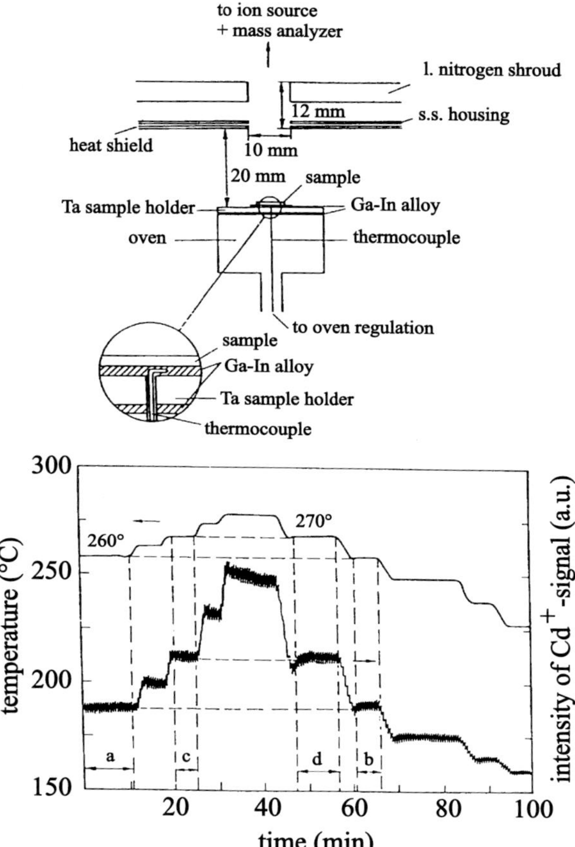


**Fig. 12.9.** Schematic drawing of the experimental setup for re-evaporation with the attached quadrupole mass spectrometer (upper panel), and the mass spectrometer signal for $Cd^{+}_{114}$ as a function of time with the temperature variation of the bulk Cd deposit (lower panel) (taken from [12.38])

signal of the QMS as a function of time. The sensitivity of the QMS signal on temperature variations was tested separately. A clear change in this signal at temperature variations of the sample larger than 1°C could be resolved. Also the reproducibility of the measurements can be seen from these typical results depicted in Fig. 12.9. There are two pairs of time intervals (a, b) and (c, d) with temperatures of 260°C and 270°C, respectively. The time intervals are separated by 20 min (c, d) or 50 min (a, b) and reveal the same intensity of the signal within one pair. Based on this fact, one may conclude that the QMS signal is only a function of the source temperature and there is no memory effect superimposed in the experimental set-up.

Figure 12.10 shows the intensities of the QMS signals plotted as functions

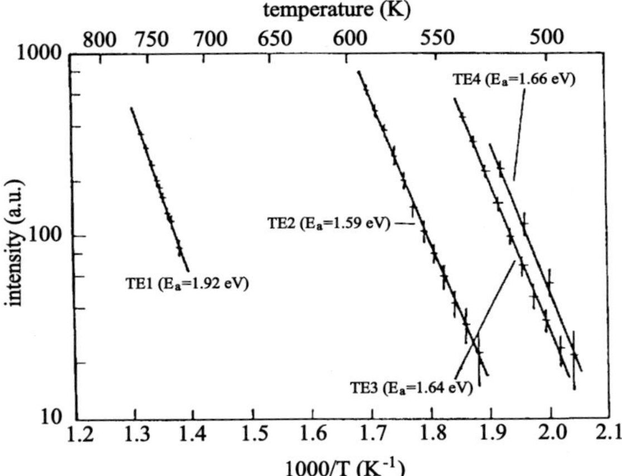


**Fig. 12.10.** Mass spectrometer signals of $Te_2^+$ as a function of inverse temperature of the deposits for free Langmuir-type evaporation. TE1, bare CdTe(111) surface; TE2, bulk Te pieces; TE3, thick Te film deposited on CdTe(111) substrate; TE4, thick Te film deposited on GaAs(100) substrate (taken from [12.38])

of inverse temperatures for $Te_2^+$ ions. The given error bars originate from the slight temperature variations mentioned above. The data show a clear exponential dependence on temperature, in accordance with the Arrhenius law (see (12.1)). From the slope of the lines the activation energies $E_a$ of thermal desorption for Te in the performed experiments have been calculated. The activation energies for thermal desorption of Cd and Te from zincblende CdTe substrate crystal were equal to: $1.13 \pm 0.06$ eV/atom and $1.92 \pm 0.13$ eV/molecule, respectively, while the relevant energies for Cd and Te desorbed from the bulk noncrystalline deposits were equal to $1.13 \pm 0.12$ eV/atom and $1.64 \pm 0.18$ eV/molecule, respectively [12.38]. Consequently, the strengths of the individual atom–atom bonds are given by

$$\begin{aligned} F_{\text{Cd–Cd}}(\text{bulk}) &= 1.13/12 = 0.0942 \approx 0.09\,\text{eV/atom, and} \\ F_{\text{Te–Te}}(\text{bulk}) &= 1.64/12 = 0.137 \approx 0.14\,\text{eV/atom,} \end{aligned}$$

for the thick bulk-like non-crystalline phase, while

$$\begin{aligned} F_{\text{Cd–Te}}(\text{CdTe(111)}) &= 1.13/4 = 0.283 \approx 0.28\,\text{eV/atom, and} \\ F_{\text{Te–Cd}}(\text{CdTe(111)}) &= 1.92/4 = 0.478 \approx 0.48\,\text{eV/atom,} \end{aligned}$$

for the single-crystalline CdTe phase.

To distinguish between the species in the last desorbing monolayers of the deposits and the first monolayers of the substrate crystal, GaAs substrates were chosen instead of CdTe to study further with QMS the thermal desorption processes of Cd and Te deposits [12.39]. Figure 12.11 shows the $Te_2$ QMS signal with the GaAs substrate temperature as a function of time. The oscillations of the QMS signal are owing to slight variations in the sample

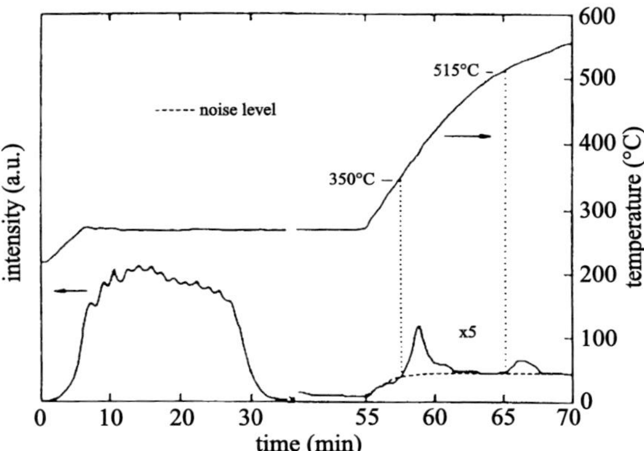


**Fig. 12.11.** Intensity of the $Te_2^+$ QMS signal together with the GaAs(100) substrate temperature as functions of time (taken from [12.39])

temperature not resolved in the temperature plot. In the second part of the plot, the time axis is extended by a factor of 2 and the QMS signal is multiplied by 5. One can easily correlate the increase in the noise levels with the moment of the increase in temperature.

The GaAs samples covered with Te were heated to temperatures between 220 and 280°C, while the intensity of the re-evaporating $Te_2$ and the temperature were monitored using an $x-t$ recorder. When the signals had dropped to the noise level the samples were kept at constant temperature for $5-40$ min. Then they were heated either stepwise or continuously to $500-600$°C. Raising the temperature caused a small but detectable increase in the noise level owing to background effects. In addition to that, a distinct increase of the $Te_2$ intensity above the noise level at about $350-360$°C for all of the GaAs samples covered with Te which have been investigated could be observed. Some minutes later the signal dropped to the noise level again but at $510-520$°C another smaller amount of re-evaporating $Te_2$ was detected. The latter $Te_2$ signal could be observed only if the GaAs sample had been preheated prior to the Te deposition, which is quite reasonable considering that the Te should be bound with bonds of different strength to GaAs than to an oxide.

The performed thermal desorption experiments concerning $Te_2$ deposits on GaAs(100) surfaces, using QMS [12.39], show that two regions of different surface adsorption may be clearly distinguished. The first region, nearest to the substrate surface, is created by adparticles strongly bound to the surface, apparently chemisorbed, while the second is created by adparticles more weakly bound to the surface, apparently physisorbed. The Te species belonging to near surface adsorbate areas (left on the surface after the desorption of the bulk deposits has been completed at lower desorption temperatures) re-evaporate from the GaAs(100) substrates at different temperatures, accordingly to the strength of their bonds to the substrate surface atoms.

## 12.3 Step Advancement and Bunching Processes

The importance of step kinetics for crystal growth under low supersaturation conditions has been indicated a long time ago in the now classic paper of Burton, Cabrera, and Frank (BCF) [12.40]. These kinetic processes are especially important for epitaxy, when the step-flow growth mode (SF-mode) is considered (this means growth on vicinal surfaces, which proceeds through advancement of pre-existing steps (see Fig. 1.7)).

### 12.3.1 Growth Conditions on Vicinal Surfaces

When an adatom during epitaxial growth reaches a descending atomic step on the surface, it typically has to overcome a higher energetic barrier in order to jump onto a different atomic plane, and thus it has a higher probability to be reflected back (Fig. 12.12a) [12.41]. This additional barrier was discovered by Ehrlich and Hudda [12.42] and, soon afterwards, Schwoebel and Shipsey [12.43] pointed out that it has a stabilizing effect on growth in the SF-mode. Because diffusing adatoms preferably attach to steps from the terrace below, rather than from the terrace above, a terrace wider (narrower) than its upper neighbor receives more (less) mass out of the flux of particles impinging onto the surface, and this mass diffuses to the upper step that advances faster (slower) (Fig. 12.12b). As a result, there is a tendency to maintain the average terrace size. This is a very important effect which in usual epitaxial growth procedures is reached by growing a buffer layer, i.e., a layer of the same material as the substrate [12.44].

On a singular surface (low index atomic plane of the substrate crystal), these Ehrlich–Schwoebel (ES) barriers lead to unstable growth: large pyramidal features (mounds) are created on the surface (see Fig. 12.13). The origin

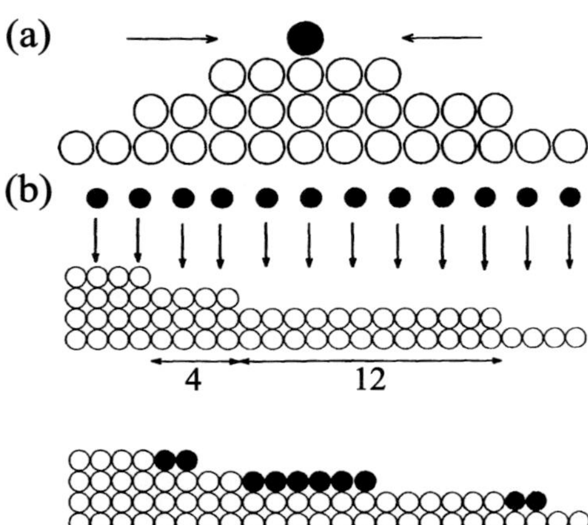


**Fig. 12.12.** Schematic illustration of the Ehrlich–Schwoebel barrier effect on: (**a**) singular and (**b**) vicinal surfaces (taken from [12.41])

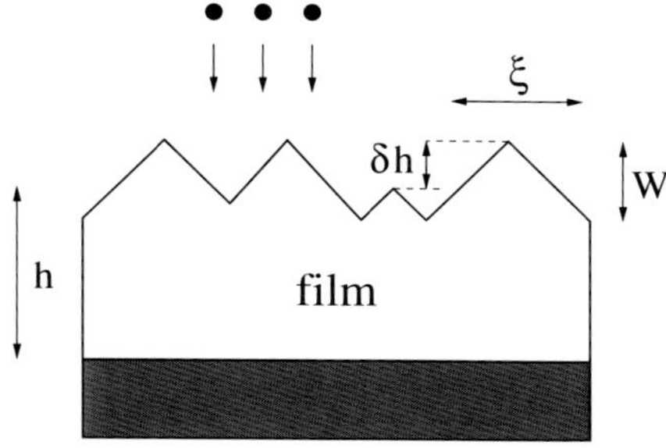


**Fig. 12.13.** Schematic illustration of the mound morphology during epitaxial growth. The typical mound size $\xi$ and the surface width $W$ increases with the film thickness $h$ according to a power law (taken from [12.41])

of this instability is a growth–induced surface current of adsorbed atoms [12.41]. Because the adatoms are reflected from descending steps and attach preferentially to ascending ones, the current is uphill and destabilizing. The concentration of diffusing adatoms is maintained by the incoming particle flux; thus, the surface current is a non-equilibrium effect.

The average lateral size $\xi$ of the mounds is found to increase with the film thickness $h$ according to a power law, $\xi \sim h^{1/z}$ with $z \approx 2.5-6$ depending on the material, and growth conditions used [12.45]. The slope $s$ of the mounds' hillsides is either observed to remain constant or increases with the film thickness as $s \sim h^{\lambda}$ (see [12.46]). The surface width $W$ increases due to both the lateral coarsening and the increase of the slope, $W \sim h^{\beta}$, where $\beta = 1/z + \lambda$. Both scenarios were found in computer simulations and in experiments [12.45].

Growth on vicinal surfaces is also unstable, despite the stabilizing effect of the step-edge barriers leading to terrace size equalization during the SF growth. As was first pointed out in [12.47], the steps are unstable towards transverse meandering since matter is more likely to attach to already advanced parts in the steps. Thus a vicinal surface is unstable with respect to fluctuations perpendicular to the direction of the slope of this surface. Consequently, completely stable epitaxial growth on the vicinal surface is impossible when the ES barriers are present there. Theoretical treatment of the crystal growth problems related to growth on vicinal surfaces are presented in a series of papers in [12.48].

### 12.3.2 Step Advancement Kinetics

Let us now discuss briefly the attachment kinetics to a step with the ES barrier [12.3]. The incorporation rate of an adatom from an upper terrace is usually not the same as that from a lower terrace as shown in Fig. 12.14. From the upper terrace an adatom has to break many chemical bonds with the underlying substrate atoms when it crosses over the step down to the crystallization position. From the lower terrace, on the other hand, an adatom can simply make additional bonds with the step atoms before it reaches kink sites. The additional energy barrier for the crystallization makes the kinetic

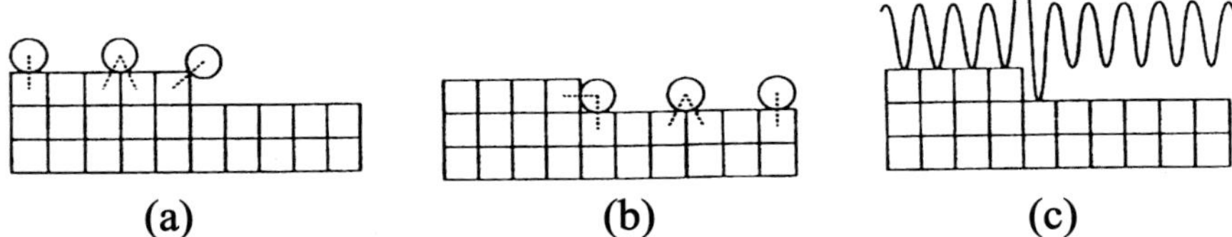


**Fig. 12.14.** Schematic illustration of the asymmetry in incorporation of an adatom: (**a**) from the upper and (**b**) from the lower terraces, called the Schwoebel effect. (**c**) The potential energy profile of the adsorbed atom. The ES barrier appears at the step (taken from [12.3])

coefficient from the upper terrace $k_u$ smaller than that from the lower terrace $k_l$. This asymmetry in attachment kinetics at the step, which is expressed by the inequality $k_l > k_u$, was first studied by Schwoebel and Shipsey [12.43], and called afterwards the Schwoebel effect [12.3].

We consider now the step down configuration of the vicinal surface in the so-called terrace–step–kink (TSK) model [12.49, 50], when the step is running on average in the $x$-direction at $y = 0$, and the terrace in front at $y > 0$ is lower than the terrace in the back ($y < 0$) (see Fig. 12.15). The step advance rates, $\nu_u$ and $\nu_l$, by the adatom incorporation from the upper and the lower terraces, respectively, are linearly proportional to the concentrations of adatoms at the step coming from these terraces. According to the Schwoebel effect these advance rates are different.

Let us now restrict to the extreme case when $\nu_u = 0$ and $\nu_l = \infty$. In this case, there is no crystallization from the upper back terrace. Since the crystallization takes place by the atom incorporation only from the lower terrace, this case is called a one-sided model. Furthermore, the kinetics from the lower front terrace is assumed to be extremely fast, such that the local

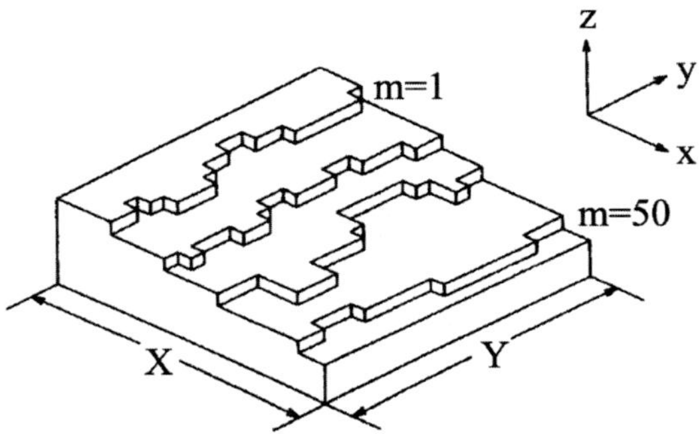


**Fig. 12.15.** Schematic illustration of a vicinal surface presented according to the TSK model. Step bunching is indicated by 5 terraces which are separeted by 4 steps comprising 50 monolayers altogether (taken from [12.51])

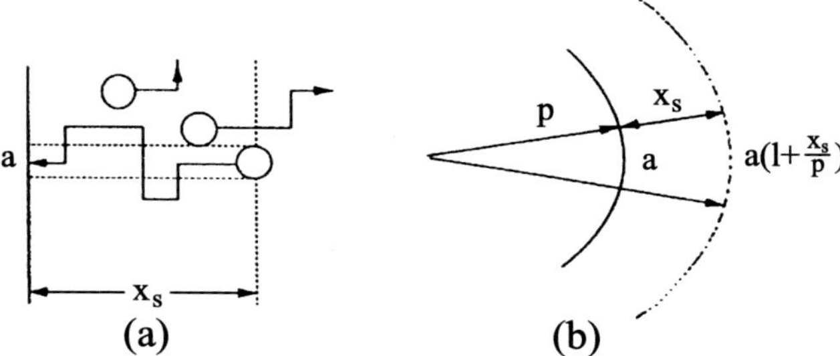


**Fig. 12.16.** Capturing region of adsorbed atoms of a one-sided model in front of a step: (**a**) for a straight step and (**b**) for a curved step (taken from [12.3])

equilibrium is reached. As in the BCF model, the atoms are deposited on the crystal surface with a flux $\phi$. The adsorbed atoms then diffuse on the terrace surface with a surface diffusion constant $D_s$, and subsequently become incorporated (crystallized) into the step, or evaporate back into an ambient vapor after a life time $\tau$. A straight step exhibits an advancement velocity $\nu_0$ resulting from the incorporation of atoms (deposited on the terrace) in the range of the surface diffusion length $x_s = \sqrt{D_s\tau}$ in front of the step (see Fig. 12.16).

The stability of the straight step depends on the advancement kinetics [12.47]. If the step is pushed forward at some part by fluctuation, the region to incorporate adatoms expands radially, as shown in Fig. 12.16b. The area of the capturing region increases approximately by a factor of $1+x_s/2\rho$, and the velocity increases with the same factor. Thus, the step with a curvature $\kappa = 1/\rho$ has a velocity higher than the straight one by $\delta\nu_d = \nu_0(x_s/2)\kappa$. The bump is accelerated as compared to the straight part and is pushed further forward. Thus, the diffusion causes a destabilization of the step profile.

Competing with this destabilization is the stabilization effect by the step stiffness. Since the equilibrium concentration of adatoms increases at a curved part, the deposition flux to maintain equilibrium at a curved step should be increased, which causes the relevant decrease in the supersaturation, and the corresponding decrease in the velocity of the bump. Both of these effects, the destabilizing and the stabilizing one, are in the first order approximation proportional to the curvature appearing in the step [12.47]. It is important to notice that since the diffusional instability increases with the velocity $\nu_0$, the instability probability increases as well by increasing the deposition flux $\phi$.

Interesting results concerning the equilibrium step dynamics on vicinal surfaces have been obtained in experiments with direct current (DC) heated silicon substrates and MBE grown epilayers [12.52–54]. The influence of DC heating of the substrate crystal on the behavior of monoatomic steps during sublimation and during epitaxial growth on Si(111) and Si(100) vicinal surfaces (created by slight off-orientation from the nominal atomic planes) was observed by *in situ* reflection electron microscopy. The conditions of step bunching (formation of multiple atomic height steps) and debunching are found. The behavior of monoatomic steps under the influence of electric DC has been explained in terms of diffusion and drift of charged adatoms and their interaction with steps.

### 12.3.3 Mass Transport Between Steps; Step Bunching

The recent development of scanning tunneling microscopy (STM) and atomic force microscopy (AFM) techniques [12.55] has allowed direct observation of real surfaces of different solids. Among other effects, also step bunching has been observed in many cases [12.56–58]. This phenomenon, destabilizing the step geometry on vicinal surfaces, has a deteriorating effect in epitaxial growth, especially of low dimensional heterostructures (see Sect. 3.2 and Sect. 11.5). Therefore, it is worthwhile discussing it in more detail.

Step bunching is brought about provided that a wider terrace than an average width $l_{\mathrm{av}}$ exhibits a tendency to extend further [12.51]. This effect may occur in the case of sublimation (dissolution) of the crystal vicinal surface as well as in the case of epitaxial growth on such a surface [12.59]. To gain an insight into the mechanism of the step bunching process, let us consider the simple stepped surface shown in Fig. 12.17.

The mass transport between two steps in a straight step array parallel to the $y$-axis (see Fig. 12.15) is carried out by a lateral flux $J$ of adatoms in the $x$-direction. This flux, generated by surface diffusion, can be expressed as

$$J = D_s \frac{\partial \Theta}{\partial x} . \tag{12.30}$$

Here $D_s$ is the surface diffusion constant, and $\Theta$ the concentration of adatoms on the terraces. The conservation law of adatom concentration and the steady state condition lead to the following diffusion equation on $\Theta$:

$$\left( \frac{\partial J}{\partial x} \right) - \frac{\Theta}{\tau} = D_s \left( \frac{\partial^2}{\partial x^2} \right) - \frac{\Theta}{\tau} = 0, \tag{12.31}$$

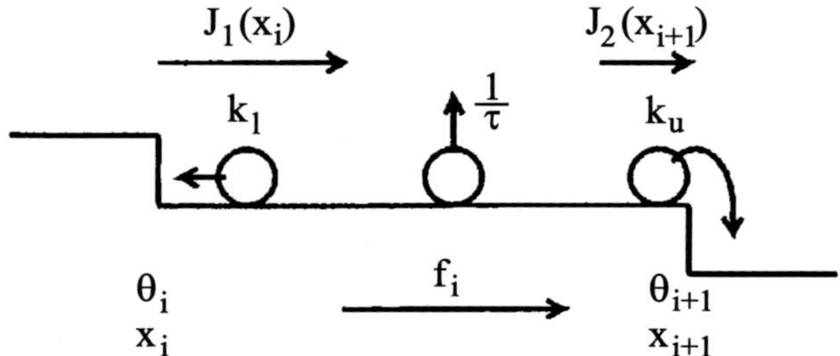


**Fig. 12.17.** Schematic illustration of the elementary adatom processes on a monoatomic vicinal surface (taken from [12.51])

where $\tau$ is the evaporation lifetime of an adatom. As for the boundary condition imposed on $\Theta(x)$ on each terrace, the continuity of the flux at an upper edge $x_i$ and the lower edge $x_{i+1}$ may be adopted.

$$J_1(x_i) = k_l\left(\Theta_i - \Theta(x_i)\right), \text{ and } J_2(x_{i+1}) = k_u\left(\Theta(x_{i+1}) - \Theta_{i+1}\right). \quad (12.32)$$

Here $\Theta_i$ is the equilibrium adatom concentration at the $i$th step edge; the positive direction of $J_i$ and $x$ are taken to be the step-down direction. If the adatoms kinetic coefficients $k_l$ and $k_u$ are sufficiently large, $\Theta(x_i)$ approaches the equilibrium value $\Theta_i$ at the $i$-th step edge. This value is, however, affected by the step–step interaction [12.60] what is expressed by the equation

$$\Theta_i = \Theta_0 \exp\left(-\frac{f_i \Omega}{kT}\right). \quad (12.33)$$

Here $\Theta_0$ is the equilibrium concentration of adatoms at an edge of isolated step, $f_i$ stands for the force acting on the $i$-th step per unit length due to the step–step interaction, and $\Omega$ stands for the surface area per atom. The positive direction of $f_i$ is also taken to be the step-down direction (see Fig. 12.17). The force $f_i$ acting on the $i$-th step can be calculated from the step–step interaction energy $U$ per unit length as

$$f_i = -\left(\frac{\partial}{\partial x_i}\right)\left(U(x_{i+1} - x_i) + U(x_i - x_{i-1})\right). \quad (12.34)$$

For step–step interactions which are due to elastic interactions or dipole–dipole interactions the interaction energy falls of as the square of the distance between the steps, $U(x) = A/x^2$ [12.60]. On the other hand, the movement of each step position $x_i$ is determined by the lateral fluxes at the step edge from both sides.

$$\frac{dx_i}{dt} = \Omega\left(-J_1(x_i, \Theta_i, \Theta_{i+1}) + J_2(x_i, \Theta_{i-1}, \Theta_i)\right). \quad (12.35)$$

The mechanism of step bunching consists of the following steps:

(i) For $f_i > 0$ (attractive step–step interaction), $J_1$ at an upper step edge and $J_2$ at a lower step edge are positive (as shown in Fig. 12.17; a positive flux $J_1$ means here a dissolution process of the upper step). $J_1$ is a monotonically increasing function of the terrace width, while $J_2$ is almost independent of the terrace width around $l_{\mathrm{av}}$ due to competition of evaporation flux and diffusion flux. The extension velocity of the terrace width is proportional to $J_1 + J_2$, and it becomes a monotonically increasing function of the terrace width in the vicinity of $l_{\mathrm{av}}$. Thus, a wider terrace than an average will broaden further, and step bunching occurs.

(ii) In the case of $f_i < 0$ (repulsive step-step interaction) on the other hand, $J_1$ is a monotonically increasing positive function of the terrace width, while $J_2$ is negative and almost independent of the terrace width near $l_{\mathrm{av}}$ due to saturation of the flux to the capture rate at the step edge. Thus, $J_1 + J_2$ becomes a monotonically increasing function of the terrace width in the vicinity of $l_{\rm av}$, and a wider terrace will broaden further, which again causes the step bunching process.

An important step-bunching-related effect, which was theoretically predicted a long time ago [12.61], is related to the behavior of monoatomic steps on the crystal surface where multiple atomic height steps (bunches) appear. It consists of the following rule: no monoatomic steps can move through the bunch of steps; this means that bunches completely block monoatomic steps on the vicinal surfaces. From the point of view of epitaxial growth technology, one has to notice that the SF-mode of epitaxy is excluded if step bunches are present on the vicinal surface of the substrate crystal.

# 13. Quantum Mechanical Aspects

Crystallization, seen as a microscopic, atomic-level process, means incorporation of atoms belonging to a disordered, metastable phase (solid, liquid or vapor) into the ordered structure of a solid crystal (stable phase). This process is accomplished by the creation of chemical bonds between atoms of the metastable phase, which occur in the nearest vicinity to the surface of the crystal, and the atoms of the crystal, which are located in the outermost atomic plane of the surface. Understanding the processes of bonding atoms into a crystal lattice requires a quantum mechanical approach. This concerns crystallization of the bulk crystals as well as the fundamental processes of epitaxial growth.

The most concise introduction to the quantum mechanical framework needed for the subsequent discussion is given in the fundamental book by W.A. Harrison [13.1]. In order to be sufficiently brief, we will present here the main items of this introduction.

## 13.1 Framework of Quantum Mechanics

The state of an electron is represented by a wave function, designated as $\Psi(\boldsymbol{r})$. A wave function can have both real and imaginary parts. To say that an electron is represented by a wave function means that specification of the wave function gives all the information that can exist for that electron except information about its spin. The representation of each electron in terms of its own wave function is called the one-electron approximation. Physical observables (quantities that can be measured) are represented by linear operators acting on the wave function. The operators corresponding to the two fundamental observables, position and momentum, are defined as: position $\leftrightarrow \boldsymbol{r}$, momentum $\leftrightarrow \boldsymbol{p} = (h/2\pi\mathrm{i})\nabla$, where $h$ is Planck's constant, and $\nabla$ is the gradient operator. The operator $\boldsymbol{r}$ means simply multiplication (of the wave function) by a position vector $\boldsymbol{r}$. Operators for other observables can be obtained from $\boldsymbol{r}$ and $\boldsymbol{p}$ by substituting these two operators in the classical expressions for other observables. For example, the potential energy is represented by multiplication by $V(\boldsymbol{r})$, while the kinetic energy is represented by $p^2/2m = -((h/2\pi)^2/2m)\nabla^2$. A particularly important observable is the energy of an electron. This energy is represented by a Hamiltonian operator of the form:

$$H = -\frac{(h/2\pi)^2}{2m}\nabla^2 + V(\boldsymbol{r}). \tag{13.1}$$

In quantum mechanics one has to deal with average values of observables. Even though the wave function $\Psi$ describes an electron fully, different values can be obtained from a particular measurement of some observable. The average value $< |O| >$ of many measurements of the observable $O$ for the same $\Psi$ is determined by

$$< |O| > = \frac{\int \Psi^*(\boldsymbol{r}) O \Psi(\boldsymbol{r}) \mathrm{d}^3\boldsymbol{r}}{\int \Psi^*(\boldsymbol{r})\Psi(\boldsymbol{r})\mathrm{d}^3\boldsymbol{r}}. \tag{13.2}$$

The integral in the numerator on the right side of the equation is a special case of a matrix element; in general the wave function appearing to the left of the operator $O$ may be different from the wave function to the right of it. In such a case, the Dirac notation for the matrix element is used, as given by

$$< \Psi_1 | O | \Psi_2 > \equiv \int \Psi_1^*(\boldsymbol{r}) O \Psi_2(\boldsymbol{r}) \mathrm{d}^3\boldsymbol{r}. \tag{13.3}$$

It should be noticed that if $\Psi$ depends on time, then so also will $< |O| >$. Equation (13.2) is the principal assertion of quantum mechanics because it makes a connection of wave functions and operators with experiment. For example, the probability of finding an electron in a small region of space, $\mathrm{d}^3\boldsymbol{r}$, is given by $\int \Psi^*(\boldsymbol{r})\Psi(\boldsymbol{r})\mathrm{d}^3\boldsymbol{r}$, which means that $\Psi^*(\boldsymbol{r})\Psi(\boldsymbol{r})$ is the probability density for the electron.

Equation (13.2) leads also to the conclusion that there exist electron states having discrete or definite values for the relevant energy operators. Since any measured quantity must be real, (13.2) suggests that the operator $O$ is Hermitian. It is known from mathematics that it is possible to determine eigenstates of any Hermitian operator. In the case of the Hamiltonian operator (it is Hermitian, too), eigenstates are obtained as solutions of the differential equation

$$H\Psi(\boldsymbol{r}) = E\Psi(\boldsymbol{r}), \tag{13.4}$$

which is called the time-independent Schrödinger equation. The quantity $E$ in this equation is the eigenvalue of the Hamiltonian. It is known also that the existence of boundary conditions (such as the condition that wave functions vanish outside a given region of space) will restrict the solutions to a discrete set of eigenvalues $E$. Eigenstates determined by solving (13.4) are wave functions which an electron may or may not have. If an electron has a certain eigenstate, it is said that the corresponding state is occupied by the electron. However, the various states exist whether or not they are occupied. We see that a measurement of the energy of an electron represented by an eigenstate will always give the value $E$ for that eigenstate.

The potential energy $V(\boldsymbol{r})$ of an electron in a free atom is spherically symmetric. This means that one can expect the angular momentum of an orbiting electron not to change with time. In the quantum-mechanical context this means that electron energy eigenstates can also be chosen to be angular momentum eigenstates [13.1]. The spatial wave functions representing these states are called orbitals since we can imagine the corresponding classical (not quantum-mechanical) electron orbits as having fixed energy and fixed angular momentum around a given axis. The term orbital will be used to refer specifically to the spatial wave function of an electron in an atom or molecule. We will also use the term orbital for electron wave functions representing chemical bonds where the corresponding electron orbits would not be so simple. The angular momentum of the electron orbiting in an atom takes on discrete values $|L| = \sqrt{l(l+1)(h/2\pi)}$, where $l$ is an integer greater than or equal to 0. For each value of $l$ there are $2l+1$ different orthogonal eigenstates; that is, the component of angular momentum along any given direction can take on the values $m(h/2\pi)$ with $m = -l, -l+1, \ldots, l-1, l$. $m$ is called the magnetic quantuum number. The $2l+1$ orthogonal eigenstates, with different $m$ values, all have the same energy.

In the one-electron approximation, electron orbitals in atoms may be classified according to angular momentum. Orbitals with zero, one, two, and three units of angular momentum are called s, p, d, and f orbitals, respectively. Electrons in the last unfilled shell of s and p electron orbitals are called valence electrons. The principal periods of the periodic table of elements contain atoms with differing numbers of valence electrons in the same shell. The properties of the atom depend mainly upon its valence, that is upon the number of its valence electrons. When atoms are brought together to form molecules, their energies are shifted because the atomic states become combined. Mathematically, they are represented by linear combinations of atomic orbitals (LCAOs) of the atoms constituting the molecule. The combinations of valence atomic orbitals with lowered energy are called bond orbitals, and their occupation by electrons bonds the molecules together. Bond orbitals are symmetric or nonpolar when identical atoms bond but become asymmetric or polar if the atoms are different. The LCAOs are used as a basis for studying covalent and ionic solids, however, for metals the basis consists of plane waves [13.1].

For any given value of $l$ and $m$ there are many different energy eigenstates; these are numbered by a third integer, $n$, in order of increasing energy, starting with $n = l + 1$. This quantum number $n$ is called the principal quantum number [13.2].

The s-orbitals have vanishing angular momentum; $l = 0$ (and $m = 0$, since $|m| \leq l$). The wave function for an s-orbital is spherically symmetric, as shown in Fig. 13.1a. The lowest energy state, $n = 1$, is called a 1s state, and the next state is the 2s state.

**(a) atomic orbitals**

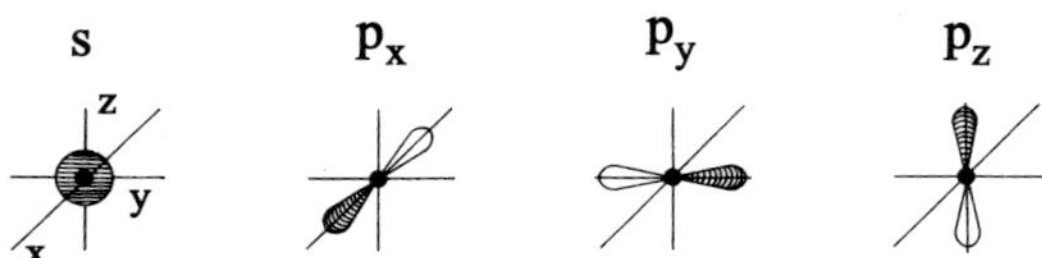


**(b) binding orbital pairs**

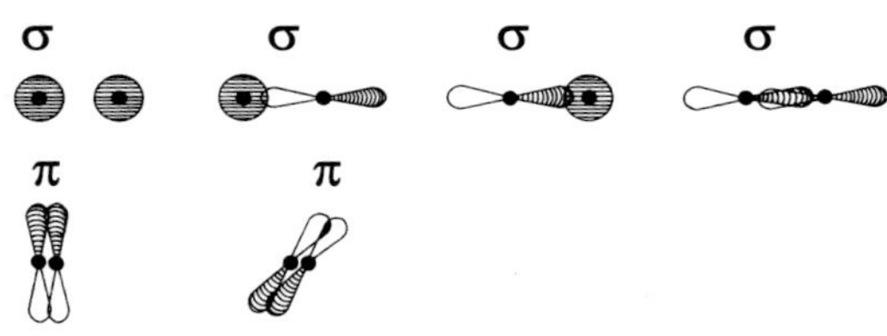


**(c) hybrid orbitals**

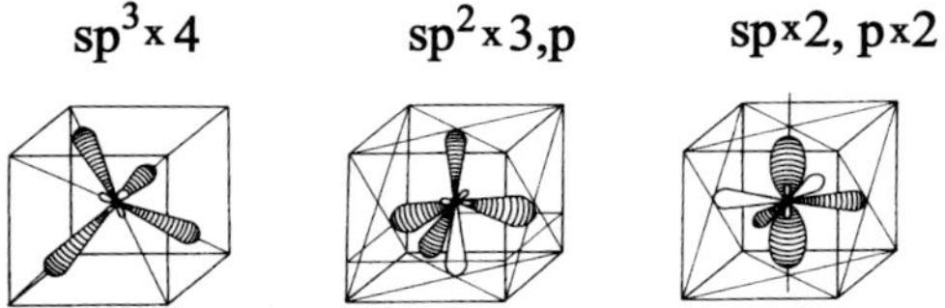


**Fig. 13.1.** (**a**) Schematic illustration of atomic s- and p- orbitals. (**b**) Different pairs of s- and p-orbitals in $\sigma$- and $\pi$-bonding states. (**c**) Hybrid orbitals of three p- and one s- orbital ($sp^3$ hybrid), two p- and one s- orbital ($sp^2$ hybrid), and one p- and one s-orbital (sp hybrid) (taken from [13.3])

The p-orbitals have one unit of angular momentum, $l = 1$; there are three orbitals corresponding to $m = -1$, $m = 0$, and $m = +1$ (see Fig. 13.1a). Any orbital, including those of the p series, can be written as a product of a function of radial distance from the nucleus and one of the spherical harmonics $Y_l^m$, which are functions of angle only (for more details see [13.4], p. 79):

$$\Psi_{nlm}(r) = R_{nl}(r) Y_l^m(\Theta, \phi), \tag{13.5}$$

where $r, \Theta, \phi$ are here the spherical coordinates. For a given $l$, the radial function is independent of $m$. For s-orbitals, the spherical harmonic is $Y_0^0 = \sqrt{1/4\pi}$. For p-orbitals, the spherical harmonics are given by: $Y_l^{-1} = \sqrt{3/8\pi} \sin\Theta \exp(-\mathrm{i}\phi)$, $Y_1^0 = \sqrt{3/4\pi} \cos\Theta$, and $Y_1^{+1} = -\sqrt{3/8\pi} \sin\Theta \exp(\mathrm{i}\phi)$.

In solid state physics it is frequently more convenient to take linear combinations of the spherical harmonics to obtain angular dependences proportional to the component of radial distance from the nucleus along one of the three orthogonal axes $x$, $y$, or $z$ [13.5]. In this way, the three independent p-orbitals may be written as

$$\Psi_{nlm}(r) = \sqrt{\frac{3}{4\pi}} R_{n1}(r) Y_1^m(x, y, z), \tag{13.6}$$

where $Y_l^{-1} = x/r, Y_1^0 = y/r$, and $Y_1^{+1} = z/r$. For each $n$ when $l = 1$, there are three p-orbitals oriented along the three Cartesian axes. Diagrams such as those shown in Fig. 13.1a illustrate the three angular forms.

The d-orbitals have two units of angular momentum, $l = 2$, and therefore five $m$ values: $m = -2$, $m = -1$, $m = 0$, $m = +1$, and $m = +2$. They can be conveniently represented in terms of Cartesian coordinates in the form

$$\Psi_{n2m}(r) = \sqrt{\frac{15}{4\pi}} R_{n2} Y_2^m, \tag{13.7}$$

where $Y_2^{-2} = yz/r^2, Y_2^{-1} = zx/r^2, Y_2^0 = xy/r^2, Y_2^{+1} = (x^2 - y^2)/(2r^2)$, and $Y_2^{+2} = (3z^2 - r^2)/(2r^2\sqrt{3}$. A very important feature of d-orbitals is that they are concentrated much more closely at the nucleus than are s and p-orbitals.

The f orbitals, which have three units of angular momentum, $l = 3$, and therefore seven values of $m$, are even more strongly concentrated near the nucleus and isolated from other neighboring atoms than the d-orbitals are. They are, however, important in studying properties of the rare-earth metals. Therefore, we will introduce the relevant mathematical formulae in the further text in places related to these materials.

Discussing states of minimum energy, which is the main goal in determination of crystal lattice bonding, one usually is not interested in how the wave function changes with time. However, for the cases in which that information is wanted, one has to use the time-dependent Schrödinger equation, exhibiting explicitly the time dependence of $\Psi$

$$\mathrm{i}\frac{h}{2\pi}\frac{\partial \Psi}{\partial t} = H\Psi. \tag{13.8}$$

### 13.1.1 Interatomic Bonds in Small Molecules

The elementary process of bonding atoms in solid crystals, or at their surfaces, is in principle similar to the process of bonding atoms in small molecules. Therefore, let us first introduce the relevant terminology used in the quantum mechanics of this subject [13.1]. In describing states of a small molecule (the simplest example of such a molecule is the hydrogen molecule $H_2$, with two electrons which are represented by orbitals $|1>$ and $|2>$ of the 1s states of hydrogen atoms constituting the molecule, respectively) one has to determine the electronic states of the molecule. This can be done in the simplest way by representing the molecular state $|\Psi>$ as a linear expansion of the electronic states in the constituent atoms

$$|\Psi>=\sum_{\alpha=1}^{n} u_\alpha\,|\,\alpha>, \tag{13.9}$$

where $\alpha$ enumerates the possible electronic states, so called basic states, which are selected to be normalized ($<\alpha|\alpha>=1$) and orthogonal ($<\beta|\alpha>=0$, if $\beta\neq\alpha$), while $u_\alpha$ are the expansion coefficients. Solving the relevant Schrödinger equation (13.4), one may find the eigenstates of the molecule and the eigenvalues of the Hamiltonian operator for this molecule. The resulting two simple algebraic equations have the form

$$(\varepsilon_s - E)u_1 - V_2 u_2 = 0 \quad \text{and} \quad -V_2 u_1 + (\varepsilon_s - E)u_2 = 0 \tag{13.10}$$

where the 1s state energy is $\varepsilon_s = <1|H|1> = <2|H|2>$. The matrix element $V_2 = -H_{12} = -H_{21}$ is called the covalent energy, and is defined to be greater than zero; $V_2$ will generally be used for interatomic matrix elements, in this case between s-orbitals. Equations (13.10) are easily solved to obtain a low-energy solution, the bonding state, with energy

$$\varepsilon_b = \varepsilon_s - V_2 \tag{13.11}$$

as well as a high-energy solution, the antibonding state, with

$$\varepsilon_a = \varepsilon_s + V_2. \tag{13.12}$$

Substituting the eigenvalues given in the last two equations back into (13.10) gives the coefficients $u_1$ and $u_2$. For the bonding state, $u_1 = u_2 = 1/\sqrt{2}$ and for the antibonding state, $u_1 = -u_2 = 1/\sqrt{2}$. The conventional depiction of these bond orbitals and antibond orbitals is illustrated in Fig. 13.2a for the case of the $H_2$ molecule. In molecular systems in which the energies of the constituent atoms are different, one expects that the relevant LCAOs will consist of the orbitals of the constituting atoms, too; for example in the case of LiH molecules, the linear combinations are those of hydrogen 1s-orbitals and the lithium 2s-orbitals. Such molecules like LiH are said to have a heteropolar bond.

In calculating the energy of heteropolar bonds, (13.10) must be modified so that $\varepsilon_s$ is replaced by two different energies: $\varepsilon_s^1$ for the low-energy state (for the energy of the anion) and $\varepsilon_s^2$ for the high-energy state (for the energy of the cation). This modification leads to the form

$$(\varepsilon_s^1 - E)u_1 - V_2 u_2 = 0 \quad \text{and} \quad -V_2 u_1 + (\varepsilon_s^2 - E)u_2 = 0. \tag{13.13}$$

The value of one half of the anion–cation energy difference $V_3 = (\varepsilon_s^2 - \varepsilon_s^1)/2$ is the polar energy; it is convenient to define the average of the cation and anion energy as $\varepsilon_{\text{av}} = (\varepsilon_s^1 + \varepsilon_s^2)$. Then (13.13) become

$$\begin{gathered}(\varepsilon_{\text{av}} - V_3 - E)u_1 - V_2 u_2 = 0 \\ \text{and} \\ -V_2 u_1 + (\varepsilon_{\text{av}} + V_3 - E)u_2 = 0\end{gathered} \tag{13.14}$$

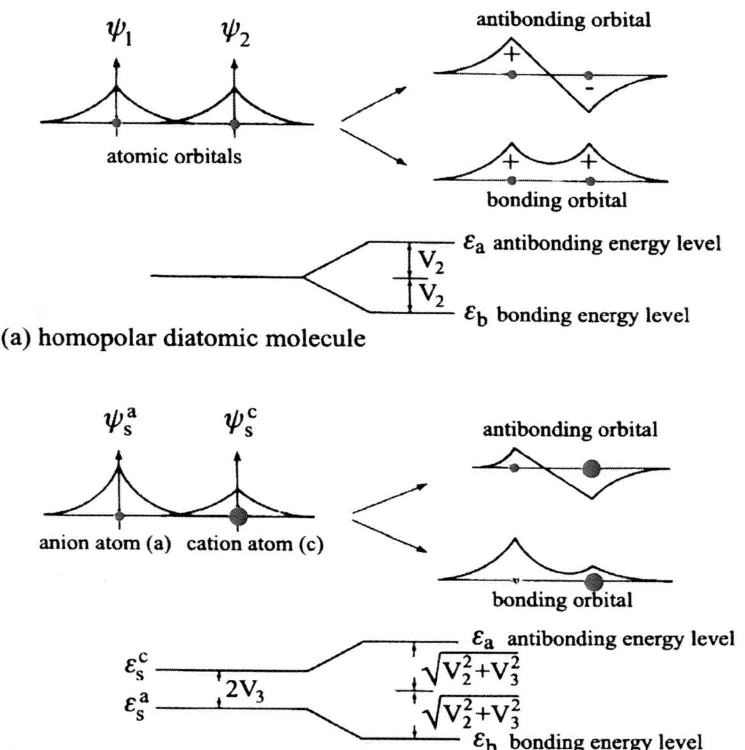


**Fig. 13.2.** The formation of bonding and antibonding combinations of atomic orbitals in diatomic molecules, and the corresponding energy-level diagram: (**a**) for homopolar molecules, like $H_2$, (**b**) for heteropolar molecules, like LiH (taken from [13.1])

which has the solution given by $\varepsilon_b = \varepsilon_{\text{av}} - (V_2^2 + V_3^2)^{1/2}$, and $\varepsilon_a = \varepsilon_{\text{av}} + (V_2^2 + V_3^2)^{1/2}$. The energies $\varepsilon_b$ and $\varepsilon_a$ are called bonding and antibonding energies, respectively. The splitting of these levels is shown in Fig. 13.2b. In looking at the energy-level diagram of that figure, it is easy to recognize that the interaction between the two atomic levels, represented by $V_2$, pushes the levels apart. It is also shown in this figure that the charge density associated with the bonding state shifts to the low-energy side of the molecule (the direction of the anion). This means that the molecule has an electric dipole moment.

Polarity of bonding is an important concept in bulk solids and at their surfaces. To describe polarity mathematically, first one obtains $u_1$ and $u_2$ values for the bonding state by substituting $\varepsilon_b$ for the energy $E$ in (13.14). The first equation of these can be rewritten as $u_1 = V_2 u_2/(\sqrt{V_2^2 + V_3^2} - V_3)$. If the individual atomic wave functions do not overlap, the probability of finding the electron on atom 1 will be equal to $u_1^2/(u_1^2 + u_2^2)$, and the probability of finding it on atom 2 will be $u_2^2/(u_1^2+u_2^2)$. These probabilities may be expressed by $(1+\alpha_p)/2$ and $(1-\alpha_p)/2$, for the electrons appearing on the atom 1, and on the atom 2, respectively. Here, $\alpha_p$ is the polarity defined by

$$\alpha_p = u_1^2 - u_2^2 = \frac{V_3}{\sqrt{V_2^2 + V_3^2}}. \tag{13.15}$$

One may expect the dipole of the bond to be proportional to the polarity $\alpha_p$. A complementary quantity to this is the covalency, defined by

$$\alpha_c = \frac{V_2}{\sqrt{V_2^2 + V_3^2}}. \tag{13.16}$$

### 13.1.2 Chemical Bonds in Solid Crystals

It is well known that chemical bonds in solid crystals are caused by two, essentially different, types of forces, namely, Coulomb forces describing electrostatic interactions of point charges, and quantum mechanical forces related to the tunneling effect in the interatomic charge transfer process [13.3]. The forces of the first type cause ionic bonding and Van der Waals bonding. Ionic bonding occurs when forces acting between individual negative and positive point charges are involved in the interatomic interaction process. On the other hand, Van der Waals bonding means that higher moments of the atomic charge distribution, in particular dipole or quadrupole moments, are responsible for the forces.

Two previously neutral atoms A and C can only form an ionic bond if they become electrically charged. This may occur if a state of lower energy can be attained by the transfer of one or more electrons from atom C (cation) to atom A (anion), i.e., when the electron affinity of atom A is greater than the ionization energy of atom C. This is most likely the case when a closed shell configuration of both of these atoms is reached by transferring one electron from C to A. However, one has to notice that the energy gain obtained by electron transfer is only one contribution to the lowering of energy involved in ionic bonding. The second contribution, which is superimposed on the first one, arises from the electrostatic interaction between the ions created by the charge transfer. This contribution depends on the relative positions of the ions in the crystal lattice, and is thus also dependent on the crystal structure. Usually, it is assumed that a change in the mutual positions of the ions does not influence the electron transfer between the atoms C and A.

The second type of binding forces, namely the quantum mechanical forces, produce covalent and metallic bonding in the crystal lattice. Let us suppose that the atoms forming a crystal are at first separated by a large distance, and then the distance is stepwise shortened. The electrons of the outermost atomic shells (the valence electrons), which at first were localized only at their native atoms become then more delocalized, spreading finally over more than only one atom. Consequently, the energy of the electrons lowers, too, due to their delocalization, which means the occurrence of chemical bonding, by covalent bonds. One necessary condition for creation of covalent bonds is that the atoms are close enough to each other to ensure that electrons will tunnel through the potential barrier separating the atoms. On the other hand, however, a second condition has to be fulfilled, namely that the atoms are not too close, because then valence-electron delocalization would not occur, or would be strongly limited to the nearest vicinity of their native atoms. For very small separations the quantum mechanical forces cause the repulsion of atoms. Beside the electrostatic repulsion these repelling quantum forces are mainly responsible for keeping the atoms in the crystal lattice at a certain distance apart, thus preventing the crystal from collapsing.

Delocalization of the valence electrons may cause them to spread either over all atoms in the crystal lattice or primarily only over the nearest neighbors in the lattice. The first case occurs in crystal lattices of metals, and leads to metallic bonding, while the second case, occurring mostly in some of the semiconductor crystals, is called covalent bonding. The latter case arises when the neighboring atoms exhibit different electron affinities, which in contrast to ionic crystals, are not disparate enough to make the transfer of one whole electron energetically advantageous. Consequently, only a partial electron transfer then takes place. The fraction of the electron transferred to the neighboring atom depends on the positions in which both atoms align themselves in the crystal lattice with respect to each other. The bond is now no longer purely covalent but instead is partially ionic in its physical nature.

The fact that a particular bonding type is dominant in a given crystal has decisive consequences for its structure. This statement holds for the surface structure, too.

In the case of ionic bonding, the structure which predominates in the bulk crystals is that which yields the largest energy of electrostatic interatomic interaction (the Madelung energy). If one orders the various structures of an ionic crystal (ionic bonds occur there) according to the absolute values of their Madelung energies, the first place takes the rocksalt structure followed by the cesium chloride, wurtzite and zincblende structures [13.3]. Thus, the structure that one expects in the case of pure ionic bonding is the cubic rocksalt structure involving six nearest neighbors. If instead of a 3D arrangement of atoms one considers a 2D arrangement in an atomic plane of the crystal, then the largest value of the Madelung energy is attained by a quadratic structure, such as that corresponding to a (100) lattice plane of a rocksalt crystal.

In the case of metallic bonding, the valence electrons are delocalized to the same extent in all directions, and the directions in which the nearest neighbors are located are thus arbitrary. Since the bond forces have the tendency to draw the atoms closer together until repulsion prevents further contraction one expects an arrangement as exhibited by densely packed spheres with 12 nearest neighbors. It is well known that this is indeed the case for most of the metals and that in these materials usually the dense packing of hexagonal and cubic type is observed.

In covalent crystals, the electrons participating in bonding are, before and after formation of the bonds, located (partially or completely) in orbitals extending in characteristic directions [13.1]. These directions are significant because they determine the arrangement of neighboring atoms in the crystal lattice. For covalent bonding s- and p-type atomic orbitals (see Fig. 13.1a) are the most important orbitals. In the cases when two p-orbitals of adjacent atoms point in each other's direction, delocalization of valence electrons is particularly pronounced, and the bonds are specially strong. These types of bonds are called $\sigma$-bonds (see Fig. 13.1b). If the two p-orbitals are still parallel to each other but perpendicular to the line connecting both atoms then they create a $\pi$-bond (pairs of p-orbitals which are perpendicular to each other do not contribute to binding). In the case of $\pi$-bonds delocalization of electrons occurs vertically to the direction of the orbitals and is not as effective as that which occurs in the direction of orbitals. Generally speaking, $\pi$-bonds are weaker than the $\sigma$-bonds.

Covalent bonding occurs primarily between nearest neighbors in the crystal lattice (bonding which occurs between the second nearest neighbors is significantly weaker, and consequently, is usually omitted in structural calculations). In treating this kind of bonding, one has to be clear at first which atomic orbitals are the constituents of crystal orbitals; are these the direct s- and p-orbitals or are these linear combinations of them, so-called hybrid orbitals [13.1]? Answers to these questions depend essentially on the number of s- and p-electrons to be involved in bonding interaction. With eight electrons per atom pair, which corresponds to the covalent semiconductors of group IV and to partially ionic semiconductor compounds of groups III–V and II–VI, respectively, the most advantageous combinations are the four orthogonal linear combinations of one s- and three p-orbitals, i.e., the $sp^3$ hybrid orbitals (see Fig. 13.1c). Bonding through $sp^3$ hybrids is in general a mixture of $\sigma$-bonding and $\pi$-bonding. This means that a binding pair of two hybrid orbitals at two nearest neighbor atoms decomposes into several binding pairs of s- and p-orbitals, both of $\sigma$- and of $\pi$-type. If the two hybrid orbitals point in each other's direction then only $\sigma$-binding pairs contribute. In that sense one can speak about $\sigma$-like bonding with respect to $sp^3$ hybrid orbitals. Other, less commonly occurring, bonding cases (met mostly in chalcogenides or group IV–VI semiconductors like e.g. PbTe or PbS) are shown in Fig. 13.1c, too.

### 13.1.3 Bonding at Surfaces

Let a surface be generated by cutting off one half of an infinite crystal. Then the atoms of the first monolayer of this surface are not as strongly bound to

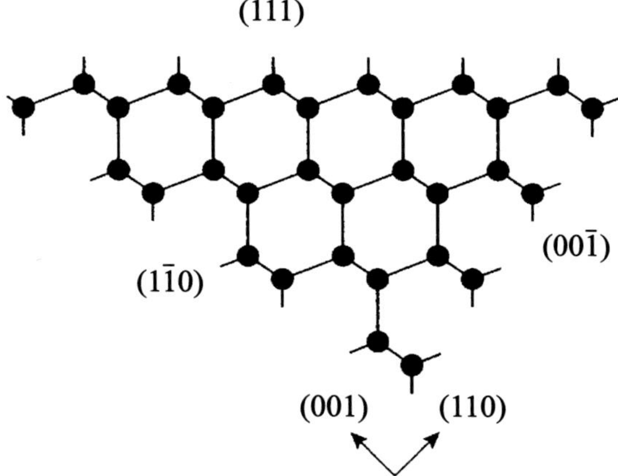


**Fig. 13.3.** Plan view of an unreconstructed structure of a homopolar semiconductor (diamond structure). Solid circles indicate atoms, and solid lines indicate bonds. Viewing at each of the indicated surface edges one can recognize the geometry of the relevant dangling bonds (dangling hybrides). At the (111) surface, for example, single dangling hybrids occur in the plane of the figure. They are perpendicular to the (111) surface and originate from each surface atom in this plane

the crystal lattice as they were before cutting off the crystal. As a matter of fact, one or more nearest neighbors of the surface atoms are missing. In the case of covalent bonding this means that one or more bonds are broken, and that unsaturated bonds occur, which are called dangling bonds. Let us consider, as an example of bonding atoms at surfaces, the case of bonding through a dangling hybrid. Such dangling bonds are schematically shown in Fig. 13.3, on the differently oriented surfaces of an unreconstructed (ideal) homopolar semiconductor.

The simplest case one may consider is the adsorption of atomic hydrogen on such surfaces. The energy of the electron state of atomic hydrogen is very low compared to the energy of the dangling hybrid, so the bonding-state energy will be near the hydrogen level and the corresponding antibonding state will be above the dangling hybrid level. The same should be true about the addition of halogen atoms, which should adsorb in just the same way. The adsorption of an alkali metal atom should be similar in all respects, except that since the energy of the alkali metal state is presumably well above the hybrid energy, the energy of the bond formed should be near, and determined by the energy of the dangling hybrid.

In the case of partial coverage, that is a coverage by less than one adsorbed atom per surface atom, the density of upper and lower hybrid states on the surface should decrease together with increasing coverage. Forming a bond with a dangling hybrid forms a neutral surface "molecule" at one of the otherwise polar sites. Neutrality can, thus, occur at either type of site where the adsorbate atom is bound to the surface.

Much more complicated surface bonding appears usually in molecular beam epitaxy (MBE) and metalorganic vapor phase epitaxy (MOVPE) growth processes of compound semiconductor thin films. Let us present a brief discussion of bonding mechanisms relevant to MBE growth of GaAs as presented by Farrel et al. [13.6]. This represents a good example for the discussion of surface bonding in terms of bonding orbitals and in addition was for more than a decade the only detailed model describing the MBE growth process. We will follow here the results and the discussion as published and presented in Figs. 13.4–13.7 [13.6].

When the Ga flux is turned on, the Ga atoms presumably arrive at the substrate surface which was assumed to be the As-rich GaAs(100)-$\beta(2\times4)$ surface (Fig. 13.4a). A single Ga atom might attempt to bond to one As atom via an As dangling hybrid orbital. However, this would be a rather short-lived species having only one covalent bond and three electrons in the Ga nonbonding orbitals (a total of five electrons, two from the As dangling orbital and three from the Ga itself, are shared among one bonding orbital and two dangling orbitals). A somewhat more stable species is produced, however, when a Ga atom is inserted into an As-dimer bond. Here, two covalent bonds are formed and only one electron is left in a Ga nonbonding orbital. Both species are expected to be more or less labile, and a given Ga atom may enter into a number of these unstable surface complexes before it reaches its final bonding site.

The insertion of two Ga atoms into two adjacent As-dimer bonds, however, does allow the formation of a stable Ga species on the surface. This is a Ga dimer with five covalent bonds and no electrons in the Ga dangling orbitals. (The two single electrons left in the nonbonding Ga orbitals of each of the two inserted Ga atoms combine to form a covalent bond between the two inserted atoms.) Note that there are two sites per unit cell where such a Ga dimer could form. These are the a and b sites shown in Fig. 13.4a. Unlike the As dimer, the Ga dimer has approximately $sp^2$ bonding, with the consequence that it is relaxed inward toward the bulk. This will relax the underlying As atoms, too. The configuration is shown in Fig. 13.4b, and the relaxation of the As atoms serves an important function in setting up the following stages. (Note that at this stage the insertion of either one or a pair of Ga atoms into the As-dimer bonds is more stable than chemisorption across As dangling orbitals [13.6].)

When approximately one-quarter of a monolayer of Ga has been deposited, there will be, on the average, one Ga dimer per unit cell. Since there are two sites per unit cells, these Ga dimers may be either next to those in the adjacent cell, as in Fig. 13.4b, or offset by one atomic spacing as would be the case if the dimer at point a had been inserted at point b instead. Because of the As-stabilized conditions used in MBE growth, while these Ga dimers

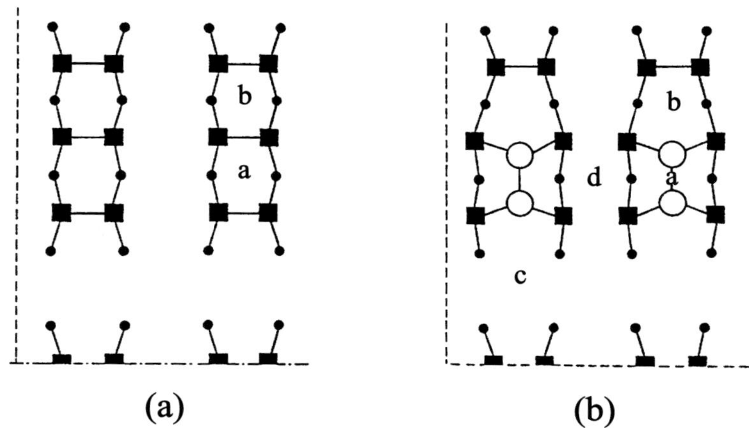


**Fig. 13.4.** (**a**) Planar view of the As-rich GaAs(100)-(2×4) surface unit cell. The As atoms (■) form three dimers per unit cell. The dimer vacancy exposes four Ga atoms (●) in the second layer. A Ga dimer can be inserted at either a or b sites. The size of the symbol indicates the proximity of the layer to the surface. (**b**) After the chemisorption of the $\frac{1}{4}$ monolayer of Ga (○), one Ga dimer per unit cell initiates growth of the second layer. In the right-hand cell the dimer may be aligned with that in the left-hand adjacent cell, as shown, inserted at a, or staggered if inserted at b. Insertion of a second Ga dimer at point d necessitates the filling of the As vacancy at point c (taken from [13.6])

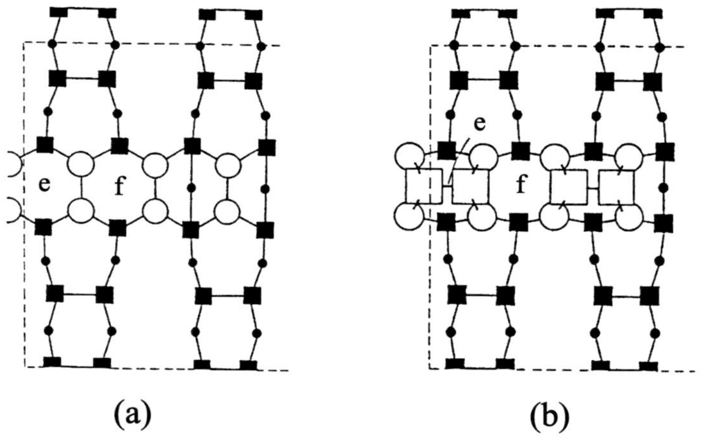


**Fig. 13.5.** (**a**) Coupled with the filling of the As dimer vacancy in the lower As layers, the chemisorption of a second Ga dimer per unit cell opens up a new site for the chemisorption of an As dimer centered at either e or f sites. (**b**) The chemisorption of an As dimer (□ – □) initiates growth of the top of the surface bilayer and completes the second stage of the growth cycle (taken from [13.6])

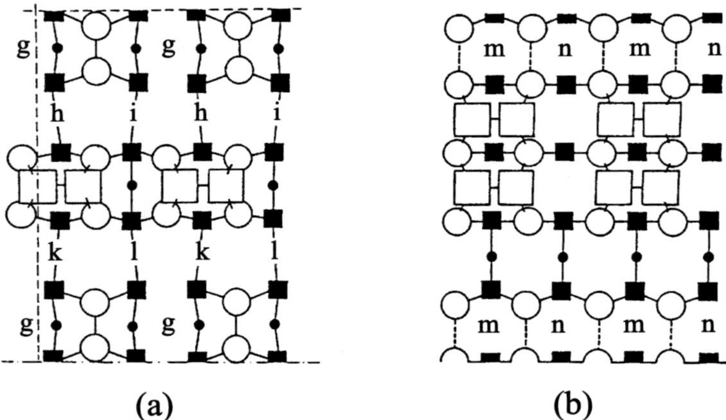


**Fig. 13.6.** (**a**) The chemisorption of a third Ga dimer per unit cell completes the third stage of the growth cycle. Note that this structure is stabilized relative to a $(4 \times 2)$ unit cell by the presence of $\frac{1}{4}$ of a monolayer of As in the outermost layer. Further growth necessitates the simultaneous filling of the Ga dimer vacancy site at point g and the chemisorption of one As dimer per unit cell at either points h or i or at points k or l. (**b**) The correlated chemisorption of a Ga dimer at point g and an As dimer at point h. This opens up one additional site for the adsorption of an As dimer. Site m is favored over site n on the basis of back-bonding (taken from [13.6])

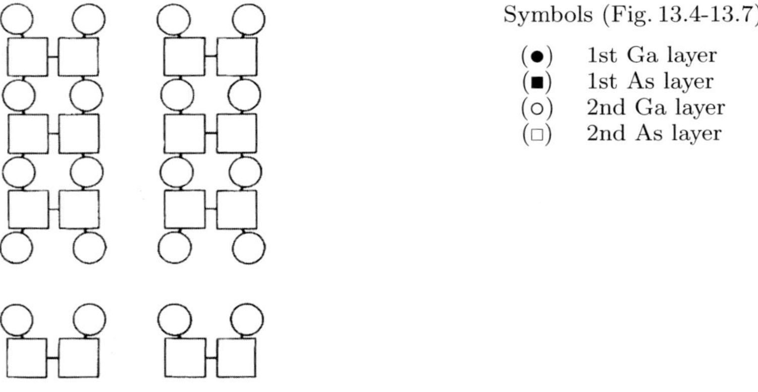


**Fig. 13.7.** The growth cycle is complete and the As-rich structure shown in Fig. 13.4a is recovered. This occurs when an As dimer and a Ga dimer are simultaneously chemisorbed on the structure shown in Fig. 13.6a, followed by the chemisorption of the final As dimer (taken from [13.6])

are being formed there is still an As flux to the surface that greatly exceeds the Ga flux. However, at this stage of growth there are essentially no stable binding sites for As, and short-lived species presumably form and decay with the subsequent diffusion and desorption of $As_2$.

For the growth process to proceed further, four Ga atoms must be chemisorbed contiguously as two adjacent dimers in order to provide a bonding site for an $As_2$ dimer. In addition, the dimer vacancy in the first layer must be filled to provide a basis for the chemisorption of further Ga atoms. As will be seen, these two steps (the chemisorption of an As dimer to fill the vacancy and the chemisorption of a second pair of Ga atoms) must occur essentially simultaneously within a given unit cell to achieve an electrically stable situation. This step, and a similar one near the end of the cycle, appear to be the limiting steps for epitaxial growth to occur. In addition, the necessity of having an $As_2$ dimer (or its monoatomic constituents) available for concurrent chemisorption may explain, in part, the requirement of a much larger As than Ga flux to the surface for successful growth to occur. The further steps of the homoepitaxial growth of GaAs on the GaAs(100) nominally oriented surface are shown in Figs. 13.5–13.7.

## 13.2 Surface Structure

As already mentioned, the most characteristic feature of epitaxy is the strong influence which the surface of the substrate crystal exerts on the growth process. Knowing the parameters characterizing this surfaces is, therefore, indispensable for getting a control over the process of epitaxy. Accordingly, we will proceed with a discussion of the most important features of the substrate surfaces with emphasis on the geometric properties (order and symmetry) of these surfaces.

### 13.2.1 Physical Principles

The term "surface", when used in the macroscopic sense, means the outermost face of the solid, which can be produced and maintained under normal external conditions by conventional methods like cutting, polishing and etching. This macroscopic approach implies that the surface is in contact with the ambient atmosphere, and that oxide layers, as well as water vapor, carbon or other chemical elements, can be deposited on it [13.3]. On the other hand, in the microscopic sense, such a surface is not a true surface but rather a transition region, separating the bulk of the solid from the adjoining environment. The term "surface with coverage" is more realistic when describing this situation.

In the microscopic sense, the term "surface" means an atomically clean and atomically smooth crystal plane, which creates an abrupt transition from the bulk solid to an ideal vacuum or gas. Thus, if the adjoining environment is not an ideal vacuum, but rather a vapor, then a solid–solid interface is created by the adparticles and the plane atoms. Producing atomically clean and atomically smooth surfaces or abrupt interfaces, and keeping them in ideal conditions, is an extremely difficult task. Thus, in the real physical situation one deals rather with covered surfaces and with interface regions. A real surface has been found to exhibit a more or less unknown atomic order. Thus, the task of determining this order (the crystallographic structure) became a considerable research problem. In order to solve this problem the following research philosophy has been adopted. It is based on three sequential steps, namely, on radical reduction of the complexity of real systems, studying then models created in this way, and finally on stepwise extension of the complexity (reality) of the studied system to a degree, required by the problem being under consideration. The models used in such research procedures are primarily the atomically clean and atomically smooth surfaces and abrupt interfaces.

The most surprising discovery made in relation to surface science was connected with the geometrical structure of clean surfaces. Experimentally and theoretically it has been shown, that the atoms of a clean surface undergo relatively large displacements compared to their positions in an infinite bulk crystal.

Structural changes at the surface are a consequence of altered chemical bonds as compared to the bulk of the crystal. Information on these changes may be gained by studying the bonding forces at the surface in detail. The atoms in the surface layer experience forces different from those, that are acting in the bulk of the crystal. Consequently, they are subjected to displacements from their original sites in the bulk [13.7]. Since the forces acting on atoms of the second monolayer are partly determined by the positions of the atoms of the first (outermost towards the vacuum) monolayer, these forces are also subjected to changes accompanied by displacements in the second layer. Such interaction concerns each successive atomic monolayer near the crystal surface; however, the relevant displacements decrease from one monolayer to the next until vanishing at a certain depth from the surface. Let us now discuss the so-called surface-induced atomic displacements in more detail.

We denote these displacements by $\mathrm{d}\boldsymbol{r}_{\boldsymbol{s}_1,\boldsymbol{s}_2}(i)$ and the new positions of atoms in the surface of the crystal by $\boldsymbol{r}'_{\boldsymbol{s}_1,\boldsymbol{s}_2}(i)$. Then one gets

$$\boldsymbol{r}'_{\boldsymbol{s}_1,\boldsymbol{s}_2}(i) = \boldsymbol{r}_{\boldsymbol{s}_1,\boldsymbol{s}_2}(i) + \mathrm{d}\boldsymbol{r}_{\boldsymbol{s}_1,\boldsymbol{s}_2}(i), \tag{13.17}$$

where $\boldsymbol{s}_1$ and $\boldsymbol{s}_2$ represent the displacements of the atom $(i)$ along the two directions of the 2D lattice's primitive vectors $\boldsymbol{\tau}_1$ and $\boldsymbol{\tau}_2$. The displacements $\mathrm{d}\boldsymbol{r}_{\boldsymbol{s}_1,\boldsymbol{s}_2}(i)$ may be divided into two classes with regard to their effect on translational symmetry of the surface crystal lattice. If this symmetry is not affected, the displacements cause surface relaxation, and are thus called relaxing displacements. In this case, equivalent atoms in different elementary

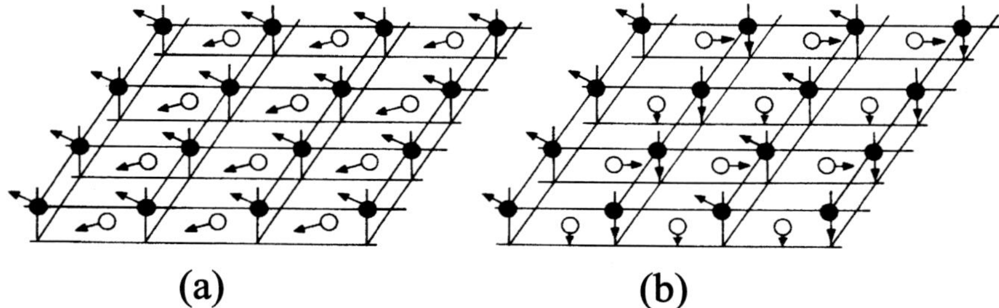


**Fig. 13.8.** Schematic illustration of lattice relaxation (**a**), and lattice reconstruction (**b**) for the case when surface-related atomic displacements extend to the first and second atomic monolayers, only. In (**b**) a 2×2 reconstruction is shown (taken from [13.3])

cells of the lattice are displaced in the same way (see Fig. 13.8a). If, on the other hand, the translational symmetry is altered by the displacements, then the surface becomes a reconstructed one. In this case, equivalent atoms in different elementary cells of the lattice are not all displaced in the same manner, i.e., the $\mathrm{d}\boldsymbol{r}_{\boldsymbol{s}_1,\boldsymbol{s}_2}(i)$ depends on $\boldsymbol{s}_1$ and $\boldsymbol{s}_2$ (see Fig. 13.8b).

For an ideal crystal it is easy to construct primitive surface translations, which are the smallest translations in the plane of the surface by which each atom is replaced by another atom. Let these translations be equal to $\boldsymbol{\tau}_1$ and $\boldsymbol{\tau}_2$, i.e., let them be defined by the primitive vectors of the 2D crystal lattice of the surface. If these primitive translations repeat on the whole surface, one calls this surface unreconstructed. If the surface geometry exhibits the same symmetry as the truncated bulk solid (as an unreconstructed surface), being however relaxed, then it is referred to as exhibiting a 1×1 (one-by-one) reconstruction. This is so even though the atoms at the surface may lie as much as an Ångstrom away from the truncated bulk lattice sites. The actual translational symmetry of a real surface frequently exhibits primitive translations of $2\tau_1$ and $\tau_2$, or generally speaking, of $n\boldsymbol{\tau}_1$ and $m\boldsymbol{\tau}_2$. For such surfaces the reconstructions are called 2×1 (two-by-one) or $n \times m$ ($n$-by-$m$), respectively. These notations are in common with the Wood notation system (for more information see [13.8, 9]), most frequently used for identification of the reconstructed surfaces.

Experimental data, most frequently based on electron diffraction, do not show the nature of the reconstruction, only its symmetry. Thus, a series of works have speculated on a variety of patterns of distortion or missing atoms. The currently available data, concerning surface reconstruction of different materials, are enormous in amount, and are related to both the theory of crystal surface structure and the experimental results. For a review one may see the references [13.9–14].

### 13.2.2 Reconstructed Surfaces; Theoretical Methodology

The quantitative description of the surface structure is based on the adiabatic principle of solid state physics [13.3]. According to this principle, it is possible to average the bonding forces in a crystal due to the electronic interactions over the states of electrons involved in these interactions. These electron states may be calculated separately for a fixed configuration of the cores of atoms creating the crystal lattice. The total energy $E_{\text{tot}}$ of the crystal in its ground state may be, thus, expressed as a function of the positions $\boldsymbol{r}(i)$ of all the atom cores in the crystal; $E_{\text{tot}}(\boldsymbol{r}(1), \boldsymbol{r}(2), \ldots, \boldsymbol{r}(N))$. The force $\boldsymbol{F}(i)$ acting on the $i$-th atom is then given by the gradient over the total energy; $\boldsymbol{F}(i) = -\nabla_{\boldsymbol{r}(i)} E_{\text{tot}}$. In equilibrium the geometric sum of all forces acting on the $i$-th atom must be zero, or equivalently, the total energy of the crystal lattice attains a minimum as a function of $\boldsymbol{r}(i)$. Solving the equation

$$F(i) = -\boldsymbol{\nabla}_{\boldsymbol{r}(i)} E_{\text{tot}}(\boldsymbol{r}(1), \boldsymbol{r}(2), \ldots, \boldsymbol{r}(N)) = 0 \tag{13.18}$$

one may, in principle, determine the structure of the crystal (the positions of the atom cores in the crystal lattice), as well as the structure of its surface. Materials-specific structure differences are in a formal sense described by the differences of the functional dependence of $E_{\text{tot}}$ on the position vectors $\boldsymbol{r}(i)$. Let us now concentrate on surface structures, only.

The theoretical methodologies used in the procedures of determination the structures of surfaces can be roughly classified into two groups: those that use quantum-mechanical potentials, and those that use empirically determined classical potentials [13.11]. The quantum-mechanical methods can be further subdivided into three major types, namely: self-consistent field, linear- combination of atomic orbitals (SCF-LCAO), density functional theory (DFT) [13.13, 15], and tight-binding depending upon how the interactions between the electrons are treated. The empirical classical-potential models have primarily been developed and applied to “ionic” insulators because of the presumed predominance of the Coulombic interactions between the ions in the lattice [13.15]. They have, however, found various applications in studies on surfaces of other materials. These techniques can be subdivided into three categories as well: Green's function techniques, slab calculations, and cluster calculations. The main difference between these methods is the boundary condition used to model the semi-infinite surface. The Green's function techniques treat the semi-infinite nature of the surface exactly (and as a result are the most complex). Slab models treat the infinite 2D nature of the surface property, but have a finite thickness in the third dimension perpendicular to the surface. Finally, the cluster methods model the surface with a finite set of atoms. For the details of a particular method, the reader is referred, primarily, to the review literature and to graduate-level texts (see the listing presented in [13.11]). Here we will emphasize only the basic terminology; the goal is to enable the reader to evaluate computational results.

One feature, common to all of the mentioned methods, is the way in which the equilibrium surface atomic geometry is obtained. That is, once the method is chosen, and the total-energy functional specified, atomic forces are computed, and the total energy minimized as a function of the atom core coordinates (see (13.18)). The methods differ, of course, in the way in which the total energy functional is specified.

In order to characterize in more detail the quantum mechanical approach to surface structures, we will discuss now the main calculation procedures used in this approach. The differences between the various quantum mechanical methods lie in the way in which the many-body electron–electron interactions are reduced to effective one-particle, or one-electron interactions. Perhaps the most straightforward way of illustrating these differences is to partition the one-electron energy into component parts as follows [13.11].

$$E_{EL}(\boldsymbol{r}(N)) = E_{KE}(\boldsymbol{r}(N)) + E_H(\boldsymbol{r}(N)) + \\ + E_X(\boldsymbol{r}(N)) + E_C(\boldsymbol{r}(N)). \tag{13.19}$$

Here, $E_{KE}$ is the one-electron kinetic energy and electron–ion attraction which results from (13.19) with the potential $V(\boldsymbol{r})$ in the form of Coulomb attraction energy. The remaining terms stands for different electron–electron interactions; $E_H$ represents the two-electron Hartree screening (or Coulomb) interaction, $E_X$ represents the two-electron exchange (or Fock) interaction, arising from the indistinguishability of the electrons, which must be reflected by the wave functions; and $E_C$ is the electron correlation energy.

Let us start with the SCF-LCAO method, which simplifies the computation of the electronic energy by restricting the one-electron wave function to a single Slater determinant [13.16]. This requirement is one way to assure that the wave function is antisymmetric and describes electrons as indistinguishable particles. It also results in the neglect of electron correlation effects (although a limited amount of correlation is included since the method also obeys the Pauli exclusion principle [13.17]). This level of calculation is called the "Hartree–Fock" level. Equation (13.19) reduces at this level to the form

$$E_{EL}(\boldsymbol{r}(N)) = E_{KE}(\boldsymbol{r}(N)) + E_H(\boldsymbol{r}(N)) + E_X(\boldsymbol{r}(N)), \tag{13.20}$$

where all of the terms are now evaluated explicitly. To do this, the one-electron wave functions are computed using the effective one-electron Hamiltonian operator, referred to in this context as the Fock operator, obtained from the variational principle. Note that the one-electron wave functions, $\psi_i(\boldsymbol{r})$, appear in the definition of the Fock operator, requiring an iterative, or "self-consistent" solution. Once the one-electron wave functions are found, the energy terms in (13.20) can be evaluated [13.11].

In summary, one has to notice that the SCF-LCAO method neglects electron correlation effects by limiting the wave function to a single Slater determinant. The remaining energy terms of (13.20), i.e., the kinetic, electron–ion, Coulomb, and exchange energies are then evaluated explicitly.

The next method, the density functional method (DFT), differs from the SCF-LCAO method in that the electron density, $\rho(\boldsymbol{r})$, is used as the variable of interest [13.18]. It has been shown [13.19] that the ground state electronic energy can be expressed as a functional of the external (or nuclear) potential, $\nu(\boldsymbol{r})$, and the electron density as

$$E(\rho(\boldsymbol{r})) = \frac{e^2}{4\pi\varepsilon_0}\left(T(\rho(\boldsymbol{r})) + \int \rho(\boldsymbol{r})\nu(\boldsymbol{r})\mathrm{d}\boldsymbol{r} + V_{ee}(\rho(\boldsymbol{r}))\right), \tag{13.21}$$

where $T(\rho(\boldsymbol{r}))$, is the kinetic energy, the second term (integral) is the electron–ion energy, and the third term, $V_{ee}$, represents all of the electron–electron interactions, including the Hartree, exchange, and correlation energies. The only constraint on the electron density is that it has to be "$N$-representable", i.e., that it can be obtained from an antisymmetric wave function, and thereby the electrons behave as indistinguishable particles. Most importantly, this electron density can be a "one-electron" density, that is, an electron density constructed from some set of one-electron functions. This is a remarkable result since it exactly transforms the many-body problem into an one-electron problem, provided that the terms in (13.21) can be evaluated. This, of course, is the difficult part because the exact forms of the kinetic energy and the electron-electron interaction terms are unknown.

The most commonly used approach is the Kohn–Sham method [13.20]. In this approach, the kinetic energy term, $T(\rho(\boldsymbol{r}))$, is replaced by the kinetic energy of a system with no electron–electron interactions, $T_s(\rho(\boldsymbol{r}))$, but at the same ground state electron density of the original system (with electron–electron interactions). In this way the (newly defined) kinetic and electron–ion interaction energies can be computed from the one-electron eigenvalues, $e_i$, of a system of noninteracting electrons moving in the new external potential, $\nu_s(\boldsymbol{r})$, of the noninteracting system

$$\begin{aligned} E_{KE}(\boldsymbol{r}(N)) &= \frac{e^2}{4\pi\varepsilon_0}\left(T_s(\rho(\boldsymbol{r})) + \int \rho(\boldsymbol{r})\nu_s(\boldsymbol{r})\mathrm{d}\boldsymbol{r}\right) \\ &= \frac{e^2}{4\pi\varepsilon_0}\sum_{1}^{n/2} 2e_i \end{aligned} \tag{13.22}$$

with $\nu_s(\boldsymbol{r}) = \nu(\boldsymbol{r}) + \int(\rho(\boldsymbol{r}'))/(|\boldsymbol{r}-\boldsymbol{r}'|)\mathrm{d}\boldsymbol{r}' + V_{XC}(\boldsymbol{r})$. The electron density is computed from the associated one-electron eigenfunctions, $\psi_i(\boldsymbol{r})$, as

$$\rho(\boldsymbol{r}) = 2\sum_{1}^{n/2} |\psi_i(\boldsymbol{r})|^2 \tag{13.23}$$

and the Hartree energy (also referred to as $J(\rho(\boldsymbol{r}))$) is computed as in the SCF-LCAO method

$$E_H(\boldsymbol{r}(N)) = J(\rho(\boldsymbol{r})) = \frac{e^2}{8\pi\varepsilon_0}\iint \frac{\rho(\boldsymbol{r})\rho(\boldsymbol{r}')}{|\boldsymbol{r}-\boldsymbol{r}'|}\mathrm{d}\boldsymbol{r}\mathrm{d}\boldsymbol{r}'. \tag{13.24}$$

This leaves only the exchange and correlation energies to be evaluated, along with the correction needed to account for the neglect of the electron–electron interactions in the kinetic energy computation. All of these terms are collected together as an effective one-electron term, referred to as the "exchange-correlation" energy, and given by

$$\begin{aligned} E_{XC}(\boldsymbol{r}(N)) &\equiv T(\rho(\boldsymbol{r})) - T_s(\rho(\boldsymbol{r})) + V_{ee}(\rho(\boldsymbol{r})) - J(\rho(\boldsymbol{r})) \\ &= \frac{e^2}{4\pi\varepsilon_0}\int V_{XC}(\rho(\boldsymbol{r}))\rho(\boldsymbol{r})\mathrm{d}\boldsymbol{r}, \end{aligned} \tag{13.25}$$

where $V_{XC}(\rho(\boldsymbol{r}))$ is the exchange-correlation potential.

In summary, the density functional method uses the electron density, rather than the wave function, as the system variable of interest. As a result, it is possible to transform exactly the many-body problem into an effective one-electron problem. In this effective potential, the kinetic energy is computed for a system of noninteracting electrons. The kinetic energy correction (due to the fact that real electrons interact) is then lumped together with the exchange and correlation interactions into an effective, one-electron exchange-correlation potential $V_{XC}(\rho(\boldsymbol{r}))$. The Hartree energy is computed explicitly as in the SCF-LCAO method.

We will conclude this short review of quantum-mechanical methods with the tight-binding method. The ideology of the empirical tight-binding method [13.21] is simple: none of the terms in (13.19) are evaluated explicitly. Instead, the Schrödinger equation (13.4) is recast into matrix form

$$[\boldsymbol{H}][\boldsymbol{C}] = [\boldsymbol{S}][\boldsymbol{C}][\boldsymbol{E}], \tag{13.26}$$

where $[\boldsymbol{E}]$ is the diagonal matrix of one-electron eigenvalues and $[\boldsymbol{C}]$ is the matrix of expansion coefficients (which define the eigenvectors); $[\boldsymbol{S}]$ is the overlap matrix, accounting for the spatial overlapping of the basis functions, whose elements are given by

$$S_{ij} = \iint \psi_i^*(\boldsymbol{r})\psi_j(\boldsymbol{r}')\mathrm{d}\boldsymbol{r}\mathrm{d}\boldsymbol{r}', \tag{13.27}$$

and $[\boldsymbol{H}]$ is the Hamiltonian matrix, whose elements are given by

$$H_{ij} = \iint \psi_i^*(\boldsymbol{r})\boldsymbol{H}\psi_j(\boldsymbol{r}')]\mathrm{d}\boldsymbol{r}\mathrm{d}\boldsymbol{r}'. \tag{13.28}$$

The elements of the Hamiltonian matrix are treated as adjustable parameters, fitted at the high symmetry points of the first Brillouin zone to either experimental information, or to the results of *ab initio* calculations for the bulk system. These parameters are then assumed to be transferable for use in computing the properties of surfaces. The range of these interactions is usually assumed to be the nearest-neighbor or the next-nearest-neighbor only, and the interaction matrix elements are assumed to have some parametric dependence upon the internuclear separation $d$ (in the crystal lattice). Commonly a $d^{-2}$ dependence for sp-bonded semiconductor systems is accepted [13.22].

The assumption of transferability (from bulk to surface) for the Hamiltonian matrix elements will be valid provided that the charge density at the surface is not significantly different from that in the bulk. The success of this assumption (see [13.23] and [13.24]) for the covalently bonded semiconductor systems is an *a posteriori* justification for its use. For the more "ionically" bonded insulating systems, this assumption should be equally valid [13.11].

The most important aspect of the empirical tight-binding method, however, is that the electron–electron interactions are never computed, but included empirically through the parametrization of the Hamiltonian matrix elements. This has consequences in the way the total energy is evaluated, since the electronic energy can only be expressed as the sum of the one-electron eigenvalues. This sum, however, overestimates the electronic energy by double-counting the Hartree and exchange interactions. Because this extra energy cannot be explicitly accounted for, it is usually lumped together with the nuclear–nuclear repulsion, and the total energy is rewritten as

$$E_{\text{tot}}(\boldsymbol{r}(N)) = E_{\text{bs}}(\boldsymbol{r}(N)) + U(\boldsymbol{r}(N)), \tag{13.29}$$

where $E_{\text{bs}}$ is the sum of the occupied one-electron eigenstates (commonly termed the "band structure" energy), and $U$ is a pair potential (i.e., a potential that depends only upon the pairwise interactions between the atoms in the system) representing the nuclear repulsion and electron-double counting terms [13.25].

Summarizing, it has to be emphasized that in the empirical tight-binding method none of the interactions in the system are computed explicitly, in contrast to the SCF-LCAO and density functional methods. Instead, they are included empirically, through the parametrization of the Hamiltonian matrix and the pair potential $U$ [13.11].

### 13.2.3 Reconstructed Surfaces; Materials-Related Examples

The first insight into the structure of crystal surfaces can be obtained by applying a simple set of chemical and physical principles. In fact, these principles can be cast as a set of five rules [13.11]:

(i) saturate the dangling bonds;
(ii) form an insulating surface;
(iii) take into consideration the kinetics of surface ordering/re-ordering processes;
(iv) form an electrically neutral surface;
(v) conserve the bond lengths.

Let us briefly comment on these rules.

As already mentioned above, the creation of the surface causes dangling bonds to appear there, that is, generates bonds which are used to bind surface atoms to their, now missing, bulk neighbors. This is an energetically unfavorable situation, because these dangling bonds are only partially filled with electrons. Therefore, a driving force exists at the surface to redistribute the dangling bond electrons (the charge density) into a more energetically favorable configuration. In general, this occurs in a way which satisfies the local chemical valences of the surface atoms.

In order to ensure an energetically favorable state of the surface, one has to eliminate the dangling bonds. This can be done by creating an "insulating" surface, i.e., by forming new bonds at the surface, either between surface atoms, or between the surface atoms and atoms of adsorbates (adatoms). This means opening a gap between the occupied and nonoccupied surface states. The surface can, however, become insulating also as a result of surface structural rearrangements, i.e., by surface relaxation or reconstruction, that transfers electrons between surface atoms. Finally, the energy of the surface can be lowered electronically, through strong electron correlation effects which also open a gap between the occupied and unoccupied surface states.

In general, and especially for the fracture surfaces of the crystalline metal oxides, the structure exhibited by any surface is dependent upon the processing history of the sample. That is, the "structure observed will be the lowest energy structure kinetically accessible under the preparation conditions" [13.24, 26, 27]. This applies to the cleavage surfaces as well, where the exhibited surface structures are "activationless" in the sense that the activation energy for the relaxation or reconstruction is less than the energy provided by the cleavage process.

The energetically most favorable way of pairing up dangling bond electrons, or equivalently to form an electrically neutral surface, is to fully occupy the anion dangling bonds and simultaneously to empty the cation dangling bonds. Alternatively this can be thought of as completely filling the valence band orbitals while completely emptying the conduction band orbitals, which is the situation characteristic for the bulk of the crystal. This process is also referred to as autocompensation.

All of the factors described above provide potential driving forces for the atoms to move away from their bulk atomic positions and lower the surface energy. If this movement creates local strain in the surface or subsurface region, which results from the distortion of the local bonding environment, then the surface energy will be raised and the movement of the surface atoms is resisted. However, not all surface strains are created in equal ways. For example, distortions of bond angles typically cost an order of magnitude less energy than distortions in near-neighbor bond lengths. Consequently, surface atomic motions which move the atoms into electronically favorable configurations, while (nearly) conserving near-neighbor bond lengths, are the most favorable motions.

In conclusion it is worth emphasizing that the topology of the surface is the factor that controls which atomic motions will be energetically favorable. Hence, it is the balance between the surface energy lowering due to the elimination of surface dangling bonds and the creation of an insulating surface, and the energy cost due to induced local strain that primarily determines nature of surface relaxation and reconstructions.

Let us now discuss some selected, materials-specific surface reconstructions, as examples of how nature orders the crystal surfaces according to the principle of minimization of the surface energy.

The best known, and probable most deeply studied materials group is the group of tetrahedrally coordinated elemental (Si, Ge) and compound (GaAs, InAs, CdSe, etc.) semiconductors [13.13]. The structure of most of these semiconductors is illustrated in Fig. 13.9, in which a ball-and-stick model of the zincblende (compound semiconductors) and diamond (elemental semiconductors) atomic geometry is given. The tetrahedral nature of the local atomic coordination is indicated in this figure by heavy lines. The typical low-index surfaces of these semiconductors are shown in Fig. 13.3 for the diamond lattice.

Among the elemental semiconductors, Si surfaces are the most thoroughly studied. This makes Si an ideal example for the present discussion. The Si(111) surface is the cleavage face of Si crystals. This surface may exhibit, however, two possible cleavage terminations. To obtain the one illustrated in Fig. 13.3, the surface is cleaved so that only one bond per surface atom is broken ("single bond scission"). As noted earlier, the truncated bulk geometry is not stable because the surface atoms relax to saturate the dangling bonds. For Si(111) this is believed to occur via the formation of two surface layers

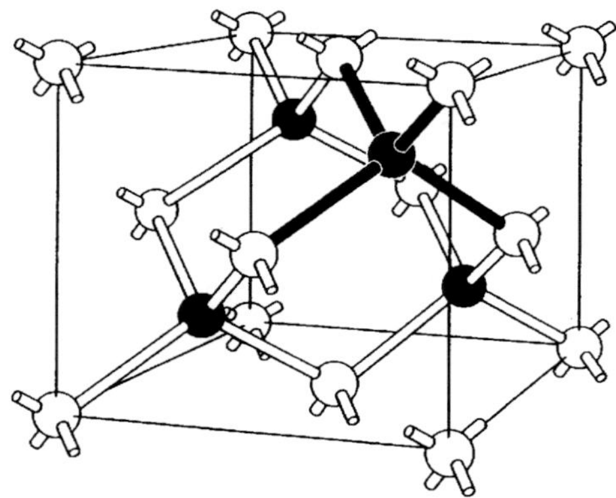

**Fig. 13.9.** Ball-and-stick model of the zincblende (ZnS) atomic geometry. Open circles represent cations (e.g., Zn atoms) and closed circles anions (e.g., S atoms). If both species are identical (e.g., C, Si, Ge), then this structure becomes the diamond atomic geometry. Balls represent atomic species and lines (i.e., "sticks") the bonds between them. The heavy shadowed lines around the second-layer atom indicate the tetrahedral coordination of the individual atomic species in these structures (taken from [13.13])

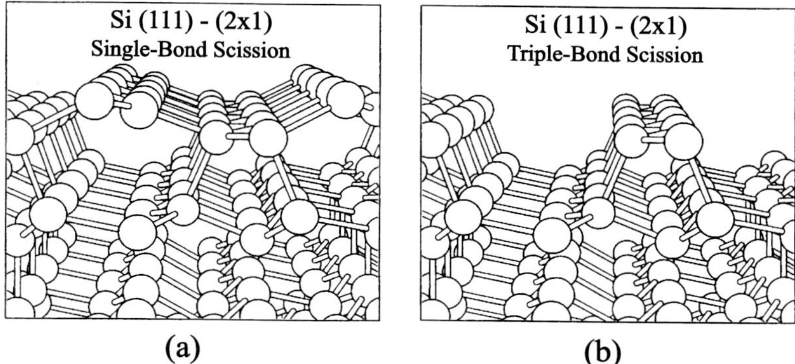


**Fig. 13.10.** Ball-and-stick model of the (2×1) $\pi$-bonded chain structure resulting from (**a**) the single-bond scission cleavage of Si, (**b**) the triple-bond scission cleavage of Si (taken from [13.13])

composed of $\pi$-bonded chains of locally $sp^2$ coordinated Si atoms as shown in Fig. 13.10a. The $sp^2$ chain is a recurring structural motif in the surface structure of tetrahedrally coordinated semiconductors. It is an example of how the surface atoms relax so that the normally tetravalent, fourfold coordinated Si atoms can satisfy their fourfold valence with threefold coordination. Such $\pi$-bonds occur in both dimer and chain structural motifs on Si surfaces, thereby permitting the "dangling" $sp^3$ electrons of the truncated bulk surface to participate in chemical bonds which stabilize the reconstructed surface. For further information on this subject the reader is referred to [13.28].

The second possible cleavage termination of the Si(111) surface has been proposed by Haneman and coworkers [13.29]. According to this work, the (111) surface cleaves by breaking three bonds per Si atom, instead of one bond, thereby uncovering a different atomic plane parallel to the (111) surface. In this case the $sp^2$ surface chain would appear as shown in Fig. 13.10b. The $sp^2$ chain is a common structural motif in both models, and by itself accounts for many features of the experimental surface characterization measurements.

Upon annealing above about 300°C, the $(2 \times 1)$ cleavage structure converts irreversibly into the $(7 \times 7)$ structure, although the precise nature of this conversion is still being discussed [13.30]. A schematic diagram of the dimer–adatom stacking fault model for this structure is given in Fig. 13.11. Three new surface motifs occur in this structure. Dimers appear along the outside edges of the unit cell. Adatoms (which are threefold coordinated) appear over some of the atoms in the second layer but not over others (called rest atoms). By the third layer all of the atoms are fourfold coordinated as in the bulk. On one half of the unit cell the atoms in the second layer lie directly over those in the fifth layer as in the wurtzite rather than the zincblende packing sequence. Thus, they constitute a stacking fault in the diamond lattice.

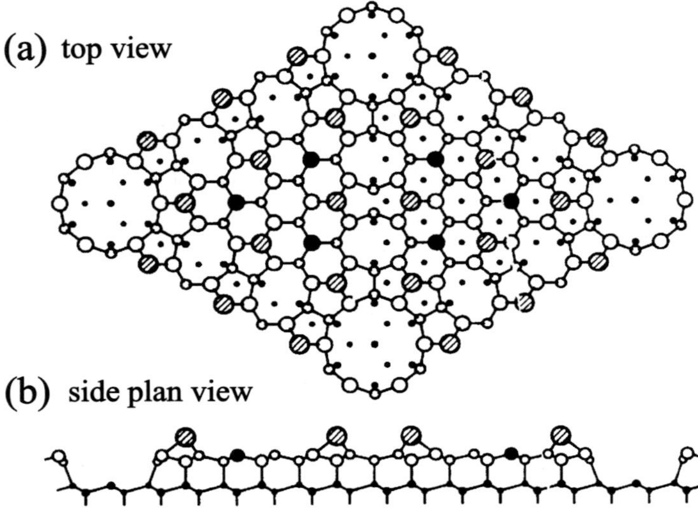

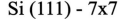

**Fig. 13.11.** Schematic illustration of the top and side views of the dimer–adatom stacking fault model of the Si(111)-(7×7) structure. The side view is given along the long diagonal of the unit cell. In the top view, the large shaded circles designate the adatoms in the top layer of the structure. The large solid circles designate "rest atoms" in the second layer which are not bonded to an adatom. Large open circles designate triply bonded atoms in this layer, whereas small open circles designate fourfold coordinated atoms in the bilayer beneath. Smaller solid circles designate atoms in the fourth and fifth bilayers from the surface. The size of all circles is proportional to the proximity to the surface. The side view is a plan view of nearest neighbor bonding in a plane normal to the surface containing the long diagonal of the surface unit cell. Smaller circles indicate atoms out of the plane of this diagonal (taken from [13.13])

The (100) surface of Si is even more important because it is a widely used substrate for microelectronic fabrication. It is also a common template for MBE. This surface is prepared by ion bombardment and annealing. It exhibits a (2×1) structure. The current consensus about the atomic geometry of this surface is that it consists of tilted dimers as shown in Fig. 13.12 [13.18]. Thus, one again finds a dimer motif as a means to achieve threefold surface coordination of a group IV atom. More on Si surface reconstructions may be found in [13.15].

Compound semiconductors represent a more complicated case with respect to surface reconstructions since the surface stoichiometry gives an additional degree of freedom in order to minimize the total energy of the surface. Thus in general a large variety of surface reconstructions is possible depending on the supply of constituents to the surface. Their densities define the

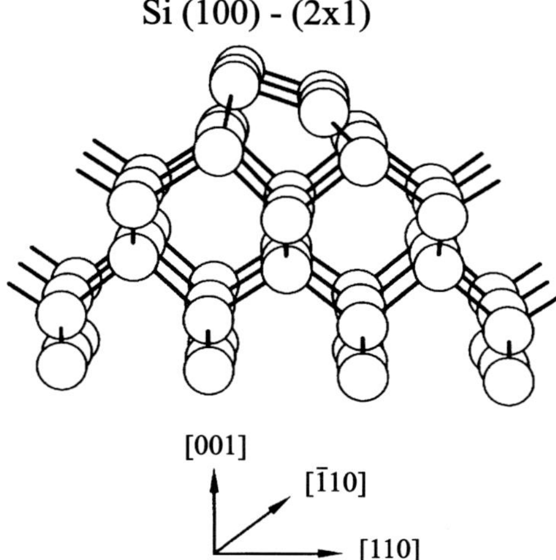


**Fig. 13.12.** Ball-and-stick model of the buckled dimer structure of the Si(100)-(2×1) surface (taken from [13.13])

chemical potential at the surface which is the driving force for the appearance of the different surface reconstructions. The best studied example in that respect is the GaAs(001) surface which is the starting surface for many electronic or optoelectronic device structures. The total surface energy has been calculated by DFT for many surface models. A summary of such results is shown for GaAs in Fig. 13.13 [13.34]. In the latter reference total energy diagrams and corresponding structural models for other III–V semiconductors can also be found. In epitaxial growth the proper surface is adjusted then by changing the chemical potential via the flux from the sources; in MBE by varying the temperature of the effusion cells, and in MOVPE via the flow rates of the precursors. Thus the starting surface for growth can be experimentally very well controlled especially when a monitoring device like RHEED or RAS is available (see Chap. 10) for excact calibration.

More complicated cases are constituted in gas phase epitaxial growth because the gas phase may contain many radicals which might adsorb on the surface and influence the surface reconstruction. Examples have been discussed in Chap. 8 where growth with TMGa at low temperatures produces new reconstructions via the adsorption of $CH_3$ or $CH_2$ (Fig. 8.28). While in this situation the growth conditions are not at all standard and the hydrocarbonated surfaces are easily recognized by RAS the case of atomic hydrogen, being always present on the growing surface in MOVPE, is more difficult to detect. However, the recent studies on the InP(001) surface grown under P-rich conditions showed that InP(001)-(2×1)/2×2) observed in diffraction could not be built up just from P (In) atoms [13.35]. Instead a InP(001)-H-(2×2) reconstruction containing two hydrogen atoms per unit cell was

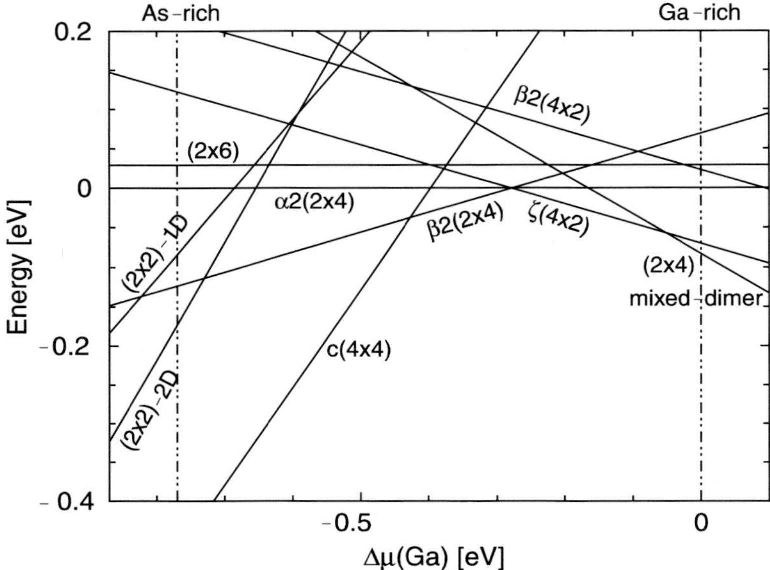


**Fig. 13.13.** Relative formation energy per (1×1) unit cell for different reconstructions of the GaAs(001) surface as a function of the Ga chemical potential. The vertical dashed lines mark the approximate As- and Ga-rich limits of the thermodynamically allowed range. (taken from [13.34] where also the structural models can be found).

suggested from DFT calculations and RAS experiments as the main reconstruction [13.36]. Such results can have major implications for understanding the many differences in physical properties observed between similar surfaces in MBE and MOVPE. One such example is the diffusion length of Ga atoms on GaAs(001) which is at least one order of magnitude larger in MOVPE than in MBE [13.37].

## 13.3 Substrate Surface Structure and the Epitaxial Growth Processes

We will discuss three growth examples on the basis of the microscopic surface structure.

### 13.3.1 GaAs(001) Homoepitaxy

The most important process influenced by the substrate surface structure is homoepitaxial growth itself. Much progress has been achieved recently towards a microscopic understanding with the example of GaAs(001) with the help of *ab initio* density functional theory in combination with kinetic Monte Carlo (KMC) simulations [13.38]. This approach can help to bridge the time scales (seconds) and length scales (micrometers) of epitaxial growth to those of the ruling microscopic processes which operate in the subnanometer and the femto- to picosecond range. The use of DFT alone of course has been hampered by the need to bridge computationally these many orders of magnitude in scales. In the KMC simulations time evolution proceeds by the discontinuous changes of the occupation of discrete lattice sites. These events may be either adsorption or desorption of molecules or the hopping process of an atom to another lattice site. The rates of the events are determined parameter free on the basis of the energies obtained from DFT calculations.

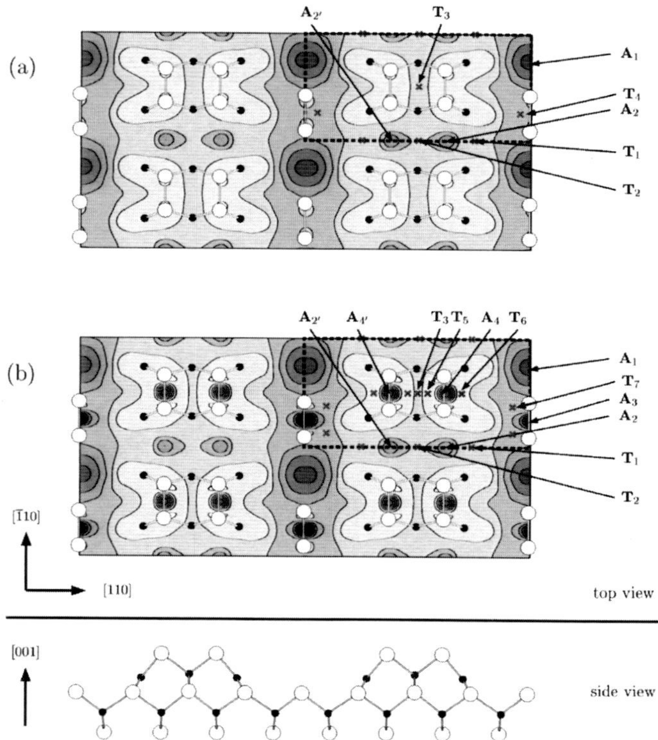


**Fig. 13.14.** Potential energy surfaces (PES) for a Ga adatom on the GaAs(001)-$\beta 2(2\times 4)$ surface. (**a**) PES obtained when the adatom is relaxed from 0.3 nm above the surface. (**b**) PES obtained when the adatom is relaxed from 0.05 nm above the surface with the surface dimers initially broken. The dahed lines represent the unit cell. The contour line spacing is 0.2 eV. The atomic positions of the clean surface are indicated for atoms of the upper two layers and for the As dimers in the third layer (As: empty circles. Ga: filled circles) (taken from [13.41])

The following events have been included in the work published by Kratzer et al. [13.38–40] on the GaAs(001)-$\beta2(2\times4)$ surface:

(i) **Ga adsorption** occurs with a sticking coefficient of unity and is thus proportional to the flux of Ga atoms to the surface.

(ii) **diffusion!Ga** is performed on the basis of potential energy surfaces [13.41] over the configurational space spanned by the adatom and the substrate atoms (Fig. 13.14 gives an example). For computational simplicity, however, only the subspace containing the lateral coordinates of the adatom is considered. The other coordinates are fully relaxed. The migration of the adsorbed adatom is then described by hopping between minima in the potential energy surface (binding site) to transitions sites (saddle points in energy), the corresponding energy difference being the activation energy. Altogether 29 different site specific hopping transitions have been included in the calculations of [13.38]. Ga hopping ($10^{-12}$ to $10^{-9}$ s) along the trench $[\bar{1}10]$ turns out to be faster than from trench to trench and is thus strongly anisotropic.

(iii) **Ga incorporation** is given through the binding energies calculated in each configuration by DFT. It is strong when As is adsorbed above the Ga atom, if it forms a Ga dimer with a neighbouring atom or if it sits in a string of Ga atoms in the $[\bar{1}10]$ direction.

(iv) $As_2$ **molecules** only stick to the surface after Ga has been deposited. Otherwise they are only weakly bounded. Because of the high binding energy of $As_2$ it is incorporated as a whole unit at sites where it can become an As surface dimer.

Figure 13.15 gives examples of adsorption at one or two Ga adatoms in the trench of the GaAs(001)-$\beta2(2\times4)$ surface. These very stable complexes are found to have life times on the surface beyond 0.1 s at temperatures below 800 K and thus have sufficient time to interact with other Ga atoms. In such a way the GaAs(001)-$\beta2(2\times4)$ surface with two dimers in the top layer per unit call is transformed locally into a GaAs(001)-$\beta(2\times4)$ with three dimers in the top layer per unit cell. These local $\beta(2\times4)$ structures (Fig. 13.16b) act then as a precursor for nucleation of the new layer (Fig. 13.16c or d).

### 13.3.2 Quantum Dots Grown on Surfaces of Different Reconstruction

We will discuss the example of the InAs/GaAs system, which is the most interesting system from the point of view of self organization during epitaxy of quantum dot arrays (see Sect. 11.5.1). The growth process of dots (3D islands) is strongly dependent on surface reconstruction of the substrate crystal, because this reconstruction reflects the configurational energy of the surface. On the other hand, the equilibrium shape of the quantum dots results from the competition between the surface and elastic energies at the dot–substrate

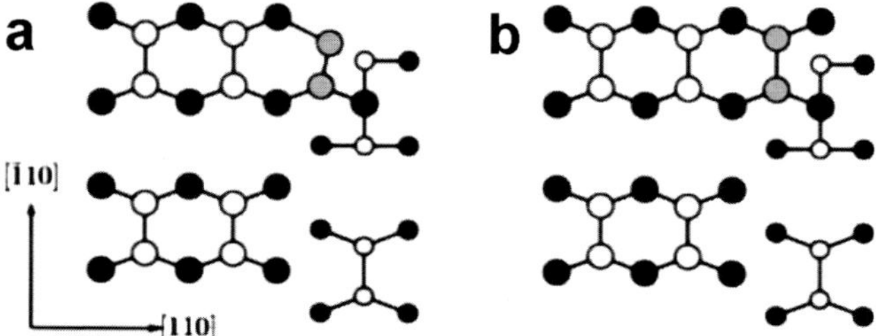


**Fig. 13.15.** (**a**)$As_2$ molecule (grey) adsorbed on a single Ga adatom that has split a trench As dimer establishing three As-Ga backbonds, thereby forming a stable Ga-As-As-$Ga_2$ complex; (**b**) $As_2$ molecule adsorbed on two Ga atoms in the trench, establishing four As-Ga backbonds and thereby forming stable $Ga_2$-As-As-$Ga_2$ complexes. As (Ga) atoms of the GaAs(001)-$\beta 2(2\times4)$ surface are shown as empty (filled) circles. (taken from [13.38])

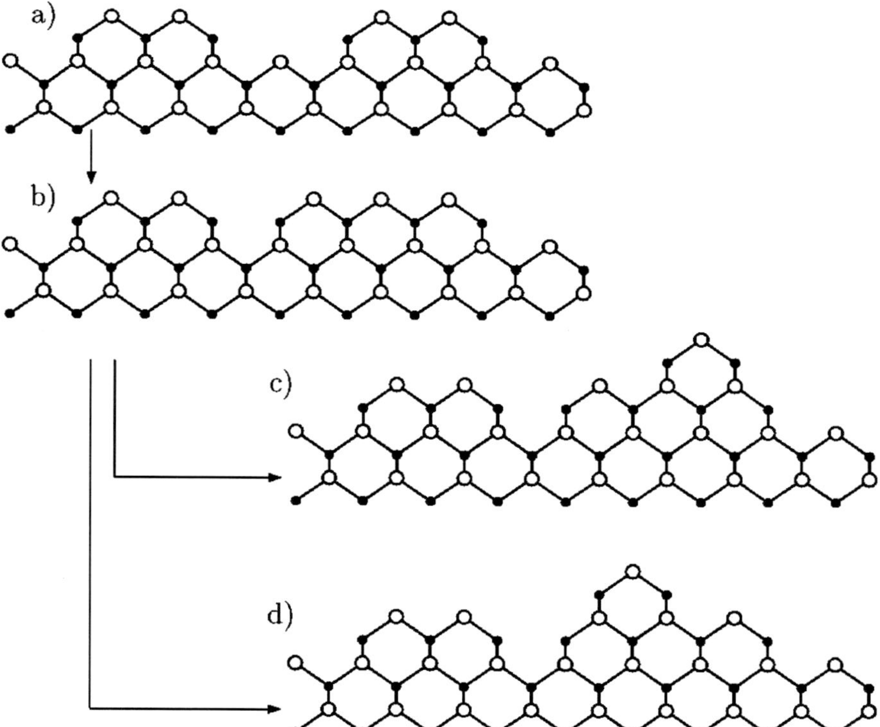


**Fig. 13.16.** Growth scenario as proposed in [13.39]: (**a**) The starting GaAs(001)-$\beta 2(2\times4)$ surface is (**b**) locally transformed into a GaAs(001)-$\beta(2\times4)$ (Fig. 13.15) from where (**c** or **d**) nucleation of a new layer takes place (taken from [13.39])

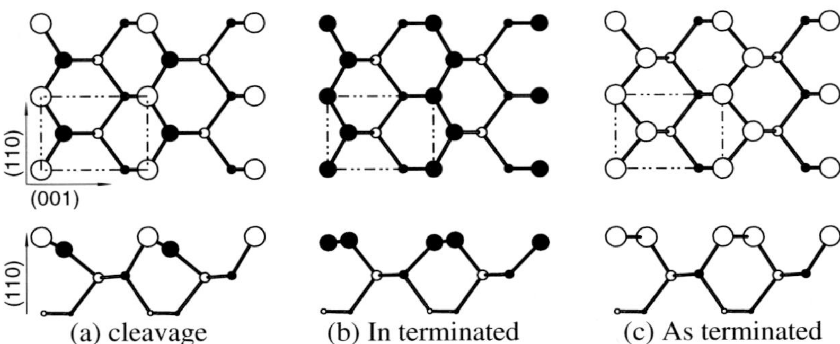


**Fig. 13.17.** Atomic structure models for the InAs(110) surface, top and side views. Open and filled circles denote As and In atoms, respectively; the size of all of the circles is proportional to their proximity to the surface (taken from [13.31])

interface (the latter energy reflects the lattice mismatch-generated stress at this interface). The theoretical analysis of the shape and stability of quantum dots in this material system, taking into consideration the different possible surface reconstructions, has been presented in [13.31].

The stable surface reconstruction is that with the lowest surface free energy. In the case of a two-component system, like surfaces of III–V compounds, also the stoichiometry of the surface enters in the energetic balance of the surface. In the optimization procedure of [13.31], leading to the equilibrium shape of the InAs dot grown on the GaAs surface, stoichiometry enters as an additional chemical degree of freedom into this procedure. The local-density approximation is applied there to the exchange-correlation energy functional by using parametrization of the correlation energy of the homogeneous electron gas. Surfaces are approximated by periodically repeated slabs (see [13.11] for details on the slab model used to simulate a surface). For example, (111)-slabs are built of stacks of In-As double layers, which are cation or anion terminated, respectively, on the opposite sides of the slab.

The simplest approach to calculate surface energies is realized by taking the same reconstruction on both sides of the slab. In that case the surface energy can directly be inferred from the total energy of the slab by subtracting the bulk energy. In the considered case of InAs dots grown on GaAs, the InAs surfaces are assumed to display surface reconstructions equivalent to those of GaAs.

For the (110) orientation in the case of InAs, shown in the atomic structure models in Fig. 13.17, always the relaxed $(1\times1)$ cleavage plane yields the lowest surface energy, independent of the As chemical potential. The As-terminated (110)-$(1\times1)$ surface, which for GaAs becomes stable under As-rich conditions, remains unstable for InAs. The relaxed cleavage surface displays the well-known outward rotation of the As atom.

For the (100) orientation, the calculations of [13.31] yield the $\beta2(2 \times 4)$

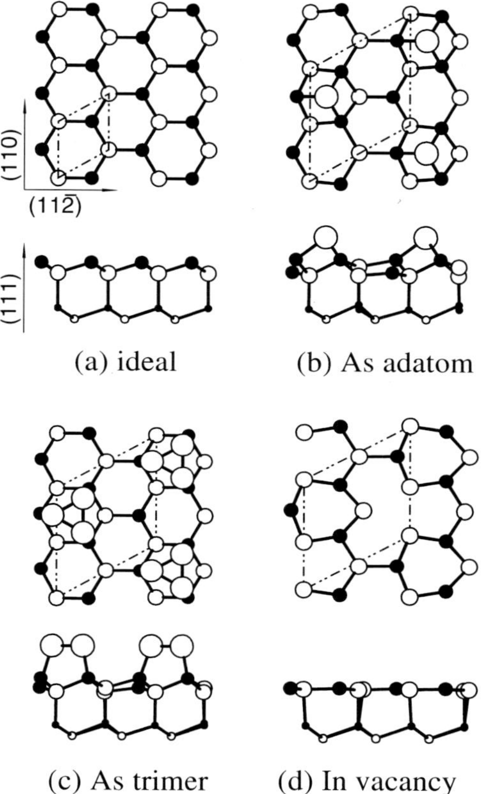


**Fig. 13.18.** Atomic structure models for the InAs(111) surface. Open and filled circles denote As and In atoms, respectively; the size of all of the circles is proportional to their proximity to the surface (taken from [13.31])

reconstruction as the energetically preferred surface structure under As-rich conditions. Again, the most As-rich c$(4 \times 4)$ reconstruction, which becomes stable in the As-rich environment in the case of GaAs, does not become stable for InAs. Experimentally the surface reconstruction has been reported to change from $(2 \times 4)$ to $(4 \times 2)$ as a function of As chemical potential.

Both for the (111) (see Fig. 13.18) and the $(\bar{1}\bar{1}\bar{1})$ (see Fig. 13.19) orientation the predicted equilibrium reconstructions are consistent with experimental data. While the As-trimer structure becomes the energetically preferred reconstruction of the GaAs(111) surface in the As-rich environment, for InAs the stoichiometric In vacancy reconstruction is stable independent of the As chemical potential and the As-trimer structure always remains unstable. For the $(\bar{1}\bar{1}\bar{1})$ orientation on the other hand both GaAs and InAs display equiv-

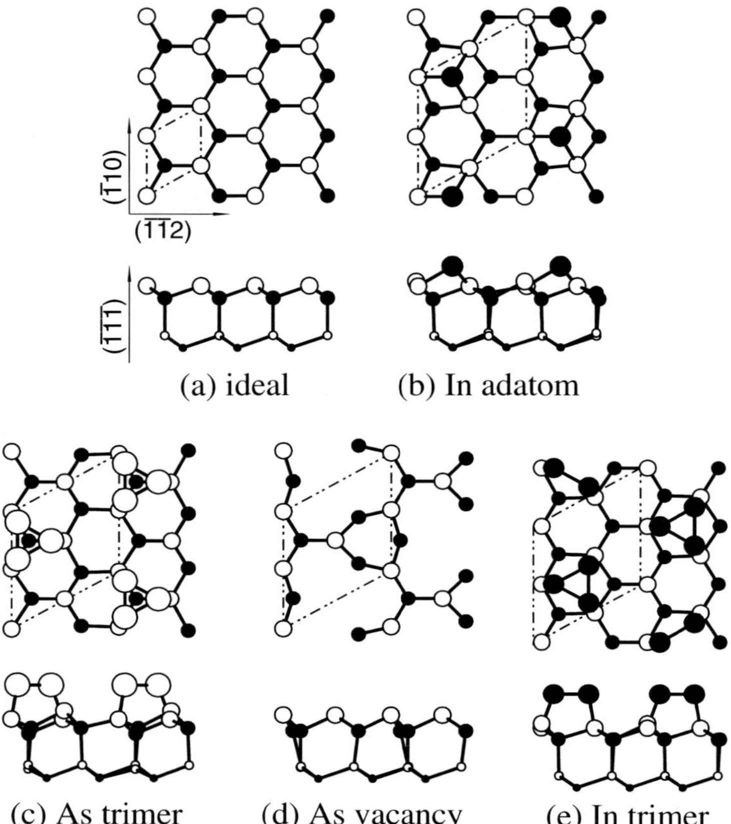


**Fig. 13.19.** Atomic structure models for the InAs $(\bar{1}\bar{1}\bar{1})$ surface. Open and filled circles denote As and In atoms, respectively; the size of all of the circles is proportional to their proximity to the surface (taken from [13.31])

alent surface reconstructions, with an As-trimer reconstruction being stable under As-rich conditions.

Since epitaxial growth of InAs quantum dots is most often performed under As-rich conditions, only the surface energies for surfaces in equilibrium with bulk arsenic have been taken into consideration when comparing the calculated dot shapes to experiment. These surface energies are listed below:

| *Orientation* | *Reconstruction* | *Surface energy (meV/Å²)* |
|---|---|---|
| (110) | $(1 \times 1)$ relaxed cleavage plane | 41 |
| (100) | $\beta(2 \times 4)$ | 44 |
| (111) | $(2 \times 2)$ In vacancy | 42 |
| $(\bar{1}\bar{1}\bar{1})$ | $(2 \times 2)$ As trimer | 36 |

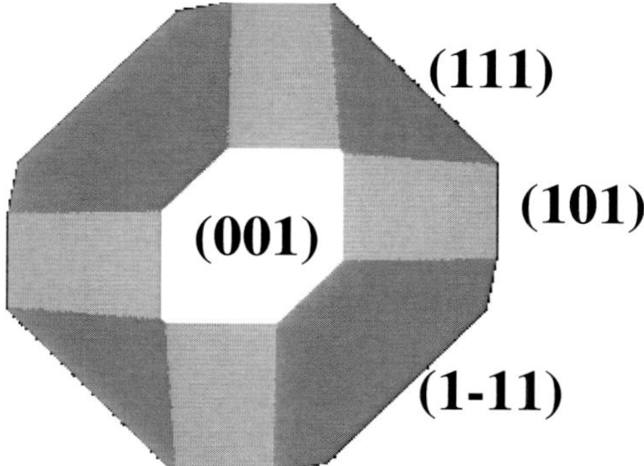


**Fig. 13.20.** Equilibrium shape of InAs quantum dots in an As-rich environment. Surfaces are labelled by their Miller indices (taken from [13.9])

The theoretical shape of the quantum dots (3D islands) of InAs calculated with these data exhibit all four orientations (see Fig. 13.20), which is in agreement with experimental results obtained when InAs dots were grown on GaAs by MOVPE and MBE [13.32, 33].

### 13.3.3 Ordering in InGaP

Like many ternary semiconductors InGaP shows the $CuPt_8$-type ordering effect on the group-III sublattice. This ordering is described by an ordering parameter $\eta$ which is zero for no ordering and 1 for complete orderinh. The ordering results in a reduction of the bandgap and a reduced bulk symmetry

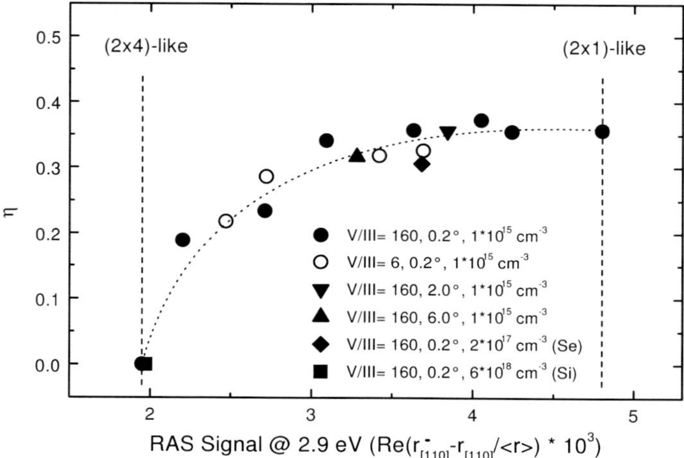


**Fig. 13.21.** Correlation of ordering parameter to the RAS amplitude at 2.9 eV (surface contribution only) which scales with the (2×1)/(2×4) surface domain ratio (taken from [13.43])

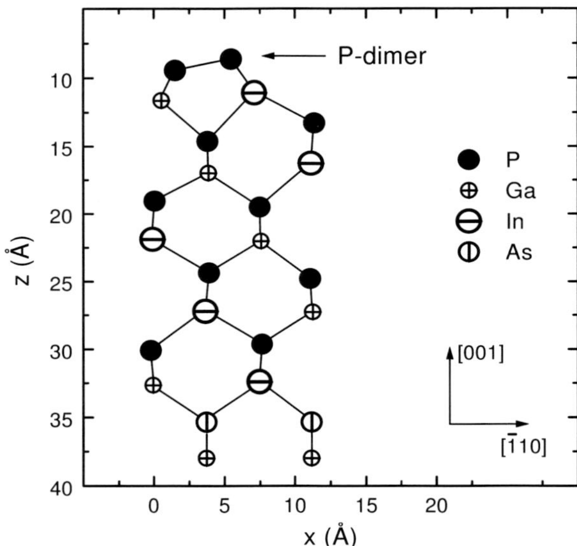


**Fig. 13.22.** Calculated atomic positions of the P-rich subsurface region of InGaP(001) using DFT-LDA based TE minimization. The resulting buckling of the dimers can be clearly seen. (taken from [13.43])

depending on the actual ordering. For technological applications these properties are not at all desired. Growth under many different conditions, fluxes and temperatures had been done in order to minmize this ordering. By measuring simultaneously the ordering and the surface reconstruction in MOVPE it was found that all the data could be mapped on one curve in a diagram of the ordering parameter versus the domain ratio of (2×1)/(2×4)[13.43] as given in Fig. 13.21. This clearly states that the surface reconstruction determines the bulk growth. Maximum ordering occurs in connection with the P-rich InGaP(001)-(2×1)-like surface and no ordering on the group-III-(2×4)-like surface. Since the former reconstruction has P-dimers on the surface while the latter has not it was concluded that the P-dimers are responsible for the ordering to occur. The buckling of the dimers gives two inequivalent positions for the group III atoms (Fig. 13.22). The DFT calculation shows moreover that the configuration with the smaller atom (Ga) below the strained dimer rows and the larger (In) atom between the dimers rows is energetically more favorable (0.24 eV per surface atom) than the opposite case (In below P dimer, Ga in between dimer rows). This case of InGaP ordering therefore constitutes a very nice example where the influence of microscopic structure on macroscopic structure can be very directly observed and moreover constitutes also a very convincing example for *ab initio* theory in describing epitaxial growth.

# Part V

# Heteroepitaxy

# 14. Heteroepitaxy; Growth Phenomena

The most frequently used, and most important, epitaxial growth process is heteroepitaxy, namely, the epitaxial growth of a layer or a thin film with a chemical composition, and usually also structural parameters, different from those of the substrate. Different aspects of heteroepitaxy have already been presented and discussed in this book. The definition of this growth process is given in Sect. 1.1, which is followed by an introduction to heterogeneous nucleation in Sect. 2.1. and a detailed presentation of interface dislocations occurring in heterostructures. A concise survey of the specific features of heteroepitaxial layers is presented in Sect. 2.3.2, while the most interesting application areas of heteroepitaxial multilayer structures are discussed in Sect. 3.1. The thermodynamics of heterointerface formation processes is described in Sect. 11.4, while the most fascinating problem of self organization in heteroepitaxial growth is discussed in detail in Sect. 11.5. Finally, the atomistic approach to heteroepitaxy, including growth on vicinal surfaces, and the important role of surface structure for heteroepitaxy, are the subjects of Chaps. 12 and 13, respectively.

Keeping in mind the above listed items related to heteroepitaxy, already presented and discussed in this book, we will concentrate in this chapter on specific features and implementations of heteroepitaxial growth phenomena. We will start with discussing heteroepitaxy of nearly lattice-matched heterostructures.

## 14.1 Nearly Lattice-Matched Heterostructures

The crucial problems of heteroepitaxy are connected with lattice mismatch, or misfit (for definition see Sect 2.3.2). These terms refer to the disregistry of the equilibrium interfacial atomic arrangements of the substrate and the overgrown epilayer. Differences in atomic spacing of lattice symmetries, which are characteristic of the two crystals in the absence of interfacial interaction between them, are responsible for this disregistry.

Lattice mismatch between the substrate and the growing film has significant effects on epitaxy. The most important are the following three:

(i) inducing structural defects (e.g. misfit dislocations) in the transition region between the substrate and the growing film,

(ii) affecting the growth morphology of the film in both 3D island and 2D layer growth modes,

(iii) influencing the epitaxial orientation of the grown film by affecting the heterogeneous nucleation process.

These effects are due to the interfacial energy which is related to the strength of bonding at the interface as well as to the degree of lattice mismatch. As the bonding strength increases, the system tends towards pseudomorphic (coherent, perfectly matched) growth [14.1] with the resulting homogeneous strain which depends on the bonding strength, the lattice mismatch, the thickness of the film, the pre-existence of dislocations, the material parameters (e.g., the elastic properties) and the growth conditions [14.2]. However, as the layer thickness increases, the homogeneous strain energy $E_H$ becomes so large that a thickness is reached when it is energetically favorable for misfit dislocations to be generated at the interface (see Fig. 2.17). The overall strain will then be reduced, but at the same time the dislocation energy $E_D$ will increase from zero to a value determined by the lattice mismatch. The existence of this characteristic thickness of the epilayer, called the critical thickness, was first indicated in the theoretical study by Frank and van der Merwe [14.3]. It has been subsequently treated theoretically by others and confirmed by various experimental observations [14.4].

If the lattice mismatch between the substrate and the growing epilayer is sufficiently small ($\leq$2%) the epilayer usually grows with a 2D, layer-by-layer mode (FM mode see Sect. 1.2) and the relaxation of the homogeneous strain arises through generation of misfit dislocations leading to the plastic relaxation of this layer. For larger mismatch a 3D, island growth mode (VW mode) is generally developed before the plastic relaxation of the layer occurs. Thus, the strain relaxation with increasing thickness of the layer is accomplished first through an elastic relaxation into coherent 3D islands and continues then for thicker layers through misfit dislocations generated in the islands [14.5]. Consequently, one may distinguish in this case two characteristic thicknesses of the growing epilayer, namely, the "transition thickness" $t_{\text{3D}}$ (for the 2D$\rightarrow$3D transition) [14.6–9] and the "critical thickness" $t_{\text{MD}}$ (for the generation of misfit dislocations), which satisfy the inequality $t_{\text{3D}} < t_{\text{MD}}$ [14.4].

One may now define the relevant lattice mismatch conditions for calling a definite heteroepitaxy pair of materials to be "nearly-lattice-matched". This is an arbitrary choice; however, we will apply here as the criterion the nonoccurrence of the $t_{\text{3D}}$ transition thickness. Consequently, we define as "nearly-lattice-matched" these heteroepitaxy pairs which are characterized by a lattice mismatch smaller than 2%.

### 14.1.1 Critical Thickness; Theoretical Treatment

In discussing the phenomenon of critical thickness of an epilayer, we will follow here the theory of Ball and van der Merwe [14.9]. The basic assumption of this theory is that the configuration of the substrate–epilayer system (the $s/e$ system (see Sect. 1.1 and Fig 1.1)) is that of minimum energy. For a particular $s/e$ system consisting of a semi-infinite substrate $A$ and an epitaxially grown layer $B$ of thickness $t$, the interfacial energy per unit area $E_I$ is given as the sum of the homogeneous strain energy $E_H$ and the dislocation energy $E_D$:

$$E_I = E_H + E_D. \tag{14.1}$$

The strain energy is defined by the interfacial shear modulus $\mu$ and Poisson's ratio $\nu$, as well as by the thickness $t$ and the strains $e_i$ in the overgrown layer. This energy may be expressed in the form

$$E_H = \frac{\mu t}{1-\nu}\left(e_1^2 + 2\nu e_1 e_2 + e_2^2\right) \tag{14.2}$$

where $e_i$ are defined by (2.2). The dislocation energy is expressed in more complicated form

$$E_D = \frac{\mu b}{4\pi(1-\nu)} \sum_{i=1}^{2} \frac{|e_i + f_{0,i}|}{\cos\gamma_i \sin\beta_i}(1-\nu\cos^2\beta_i)\ln\left(\frac{\rho R_i}{b_i}\right), \tag{14.3}$$

where $b$ is the magnitude of the Burgers vector characterizing the dislocation at the interface, $\beta$ and $\gamma$ are the angles between the Burgers vector and the dislocation line, and between the glide plane of the dislocation and the interface, respectively. The natural misfit $f_{0,i}$ between the epilayer and the substrate is defined by (2.5). The cut-off radius of the dislocation $R_i$ defines the outermost boundary of the dislocation's strain field, while $\rho$ is a numerical factor used to take the core energy of the dislocation into account (usually $\rho = 4$ [14.9]).

In some cases, on (001) interfaces the misfits and lattice parameters will be identical with respect to the two perpendicular interfacial directions [110] and $[1\bar{1}0]$ in which misfit dislocations are observed to lie [14.10]. For such a situation (14.2) and (14.3) can be simplified because the homogeneous strains $e_1$ and $e_2$ will be the same, equal to $e$. Consequently

$$E_H = 2\mu t e^2 \frac{1+\nu}{1-\nu} \tag{14.4}$$

and

$$E_D = \mu b \frac{|e + f_0|(1-\nu\cos^2\beta)}{2\pi(1-\nu)\cos\gamma\sin\beta}\ln\left(\frac{\rho R}{b}\right). \tag{14.5}$$

One has to notice that the strain energy $E_H$ is zero at zero strain ($e = 0$), while the dislocation energy $E_D$ falls to zero at $f = e + f_0 = 0$ which is the

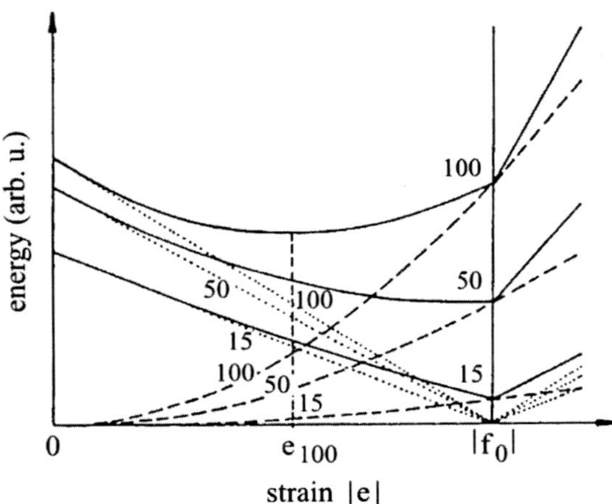


**Fig. 14.1.** Curves showing the dependence of (i) homogeneous strain energy per unit area $E_H$ (- - -); (ii) dislocation energy per unit area $E_D(\cdots)$ and (iii) interfacial energy $E_I$(——), for layer thicknesses of 100, 50 and 15 times the lattice constant of the substrate $a_s$. The presented curves correspond to a film of $GaAs_{0.90}P_{0.10}$ on GaAs with a natural misfit of $-0.36\%$ (taken from [14.9])

condition of pseudomorphism in the $s/e$ system, or of coherency of the layer with the substrate.

The actual configuration of the considered system will be given by the strain for which the interfacial energy $E_I = E_I(e)$ is a minimum [14.9]. It can be seen from Fig. 14.1 that for a film thickness of 15 and of 50 times the substrate lattice constant $a_s$, this minimizing strain is $|e| = |f_0|$, corresponding to a coherent epilayer which is free of misfit dislocations. However, in the case of a film of thickness 100 times $a_s$ the minimizing strain $e_{100}$ is somewhat smaller, indicating that misfit dislocations will be present [14.11] and spaced at intervals of

$$p_{100} = \frac{a_s}{f_0 + e_{100}}. \tag{14.6}$$

Looking at the curves in Fig. 14.1, one may conclude that when the thickness $t$ of the epilayer is small, the curve of $E_I$ has a negative slope at $|e| = |f_0|$ and this is the condition for a coherent film with no misfit dislocations. As the thickness increases, the slope of $E_I$ at $|e| = |f_0|$ increases to zero and then becomes positive. The thickness at which the slope is zero is the critical thickness $t_{MD}$, above which misfit dislocations will occur. The criterion for the critical thickness is thus the following relation

$$\frac{\partial E_I}{\partial |e|} = 0, \quad \text{evaluated at } |e| = |f_0|, \tag{14.7}$$

which leads to the expression

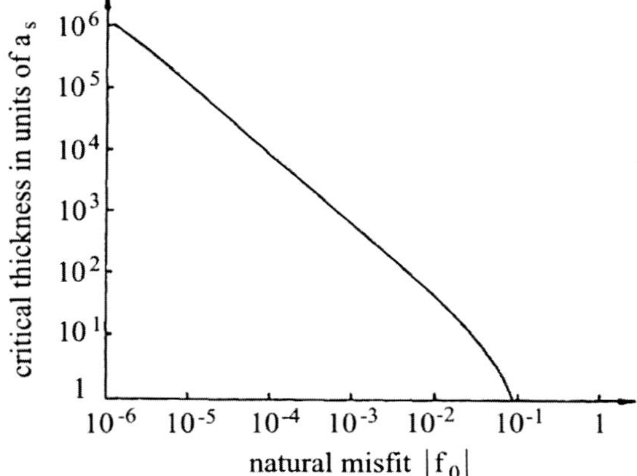


**Fig. 14.2.** Critical thickness in units of lattice parameter $a_s$ for the introduction of $60^\circ$ misfit dislocations on (111) glide planes in a (001) interface ($\nu = 0.25$) (taken from [14.9])

$$t_{\mathrm{MD}} = \frac{b(1 - \nu \cos^2 \beta)}{8\pi |f_0| (1 + \nu) \sin \beta \cos \gamma} \ln \left( \frac{\rho R}{b} \right) \tag{14.8}$$

from which $t_{\mathrm{MD}}$ may be calculated for a given natural misfit $f_0$. Since the introduction of the first dislocations is being considered, the cut-off radius $R$ can be set in this expression as equal to the thickness of the film.

Figure 14.2 shows a plot of $t_{\mathrm{MD}}$ as a function of the natural misfit $f_0$ for the case of $\nu = 0.25$ where the misfit dislocations are of the $60^\circ$ type lying on (111) glide planes in an (001) interface. Again, $\rho$ is taken as 4.

The (001) interface of diamond and sphalerite $s/e$ systems has eight different possible misfit dislocation slip systems of the $60^\circ$ type which will produce interfacial dislocations, and all of these are equivalent in the relief of misfit. The assumption was made that the misfit is relieved simultaneously in both of the interfacial directions 1 and 2 since the critical thicknesses for the $60^\circ$ dislocations along these two directions will be the same.

In the case of interfaces of lower symmetry the dislocation slip systems are generally not equivalent in the relief of misfit and each system will have its own critical thickness. The dislocations which are in fact introduced will, according to the minimum energy principle, be those with the lowest critical thickness. The critical thickness for each system may be calculated from (14.7) with $e = e_1$ and $E_I$ given by (14.1)–(14.3), and the interfacial direction 1 is perpendicular to the interfacial dislocations being considered.

The comparison between the theoretically predicted values of the critical thickness and experimental observations has been discussed by Matthews [14.12]. In metals the predicted and the experimental values agree fairly well. In some materials, however, the thickness to which coherency persists is very much larger than the predicted value. For example, germanium films on gallium arsenide are coherent to a thickness of approximately 2 μm, while the theoretical result is approximately 300 nm [14.13]. Despite such discrepancies, the calculated values of critical thicknesses can serve as a useful indication of the lower limit of the thickness at which misfit dislocations are introduced. This conclusion is justified by the fact that there is no case recorded, so far, of misfit dislocations to occur at thicknesses below the theoretically predicted critical thickness.

A comprehensive theoretical treatment of the conditions under which coherency between an epilayer and its substrate can be maintained may be found in [14.14]. In this presentation the energetics of misfit dislocations at the $s/e$ system interface is discussed in detail, and is followed by an analysis of the forces acting on dislocations in strained heterostructures. The treatment is concluded by a discussion of the kinetics of creation of misfit dislocations leading to relaxation of strain in the interface region. The mechanisms of formation of misfit dislocations and of strain relaxation, as studied by high resolution transmission electron microscopy (HRTEM), are presented in [14.15, 16] with a concise theoretical estimation.

An important item concerning the critical thickness of epitaxial strained layers is its dependence on growth orientation. This effect is caused by both the crystallographic geometric factors and the anisotropy of elastic parameters of the $s/e$ materials system [14.17]. In a standard treatment of critical thickness, these two effects reduce it in all non-{001} growth orientations except the {111} orientations. Close to (111), where the (111) slip system is inactive, the critical thickness is slightly increased to 1.23 times the {001} value for GaAs and 1.37 times for Si. The analysis presented in [14.17] also shows that the critical thickness is decreased for any off-cut away from the (001)- or (111)- growth orientations and that the off-cut directions for the smallest decreases are $\langle 100 \rangle$ and $\langle 2\bar{1}\bar{1} \rangle$, respectively.

### 14.1.2 Critical Thickness; Experimental Data

The most flexible growth technique which enables the experimental studies on the critical thickness phenomenon is MBE. The UHV environment characteristic of this technique ensures the required cleanness of the substrate surface and allows for application of surface analytical techniques like RHEED or optical reflectance spectroscopies to study the $s/e$ material system. Therefore, the presentation of experimental data concerning the critical thickness will be based on MBE experiments [14.4].

The theoretical treatment presented in Sect. 14.1.1, is based on energetic arguments similar to the Matthews–Blakeslee (MB) model [14.18]. This is the most realistic approach for making an estimate of $t_{MD}$ at thermodynamic equilibrium. However, in non equilibrium conditions, which may be achieved with MBE or MOVPE, the experimentally observed critical thicknesses of metastable pseudomorphic layers are much larger than the equilibrium values of $t_{MD}$. Moreover, these non-equilibrium critical thicknesses are strongly dependent on the growth conditions, especially the growth temperature [14.19–21].

RHEED is the experimental technique most frequently used in MBE (for a detailed description see Chap. 10) to measure *in situ* the $t_{MD}$ and the $t_{3D}$ thicknesses during a heteroepitaxial growth process. As an example of experimental data gained with RHEED the results of the work [14.6] concerning $In_xGa_{1-x}As$ epilayers grown on pseudomorphic InGaAs/InP substrates will be presented and briefly discussed.

RHEED oscillations (characteristic for 2D growth), 3D growth mode onset (deduced from intensity variations of 3D spots), and plastic relaxation onset (deduced from variations of surface lattice parameters) have been measured in this work for strained $In_{0.82}Ga_{0.12}As$ (2% lattice mismatch to the substrate) and InAs (3.2% mismatch) layers grown at a temperature of 525°C. The results are presented in Fig. 14.3. The transition from plastic relaxation through misfit dislocations to plastic relaxation through 3D island generation occurred at the composition $x_{TR}$, of the ternary compound, at which the lattice mismatch of the layer to the substrate was equal to 1.7%.

Looking at the curves presented in Fig. 14.3 one may conclude that there is competition between the two relaxation mechanisms. For strains (lattice mismatch) somewhat less than 2%, curves $t_{MD}$ and $t_{ID}$ coincide (at the composition $x_{TR}$ of the ternary compound layer), which means that a transition in relaxation mechanisms occurs. For lower strains, i.e., at compositions of the ternary compound $x < x_{TR}$, smooth surfaces of the layer growing in 2D mode and misfit dislocations at the $s/e$ interface occur. However, for higher strains, the layer grows predominantly in a 3D mode and relaxes through 3D island dislocations (the MB-type models for critical thickness calculation are no longer applicable in this case (see Fig. 14.3)). Consequently, two completely different relaxation mechanisms (through misfit dislocations or through 3D island dislocations) may occur during heteroepitaxial growth, depending on whether the growth mode is 2D or 3D. This effect explains also the temperature dependence of the critical/transition thicknesses. Namely, growing at high temperatures favors near equilibrium conditions and the appearance of a 3D growth mode for highly strained layers (the transition thickness 2D → 3D and the related relaxation of the strain through 3D islands is smaller than the critical thickness related to misfit dislocations). In contrast, lowering the temperature forces a 2D growth mode, due to a decrease of the interplane mobility of surface adatoms (relaxation of the strain occurs now at the critical thickness $t_{MD} > t_{3D}$ through misfit dislocations).

An increase in critical thicknesses for MBE growth of coherent interfaces by at least a factor of seven for the composition of InGaAs corresponding to less than 45% indium was observed in [14.22], when the growth temperature was lowered from the ordinary values for this compound of 500-600°C to 460°C. A similar effect exerted on the critical thickness by lowering the MBE growth temperature in single quantum well structures of $In_xGa_{1-x}As/GaAs$ has been observed in [14.21]. The results of the investigations performed with photoluminescence in highly strained ($0.36 \le x \le 1$) SQWs are shown in Fig. 14.4.

### 14.1.3 Epitaxy on Compliant Substrates

An interesting approach to grow pseudomorphic (coherent) structures over the critical thickness has been suggested by Lo [14.23], and then demonstrated in [14.24, 25]. The idea is to grow epitaxially strained layers on free-standing, thin substrates, what is a direct conclusion of the following arguments.

For conventional pseudomorphic heterostructures, the strain predominantly occurs in the epitaxial thin film because the substrate is too thick

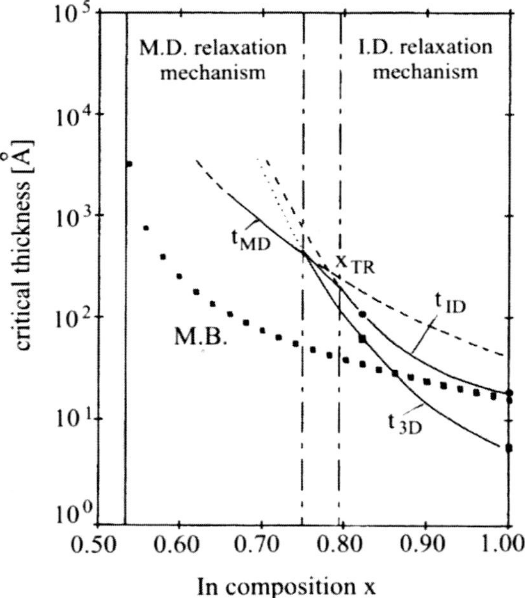


**Fig. 14.3.** Critical/transition thickness curves for InGaAs/InP corresponding to the onset of 3D growth ($t_{3D}$), plastic relaxation through misfit dislocations ($t_{MD}$) and plastic relaxation through 3D island generation ($t_{ID}$). These experimental data correspond to a MBE growth temperature of 525°C for which the transition from plastic relaxation through misfit dislocations to plastic relaxation through 3D island generation occurred at composition $x_{TR}$ of the ternary compound at which the lattice mismatch to the substrate was equal to 1.7%. The black square-dotted line gives the Matthews–Blakeslee (M.B.) equilibrium critical thickness curve which characterizes plastic relaxation through misfit dislocations generated in layers grown with a 2D growth mode (taken from [14.6])

to be compliant. However, for thin enough substrate platelets, the elastic energy is more evenly distributed between the epilayer and the substrate so that the total energy is considerably reduced. In the extreme case, if a free-standing substrate (e.g., a diaphragm structure) is used, which is thinner than the critical thickness of the substrate material, then the overall strain energy will never be large enough to generate misfit dislocations, regardless of how thick the epitaxial layer is. The easiest way to understand this argument is to consider the epitaxial layer as a "substrate" and the thin substrate as an "epitaxial layer" in the conventional sense. Since the "epitaxial layer" is thinner than the critical thickness, no misfit dislocations will occur under whatever "substrate" thickness.

To establish the quantitative relationship between the substrate thickness and the maximum allowable pseudomorphic epilayer thickness, let us consider a model structure made of two {100} semiconductors $A$ and $B$, as shown in Fig. 14.5a. Lattice mismatch for the coherent {100} interface gives rise to equal biaxial stresses ($\sigma_{11} = \sigma_{12} = \sigma$) and strains ($e_{11} = e_{12} = e$) in both materials [14.23]. The condition of mechanical equilibrium requires that the net force on the material system must be zero, which leads to the relation

$$\sigma_a h_a - \sigma_b h_b = 0. \tag{14.9}$$

The misfit strains must also partition, such that

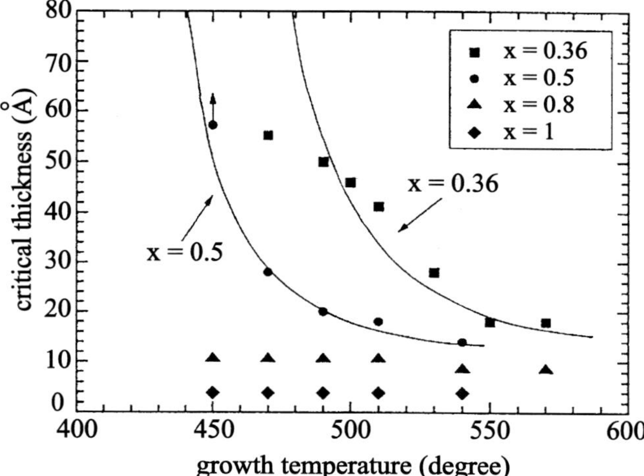


**Fig. 14.4.** The critical thickness data as a function of MBE growth temperature for four different values of the In content in InGaAs/GaAs. The value of the critical thickness for $x = 0.5$ at 450°C is larger than 55 Å as marked by the arrow. The two solid lines present the results of theoretical calculations of the critical thicknesses (taken from [14.21])

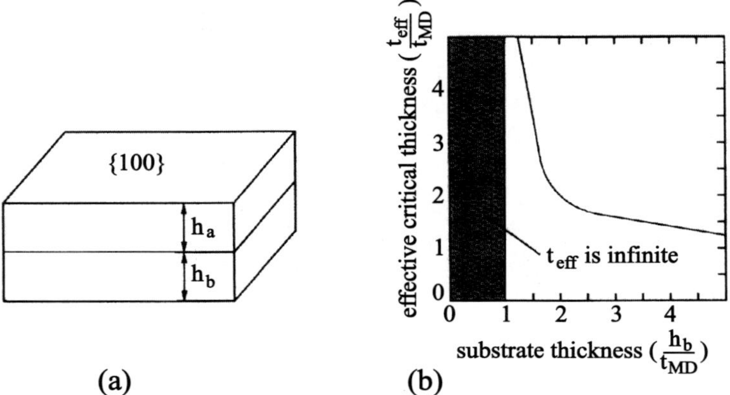


**Fig. 14.5.** Semiconductor $A$ grown on a lattice-matched thin substrate (**a**). Dependence of effective critical thickness on the substrate thickness; both axes are normalized to the critical thickness for an infinitely thick substrate (**b**) (taken from [14.23])

$$e_0 = e_a + e_b = \frac{\Delta a}{\langle a \rangle}, \tag{14.10}$$

where $h$ is the epilayer thickness and $\Delta a / \langle a \rangle$ is the lattice mismatch (see (2.5)) between $A$ and $B$. Neglecting the bending stresses, the strain energy per unit area in the system can be represented as

$$E = K_a(h_a e_a^2) + K_b(h_b e_b^2) \tag{14.11}$$

where $K_a$ and $K_b$ refer to the relevant elastic constants for the two materials. For simplicity, we assume that all the elastic constants for both materials are identical, so the subscripts of $K$ can be neglected. This is often a good approximation for most pseudomorphic structures such as InGaAs/GaAs and SiGe/Si. Combining (14.9) and (14.11), and using $\sigma = Ke$ to relate the stresses and the strains, the elastic energy can be written as

$$E = Ke_0^2 \left( \frac{h_a h_b}{h_a + h_b} \right). \tag{14.12}$$

If the strain energy reaches the critical energy $E_{\mathrm{MD}}$, misfit dislocations will be generated. For conventional structures where the substrate is much thicker than the epilayer, the critical thickness for infinite substrates is found as

$$t_{\mathrm{MD}} = \frac{E_{\mathrm{MD}}}{Ke_0^2}. \tag{14.13}$$

Defining $h_a$ as the effective critical thickness $t_{\mathrm{eff}}$, when $E$ in (14.12) is equal to $E_{\mathrm{MD}}$, one obtains the dependence of $t_{\mathrm{eff}}$ on the substrate thickness as

$$\frac{1}{t_{\mathrm{eff}}} = \frac{1}{t_{\mathrm{MD}}} - \frac{1}{h_b}. \tag{14.14}$$

If the substrate is thinner than the critical thickness ($h_b < t_{\mathrm{MD}}$), then the effective critical thickness cannot be found from (14.14). This means that the strain energy in (14.12) is never able to reach the critical strain energy $E_{\mathrm{MD}}$. Therefore, infinitely thick pseudomorphic structures can be grown without misfit dislocations. Even if the substrate is slightly thicker than $t_{\mathrm{MD}}$ a pseudomorphic layer thicker than $t_{\mathrm{MD}}$ can still be obtained. The effective critical thickness as a function of the substrate thickness is plotted in Fig. 14.5b.

Besides misfit dislocations, another critical problem for many heteroepitaxial devices is the density of threading dislocations which propagate from the interface into the epitaxial layer (see Sect. 2.2). For thick substrates, misfit dislocations at the interface always experience a pulling force attracting them towards the film surface. Consequently, misfit dislocations have a tendency to move to the surface as threading dislocations by climbing or slipping. This is the effect of image force, similar to the image force experienced by a charged particle close to a metal boundary. However, if a sufficiently thin substrate is used, the image force tends to pull the interface dislocations away from the film and into the substrate. This effect can be considered as dislocation gettering, similar to the point defect gettering where defects are diffused away from the surface layer. The dislocation gettering effect by thin substrates has great promise in improving the quality of heteroepitaxy.

Concluding these considerations one may state that the use of free-standing, thin substrates can improve the heteroepitaxial, pseudomorphic materials by two effects, the increase of effective critical thickness and the gettering of threading dislocations.

On the way to practical realization of pseudomorphic structures grown on free-standing substrates, Teng and Lo [14.24] have suggested, on the basis of a dynamic strain analysis of the critical thickness, the use of semiconductor membranes as compliant substrates. In the example of the InGaAs/GaAs material system they have shown that a membrane should be thinner than 120 nm for a strain of 1% and can be 1 μm thick for a strain of 0.5%. Such thicknesses can be achieved by existing semiconductor device technology. The practical realization of the pseudomorphic MBE growth on a semiconductor membrane has been demonstrated by fabricating first a 80 nm thick compliant platform of GaAs and then growing on it an exceedingly thick, high-quality pseudomorphic $\mathrm{In_{0.14}Ga_{0.86}As}$ layer. The layer that has been grown on this membrane exceeded its usual critical thickness by about 20 times without strain relaxation [14.25]. Moreover, X-ray analysis confirmed a shift in the InGaAs peaks, indicating an unrelaxed strain of 0.9%, while atomic force microscopy profiles verified that the layer grown on a compliant substrate is much smoother than the layers grown on ordinary plain substrates.

An example of a bench-like compliant GaAs platform prepared for MBE regrowth with InGaAs pseudomorphic film is shown in Fig. 14.6 [14.25]. To create this platform, first an MBE sample consisting of a 100 nm thick $\mathrm{Al_{0.8}Ga_{0.2}As}$ layer followed by an 80 nm GaAs layer has been grown on a

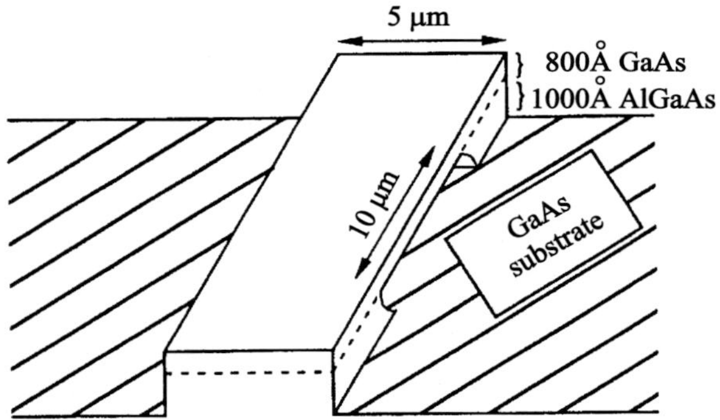


**Fig. 14.6.** Schematic illustration of a bench-like compliant GaAs platform prepared for MBE regrowth (taken from [14.25])

plain standard GaAs substrate. Then photolithography was used to define 5 μm-wide stripe patterns that are separated by a center-to-center distance of 10 μm. Raised mesas were then created by etching through both the 80 nm GaAs top layer and the 100 nm AlGaAs layer. After forming the mesas, a second layer stripe pattern running perpendicular to the mesa stripes was defined. The bench-like platform illustrated in Fig. 14.6 resulted from selectively undercutting the unprotected AlGaAs with $1\,HF/5\,H_2O$ etching solutions. This solution was gently agitated to allow $H_2$ reaction products to escape. The HF solution also enriched and passivated the GaAs surface with As, thus preparing the surface for MBE regrowth.

The idea of using compliant substrates in heteroepitaxial growth has received wide attention because of its potential to overcome the serious problems related to growing pseudomorphic heterostructures of lattice mismatched materials. This idea, also called "lattice engineered compliant substrates" [14.26] allows for pseudomorphic growth of any III–V group or related semiconductor compound on a universal substrate like Si or GaAs [14.27]. It is worth the noticing that MBE growth technique is not the only way of creating applications of compliant substrates in heteroepitaxial technology. In a report concerning growth of InGaAs multi quantum well structures with a lattice mismatch to GaAs compliant substrate equal to 1.5%, the MOVPE growth technique was applied with success [14.28]. The compliant GaAs substrate used in this work consisted of a 3 nm GaAs layer weakly bonded on a bulk GaAs substrate with a 22° angle between their [110] axes (the fabrication process of such twist-bonded compliant substrates has been described and discussed in detail in [14.29]). The basic idea was to have the thin seed layer behaving like a free-standing membrane and being sufficiently thin to absorb the overall strain energy of a thick deposited mismatched layer without generating any structural defects in this layer.

Theoretical work devoted to growth on compliant substrates has been published recently. A phenomenological mean-field theory is presented in [14.30] for the kinetics of strain relaxation due to misfit dislocation generation in the strained-layer growth of epitaxial semiconductor films on thin compliant substrates. This theory provides a generalized dislocation kinetic framework by coupling the mechanics of an epitaxial film on a compliant substrate with a standard description of plastic deformation dynamics in semiconductor bulk crystals. It has been shown that the theoretical results gained with this theory reproduce successfully the experimental data for strain relaxation in the InAs/GaAs(110) heteroepitaxial system (for this heterosystem high compressive strain of 6.64% is characteristic).

At the end of 1999 a new, so-called paramorphic approach to growing high quality, fully relaxed heteroepitaxial layers on compliant substrates has been demonstrated, in which surface micromachining is involved in the fabrication process [14.31]. Thin and thick fully relaxed $In_{0.65}Ga_{0.35}As$ layers have been grown on InP substrates (0.81% lattice mismatch) with high structural and high optoelectronic quality at an operating wavelength of photodetectors of $\approx 2.0\,\mu m$. Full relaxation has been achieved using the paramorphic approach by growing $In_{0.65}Ga_{0.35}As$ layers lattice matched to an $InAs_{0.25}P_{0.75}$ seed membrane of predetermined lattice parameter. The $InAs_{0.25}P_{0.75}$ layer was originally grown pseudomorphically strained on the InP substrate before being separated and elastically relaxed using a surface micromachining procedure to prepare seed platforms with sizes ranging from $40\times40$ to $300\times300\,\mu m^2$.

### 14.1.4 Highly Strained Heterostructures

When the natural lattice mismatch of the heteroepitaxy-pair materials exceeds 2%, then the epitaxially grown pseudomorphic structures are usually called highly strained heterostructures. Among the III–V semiconductors the $In_xGa_{1-x}As$/GaAs(001) system (lattice mismatch ranges from 0 to 7.2%) may serve as an example of such $s/e$ material system. It is still intensively studied for its potential application in both microwave and optoelectronic device fabrication [14.9]. The Si/SiGe system, very important for microelectronic device technology, is the next example with the lattice mismatch which may reach 4.17% for the extreme case of the Si/Ge pair [14.32]. Finally, let us mention the very important material systems, creating the basis for blue optoelectronics namely, the (Ga,Al,In)-nitrides [14.33]. Here, the lattice mismatch reaches 3.5% in the case of the AlN/GaN pair.

One of the characteristic features of the $s/e$ materials with high lattice mismatch is the strain relaxation mechanism, which in this case occurs through generation of 3D islands. Thus, the characteristic layer thickness for heteroepitaxy of the highly strained heterostructures is the transition thickness $t_{3D}$. A complete model for strained epitaxial island growth has been given in [14.34]. This model, assuming linear elastic behavior, has been used to analyze an isolated arc shaped island with elastic properties similar to those of the substrate. The finite element analysis performed in this work has shown that in order to minimize the total energy, which consists of strain energy, surface energy, and film/substrate interface energy, a coherently strained island will adopt a particular height-to-width aspect ratio that is a function of only the island volume. Consequently, for an island with volume greater than a certain critical size, the inclusion of a mismatch strain relieving edge dislocation is favorable. The proposed criterion for the critical size is based on a comparison of the configurational force acting on the edge of the island in the presence of an edge dislocation. A finite element calculation combined with an analytical treatment of the singular dislocation fields has been used to determine the minimum energy island aspect ratio for the dislocated island/substrate system.

### 14.1.5 Surfactant-Mediated Heteroepitaxy

One of the main goals of current research in the field of highly strained heterostructures is to increase the transition thickness $t_{3\mathrm{D}}$ below which the layer-by-layer growth can be maintained. A reasonable way to avoid the formation of 3D islands is to use the kinetic limitations in order to prevent the $s/e$ system from reaching equilibrium. In other words, this means that the mass transport at the substrate surface, and thus the surface diffusion length, should be reduced. Usually, the surface diffusion length, which can be written as $\lambda = \sqrt{D\tau}$, where $\tau$ is the mean residence time of atoms at the surface and $D$ is the surface diffusion coefficient, has been reduced through the decrease of $D$ by lowering the growth temperature. However, this in turn may have a deleterious effect on the device quality of the epitaxially grown material. Another approach, which seems to be very promising, is the use of surfactants [14.35–39].

One of the most frequently grown heterostructure for device application is the $A/B/A$ heterostructure, in which material $B$ is embedded in the material $A$. This structure occurs in quantum wells, superlattices and other low dimensional heterostructures (see Sects. 3.1 and 3.2). In growing an embedded layer, first a heterolayer must be formed, followed by a capping layer of the substrate species. Unfortunately, only one of these two species can have the lowest surface free energy of the two. Thus, when growing an embedded layer either the growth mode of the heterolayer, or the capping layer will be the VW-mode (immediate islanding) [14.37,38]. This is so, because if material $A$ wets material $B$, $B$ will not wet $A$. Any attempt at growing an $A/B/A$ heterostructure must overcome this fundamental obstacle.

In their pioneering work concerning this item Copel et al. [14.35] proposed the use of a surface-active species, the so-called surfactants, to reduce the surface free energies for $A$ and $B$ and suppress island formation, as demonstrated in the growth of Si/Ge/Si(001) with a monolayer of As. The idea was to gain control of growth by manipulation of surface energetics which should provide a new avenue to achieve high-quality man-made microstructures against thermodynamic odds.

Theoretical models leading to the relations (12.4) (e.g., Kossel's model) suggest that the growth mode of the epilayer is determined by the free energy of the substrate surface $E_s$, the interface free energy $E_I$, and the surface free energy of the epilayer $E_F$, neglecting the strain energy of the film. The inequality

$$E_s > E_F + E_I \tag{14.15}$$

sets the condition for the epitaxial film to wet the substrate surface. In this case the FM-mode of epitaxy may occur. If the inequality has the opposite sign, one usually obtains the VW-mode of growth, i.e., no wetting of the substrate. The SK-mode of growth generally occurs when there is wetting of the substrate but the overlayer's strain is unfavorable, or when there is the added complication of interface mixing (interdiffusion effect) and/or surface reconstruction [14.35]. For the case of the heterostructure Si/Ge/Si, Ge grows on Si in the SK-mode, and Si grows on both Ge and Ge-on-Si in the VW-mode [14.37].

A substantial modification of the film growth may be obtained by introducing a third element which lowers the surface free energy of both Ge and Si. In this case, surface segregation of the third element, the surfactant, is strongly favored during growth. As a result, islanding of the film will be kinetically inhibited. This effect has been achieved in [14.35] through the passivation of the Si(001) surface with one monolayer of As prior to growth. The As atoms of the passivating layer, which contain one extra valence electron per surface atom, fill the dangling bonds which normally occur on the clean Si(001) and Ge(001) surfaces, thereby creating a stable termination [14.40]. One has to notice that both clean Si(001) and As-capped Si(001) have (1×2) unit cells caused by the formation of Si or As dimers (see Chap. 13). However, the difference between growth on the clean and As-capped surfaces is not due to the presence or absence of a reconstruction, but to the energetically favored filling of dangling bonds. By using the As-passivated surface as a stage for epitaxial growth, one is able to alter the growth mode to induce wetting of the substrate. Since As segregates to the surface during growth, the structure itself incorporates relatively small quantities of As.

Two possible mechanisms have been proposed in [14.37] for an explanation of surfactant-mediated heteroepitaxy. The first is a dynamic mechanism based on enhanced incorporation of the growth species, while the second is a static one, based on the stress of a chemisorbed layer.

The effect of a surfactant on surface dynamics may be understood on an intuitive basis. Since dopants are energetically driven to segregate towards the surface, the surfactant will drive, during the growth, any incoming species to a subsurface site. Once the species is incorporated into the crystal lattice of the growing layer, its mobility is severely diminished. Energy-minimization calculations based on norm-conserving pseudopotentials verify the strong tendency of As to segregate. A comparison of (001) slab energies for ...Si/Ge/Si/As and ...Si/Ge/As/Si show a difference of 2.3 eV per dimer favoring the segregation of As, namely, the ...Si/Ge/Si/As structure. Likewise, a comparison of ...Si/Ge/As and ...Si/As/Ge favors As termination (the Si/Ge/As structure) by 1.4 eV per dimer [14.37]. In view of such large energy differences, it is reasonable to suppose that site exchange between the surfactant and the Si or Ge atoms occurs rapidly enough to alter the growth dynamics. In fact there is a profound difference between the dynamics of growth with and without a surfactant.

Let us imagine an atom condensing from the vapor phase onto a clean surface. In the first moment, or few moments, this atom exists as an isolated surface adatom, which is free to migrate until it reaches a step or some other defect site. There is, in fact, little to prevent the adatom from moving on the substrate surface at the growth temperature. However, in the presence of a surfactant atom on the surface, there may be some period during which an adatom is quite mobile, but once the adatom becomes incorporated into a site underneath the surfactant atom, the diffusion of it becomes strongly reduced. The incorporation into a subsurface site may not require the presence of a step or a lattice defect, but only a site exchange with the surfactant atom.

The second mechanism is based on the calculation of the stress of clean and chemisorbed (111) Si and Ge surfaces [14.41]. The results are summarized in Table 14.1. They show that both clean and Ge-covered Si(111) surfaces involve significant components of compressive stress. On the other hand, As chemisorbed surfaces involve even more substantial tensile stress. The origin of the tensile stress is, however, not lattice misfit, but the result of the bond-angle distortion.

Experimental measurements of sample bending caused by film growth on Si(001) confirm the trends predicted in [14.41], supporting the idea that an adsorbed layer of group-V atoms may compensate for the stress of a heavily compressed Ge film [14.42]. A shortcoming of this mechanism is that it does

**Table 14.1.** Stress due to clean and adsorbed surfaces of group-IV semiconductors [14.37]. Results for (111) surfaces are from theory [14.41]. Results for Si (001) are from experimental work, and represent values relative to the clean surface [14.42]

| Surface | Stress | Direction |
|---|---|---|
| Si (111) | −0.54 | Compressive |
| Si (111)/Ge | −4.45 | Compressive |
| Si (111)/Ge | −1.12 | Compressive |
| Si (111)/As | 2.27 | Tensile |
| Ge(111) | −0.73 | Compressive |
| Ge(111)/As | 2.64 | Tensile |
| Si (100)/Ge | −0.73 | Compressive |
| Si (100)/As | 1.3 | Tensile |

not predict a change in growth mode, only an increase in the thickness that can be achieved before islanding. Therefore, it seems to be reasonable to suggest that both mechanisms may assist in surfactant-mediated epitaxy.

A good test of the importance of surface stress would be to grow in the presence of a group-III adlayer, where compressive stress is predicted for both the heterolayer and the surfactant. This problem has been addressed and extensively discussed in [14.43]. A theoretical analysis for the system Ga/Si(111)/Ge(111) using mainly thermodynamic arguments has been presented there for explaining the role of the column-III surfactant (Ga) in initiating a transition from 3D growth mode into 2D mode in Si-on-Ge epitaxy. It has been shown that in equilibrium conditions the adsorption of only 1/10 monolayer of Ga on the Si(111) surface is enough to initiate this 3D$\rightarrow$2D transition.

The reverse problem of Ge-on-Si epitaxy with Ga as a surfactant has been investigated in [14.44]. A TEM image of the grown structure has revealed that the Ge epilayer consists of two domains. The first, denoted as A-type Ge, exhibits the same orientation as the substrate, while the second, denoted as B-type Ge, is rotated by 180° with respect to the substrate. This effect has been explained by the following argumentation. The key to the nucleation of both A- and B-type Ge is the bonding structure of Ga atoms. In the reconstructed Si(111)-(6.3×6.3) surface (the substrate for the epitaxial growth) the surfactant Ga atoms occupy substitutional sites in the top half of the Si double layer (see [14.4]). There, Ga atoms may be bonded either to A-type or to B-type Si bonds which gives rise to the nucleation of A- and B-type domains of the Ge epilayer [14.44].

Growth with surfactants is usually accomplished with MBE [14.45–48]. This is a consequence of the indispensable condition for surfactant-mediated epitaxy, i.e., the feasibility of a precise control of surface coverage in the monolayer range during the growth process. Sb, As and Ga have been found to be the elements best suited as surfactants in group-IV elemental and compound epitaxy, while tellurium acted well as a surfactant in growth of highly strained III–V compounds. A lot of papers have been devoted to theoretical and experimental studies on surfactant-mediated epitaxy. The reader is referred to [14.4] for more detailed information on this subject.

### 14.1.6 Heteroepitaxial Lateral Overgrowth

One of the advantages of epitaxy is that the substrate surface can be given any desired crystallographic orientation and any misorientation (off- orientation). Besides, the substrate surface can be chosen planar or non-planar, which enables epitaxial crystallization in different directions, e.g., perpendicular or lateral in relation to the substrate surface. Moreover, the substrate surface can be partially masked, which gives the possibility of growing epitaxially only on some selected areas of this surface.

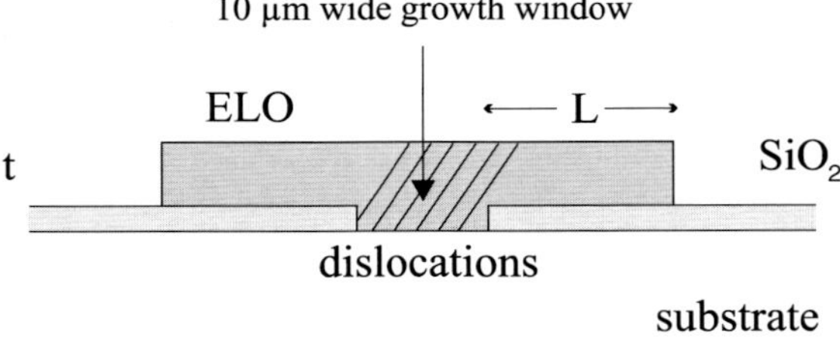


**Fig. 14.7.** Schematic illustration of geometry and principle of the ELO growth process and the characteristic phenomenon of substrate defect filtration (taken from [14.49])

Making use of the latter advantage of epitaxy, a very important epitaxial growth technique has been invented, which is known as "epitaxial lateral overgrowth" (ELO). The principle of the ELO technique is shown schematically in Fig. 14.7 [14.49]. The substrate is covered by a thin $SiO_2$ film and patterned by conventional photolithography to form on its whole area a grating of oxide-free seeding windows. Then, an epitaxial layer is grown on such substrate. It nucleates on the seeds and epitaxy proceeds in the direction normal to the substrate surface. Next the growth in lateral directions over the oxide film starts when the crystallization front exceeds the top surface of the oxide. This is partly due to a strong anisotropy in growth rate at fairly low growth temperatures [14.50].

The key feature of the ELO growth technique is the creation of a buffer layer which filtrates defects of the substrate. It is known from many experiments that dislocations present in the substrate cannot penetrate through the polycrystalline $SiO_2$ film. Consequently, if dislocations from the substrate can penetrate the epilayer, then they will be present in the epilayer only in the areas grown vertically over the seeds (see Fig. 14.7). This means, however, that a significant reduction of the average density of defects in the epilayer can be expected in the parts which grow laterally over the oxide mask stripes. Finally, when the horizontal dimension $L$ of the overgrown epilayer becomes larger than the width of the growth window, then the structural quality of these parts of the epilayer that cover the oxide areas may be extremely high (average density of structural defects nearly zero can be expected). It is important to notice that for a successful ELO process nucleation of the epilayer material on the oxide film must be avoided.

The ELO technique has been applied with success for growing different materials, e.g., Si, GaAs, GaP, and GaN, by LPE, CVD or MOVPE. Especially well suited for ELO is LPE [14.49–53], which enables a precise control of the liquid solution supersaturation at the growing interface. Thus, the nucleation of the epilayer on the oxide surface can easily be avoided. Moreover, this also allows to get a high lateral to vertical growth velocity ratio, without any polycrystalline growth on the oxide surface [14.52]. ELO may be performed as a homoepitaxial growth process [14.49–51, 54], as well as a heteroepitaxial process [14.55–57]. In the latter growth mode usually the heteroepitaxial substrate (e.g. SiC) is first covered by a standard epitaxial process with a thin buffer layer of the heteroepilayer material (e.g. GaN), which is subsequently covered by a patterned masking oxide and overgrown by homoepitaxial ELO process with the heteroepilayer material (e.g. GaN). In this way, the growth of heteroepitaxial (in relation to the thick substrate), highly strained epilayers, which are structurally perfect, is possible through ELO [14.58] (compare this with the model picture shown in Fig. 1.1d).

ELO has been the main ingredient to the success of the GaN technology [14.59]. In order to be more precise, let us discuss the ELO in the case of growth of GaN on the GaN(0001)-on-$Al_2O_3$(0001) substrates, according to the experimental data given in [14.60]. Selective area growth (SAG) of GaN was performed in this work on 1.5 μm thick undoped GaN-on-sapphire substrates. A 70 nm thick $SiO_2$ layer was deposited as mask material by RF sputtering. Patterning of the mask was carried out by laser holography (He-Cd laser; 422 nm wavelength) and $NH_4HF_2$ etching to form two types of line patterned windows; one parallel to the $<1\bar{1}00>$ and the second parallel to the $<11\bar{2}0>$ axes of the GaN substrate. The line patterns were aligned periodically at 1 μm between. SAG of GaN was performed at atmospheric pressure by MOVPE. The source gases were trimethylgallium (TMGa) and ammonia ($NH_3$) with hydrogen as the carrier gas. The growth temperature was 1070°C, and the growth time was varied from 2 to 30 min. Figure 14.8 and 14.9 show SEM images of GaN grown on the line patterns along the $<11\bar{2}0>$ and along the $<11\bar{2}0>$ directions, respectively. In the first case submicron line structures of GaN were obtained with a (0001) facet on the top and $\{1\bar{1}01\}$ facets on the sides at the growth time of 2 min, as shown on Fig. 14.8a. The area of the (0001) facet decreased with increasing growth time from 2 to 7 min, as shown in Fig. 14.8b. For the further growth time, the neighboring submicron line structures of GaN began coalescing with each other and at the growth time of 30 min a buried structure of periodically aligned $SiO_2$ masks overgrown with GaN, namely an ELO GaN layer was obtained, as shown in Fig. 14.8c. The surface of the GaN layer is uniform and grain boundaries due to the coalescence are not observed.

In the second case, submicron line structures of GaN with a (0001) facet on the top are observed at the growth time of 2 min, but $\{1\bar{1}01\}$ facets on the sides are not clear, as shown in Fig. 14.9a. Because of the lateral overgrowth on the $SiO_2$ mask the area of the (0001) facet increases at the growth time from 2 to 7 min, as shown in Fig. 14.9b. For the further growth time, the neighboring submicron line structures of GaN coalesce and an ELO GaN layer was obtained at the growth time of 30 min, as shown in Fig. 14.8c.

The process of the lateral overgrowth in the described cases is illustrated schematically in Fig. 14.10. The area of the (0001) facet on the top of the GaN line structures grown on the $<11\bar{2}0>$ line pattern decreased with increasing

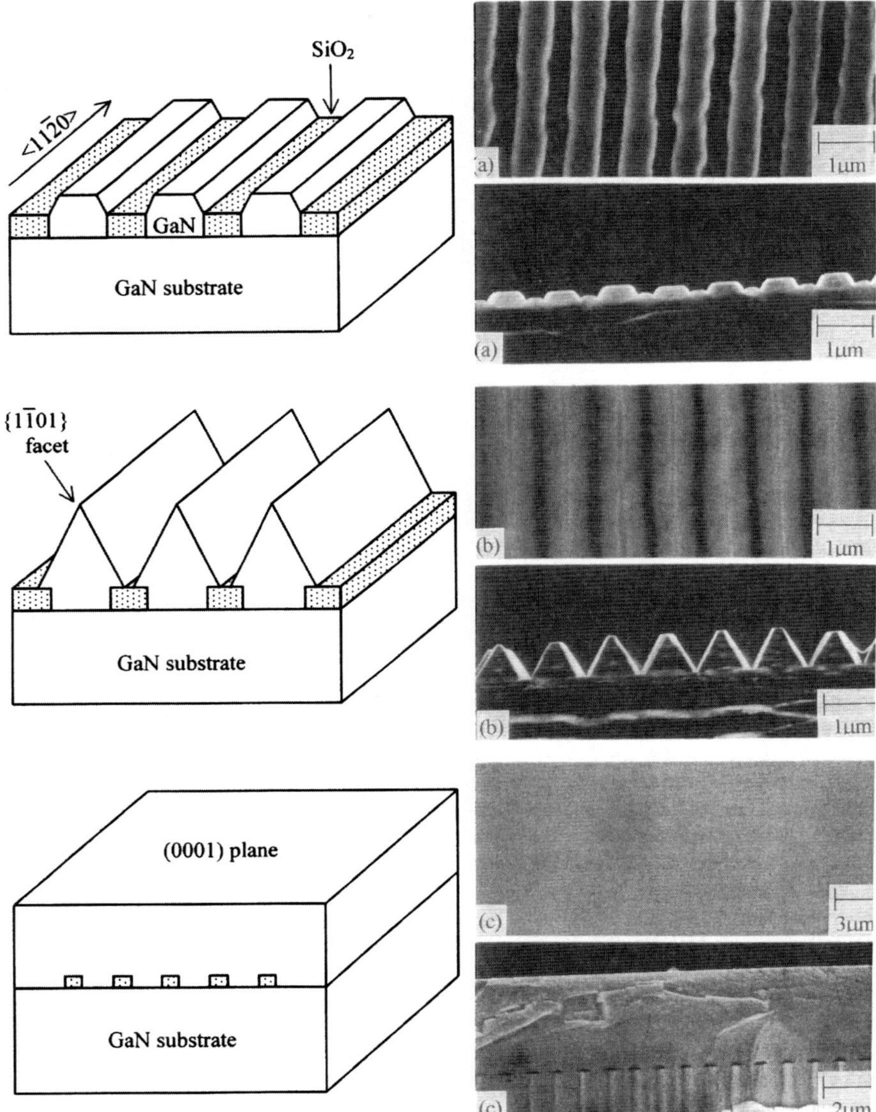


**Fig. 14.8.** Schematic illustration as well as top and cross-sectional SEM images of GaN grown on the <11$\bar{2}$0> line pattern at the growth times of 2 min (**a**), 7 min (**b**) and 30 min (**c**) (taken from [14.60])

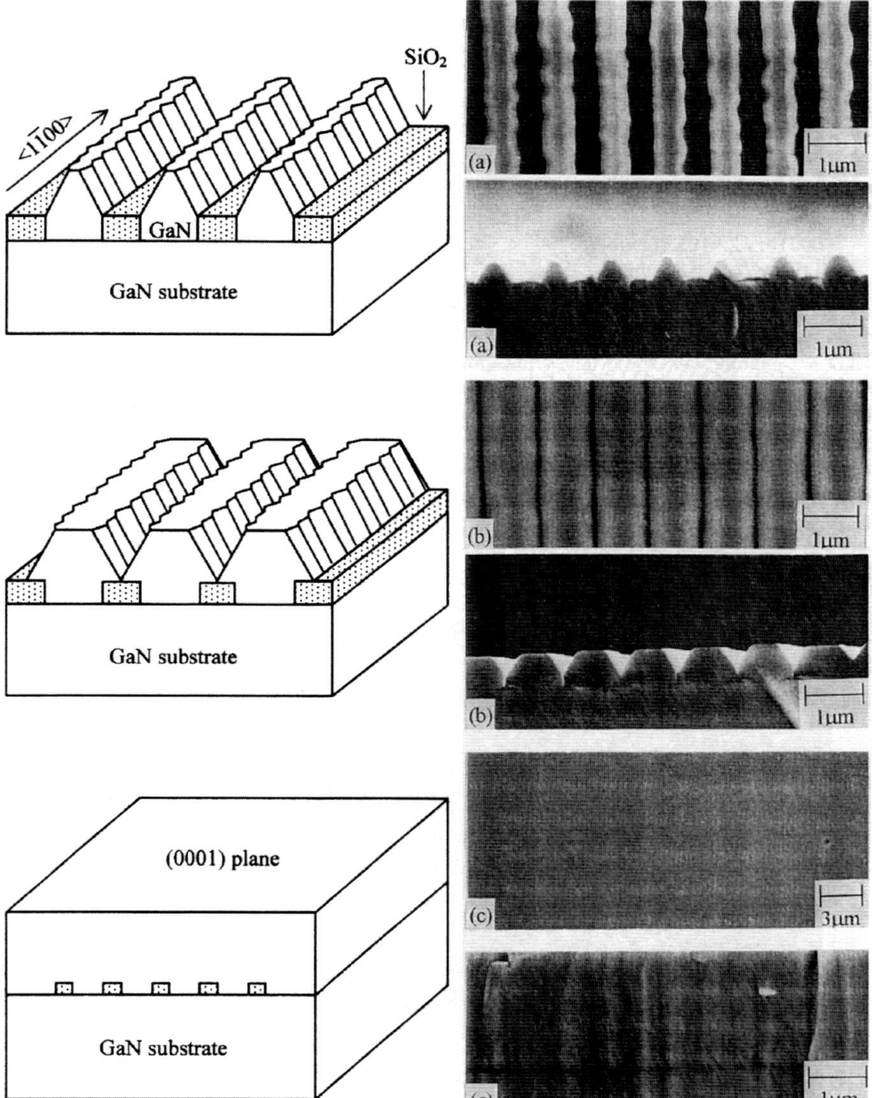


**Fig. 14.9.** Schematic illustration as well as top and cross-sectional SEM images of GaN grown on the <1$\bar{1}$00> line pattern at the growth times of 2 min (**a**), 7 min (**b**) and 30 min (**c**) (taken from [14.60])

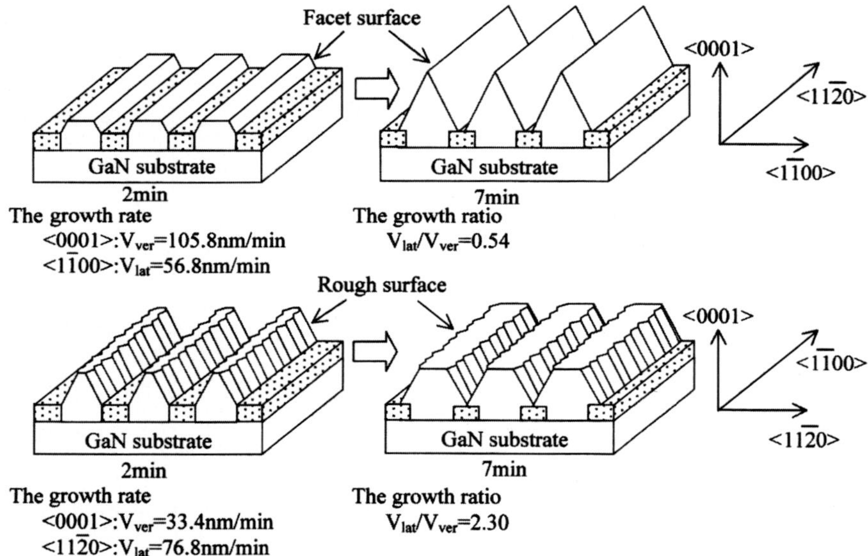


**Fig. 14.10.** Schematic illustrations indicating the differences in the ELO process on the $<11\bar{2}0>$ and $<1\bar{1}00>$ line patterns (taken from [14.60])

time of the growth from 2 to 7 min, but that on the $<1\bar{1}00>$ line pattern increased. On the $<11\bar{2}0>$ line pattern, the growth rate towards the $<0001>$ direction (105.8 nm min$^{-1}$) is faster than that towards the $<1\bar{1}00>$ direction (56.8 nm min$^{-1}$), resulting in the structures which are covered with $\{1\bar{1}01\}$ facets. On the other hand, on the line pattern $<1\bar{1}00>$, the growth rate towards the $<0001>$ direction (33.4 nm min$^{-1}$) is slower than that towards the $<11\bar{2}0>$ direction (76.8 nm min$^{-1}$), resulting in the (0001) facet on the top and non-singular surfaces on the sides. Consequently the ELO GaN layer with a smooth surface is completed earlier when grown in the $<11\bar{2}0>$ direction on the $<1\bar{1}00>$ line pattern than in the $<1\bar{1}00>$ direction on the $<11\bar{2}0>$ pattern. In addition, since the line pitch is less than 1 µm, the ELO layer can be easily obtained at less than 15 min. Cross-sectional TEM investigations on the ELO GaN structures described above showed the effective dislocation filtering in the GaN layers, occurring through the ELO growth process [14.60].

The ELO growth technique has led to a considerable breakthrough in the technology of blue and UV laser diodes based on GaN and related compounds [14.61]. The sophisticated structure of the InGaN multiquantum well laser diode, with modulation doped strained-layer superlattices as cladding layers, grown on ELO GaN substrates is shown in a schematic illustration in Fig. 14.11. This laser emits violet radiation of 401.4 nm wavelength under room temperature continuous wave single mode operation at an optimized supplying current.

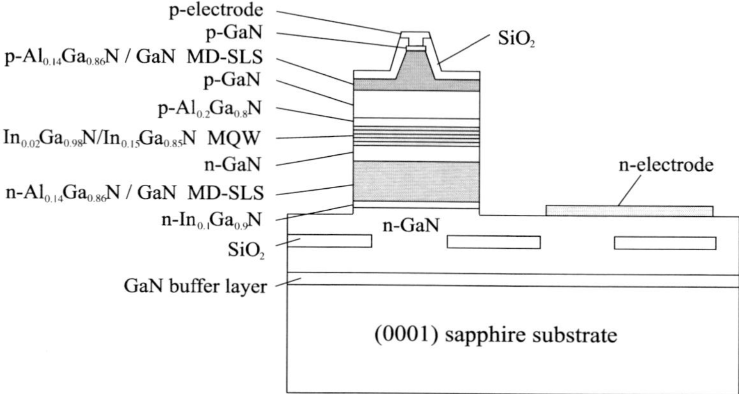


**Fig. 14.11.** Schematic illustration of the structure of the InGaN multiquantum well laser diode with modulation doped strained-layer superlattice cladding layers grown on an ELO GaN substrate (taken from [14.61])

It has also been demonstrated that the same good structural quality of the ELO GaN can be grown in a single step MOVPE process (single temperature and single nucleation step, without growing first on the sapphire substrate a GaN buffer layer by standard epitaxial process) when instead of the $SiO_2$ oxide, $Si_3N_4$ nitride is used for the mask layer [14.56]. In this case the ELO process is initiated directly on the sapphire (or SiC) substrate masked with the nitride. The nitride mask material is preferred for this process since the AlGaN used in the nucleation step does not grow on the masked surface unlike the more commonly used $SiO_2$ mask [14.62].

The one step ELO process was carried out in a MOVPE reactor at a fixed susceptor temperature of 1040°C with hydrogen and ammonia mixtures as the carrier gas at a pressure of 76 torr [14.62]. The 6H-SiC and sapphire substrates were directly coated with a 70 nm thick $Si_3N_4$ film using plasma enhanced CVD. This masking layer was then patterned using standard photolithographic techniques in arrays of stripes consisting of 2 μm openings, on a 12 μm pitch. The AlGaN nucleation layer is first deposited to a thickness of roughly 200 nm. This layer is necessary for smooth growth in the mask openings. The V/III group element flux ratio was doubled to 1800 during the GaN growth as the lateral growth rate is reported to increase with ammonia flow. Flow modulation is accomplished by rotating the wafers through a group III TEGa and TMAl rich growth zone resulting in roughly 25% duty cycle, while a background of ammonia is supplied for the remaining rotation period. A minimal Al content ($x \leq 0.3$) in the $Al_xGa_{1-x}N$ is required for successful nucleation, and differs slightly between SiC and sapphire substrates [14.63]. A schematic illustration of the structures produced with the described process is presented in Fig. 14.12.

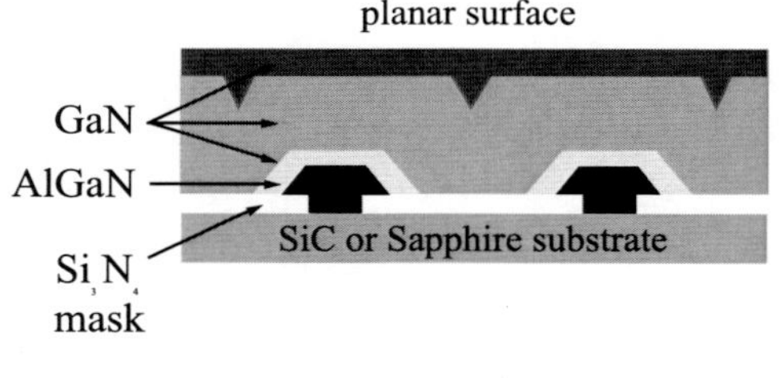


**Fig. 14.12.** Schematic illustration of the single step ELO process, with an AlGaN nucleation layer of composition x = 0.15 and thickness of 200 nm. This illustration shows the process from nucleation to planarization of the ELO GaN layer (taken from [14.64])

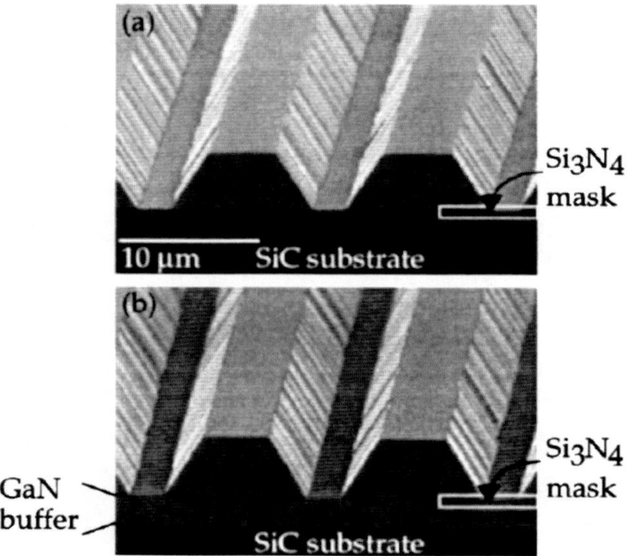


**Fig. 14.13.** ELO structures formed by direct growth on a patterned SiC substrate using a single step nucleation process (**a**), compared to conventional regrowth technique on a GaN/SiC substrate (**b**) (taken from [14.56])

When optimal nucleation conditions are met, atomically smooth AlGaN growth occurs within the first several tenths of nanometers across the entire window opening, providing the needed crystalline structure for subsequent GaN deposition. The GaN growth then continues both laterally and vertically retaining the original trapezoidal shape with a vertical to lateral growth rate of around 1:1.5. Figure 14.13a shows a SEM micrograph of the GaN/AlGaN ELO structure directly on patterned 6H-SiC, with growth suspended after deposition of roughly 2 µm of GaN in the vertical direction. The atomically flat (0001) surfaces are still present, along with the irregular $<1\bar{1}01>$ sidewalls. For comparison, a GaN homoepitaxial lateral overgrowth sample deposited on similar 6H-SiC substrate is shown in Fig. 14.13b. One may recognize that the growth features are nearly identical.

### 14.1.7 Hard Heteroepitaxy

Hard heteroepitaxy means growth of a solid film on a single crystalline substrate differing strongly from the film in at least one of the following parameters: lattice constant, crystal structure, and (or) nature of the chemical bonds [14.65]. The crystallographic orientation of the epilayer cannot easily be predicted in the case of hard heteroepitaxy, nor can simple solutions be given a priori to problems which arise usually in this case. Let us mention only two such problems as examples:

(i) How can one get a certain surface orientation by choosing suitable growth conditions if different surface orientations may grow, as for instance in the case of the CdTe/GaAs(100) interface [14.66]?

(ii) What is the volume fraction of each of the orientations occurring in the epitaxial film if different orientations grow simultaneously [14.67]?

Obviously, when one has to grow high-quality heterostructures of a material system which grows by hard heteroepitaxy, then these and related problems should be solved first for each of the materials that has to be grown [14.68].

One of the crucial problems in hard heteroepitaxy is the concept of lattice match for any pair of crystal lattices in any given crystal direction, allowing for a periodic reconstruction of the interface of the grown heterostructure. This problem was addressed in the pioneering work of Zur and McGill [14.69], in 1984.

The main idea of these authors was to compare the interface translational symmetry with that of the bulk materials on both sides of the interface, instead of comparing the bulk lattice constants. For systems which grow by hard heteroepitaxy the old lattice match criterion based on comparison of bulk lattice constants is no longer applicable. According to the definition more relevant to hard heteroepitaxy: "two lattices are crystallographically matched if the interface translational symmetry could be compatible with the symmetry on both sides of the interface, to within a given precision".

To clarify this point, let us refer to Fig. 14.14 which shows the two-dimensional translations parallel to the (101) face of rhombohedral $Al_2O_3$ (or $(1\bar{1}02)$ in the more common hexagonal notation) and the (111) face of face centered cubic (fcc) silicon. The lattice parameters of $Al_2O_3$ are $a_{\mathrm{rh}} = 5.1286\,\text{Å}$ and $\alpha = 55^\circ\ 17.36'$ [14.69]. Its (101) face has a two-dimensional (2D) rectangular symmetry with unit cell edges

$$2a\sin(\alpha/2) = 4.759\,\text{Å} \qquad \text{by} \qquad a = 5.129\,\text{Å}.$$

The (111) face of the silicon forms a 120° rhombic grid, with edges of

$$\sqrt{1/2}a = 3.840\,\text{Å}.$$

These two-dimensional lattices are depicted by the fine line grids in Fig. 14.14. One can now form superlattices by taking larger unit cells. If one takes 21 sapphire unit cells and 40 silicon unit cells in the manner seen in Fig. 14.14,

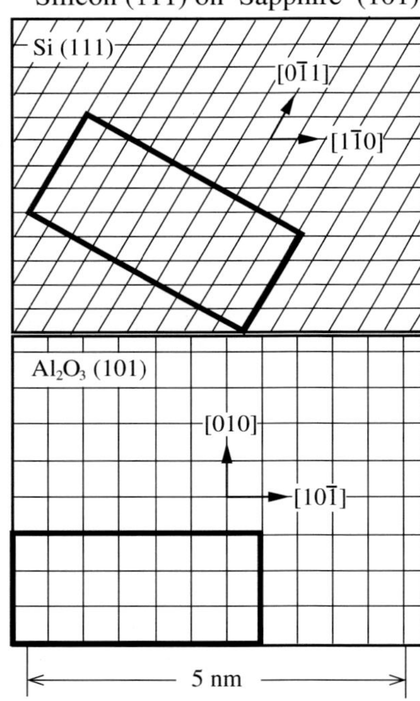


**Fig. 14.14.** Lattice translations parallel to the Si (111) and sapphire (101) faces. A cell made of 21 sapphire unite cells has almost exactly the same dimensions as a cell made of 40 silicon unit cells. Sapphire is rhombohedral with $a = 5.129$ Å and $\alpha = 55°17'$. Silicon is face-centered cubic with $a = 5.431$ Å (taken from [14.69])

then very similar superlattice cells are obtained. The silicon superlattice cell is also rectangular, with edges of

$$15.361\,\text{Å} \qquad \text{by} \quad 33.258\text{Å}.$$

The mismatch in each direction is about 0.2%. One can see that the $Al_2O_3$ [001] direction ([$\bar{1}$101] in the hexagonal notation) is parallel to one of the three equivalent Si [01$\bar{1}$], [10$\bar{1}$], [1$\bar{1}$0] directions.

The described example of lattice matching in the silicon-on-sapphire heterostructure may serve as an illustration confirming the fact that there are some possible orientations of the interface that match to within high precision [14.69]. Lattice match in this context means that the two-dimensional interface possesses translational symmetry that is compatible with that of the bulk on both sides of the interface. Such compatibility enables local structures of the interface to repeat themselves periodically over large distances. It should be stressed, however, that the effect of lattice mismatch is secondary in its importance for heteroepitaxy, in comparison to the chemistry of the interface, which will always play the major role as far as epitaxial growth is concerned.

As one may conclude from the previous example, the lattice match is characterized by two parameters: the precision of the match and the minimal unit cell area. The common superlattice unit cell in Fig. 14.14 has a mismatch of at most 0.2% in both lateral dimensions and the angle between them. The superlattice unit cell area is 511 Å$^2$. This cell is the smallest possible one that will enable a mismatch of less than 1% in the unit cell sides and angle. One can recognize that there is a trade-off between the precision of a match and the size of the superlattice cells. The larger the size of the interface unit cell, the less likely it is that the chemical forces will reinforce the lattice match condition. The precision of the match gives the lower bound to changes in lateral interatomic distances, while the common unit cell area should be limited to a value that is reasonable for interfacial periodic reconstruction. For example, limits of 1% on the match precision and 600 Å$^2$ on the common unit cell area are sufficient to reproduce the experimental results of CdTe(111)/GaAs(100) [14.66] and CdTe(111)/$Al_2O_3$ [14.70]. With the match precision and the common superlattice unit cell area taken as parameters, the problem of lattice match in hard heteroepitaxy is well defined [14.69].

## 14.2 Artificial Epitaxy (Graphoepitaxy)

Artificial epitaxy, hereafter called graphoepitaxy, is a growth process of oriented crystalline films on a substrate of different chemical composition, which is amorphous rather than a crystalline body [14.71]. In graphoepitaxy, the orienting influence on the growth of an epilayer on the substrate can be determined by various factors (forces) distinctive of the crystalline lattice. These factors of different possible nature, e.g., geometrical, mechanical, thermal, chemical or electrical, and their actions on the growth of the film can manifest themselves in both static and dynamic conditions. All these specific features of graphoepitaxy are in contrast to the usual heteroepitaxy. Despite this, and despite the fact that graphoepitaxy is based on macroscopic factors (macrosteps, macroparticles, etc.) while the usual heteroepitaxy is based on microscopic factors (monoatomic steps, atoms, etc.), the two mentioned crystal growth phenomena have much in common. This results from two facts. First, during the film growth process they can act simultaneously, and second, their mechanisms can sometimes have the same nature.

Graphoepitaxy has important practical value as an approach to preparation of well oriented films of various materials on arbitrary substrates like amorphous, ceramic or polycrystalline solids. This makes graphoepitaxy interesting for many applications in microelectronics, micromechanics, optics and optoelectronics [14.72].

Sheftal and Bouzynin [14.73] were the first researchers who succeeded in obtaining graphoepitaxial growth. In their experiments, two kinds of surface relief were made on glass plates used as substrates for oriented film growth. In the first case, two sets of parallel gratings were crossed at an angle of 90° forming a four-fold symmetrical relief. In the second case, the gratings crossed under 60° forming a three-fold symmetrical relief. These two surface reliefs correspond to two characteristic equilibrium faces, {100} and {111}, of the material to be grown i.e., to a cubo-octahedron.

As the results of the performed experiments, ammonium iodide ($NH_4I$) crystallites were deposited onto the substrates from an aqueous solution. However, different kinds of deposition were observed with the different-symmetry surface reliefs; both, however, demonstrated the effect of grapho-epitaxy.

### 14.2.1 General Principles of Graphoepitaxy

Pattern symmetry, sidewall angles of cells and surface relief topology are the main points of the crystallographic principles of graphoepitaxy [14.71]. Let us begin our considerations on graphoepitaxy with the pattern symmetry problem.

The starting criterion for an optimal choice of the orienting relief on the substrate surface is the lattice symmetry of the material to be crystallized. Depending on the type of lattice of the crystal to be grown, different growth shapes, e.g., flat, stepped or kinked, occur in graphoepitaxy.

Figure 14.15 shows two kinds of surface patterns, both having three-fold symmetry (this means that they can be self-aligned when rotated by 120°) and the one shown in Fig. 14.15a, having, in addition, six-fold symmetry. Such patterns are, in general, appropriate for graphoepitaxy of silicon and germanium, the crystallites of which are preferentially bounded by {111} faces. In Fig. 14.15a triangular cells (depressions) are separated by linear projections, whereas those in Fig. 14.15b are separated by trigonal prisms (shown as black triangles). A principal difference between these two patterns is that the first leads to twinned growth (the crystallites in twin positions are schematically shown as shaded triangles), whereas the second one allows us, in principle, to avoid such a defective growth. Consequently, the relief shown in Fig. 14.15b is preferable, if single crystalline films have to be grown by graphoepitaxy.

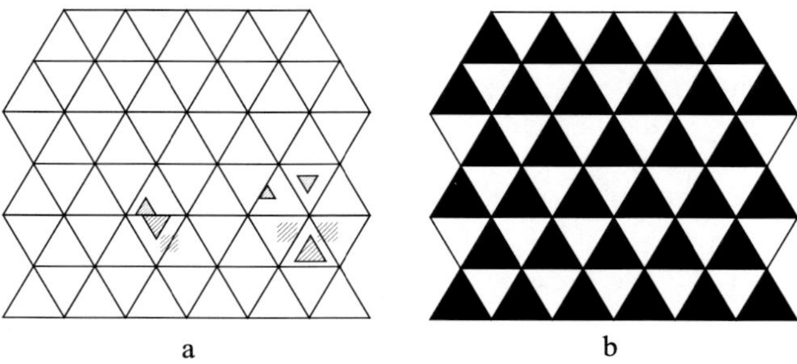


**Fig. 14.15.** Examples of microrelief suitable for graphoepitaxy of materials crystallizing in diamond-like, sphalerite, some other cubic, as well as wurtzite and hcp lattices (taken from [14.71])

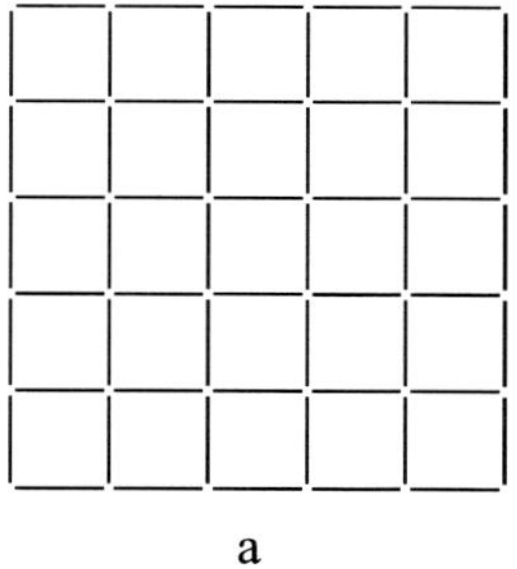


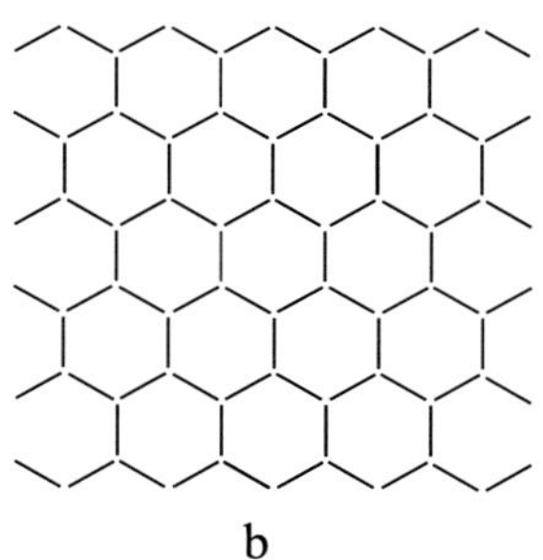


**Fig. 14.16.** Four-fold (**a**) and six-fold (**b**) symmetry surface reliefs. The relief (**a**) is appropriate for graphoepitaxy of materials that crystallize in the NaCl-type lattice (cubic habit), while relief (**b**) is for materials with hexagonal-type lattices (taken from [14.71])

Let us now consider four-fold and six-fold symmetrical reliefs shown in Fig. 14.16. The relief shown in Fig. 14.16a is evidently appropriate for materials that crystallize in the NaCl-type lattice and, accordingly, exhibit cubic habit. This relief is also suitable for materials characterized by its "relative" two-fold symmetry. The six-fold symmetrical relief shown in Fig. 14.16b is suitable for materials with hexagonal-type lattices; however, also the three-fold symmetrical relief shown in Fig. 14.15b should be effective for oriented growth of these materials because three-fold and six-fold symmetries are "relatives" in the crystallographic sense.

The next factor, which influences substantially the growth process of graphoepitaxy, i.e., the alignment of microcrystals within the surface microrelief, is connected with sidewall angles of the surface relief cells. When showing the various pattern symmetries in Figs. 14.15 and 14.16, it was implied that the sidewalls are vertical, i.e., that they form an angle of 90° with the plane of the substrate. In fact, however, it is difficult to prepare a relief with absolutely vertical sidewalls. On the other hand, in some cases it is advantageous to have non vertical walls of the relief cells.

For graphoepitaxy of materials crystallizing in cubic (e.g., NaCl) or tetragonal (e.g., Sn) growth shapes, vertical walls are evidently preferable. The same is true for materials with hexagonal-type lattices (e.g., CdS, ZnO) when they are deposited in the basal orientation. However, for crystallites with octahedral growth shapes (e.g., Si, Ge, GaAs, ZnSe), the best alignment is achieved when the walls are at the crystallographic angle of 70° 32′ or 109° 28′ to the substrate plane (these are the typical angles between two adjacent {111} faces in cubic lattices). Similarly, other crystallographic angles are appropriate for other growth shapes (i.e., for other combinations of adjacent faces). It is worth mentioning that it is difficult in practice to provide such exact (crystallographic) values of the angles, but fortunately, even angles that deviate markedly from these can be effective for graphoepitaxy [14.71].

Let us conclude the considerations on general principles of graphoepitaxy with the problem of topology of the surface relief. The topology of the relief is created by the sizes of the relief elements (cells or grooves) in the horizontal and vertical directions and by the spacing between them.

It is evident, in general, that there exists an upper limit for the horizontal size of the relief elements, which depends on the kind of the crystallization phase, on the growth temperature and other factors. For example, the higher the growth temperature and the higher the mobility of the crystallization phase particles involved in the growth process, the larger the horizontal sizes of the relief elements. On the other hand, the vertical size of the cells or grooves, i.e. the height of steps in the relief, should be optimized taking into account tentative applications of the films. For example, microrelief that is too deep is undesirable for microelectronic applications of the film, since the planarity of the structures to be grown can be destroyed. On the other hand, rather deep grooves could be desirable for applications such as waveguides.

It is clear that there is a minimum step height required to induce orientation; the smaller the horizontal sizes, the lower the height [14.71]. Klykov and Sheftal. [14.74] used, for example, steps up to 1 μm in height for cell sizes of 10–20 μm and succeeded in preparing fairly perfect single crystalline Si films, whereas a 0.1 μm height was sufficient to achieve orientation effects with a 3.8 μm spatial period of the relief [14.75].

Spacing between relief elements is an element of topology, for which no clear dependencies have been established. From a general point of view, the spacing is larger for materials which tend to form plate-type crystallites (e.g., those having the wurtzite lattice, such as CdS, or having a layered internal structure, such as GaS or $MoSe_2$).

Concluding the discussion on principles of graphoepitaxy, it should be emphasized that these principles formulate only the general conditions necessary for oriented crystallization on amorphous substrates. In particular cases, for different material systems and growth techniques, definite optimal conditions should be defined by experiment, however, taking as the basis the presented general principles [14.71].

### 14.2.2 Growth Mechanisms in Graphoepitaxy

The key issue in crystallization phenomena is the answer to the question: "How is a crystal, that has to be grown, formed in the growth process?" As applied to graphoepitaxy this question should be reformulated as follows: "What are the orientation mechanisms in the growth process?"

On the basis of current knowledge at least four different mechanisms of orientation of crystals in the film growing by graphoepitaxy should be distinguished. These are:

(i) topographic relief depending on orientation,
(ii) orientation caused by capillary forces,

(iii) oriented growth in directional crystallization, and
(iv) orientation related to anisotropic deformation.

In the graphoepitaxy growth process, these mechanisms can operate both separately and simultaneously.

The orientation by topographic relief assumes the incorporation of crystallites into a macroscopic kink or step. This is, in fact, a generalization of the classical molecular-kinetic crystal growth mechanism by Kossel and Stranski (see Sect. 12.1.1). Therefore, it is sometimes called the "macroscopic Kossel-Stranski mechanism" [14.71].

In the case of classical crystal growth (see Fig. 12.1) minimal particles (atoms or molecules) are attached to elementary kinks or steps of monatomic or monomolecular height. In the case of graphoepitaxy, "macroscopic" kinks and steps are formed on the substrate surface by the surface relief, their heights (usually $0.1-0.5\,\mu m$) being about $10^3$ times larger than that of the elementary steps. In addition, there is no relationship between the heights (and other sizes) of the crystallites and those of the macrokinks or macrosteps. However, both the microscopic (classical) and the macroscopic (relevant to graphoepitaxy) growth mechanisms are based on the principle of energy gain from the attachment of the depositing particles (atoms or molecules in the former case, and crystallites in the latter case) to the elements of the relief on the substrate/crystal surface.

Crystallites deposited in graphoepitaxy are often bound by faces of the relief that are not strictly flat. Nevertheless, if a crystallite is incorporated into a carefully prepared microrelief, a certain (usually rather large) gain in the free energy of the substrate–crystallite system will occur in comparison with the case in which the crystallite takes an arbitrary position on the substrate surface, e.g., on its flat areas. Moreover, in real systems any medium constituents, including substrate impurities, can effectively accommodate the contact between the substrate and the crystallite by forming an intermediate sublayer, e.g., a liquid-like sublayer [14.71].

The graphoepitaxy mechanism based on orientation caused by capillary forces has been proposed by Klykov et al. [14.74–76] and experimentally demonstrated for Si crystallites. According to the symmetry of Si, a microrelief similar to that depicted in Fig. 14.15b was used in the experiments. The experimental results have shown that growth of the Si film from a solution proceeds via formation of crystallites that can float on the surface of the solution. These crystallites nucleated within the solution or at the liquid–vapor interface. Once they are formed, the Archimedes force will lead to their floating provided they are larger than about $1\,\mu m$. Owing to the good wetting of both the crystallites and the cell sidewalls, the liquid in the cell forms a meniscus between the crystallite and the sidewalls. If the crystallite approaches a sidewall, the meniscus is deformed, and a restoring force arises; as a result, the crystallite is repelled from the sidewall, i.e., some inclination stabilization takes place. Azimuthal stabilization operates in a similar manner.

One may conclude that the capillary mechanism is a version of the topographic one; the meniscus can be formed only in a cell having walls, i.e., topographic elements. A principal difference between these two orientation mechanisms is that, here, in the capillary mechanism, no direct contact of the crystallites with the sidewalls of the cells is necessary. The orientation action of the microrelief is transferred via an intermediate material (the liquid phase) by means of capillary forces. It is however clear that when at a later stage of the growth process the growing crystallite reaches the sidewalls, some direct interaction with them is possible, so that the two orientation mechanisms (topographic and capillarity) can operate either simultaneously or consecutively. In this respect, graphoepitaxy of proteins could be representative [14.71].

The next mechanism of orientation of crystallites in graphoepitaxy is based on periodic thermal relief effects occurring in directional crystallization. These effects occur, for example, also in zone-melting recrystallization (ZMR) of thin films. ZMR represents directional crystallization and in particular, directional solidification of the melt [14.77]. The most prominent feature of ZMR films is the formation of subboundaries that are elongated in the direction of thermal zone propagation [14.76]. Typical distances between the subboundaries, i.e., widths of the corresponding grains, vary from several μm to about 100 μm. An example of the morphology of ZMR Si film grown on oxidized Si (an amorphous substrate) is shown in Fig. 14.17. The subgrains usually propagate from the surface of the film down to the amorphous substrate. When the thermal zone moves, new subgrains are often nucleated, or some neighboring subgrains are merged (coalesced). Misorientations between neighboring subgrains are usually about 1°.

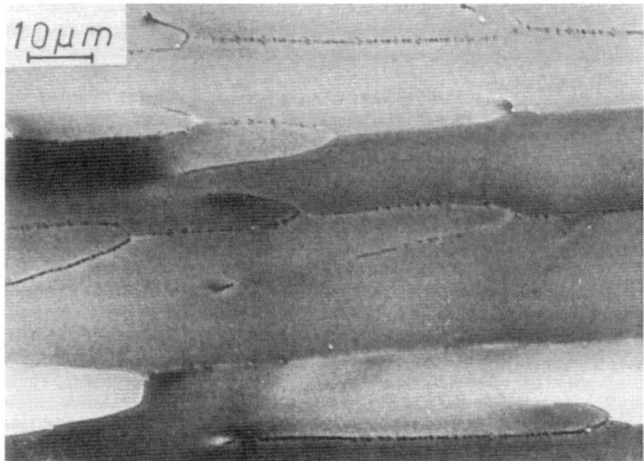


**Fig. 14.17.** SEM image (taken by a back-scattering mode of observation) of the morphology of ZMR Si film grown on an oxidized Si surface. Subboundaries are seen as a line structure. The thermal zone moved from left to right. Note the dark spots on the lines depicting the subboundaries (taken from [14.71])

The last growth mechanism which has been mentioned above is the orientation process caused by anisotropic deformations. It occurs when a polycrystalline film of the material to be grown epitaxially is deposited on an amorphous substrate and then annealed by laser irradiation performed in a special way. The laser beam passes through a mask with a crystallographically symmetrical black and white pattern (e.g., squares or triangles) and is focused onto the substrate, forming there an inhomogeneous, symmetrical temperature field. Such laser annealing up to complete melting of the film causes oriented growth of crystallites. It is important that, depending on the symmetry of the pattern, the grown films have different orientations. For example, in the case of the square pattern (four-fold symmetry axis), the films are of (100) orientation, while in the case of triangles (three-fold symmetry axis), they are of (111) orientation, as it should be in graphoepitaxy. These results can be ascribed to anisotropic mechanical strains to which the crystallites were subjected in the film during annealing.

Concluding the considerations on graphoepitaxy let us make the following statement [14.71]. "Graphoepitaxy is important, not only for crystal growth physics but also for application purposes." At least two areas of application of graphoepitaxy can be indicated at present: (a) technology of multilayer structures used in 3D microelectronics, where single crystalline (or at least highly ordered) films of active materials with different properties are separated by amorphous, passive intermediates, and (b) the emerging field of bioelectronics where strongly dissimilar materials (biological, metallic, semiconducting, etc.) are arranged on the same substrate.

# 15. Material-Related Problems of Heteroepitaxy

The basis for discussion of material-related problems of heteroepitaxy may be defined by the relevant criteria, according to which these problems can be identified. One way of formulating these criteria is connected with the conceptual tools required for formal description of epitaxy (Chaps. 11–13), i.e., thermodynamics of phase transitions, fluid dynamics of mass transport, statistical mechanics of crystal growth, and quantum mechanics of chemical bond formation. Each of these approaches, when applied to heteroepitaxy, leads to different material-related problems. Consequently, the general discussion concerning the subject of material-related problems of heteroepitaxy is an enormous task, exceeding the framework of this book.

In order to be sufficiently concise in our presentation, we have to limit substantially the scope of the discussion, taking into consideration only these problems which influence the process of heteroepitaxy in the strongest way. Thus, the basic criterion we have chosen here for the discussion of the material-related problems, is the dependence of the initial stages of epitaxial growth processes (Sect. 11.4) on the materials to be grown by heteroepitaxy. In this discussion, references will be made to the excellent reviews published in vol. 8 of the series: *The Chemical Physics of Solid Surfaces* edited by King and Woodruff [15.1].

## 15.1 Material Systems Crystallizing by the Fundamental Growth Modes

As already described in Sect. 1.2, one can distinguish five fundamental growth modes in epitaxial crystallization. These are: layer-by-layer or FM (Frank–van der Merwe) mode, layer-plus-island or SK (Stranski–Krastanov) mode, island or VW (Volmer–Weber) mode, columnar or CG mode, and step flow or SF mode. Among them, the first three modes occur most frequently in heteroepitaxy, especially when the growth proceeds on flat surfaces. Accordingly, we will limit our considerations in this section only to these three growth modes.

The possibility of occurence of a definite mode in the growth process depends on the relation between the surface energies of the substrate and the epilayer (Sect. 11.4). These relations are, however, strongly dependent on the physico-chemical properties of the materials that create the heterostructure pair, i.e. the substrate-epilayer pair. Consequently, one may expect that for some materials, only one growth mode can occur in heteroepitaxy. Let us consider a few examples, illustrating this anticipation.

### 15.1.1 Growth by the Island Mode

Extended reviews on this subject have been published by Venables et al. [15.2–4] with emphasis put on growth of noble metals, Ag, Au and Pd, on alkali halides and oxides. For silver and gold grown on a NaCl (100) surface, the adsorption energy $E_a$ of the atoms is in the range of 0.5 – 0.9 eV, with the Au values somewhat higher than the Ag values, and errors for particular deposit-substrate combination being $< 0.1$ eV. Taking into consideration that the binding energies of pairs of Ag or Au atoms in free space have the values $1.65 \pm 0.06$ eV and $2.29 \pm 0.02$ eV, respectively, one may conclude that these metals grow in heteroepitaxy on a NaCl(100) surface only by the island mode. The Ag or Au adatoms re-evaporate readily above room temperature from the substrate surface however, when they meet another adatom on this surface, they form a stable nucleus which then grows further by adatom capture [15.2].

Another experiment on metal heteroepitaxy on the NaCl(100) substrate concerns the growth of binary alloy pairs, formed from Ag, Au and Pd. It has been shown [15.5, 6] that atoms with higher value of the adsorption energy $E_a$, namely Au in the AgAu alloy or Pd in the PdAg and PdAu alloys, form on the substrate surface nuclei preferentially, and the composition of the growing alloy film is initially enriched in the element which is most strongly bound to the substrate. The composition approaches that of the sources, i.e., the composition in the atomic beams impinging from the evaporation sources of the alloys onto the substrate, only at longer times. Again the alloy crystallizes on the substrate surface by the island mode, because the measured adsorption energies of the constituent elements, equal to 0.41 eV for Ag, 0.49 eV for Au and 0.78 eV for Pd [15.2, 5, 6] are considerably less than the binding energies between atoms creating the epitaxial nuclei.

Let us now discuss an example of Pd heteroepitaxy on an oxide substrate, namely on MgO [15.7, 8]. This growth system is of interest as a model catalyst for CO molecules interacting with the grown Pd islands [15.2]. To make the subject quantitative, the shapes and size distributions were determined experimentally, as shown in Fig. 15.1. The Pd particles have (100) top faces, with different amounts of (111) and (110) inclined faces in contrast with the substrate. The nucleation density $n_s$ around $3 \times 10^{11}\,\mathrm{cm}^{-2}$ is typical for deposition at the temperature of 150 – 200°C, and the size distribution is characteristic of complete condensation, plus a small amount of coalescence. Surface diffusion around the islands is sufficient to form a polyhedral shape, but is low enough that the coalesced islands remain elongated.

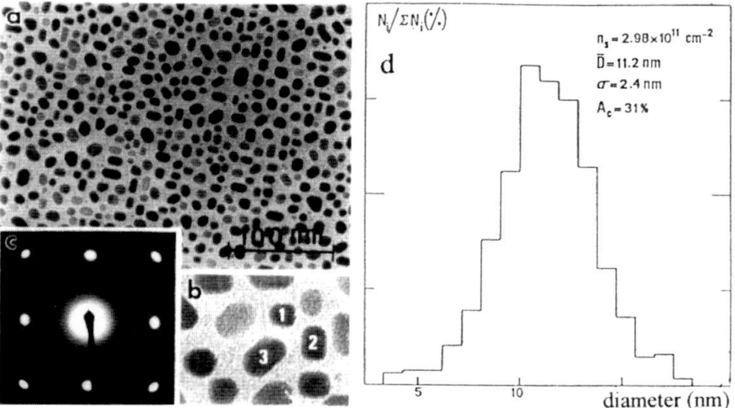


**Fig. 15.1.** Heteroepitaxial Pd islands grown on MgO: (**a**) TEM overview of the islands after some coalescence has occurred; (**b**) higher magnification view of Pd islands with different shapes numbered 1–3; (**c**) transmission diffraction pattern, giving the epitaxial orientation of all such islands; (**d**) size distribution histogram, nucleation density and other quantities derived from (**b**) (taken from [15.7])

When the adsorption energy $E_a$ is small, the adatom concentration $n_s$ can be extremely small, even at moderate growth temperature. In this situation, nucleation during heteroepitaxy is a very unlikely event on atomically smooth terraces on the substrate. In such cases, nucleation at defect sites is likely to be dominant. Several examples of nucleation on defects have been demonstrated in the literature, for example Au deposited on mica, MgO, $Al_2O_3$ and graphite, as well as many examples of metals grown on alkali halides [15.2].

### 15.1.2 Growth by the Layer-by-Layer Mode

The opposite case to island growth mode is the layer-by-layer mode. It occurs usually when on a smooth substrate surface 2D island nucleation is preferable to 3D nucleation. This mode of crystallization is usually met in homoepitaxy for different materials. In the case of heteroepitaxy some special requirements concerning the growth conditions have to be fulfilled in order to obtain this growth mode. The FM mode is often observed when metal layers grow on metal substrates [15.9]; however, it occurs for other material systems, as well.

In order to obtain the layer-by-layer growth, the surface energy of the epilayer to be grown must be less than that of the substrate surface. In addition, the lattice misfit should be small; it has to be smaller if the surface energy difference between the layer and the substrate is smaller.

For example, for metal-on-metal heteroepitaxy, the refractory metals W and Mo, with their high surface energies, are favorable as substrates for layer-by-layer growth. However, in this case two limiting situations have to be considered, namely, low temperature and the high temperature growth (or annealing). At low temperatures, diffusion limitations and high 2D nucleation rates lead to quasi-FM growth, i.e., layer growth with small grains and many multipositioning grain boundaries [15.9]. At high temperatures, misfit stress makes 2D growth energetically unfavorable above one or several monolayers; moreover, the high mobility leads then to 3D crystals on top of the 2D initial layer; the FM mode goes over into the SK mode. This means that for the FM mode to occur in the metal-on-metal heteroepitaxy, a definite temperature window exists.

Interesting material systems which can be grown by the FM or quasi-FM mode in heteroepitaxy are thin metal films grown on semiconductor substrates. Metals always grow on semiconductors by the layer-plus-island (SK) mode [15.9]; however, FM or quasi-FM growth can be obtained for these material systems when the growth process is performed at very low temperatures and is assisted with surfactants (Sect. 14.1.5) or interfactants (interface active species) [15.10]. For example, Jalochowski et al. [15.11, 12] demonstrated the FM growth of Pb films at temperatures of 16 – 110 K onto a Si(111) surface, precovered with an Au monolayer. They could measure RHEED and resistivity oscillations with a one monolayer periodicity caused by the periodic variation of the surface roughness of the Pb epilayer during monolayer-by-monolayer growth. The RHEED specular beam intensity oscillations during the growth of Pb on Si(111) surfaces at low temperatures are shown in Fig. 15.2.

At low temperatures, Au and Ag are so effective as interfactants that continuous epitaxial Pb films can be grown by the FM mode from the very beginning. Regular RHEED specular beam intensity oscillations, recorded during growth on the Si(111)-(6×6) Au surface at 16 K, start independent of the deposition rate over a wide range already at the first monolayer (Fig. 15.2, left panels). Those on the Si(111)-(7×7) uncovered substrate surface do not set in until the initial amorphous layer, whose thickness depends upon temperature, has crystallized into an epitaxial layer (asterisk in Fig. 15.2, right panels). The layers grow with perfect orientation show high electrical conductivity exhibiting so-called quantum size effects [15.13]. One may conclude from the described experimental results that epitaxial layers growing in the FM mode can be obtained even for materials which tend to grow in the SK or in the VW modes by making full use of the present understanding of the epitaxial growth phenomenon.

### 15.1.3 Growth by the Layer-Plus-Island Mode

Many combinations of heteroepitaxial material pairs grow by the layer-plus-island (SK) mode, including metals, semiconductors, gases condensed on substrate surfaces, and others [15.2]. In fact, this is the most widespread growth

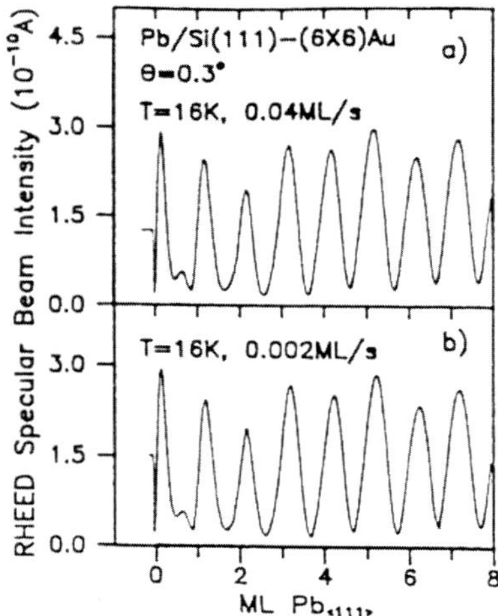


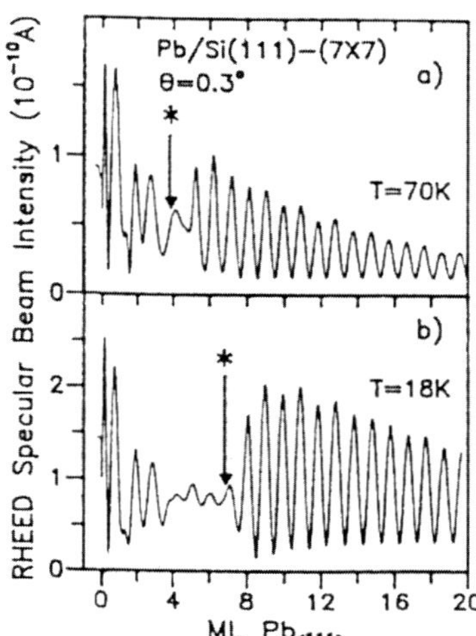


**Fig. 15.2.** RHEED specular beam intensity oscillations during heteroepitaxy of Pb on Si(111)-(6×6) Au (left panels) and on Si(111)-(7×7) (right panels) at low temperatures. Electron energy 15 keV, glancing angle of incidence 0.3° in Pb azimuth $[11\overline{2}]$ (taken from [15.13])

mode in heteroepitaxy. For the discussion of the material-related problems concerning this mode, let us select only some arbitrary examples.

We will start with the silver-on-metal growth systems in which W(110) [15.14], Fe(110) [15.15, 16], and Pt(111) [15.17, 18] are the substrate surfaces. In these systems usually two monolayers of Ag form first on the substrate surface, and then flat Ag islands grow in the (111) orientation. The growth of Ag on these substrates has been studied in ultrahigh vacuum by scanning electron microscopy (SEM) and by scanning tunneling microscopy (STM). Condensation of Ag is complete (except at the highest temperatures studied, $T > 750\,K$), while the nucleation density for creation of the islands is a strong function of substrate temperature. The critical nucleus size is in the range of 6 – 34 atoms, increasing with substrate temperature.

In the SK growth mode of the discussed heteroepitaxial pairs, strain has an important effect. This is confirmed through comparison of the experimental data on $E_d$ (surface diffusion energy) for the Ag-on-Pt(111), Ag-on-Ag(111) and Ag-on-Pt(111) covered with 1 monolayer of Ag. The respective values of $E_d$, the energy responsible for the growth of islands on the surface, increases considerably with strain. It is equal respectively to 0.16 eV, 0.10 eV and 0.06 eV, for these systems [15.2]. There are, however, some open questions concerning the Ag-on-metal systems. These concern the way in which the transition from 2D nucleation to 3D nucleation occurs. It is well known that the 2D–3D transition occurs when the nucleation processes are essentially complete. It is also known that the steps on the surface have a big effect on atomic motion within the first layer of Ag, but much less on top of this stable layer. The more perfect the substrate, the flatter the islands are. Most probably this is related to the difficulty of islands growing in height, without threading dislocations which may be generated at steps.

Among the material systems growing by the SK mode in heteroepitaxy, metals-on-semiconductors have gained the greatest interest because of their application in semiconductor device structures (mainly in microelectronic and optoelectronic devices). However, the main emphasis in the heteroepitaxial growth of these material systems is put on getting the 2D grown layer as thick as possible, which means getting the FM growth mode instead of the naturally occurring SK mode. We have already shown in Sect 15.1.2 how the interfactants Au or Ag may be used for this purpose.

It is worth mentioning that hydrogen passivation of the dangling bonds on the semiconductor surface, e.g., when growing Ag on a Si(111) surface, improves considerably the flatness of the metallic film [15.10]. Figure 15.3 shows this effect schematically [15.19]. In summary:

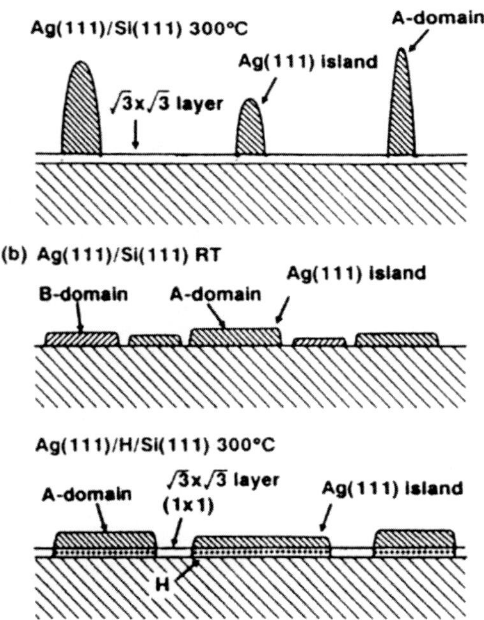


**Fig. 15.3.** Schematic illustration of the growth modes derived from heteroepitaxy of Ag on a bare Si(111) surface, at the growth temperature 300°C (top) and at room temperature (RT) (middle), and on the H-passivated Si(111) surface at 300°C (bottom) (taken from [15.19])

(i) For the growth at room temperature, hydrogen passivation significantly suppresses the rotational disorder of the Ag(111) epilayers, often observed when growth is performed on clean Si(111)-(7×7) surfaces

(ii) While on a clean surface and at higher growth temperature (300°C ) Ag growth proceeds by the SK mode, i.e., rather thick Ag islands nucleate and develop, after the completion at one monolayer of the 2D growth, exhibiting a surface structure $(\sqrt{3} \times \sqrt{3})$, growth at 300°C on the Si(111) surface terminated with atomic hydrogen (hydrogen passivated surface) changes the growth mode into the FM mode, with much thinner Ag islands predominantly of the so-called A-type (islands epitaxially oriented according to the relation Ag(111) $[11\overline{2}] \perp$ Si(111) $[11\overline{2}]$).

(iii) The persistence of a 0.5 – 0.3 monolayer thick hydrogen film at the substrate–epilayer interface has been observed even after deposition of 16 – 20 monolayers of Ag at RT and at 300°C [15.10].

The next important material system, grown by the SK mode is the semiconductor-on-semiconductor heteroepitaxial system. Growth of heterostructures by this mode, with controlled geometry of the 3D-nucleated islands, has great importance from the point of view of the technology of so-called quantum dot structures. This subject has already been widely discussed in this book, thus, the reader is referred to Sect. 11.5.1 for more information. A most recent result is described in more detail in Sect. 15.2.2.

## 15.2 Peculiarities of Heteroepitaxy of Selected Material Groups

Each pair of materials for substrates and epilayer has its special properties depending on lattice mismatch, chemical bonds, crystal structure, and thermal expansion coefficient, to mention only a few of them. So, the selection of special groups is a very subjective choice and reflects the special interest of the authors. However, we tried to choose materials which are in our opinion of wider interest since they are still under intensive investigation and discussed in the actual literature.

The first group is GaN grown on different substrate materials with different growth methods. The optimum substrate has not yet been found and the discussion between hexagonal and cubic GaN is still open. The second group are IV–VI compound semiconductors, well known as small gap materials with many applications for optoelectronic devices in the infrared. Finally, we will discuss a fairly new group of materials, the organic semiconductors, which caused a small revolution in the last decade. The challenging problem with this class of materials is their Van der Waals type of bonds, which needs a totally new concept for epitaxy, because of the extremely low sticking coefficients of the molecules. That means that usual growth techniques like MBE or MOVPE, working far from thermal equilibrium, must be replaced by methods close to thermal equilibrium supporting the Van der Waals character of the bonds.

### 15.2.1 Group III Nitrides

The number of papers published on group III nitrides has exploded in the last decade because of their technological interest. Most of the work has concentrated on the hexagonal nitrides especially GaN. Fortunately there exist a number of excellent review paperswhich deal with all the aspects of hexagonal nitrides and their application [15.20–25]. Recently also hexagonal InN, because of its revised low optical band gap, came into the focus of interest [15.26, 27]. Recent news in this field can be found in the world wide web [15.28], but a more comfortable approach for the newcomer in this field is the recently published handbook on nitride semiconductors [15.29]

All group III nitrides form, besides the hexagonal crystal structure, also the thermodynamically less stable zincblende modification, the so-called sphalerite or cubic group III nitrides [15.30, 31]. It has the advantage of the cubic structure and therefore can be grown on (001) substrates. The cubic symmetry eliminates also the piezoelectric fields occuring in the hexagonal structure, a fact which makes the interpretation and analysis of optical and electrical properties more simple. A certain disadvantage of c-GaN is however that a small fraction of the thermodynamically more stable hexagonal modification might occur in the sample. Nevertheless, c-GaN blue LEDs grown on GaAs(001) have been reported recently [15.32, 33]. Less papers have been published on the cubic GaN and therefore we will concentrate here on them in following the review by D.J. As [15.34].

For electronic devices like high electron mobility transistors (HEMTs) c-GaN has the potential of a high saturated electron drift velocity. This property could increase its applicability for high frequency devices. The large band gaps of these nitrides further predestines them also to be used at high temperatures and high powers and their chemical stability enables applications in hostile environments. Finally, cubic III nitrides are mainly grown on GaAs or Si substrates, which further enables possible future integration of III nitrides with advanced GaAs or Si technology.

The metastable zincblende structure has a cubic unit cell, containing four group III atoms (Al, Ga, In) and nitrogen atoms. A lattice picture of GaN is shown in Fig. 15.4, taking into account the different diameters of the atom species.

The position of the atoms within the unit cell is identical to the diamond crystal structure; both structures consist of two interpenetrating face centered cubic (fcc) sublattices, offset by one quarter of the distance along a body diagonal. Each atom in the structure may be viewed as positioned at the center of a tetrahedron, with its four nearest neighbors defining the four corners of the tetrahedron. The main difference from the stable hexagonal structure

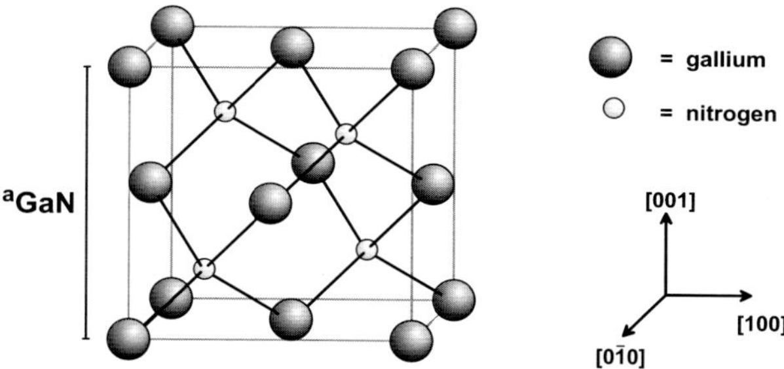


**Fig. 15.4.** Ball and stick model of the zincblende structure of cubic GaN [15.34]

is the stacking sequence of close packed diatomic planes. Whereas for the hexagonal structure, the stacking sequence of (0001) planes is ABABAB in the <0001> direction, the stacking sequences of (111) planes is ABCABC in the <111> direction for the zincblende structure. This difference occurs due to an eclipsed bond configuration of the second nearest neighbors in the wurtzite structure while in the zincblende structure the bonding configuration is staggered.

One immediately sees in Fig. 15.4 that the size of the N-atom is much smaller than that of an group III element. This has important consequences for doping since the dopant has to fit to the replaced atom. Otherwise large deformations within the crystal structure may occur, which may hinder the incorporation of the dopant species, resulting in insufficient p- or n-type conductivity of the semiconductor.

Since the natural structure is the hexagonal one, cubic substrates are needed as templates to force the III nitrides to grow in the cubic face. This fact also excludes homoepitaxy of cubic III nitrides, since no cubic bulk material exists.

Unfortunately, no suitable lattice matched substrate is available. Cubic GaN has been deposited on a number of cubic substrates, including GaAs (001) [15.35–42], Si (001)[15.43–45], 3C-SiC ($\beta$-SiC) (001) [15.46–50], GaP [15.51,52] and MgO [15.53]. All these substrates share the handicap of a very large lattice mismatch to the nitrides. In Table 15.1 the lattice parameters, thermal expansion coefficients and the calculated mismatches to c-GaN are listed.

GaAs and Si offer significant technological advantages due to the high quality materials and their commercial availability. 3C-SiC, normally prepared by deposition on (100) Si, is the closest in lattice match to the nitrides and thus should in principle produce the highest quality GaN. Unfortunately, bulk 3C-SiC has been stopped being industrially produced a few years ago. For the realization of a vertically structured optoelectronic device like a light emitting diode (LED) or a laser diode (LD) one of the most important aspects for the selection of a suitable substrate is the ability of the substrate to

**Table 15.1.** substrate lattice parameters, thermal expansion coefficient, lattice mismatch to c-GaN and thermal conductivity of prospective substrates for III-nitrides epitaxial growth. The lattice mismatch $f$ is calculated $f = (a_{\text{sub}} - a_{\text{GaN}})/a_{\text{sub}}$

| Crystal | Reference | Lattice constant $a$ [nm] | Thermal expansion coefficient $\alpha$ $[10^{-6}\text{K}^{-1}]$ | Lattice mismatch $f$ | Thermal conductivity $[\text{Wcm}^{-1}\text{K}^{-1}]$ |
|---|---|---|---|---|---|
| c-GaN* | Ponce [15.54] | 0.452 | 4.78 | - | 1.3 |
| c-InN* | Ponce [15.54] | 0.498 | 5.03 | 0.092 | 0.8 |
| c-AlN* | Ponce [15.54] | 0.438 | 4.56 | −0.032 | 2.8 |
| 3C SiC | Briot [15.55] | 0.436 | 2.7 | −0.037 | 4.9 |
| Si | Popovici [15.56] | 0.54301 | 3.59 | 0.168 | 1.5 |
| GaAs | Popovici [15.56] | 0.56533 | 6.0 | 0.20 | 0.5 |
| GaP | Strite [15.31] | 0.54512 | 4.65 | 0.17 | 0.8 |
| InP | Popovici [15.56] | 0.5859 | 4.5 | 0.229 | - |
| MgO | Popovici [15.56] | 0.4216 | 10.5 | −0.072 | - |
| $MgAlO_2$ | Popovici [15.56] | 0.8083 | 7.45 | - | - |

*) The linear expansion coefficient for cubic nitrides is not available. However, the lattice constants of cubic and hexagonal crystals are correlated by $a_{\text{cubic}} = \sqrt[3]{\sqrt{3}a^2{}_{\text{hex}}c_{\text{hex}}}$. The linear thermal expansion coefficient $\alpha_{\text{cubic}}$ is estimated from the hexagonal values for the a and c axis $\alpha_{\text{a}_{\text{hex}}}$ and $\alpha_{\text{c}_{\text{hex}}}$ by $\alpha_{\text{cubic}} \simeq 1/3 \cdot (2\alpha_{\text{a}_{\text{hex}}} + \alpha_{\text{c}_{\text{hex}}})$. The hexagonal values $\alpha_{\text{a}_{\text{hex}}}$ and $\alpha_{\text{c}_{\text{hex}}}$ are taken from [15.54].

be highly doped, both n- and p-type. From this point of view this restricts the substrates to be used to GaAs, Si and 3C-SiC.

The use of the 3C-SiC substrate is prompted by the small lattice mismatch to c-GaN (−0.037) and the high thermal conductivity of 3C-SiC ($4.9\,\text{Wcm}^{-1}\text{K}^{-1}$). However, due to the difficulties in the fabrication of cubic SiC bulk material, nowadays thick (>3 µm) 3C-SiC epilayers grown on Si (001) substrates by plasma enhanced chemical vapor deposition (PECVD) are used instead [15.46–48]. For high power devices, where the thermal conductivity of the substrate is very important, 3C-SiC will be the most appropriate material.

Growth of GaN on silicon substrates offers a very attractive way to incorporate future GaN devices onto silicon based integrated circuits. However, due to the large differences in lattice constant (mismatch 0.17) and thermal expansion coefficient, it is rather difficult to grow GaN epitaxially on Si substrates directly. The formation of amorphous $Si_xN_y$ inclusions at the interface, which act as nucleation centers for the formation of hexagonal GaN, hampers the epitaxial growth of phase-pure cubic epilayers [15.45]. Thin (<5 nm) 3C-SiC or GaAs interface layers are therefore necessary to prevent the formation of such $Si_xN_y$ inclusions.

GaAs as a substrate for cubic GaN growth is motivated by the potential for fabrication of heterostructure devices. Easy cleavage of laser facettes and the integration of GaN devices to the well developed GaAs technology makes this substrate very attractive although the thermal stability at the high growth temperatures of about 700°C may be problematic. As seen in Table 15.1 the estimated thermal expansion coefficient $\alpha_{cubic}$ of c-GaN is comparable to that of the GaAs substrate reducing cracking and thermal strain effects. The lattice mismatch $f$ of 20% is within the largest of the proposed substrates. For such large misfit systems, the assumption of elastic theory is no longer valid and a breakdown of epitaxial growth is expected, resulting in polycrystalline growth. However, recent results [15.35, 36] showed that due to the occurrence of a coincidence-lattice mismatch $f_0 = (ma_{sub} - na_{GaN})/ma_{sub}$ the residual lattice mismatch is drastically reduced and epitaxial growth enabled. In the case of c-GaN grown on GaAs the integers $m$ and $n$ are 4 and 5, respectively. Therefore, if $m = 5$ an additional lattice plan with a pure edge dislocation at the interface is incorporated into GaN. Transmission electron microscopy (TEM) measurements of the GaN/GaAs interface confirm this explanation. Thus, this heterosystem is close to true coincidence and an array of pure edge dislocations with a period of 5 GaN lattice planes can account for the entire misfit. The residual lattice misfit $f_0$ at growth temperature was estimated to be as low as $0.002 \pm 0.0002$. The various thin-film deposition methods used for the growth of cubic GaN and other group III nitrides have a number of advantages for addressing different issues in order to utilize the full potential of the cubic III nitrides for optoelectronic as well as electronic devices. However, it still requires significant progress for device applications in the areas of heteroepitaxial growth, crystal structure, substitutional doping, reduction of extended defects as well as point defects, alloying phenomena, and the formation of homo- and heterojunction structures.

Halide vapor phase epitaxy (HVPE), which uses a chloride transport method, was developed 1969 by Maruska and Tietjen [15.57] and is characterized by high growth rates. This may enable the growth of thick cubic GaN epilayers with significantly reduced misfit related defects close to the surface and can be used as pseudosubstrate material for metalorganic vapor phase epitaxy (MOVPE) and molecular beam epitaxy (MBE) [15.58]. However, this technique was abandoned in the early 1980s because of apparent difficulties in reducing the native shallow-donor concentration to nondegenerate levels and thus enabling p-type doping.

The first application of MOVPE to grow hexagonal III nitrides was reported in the mid-1980s [15.59]. Up to now, it is the most popular method to grow multilayer device structures of hexagonal III nitrides and is the only technique used for commercial III nitride production.

For cubic group III nitrides, however, only recently has MOVPE been successfully used [15.60, 61]. The precursors used for III nitride growth are trimethyl-gallium (TMGa) or triethyl-gallium (TEGa), trimethyl-aluminium (TMAl) and trimethyl-indium (TMIn) for the group III elements and ammonia ($NH_3$) or 1.1-dimethylhydrazine (DMHy) for nitrogen. The carrier gases are typically either high purity hydrogen, nitrogen or a mixture of both. In most cases GaAs substrates are used [15.60–62]; however, also 3C-SiC [15.63] are reported as substrates. Typical growth temperatures for c-GaN and c-$Al_yGa_{1-y}N$ ($y < 0.3$) are between 900°C and 950°C and for $In_xGa_{1-x}N$ between 700°C and 770°C [15.64]. Growth rates of c-GaN are in the order of about 240 nm $h^{-1}$ [15.65].

Doping in MOVPE is done *in situ* from the gas phase. For n-type doping by Si, silane ($SiH_4$) is used as precursor. In this way free electron concentrations up to $10^{20}$ $cm^{-3}$ are conveniently reached at room temperature in hexagonal GaN. Mg is the most suitable p-type dopant and can be introduced via the precursor bis-cyclopentadientyl magnesium ($Cp_2Mg$) up to a concentration of a few $10^{19}$ $cm^{-3}$. However, for activating the acceptors a post-growth heat treatment either by low-energy electron beam irradiation (LEEBI) [15.66] or in $N_2$ atmosphere at temperatures around 700°C are necessary [15.67]. Up to now no detailed studies on p- or n-type doping of cubic group III nitrides by MOVPE were reported. However, Tanaka et al. managed to grow a cubic GaN p-n homojunction [15.68].

The growth of GaN and group III nitrides by MBE (see Chap. 7) requires the development of appropriate nitrogen sources, as molecular nitrogen ($N_2$) does not chemisorb on Ga due to its large binding energy of 9.5 eV. To solve this problem different approaches are currently reported for the growth of cubic and hexagonal group III nitrides. The first approach is the use of gaseous sources like ammonia ($NH_3$) or 1,1-dimethylhydrazine (DMHy). This kind of MBE is also called reactive ion molecular beam epitaxy (RMBE). In fact, the use of $NH_3$ which dissociates at the growth front in a surface catalytic process has recently become fashionable after it was shown that high quality layers could be obtained with this method [15.69, 70]. However, this compound is quite thermally stable and as a result limits the growth temperature significantly. Therefore, lower growth temperatures, such as those needed for low temperature nucleation buffers or for layers containing In, cannot be grown as easily with $NH_3$ [15.71]. DMHy has higher reactivity than $NH_3$ and is expected to produce better quality crystals [15.39].

The second approach utilizes plasma-activated molecular nitrogen via DC plasma sources, electron cyclotron resonance (ECR) plasma sources or radio frequency (RF) plasma sources. DC plasma assisted MBE was successfully applied for the growth of cubic GaN [15.72]. However, due to the low growth rate of 10 – 30 nm $h^{-1}$, imposed by the limited nitrogen flux of the DC source, the synthesis of a 1 μm thick film would require approximately 50 hours, making it almost impossible to achieve stable growth conditions throughout such a run. In addition with increasing layer thickness the surface morphology became rougher and due to the formation of <111> facets which serve as nucleation sites for the hexagonal phase, phase purity got worse.

Compact MBE-compatible ECR sources are commercial available. ECR sources rely upon coupling of microwave energy at 2.45 GHz with the resonance frequency of electrons in a static magnetic field. Such coupling allows for ignition of the plasma at low pressures and powers, and produces a high concentration of radicals. In an ECR source approximately 10 % of the molecular nitrogen is converted into atomic nitrogen. Because these sources operate very efficiently at fairly low powers, they are usually only cooled by air. A typical growth rate of GaN achieved with an ECR source is about 200 nm $h^{-1}$ but it can be raised up to 1 μm $h^{-1}$ if exit apertures with a large number of holes are used [15.73]. A detailed description of the design and principle of operation of microwave plasma-assisted ECR sources is given by Moustakas [15.73].

Radio frequency (RF) plasma sources are among the most common sources in MBE growth of III–V materials. Nitrogen plasmas are generated by inductively coupling RF energy at a frequency of 13.56 MHz into a discharge chamber filled with nitrogen to pressures of $> 10^{-6}$ mbar. The discharge tube and the beam exit plate can be fabricated from pyrolytic boron nitride (PBN) avoiding quartz, which may be a source of residual Si or O doping of GaN. These sources are believed to produce significant concentration of atomic nitrogen. Due to the very high power used in this type of sources, up to 600 W, the plasma chambers usually must be water cooled. RF sources permit growth rates up to about 1 μm $h^{-1}$ and have become available by a number of vendors.

The different species of N produced by various kinds of plasma discharge may also have a profound impact on the growth kinetics, depending on whether the impinging species is an ionized molecule ($N_2^+$) or atomic N in an excited state ($N^*$). In a recent study by Myers et al. [15.74] using a quadrupole mass spectrometer, two different RF-plasma sources (Oxford Applied Research CARS-25 and EPI Vacuum Products Unibulb) were compared. These sources typically produce a complex mixture of active nitrogen superimposed on a background of inert molecular nitrogen. Whereas, the Oxford source produced primarily atomic nitrogen with little indication of the presence of molecular metasbles, the EPI source produced significantly less atomic nitrogen, but a significant flux of molecular nitrogen metastables. As pointed out by Newman [15.75], different compositions of the plasma are relevant for different mechanisms controlling the growth and the decomposition of GaN. Ionic and neutral atomic nitrogen can take part in growth and decomposition in contrast to the molecular nitrogen metastables, which mainly contribute to growth. This may explain the relatively low growth rates with atomic nitrogen sources. In addition, both atomic and metastable molecular nitrogen contain significantly more energy than required for GaN formation. Incorporation of atomic nitrogen releases this energy into the lattice where it can drive unfavorable reactions. In contrast, the excited molecule can incorporate one atom into the growing GaN while the other desorbs, carrying away the excess energy. Therefore, the selection of the plasma source influences the growth rate, surface morphology and optical and electrical properties of the resulting epilayer.

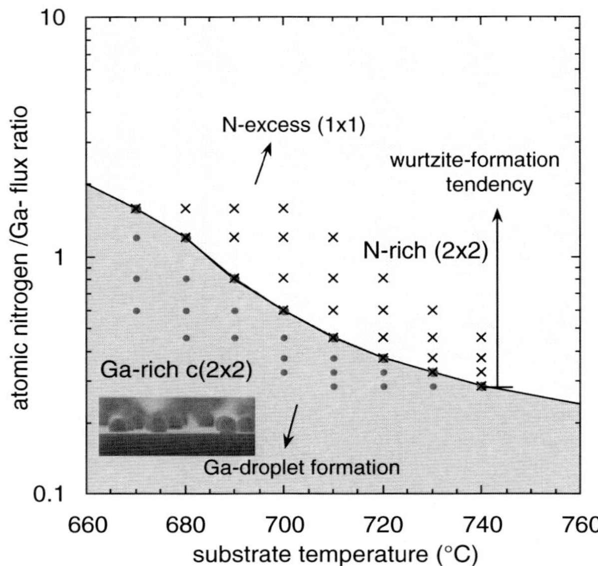


**Fig. 15.5.** Surface reconstruction of cubic GaN (001) as a function of growth temperature and N/Ga flux ratio. (taken from [15.34])

Single phase c-GaN epilayers were reported by Schikora et al. [15.36] using an MBE system equipped with elemental sources of Ga, In, Al, As, Mg, and Si. As nitrogen source a RF-activated plasma was used. Before starting the c-GaN nucleation, a GaAs buffer layer was grown first at 600°C under (2×4) reconstruction to ensure As-stabilized conditions. The nucleation of c-GaN was initiated at the same substrate temperature using an N/Ga flux ratio of about 4. After deposition of 10 – 20 monolayers (ML), the nucleation stage was stopped and the substrate temperature subsequently raised to 680 – 740°C. The static GaN (001) surface exhibits a clear (2×2) reconstruction during this period. The growth was continued at the higher temperature level varying the N/Ga flux ratio and the substrate temperature. The growth process was monitored continuously by RHEED.

The reconstruction of the growing surface is controlled mainly by surface composition. Figure 15.5 depicts the surface reconstruction diagram of c-GaN measured during growth. The V/III flux ratio is plotted versus growth temperature. In this ratio, the flux rate of atomic N is related to the Ga flux rate. It was found by mass-spectroscopic measurements that the effective flow of atomic N arriving on the surface amounts to 1 – 7 % of the $N_2$-beam-equivalent pressure, mainly dependent on the total flow rate and the RF power applied. In addition to static (2×2) reconstruction, c(2×2) and (2×2) reconstructions as well as an unreconstructed (1×1) surface were observed during growth, in accordance with other published results [15.39, 72]. At high excess of N, the unreconstructed (1×1) surface is stable. Under Ga-stabilized conditions, a c(2×2) reconstruction appears; layers grown at such conditions show n-type conductivity in Hall-effect measurements [15.76]. The (2×2) reconstruction is associated with N-stabilized conditions giving rise to p-type behavior. In a narrow range between the c(2×2) and (2×2) regime, both reconstructions occur simultaneously with different intensities of the reconstruction lines, indicating a nearly stoichiometric adatom coverage. At a temperature lower than 680°C, this requires an N/Ga flux ratio larger than 1. For substrate temperatures higher than 700°C, Ga re-evaporation becomes significant. The Ga loss from the surface must be compensated to stabilize stoichiometric conditions. Therefore, the N/Ga ratio decreases.

Due to the metastability of the cubic phase of GaN the structural perfection of epitaxial layers of this material is extremely sensitive to the growth parameters. It has been found that only slight deviations from the ideal stoichiometric growth conditions can influence the structural [15.77] and optical properties, and the type of conductivity [15.76] of undoped c-GaN. Ga excess at the surface favors the formation of Ga droplets and μm-size crystalline inclusions [15.78], which are detrimental to the epilayer surface morphology. In addition, at Ga-rich conditions different types of surface irregularities are observed as a result of successive melt-back etching in GaN and GaAs and solution growth within Ga droplets due to the change of the saturation conditions of the liquid Ga phase on the surface of the growing film [15.79]. On the other hand N-rich conditions seem to enhance the probability of the formation of inclusions with hexagonal crystalline structure [15.36, 72]. For all these reasons it is necessary to control the growth parameters during c-GaN MBE very accurately. To do this, a method which allows an accurate determination of the transition between the c(2×2) and the (2×2) surface reconstruction of c-GaN was developed [15.80]. The phase boundary which is measured with high accuracy is then used as a reference point to establish slightly Ga- or N-excess growth conditions. This method takes advantage of the fact that the transition between the c(2×2) and (2×2) surface reconstruction can be monitored by the RHEED intensity of the half-order reconstruction measured along the $[\bar{1}10]$ azimuth [15.81]. Due to the narrow window of growth conditions resulting in c-GaN it is very difficult to obtain pure single-phase layers. So, it is very important to distinguish between the cubic and the hexagonal phase when analyzing the layer structure.

The most common method to analyze structural properties of epitaxial films is high resolution X-ray diffraction. However, standard diffractometer $\omega - 2\Theta$ scans are unable to detect hexagonal subdomains, since their c-axis may be tilted with respect to the cubic main axis [15.72, 82]. Therefore, so-called "reciprocal space maps" have been used to measure hexagonal phase inclusions in c-GaN epilayers [15.72]. Layers grown under N-excess showed beside the cubic (002) reflex a pronounced diffraction peak which was attributed to an $(10\bar{1}1)$ reflection from hexagonal grains with their $[\bar{0}01]$ axis parallel to the $[\bar{1}11]$ axis of the cubic phase. This result is in good agreement with the interpretation of RHEED patterns from c-GaN grown under N-rich conditions [15.36]. However, the X-ray intensity scattered by hexagonal grains is small. Therefore, a high intensity of the primary X-ray beam and relatively long measuring times are required to detect the hexagonal grains. For cubic GaN epilayers on GaAs substrates a phase purity of better than 99.9 % has been obtained.

### 15.2.2 IV–VI Compound Semiconductors

The group of IV–VI compound semiconductors distinguishes itself from the other frequently used II–VI and III–V materials by its enormously high dielectric constant [15.83]. This results from the structure of the lead salts which is close to a phase transition to a rhombohedral modification with ferroelectric properties. As an additional effect of the large dielectric constants, the scattering of charged carriers at ionized impurities is effectively screened. This is most pronounced in PbTe, for which carrier mobilities at low temperatures of several $10^6 cm^2/Vs$ have been reported [15.84]. The effective screening due to the large dielectric constant makes the IV–VI materials quite insensitive to defects incorporated during growth or by post-growth fabrication techniques. All these properties make the IV–VI materials very much appropriate for the fabrication of Bragg mirrors to fulfill the boundary conditions, described in the next section, as closely as possible. Therefore, we will concentrate in the following on the material peculiarities of lead salts.

First attempts to grow epilayers from IV–VI materials go back to 1950 [15.85]. Vacuum deposition was used to grow on freshly cleaved NaCl substrates. However, the layers showed pronounced mosaicity and the quality was in most cases insufficient for device fabrication. A breakthrough was made by Holloway et al. [15.86, 87], by vacuum evaporation of PbTe on $BaF_2$ single crystals. Further quality improvements were achieved by Lopez-Otero [15.88], who developed the so-called hot-wall epitaxy (HWE) (see Sect. 15.2.3), where growth is carried out close to thermodynamic equilibrium. The combination of several HWE systems in a single vacuum chamber allowed the growth of quantum wells [15.89–92] and diode lasers [15.93, 94]. In spite of the remarkable success of HWE, research activities shifted in the 1980s more and more to MBE, which offers in contrast to HWE the full range of *in situ* control techniques like RHEED [15.95–97], AES [15.98, 99] and scanning tunneling microscopy (STM) [15.100]. An excellent summary of peculiarities of growth of IV–VI semiconductors can be found in [15.101].

In IV–VI MBE, the main constituents (PbTe, PbS, PbSe, SnTe or SnSe) are usually supplied from compound effusion cells due to their evaporation as binary molecules. For the growth of the ternary alloys $Pb_{1-x}M_xX$, where Pb is partially substituted by a rare-earth or alkaline-earth element (M), the alloy components M are supplied from elemental sources due to the low vapor pressures of the M chalcogenides. Elemental chalcogen sources are also required for retaining the right stoichiometry of the layers. Dopants such as Bi, Tl or Ag can be supplied from elemental as well as compound ($Bi_2Te_3$ or $Tl_2Te$) sources. For all effusion sources, operation temperatures are typically in the range of 300°C to 700°C.

The substrate temperatures in IV–VI MBE are typically in the range of 250°C to 350°C where a unity sticking coefficient of all materials except the chalcogens can be assumed. At temperatures above 380°C, a significant re-evaporation occurs from the lead salt layers, and the growth rates as well as layer composition become increasingly difficult to be controlled. At 450°C, the re-evaporation rate of the lead salt compounds is comparable to typical MBE growth rates. Since below 350°C the substrate temperature measurements with the usual thermocouple assembly is not very reliable, special substrate temperature calibration procedures are needed for precise knowledge of the growth temperatures. Technologically, the design of IV–VI MBE systems is quite similar to III–V solid source MBE systems. However, since the electrical properties of the layers are much less affected by impurities incorporated during epitaxial growth, the vacuum requirements for IV–VI MBE are considerably relaxed and high quality lead salt layers have been obtained in MBE systems with base pressures of only a few times $10^{-8}$ mbar [15.102].

There exists a big variety of substrate materials for IV–VI epitaxy. The most commonly used are summarized in Table 15.2 together with the lattice constants and thermal expansion coefficients. None of the commercially available substrates provides the ideal match to the IV–VI compounds. Therefore, the best choice depends on the given applications.

In much of the early work, cleaved alkali-halide (NaCl or KCl) single crystal plates served as substrates for lead salt epitaxial layers[15.85]. These compounds provide the same crystal structure, have lattice constants not too different from that of the lead salts, and their natural (100) cleavage plane is

**Table 15.2.** Material properties of narrow gap IV–VI semiconductors and their substrates

| material | Lattice constant at 300K [nm] | Thermal expansion coefficient at 300K [$10^{-6}K^{-1}$] |
|---|---|---|
| $BaF_2$ | 0.6200 | 18.8 |
| $SrF_2$ | 0.5800 | 18.4 |
| $CaF_2$ | 0.5463 | 19.1 |
| Si | 0.5431 | 2.6 |
| GaAs | 0.5653 | 6 |
| NaCl | 0.5640 | 40 |
| KCl | 0.6290 | 39 |
| PbS | 5.936 | 20.3 |
| PbSe | 6.124 | 19.4 |
| PbTe | 6.462 | 19.8 |

also the preferred growth direction of the lead salt compounds. However, lead salt layers nucleate in the form of 3D islands, and exhibit quite high defect densities and electrical properties much inferior to that of bulk material. This is in part due to the very large differences in thermal expansion coefficients (see Table 15.2), which leads to the formation of micro cracks or even peeling of the epitaxial layers upon thermal cycling.

Alkaline-earth fluorides, in particular $BaF_2$, are much better suited for lead salt epitaxy, in spite of their different crystal structure (calcium fluoride structure). On the one hand, the lattice mismatch between $BaF_2$ and PbSe or PbTe is not very large (−1.2 % and +4.2 %, respectively). On the other hand, in contrast to all other substrate materials, the thermal expansion coefficients above 300 K are very well matched to that of the lead salt compounds. Therefore, the cooling from the epitaxial growth temperatures to RT or below does not produce large thermal strains in the layers, i.e., the layers are mechanically stable upon thermal cycling. Furthermore, $BaF_2$ is highly insulating and optically transparent well into the far-IR region.

Like the alkali-halide substrates, $BaF_2$ substrates are usually obtained by fresh cleavage of thin plates from single crystal blocks. This produces the cleanest surface possible and besides outgassing, no further treatment is required for epitaxial growth. Although the natural (111) cleavage plane of the alkaline-earth fluorides is different from the preferred (100) growth direction of the lead salt compounds, the growth on clean (111) $BaF_2$ substrates usually yields perfectly (111)-oriented epitaxial layers with no (100)-oriented secondary phases. Practical problems with alkaline-earth fluoride substrates arise from the fact that commercially available single crystals are produced mainly for optical applications and often do not meet the high quality standards required for epitaxial growth. Based on the experience with several different commercial vendors, most large $BaF_2$ single crystals contain several small angle grain boundaries indicated by a slight splitting of the diffraction peaks in high resolution X-ray diffraction spectra. Even from the same vendor, the quality varies substantially so that often only selected crystals can be used. In addition, the cleavage of the $BaF_2$ substrates results in cleavage steps on the surface, which can be a problem for device applications. These cleavage steps can be removed by polishing [15.103].

Standard semiconductor substrates like Si and GaAs have also been tried for IV–VI heteroepitaxy. However, these materials have not only a different crystal structure, they also have much smaller lattice constants and smaller thermal expansion coefficients than the lead salt compounds (see Table 15.2). Thus, PbTe grown on 12.5 % lattice-mismatched GaAs (100) results in mixed (100)- and (111)-oriented epitaxial layers for most growth conditions [15.104], and microcracks appear in the layers when subjected to thermal cycling [15.105]. Similar observations have been made for Si (100) substrates. However, high quality lead salt layers have been obtained on (111)-oriented Si substrates [15.106, 107] using thin group IIa fluoride as buffer layers, despite the large lattice mismatch between Si and PbSe or PbTe of 11.3 % and 16 %, respectively. This is because misfit strain and strain due to differences in thermal expansion for growth in the (111) direction can be easily relaxed by gliding misfit dislocations. A different approach for lead salt growth on Si(100) substrates is the use of rare-earth chalcogenide (e.g., YbS) buffer layers [15.108]. In this case, strain is relaxed by a regular network of misfit dislocations formed by dislocation climb processes. Recently, other semiconductor substrate materials like CdTe ($a_0$ = 6.482 Å), GaSb ($a_0$ = 6.086 Å) and InSb ($a_0$ = 6.479 Å) with better matching of the lattice constants and thermal expansion coefficients to the lead salts have also become commercially available. However, up to now, no lead salt growth studies on these new substrate materials have been reported.

It is well known that three types of growth modes exist for heteroepitaxial growth, depending on the free energies of the substrate–vacuum ($\gamma_s$), substrate–overlayer ($\gamma_i$) and overlayer–vacuum interfaces ($\gamma$), as well as on the differences in the lattice constants. (A detailed discussion on heteroepitaxial growth modes can be found in Sect. 1.2 of this volume.) When $\gamma_s$ is smaller than the sum of $\gamma_i$ and $\gamma$, nucleation of 3D clusters onto the substrate occurs. This so-called Volmer–Weber growth mode is present for lead salt deposition on most non IV–VI substrates such as $BaF_2$, KCl, Si or GaAs. To obtain a smooth epitaxial surface, thick buffer layers have to be deposited on such substrate materials. When $\gamma_s$ is larger than $\gamma_i + \gamma$, homogeneous 2D layer-by-layer growth mode takes place. This type of growth mode is usually observed for IV–VI heteroepitaxy in the absence of a high misfit strain as is the case for the binary lead salts and their large bandgap ternary alloys where the ternary content is usually only a few percent. For highly mismatched heteroepitaxial systems, however, the uniform 2D epitaxial layer is not the lowest free energy configuration of the layer even if initially a 2D wetting layer is formed on the substrate. The reason is that the formation of a corrugated surface during growth allows a significant reduction of strain energy due to partial *elastic* strain relaxation perpendicular to the additional free surfaces thus created. As a result, highly strained epitaxial layers are fundamentally unstable against surface roughening, i.e., after formation of a thin wetting layer coherent 3D islands are formed on the epitaxial surface. This corresponds to the so-called Stranski-Krastanov growth mode. Depending on the combination of IV–VI compounds the growth mode can change from a 2D layer-by-layer growth, favorable for multilayers and superlattices, to a 3D growth resulting in the formation of quantum dot arrays. Particular examples were selected for this chapter leading to very interesting structures as well from the basic physics point of view as also from the possible application of such structures.

**Bragg Mirrors for the IR.** Thin layers of dielectric materials are very common for antireflection coatings of optical components like lenses. Multilayers of dielectric materials are mainly used as mirrors, especially for lasers.

These so-called Bragg mirrors are multilayer structures consisting of alternating pairs of material with different dielectric constants. The high reflectivity is caused by multiple interference of the light partly reflected at the interface of the alternating materials. This interference effect has the following main advantages in comparison to metallic mirrors:

(i) Bragg mirrors can be designed for a specific target wavelength and
(ii) they exhibit very small intrinsic absorption resulting in reflectivities beyond 99 %.

To obtain such excellent optical properties for Bragg mirrors the layers they consist of must fulfill quality criteria which are very challenging for the crystal grower. To prevent internal absorption, the single layers must be as transparent as possible, which means that the materials must be of high purity without any defects giving rise to absorption processes. To avoid losses by scattering effects the interfaces between the layers must be perfect in terms of interface roughness. Since the layer thickness determines the target wavelength the growth rate must be controlled with high precision. . This can be achieved by optical monitoring as discussed in Sect. 10.3.5 in connection with Fig. 10.29 Examples of devices in need of such tight are vertical-cavity surface emitting lasers (VCSEL) [15.109–111], resonant-cavity light-emitting diodes (RCLED) [15.112–114], solar cells with enhanced efficiency [15.115], wavelength selective photodetectors [15.116–118] or Fabry–Perot filters [15.119]. The main effect of the Bragg mirrors in these devices is either an intensity enhancement of electromagnetic waves within the optical active medium or the fact that the optical resonators support only certain wavelengths. The design rules for dielectric Bragg interference mirrors can be found in the review by W. Heiß et al. [15.120] which was used also partly as a guideline for this chapter.

Bragg mirrors for the IV–VI material system consisting of PbTe and $Pb_{1-x}Eu_xTe$ were grown by MBE on $BaF_2$ substrates [15.121]. The $Pb_{1-x}Eu_xTe$ ternary composition was determined by the PbTe to Eu beam flux ratio, and an excess $Te_2$ flux [15.122] was used to obtain the correct stoichiometry. The growth rate, which was calibrated by a quartz crystal micro balance moved into the sample position, is typically around 2 $\mu m\,h^{-1}$ for PbTe while it decreases with increasing Eu content to 0.77 $\mu m\,h^{-1}$ for EuTe. From *in situ* reflection high-energy electron diffraction studies, the EuTe (111) surface shows a very strong tendency for (100) facetation due to the resulting lowering of the free surface energy and due to the lattice mismatch to the $Pb_{1-x}Eu_xTe$ layers amounting to 2 %. As a result, 2D growth can be obtained only when the EuTe surface is kept close to the transition between the Eu- and Te-stabilized surface states, which can be easily distinguished because of their different surface reconstructions. For the Te/Eu beam flux ratio of 2 [15.123], this transition takes place at a substrate temperature of 260°C . Therefore, the EuTe layers were grown at substrate temperatures of $260 \pm 10$°C whereas for $Pb_{1-x}Eu_xTe$ a growth temperature of 340°C was used. As an example two different sample structures with different chemical compositions of the $\lambda/4$ layers are presented in Fig. 15.6.

Sample S1 represents a structure, consisting of 32 $Pb_{1-x}Eu_xTe$ layer pairs with Eu contents of 1 % and 6 %. The 1 % layer was grown as a $PbTe/Pb_{1-x}Eu_xTe$ ($x = 6\,\%$) short-period superlattice digital alloy with a 4 nm superlattice period and a $PbTe/Pb_{1-x}Eu_xTe$ thickness ratio of 5:1. Due to this digital alloy, the total number of layers is as large as 3230 for sample S1. The Bragg mirror was designed to match the band gap of PbTe at 77 K of 217 meV or a wavelength of 5.7 μm. To obtain maximum reflectivity for incidence of light from air the layer sequence starts with $x = 6\,\%$ on the $BaF_2$ substrate. Due to the long target wavelength and the large number of $\lambda/4$ layers required to obtain high reflectivity, the total layer thickness of S1 amounts to 15.96 μm. For sample S2, $\lambda/4$ layers of $Pb_{0.93}Eu_{0.07}Te$ and EuTe were used. This yields a much higher index contrast of 68 % and therefore, high reflectivities can be achieved by a very small number of layer pairs. The total layer thickness of sample S2, a Bragg mirror with 3.5 periods of $EuTe/Pb_{0.93}Eu_{0.07}Te$ layer pairs and a target wavelength of 3.8 μm, amounts to only 2.55 μm.

For structural characterization X-ray diffraction spectra of the samples were measured using a high resolution X-ray diffraction set-up with a Bartels primary monochromator and Cu $K_{\alpha_1}$ radiation. The $\Omega/2\Theta$ diffraction spectra recorded around the (222) reflex are shown in Fig. 15.6. For sample S1 the main characteristics of the diffraction spectrum in Fig. 15.6a are two main peaks from the $Pb_{1-x}Eu_xTe$ layers of the $\lambda/4$ pair with a different Eu content. The peak splitting of only 257 arcsec indicates a different lattice constant of about 0.25 %. In addition, widely spaced satellite peaks from the 1 % $Pb_{1-x}Eu_xTe$ digital alloy superlattice are observed, but no satellite peaks from the periodic $\lambda/4$ pair stacking are resolved due to the very large Bragg mirror periods of about 0.5 μm. The full width at half maximum (FWHM) of the diffraction peaks is about 99 arcsec. For S2 (Fig. 15.6b), the splitting between the EuTe and the $Pb_{1-x}Eu_xTe$ SL0 peaks of 1998 arcsec is about ten times larger than that for sample S1. The lattice constants derived from the peak positions indicate that each $\lambda/4$ layer is almost relaxed to its bulk lattice constant in spite of the 2 % lattice mismatch between the layers. This is due to the fact that the thickness of the $\lambda/4$ layers by far exceeds the critical thickness for strain relaxation. As a result, a high density of misfit dislocations is formed at each of the heterointerfaces, resulting also in a threefold broadening of the X-ray diffraction peaks to a FWHM of 300 arcsec. However, as shown below, this does not degrade the very high optical quality of the mirror structure.

The high lateral and vertical homogeneity of the grown mirror structures are demonstrated by the cross-sectional scanning electron micrograph (SEM) of a $Pb_{0.94}Eu_{0.06}Te/Pb_{0.99}Eu_{0.01}Te$ layer stack in Fig. 15.7. This reference sample was first mesa etched with a $Br/HBr/H_2O$ solution and subsequently

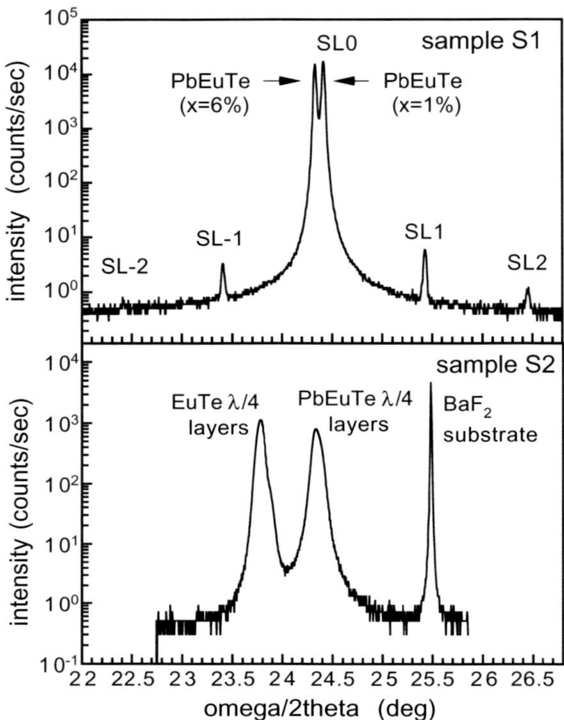


**Fig. 15.6.** $\Omega/2\Theta$ X-ray diffraction spectra (Cu $K_{\alpha 1}$) of the Bragg mirrors S1 (**a**) and S2 (**b**) around the (222) reflection. The satellite peaks in (**a**) stem from the $Pb_{0.99}Eu_{0.01}Te$ $\lambda/4$ layers, which were grown as $PbTe/Pb_{0.94}Eu_{0.06}Te$ digital alloy superlattice with a period of 40 Å (taken from [15.120])

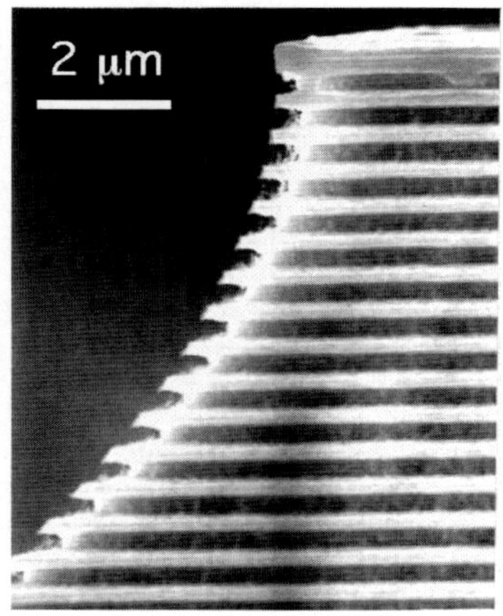


**Fig. 15.7.** Cross-sectional SEM of the cleaved edge of a 20-period $Pb_{0.99}Eu_{0.01}Te/Pb_{0.94}Eu_{0.06}Te$ Bragg mirror sample. The sample was selectively etched with a $CH_4 - H_2$ plasma to enhance the chemical contrast (taken from [15.120])

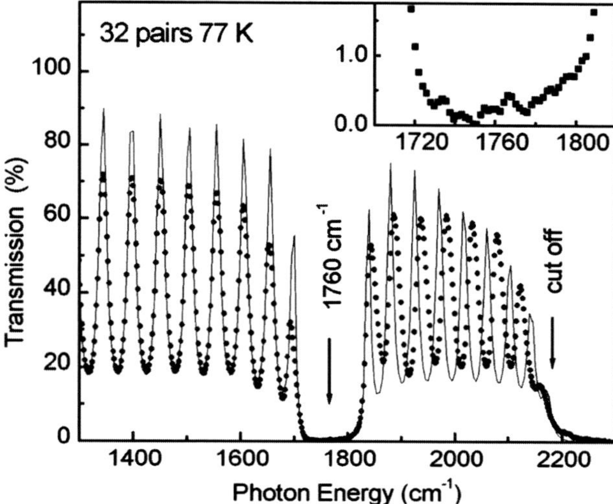


**Fig. 15.8.** Measured transmission spectrum (•) of a 32-period $Pb_{0.99}Eu_{0.01}Te$/ $Pb_{0.94}Eu_{0.06}Te$ Bragg mirror (S1) at 77 K. The inset shows the spectral region of the mirror stop band on an enlarged scale. The solid line represents the numerical result from the transfer matrix method (taken from [15.120])

selectively etched using a $CH_4/H_2$ plasma in a barrel reactor. With this method, the layers with lower Eu content are etched deeper [15.124], and hence appear darker in the SEM image than those with higher Eu content. The left edge in the picture is the side wall of the mesa stripe which is wedge shaped due to the wet chemical etching process with the Br solution.

The FTIR transmission spectrum of sample S1 at 77 K is shown in Fig. 15.8. This mirror exhibits a pronounced stop band region centered at an energy of 1760 cm$^{-1}$ ($\lambda = 5.7\,\mu$m) and exhibits a width of 80 cm$^{-1}$. As shown in detail in the inset of Fig. 15.8, a minimum transmission of about 0.3 % is observed at the center of the stop band, corresponding to a reflectivity exceeding 99 %. Such high reflectivities are well suited for VCSEL devices. The spacing of the Fabry–Perot fringes is very narrow due to the large total thickness of the 32-period sample. The transmission cut-off at 2200 cm$^{-1}$ corresponds to the fundamental absorption of the $Pb_{1-x}Eu_xTe$ layers with x = 1 %. The numerical calculation (solid line) is in very good agreement with the experimental data, apart from the slight deviation near the absorption edge region. This deviation at higher energies is indicative of a small inhomogeneity of the layer thickness (below 1 %) in the measured area of the sample.

Although the reflectivity of S1 in the stop band is sufficient for laser resonators, for probing the temperature behavior of lead salt devices, a much wider Bragg mirror stop band is desired. This can be achieved by using a

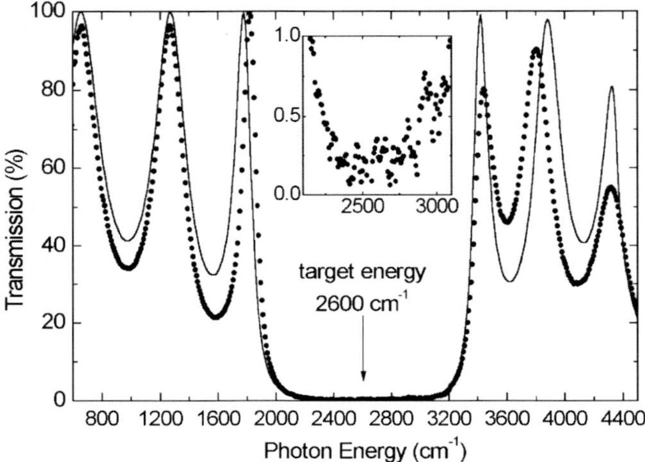


**Fig. 15.9.** Room temperature transmission spectrum of a three-period $Pb_{0.93}Eu_{0.07}Te/EuTe$ Bragg mirror (S2). The inset shows the region of the mirror stop band whose target energy coincides with the energy band gap of PbTe at room temperature. The solid line is the calculated transmission spectrum and the dots are the experimental results (taken from [15.120])

larger refractive index contrast in the case of sample S2. The room temperature transmission spectrum of S2 in Fig. 15.9 clearly shows the mirror stop band around the target energy of $2600\,cm^{-1}$ ($3.8\,\mu m$). The spacing of the Fabry–Perot interference fringes is much larger than in the case of S1, due to the reduced total thickness of the Bragg mirror stack. The transmission minimum within the stop band, shown in the inset of Fig. 15.9 on an enlarged scale, is as small as that for S1, in spite of the small number of Bragg layer pairs of 3. In contrast, the width of the stop band of about $1300\,cm^{-1}$ is 16 times larger than that of S1. In particular, the stop band width is larger than the total band gap energy shift of PbTe and PbSe between room temperature and 0 K.

**Quantum dot superlattices.** Great efforts were made to fabricate semiconductor nanostructures because of their application potential in semiconductor devices like lasers. Apart from processing techniques involving lithography and etching, the direct growth of nanostructures has evolved as a new promising approach [15.125, 126]. It is based on the natural tendency of strained heteroepitaxial layers to spontaneously form coherent, dislocation free, three-dimensional (3D) islands after formation of a uniform 2D wetting layer with monolayer thickness [15.127–129]. The driving mechanism for this Stranski–Krastanov growth mode transition is the very efficient strain energy relaxation possible within the 3D islands by lateral expansion or compression in the directions of the free side faces [15.130, 131]. When the energy gained is larger than the increase in free surface energy, this islanding transition leads to a lowering of the total free energy of the system. This is the basis of the fundamental Asaro–Tiller–Grinfeld [15.132, 133] instability of strained surfaces.

Spontaneous coherent islanding has been studied for a large number of lattice-mismatched heteroepitaxial systems ranging from SiGe/Si [15.134–136] to III–V semiconductors such as InAs/GaAs [15.125, 126, 137], InP/GaInP [15.138] and GaSb/GaAs [15.139], the wide band-gap Ga, In, and Al-nitrides [15.140–142] and the II–VI compounds [15.143–145]. However, in this chapter we will concentrate on islanding in IV–VI compounds which was reported by Springholz et al. [15.146, 150]. Because of their coherent defect-free interfaces, these quantum dots usually exhibit much better electronic properties as compared to those produced by lithography and etching techniques. In addition, very high dot densities up to $10^{11}$ $cm^{-2}$ can be easily obtained. On the other hand, the considerable variations in sizes and shapes within ensembles of self-assembled quantum dots has remained a critical issue for device applications because of the resulting inhomogeneous broadening of the electronic density of states [15.151, 152]. In addition, self-assembled growth provides only limited control of the dot sizes, spacing and positions.

Multi layering of self-assembled quantum dots can provide a possible route for improving the size uniformity and lateral order of the dots. Theoretical [15.146, 153–155] as well as experimental work [15.146, 156] has shown that within such multilayers, the elastic interaction of dots can lead to the formation of long-range spatial correlation, which may lead to a lateral ordering and size homogenization of the dots as well. Self-assembled quantum dot superlattices consist of highly strained Stranski–Krastanov island layers that are separated by spacer layers with adjustable layer thickness. These spacer layers are usually closely lattice matched to the substrate or buffer material, i.e., during dot overgrowth, the driving force for strain-induced coherent islanding is removed. As a result, a very rapid replanarization of the epitaxial surface takes place as soon as the Stranski–Krastanov islands are covered by the spacer layers. The starting surface for each new dot layer is therefore completely smooth. On the other hand, the strong lattice deformations of the matrix around the buried islands produces a nonuniform strain distribution on the epitaxial surface due to the large difference in the lattice constant between the islands and the matrix material. For a compressively strained island, the surrounding matrix is locally expanded, whereas it is locally compressed around tensile islands. During the growth of the new dot layer, preferred dot nucleation takes place at the minima of the non-uniform strain distribution that is spatially correlated to the dot positions in the previous layer. As a result, long-range *vertical* dot correlations across the spacer layers are formed [15.149]. In quantum dot superlattices of IV–VI compounds a particular efficient lateral ordering of the PbSe dots takes place with the formation of almost perfect hexagonal 2D lattices of dots on the surface already after a few superlattice periods. In addition, the layer-to-layer dot correlation for these IV–VI materials is not parallel but inclined to the growth direction resulting in a unique fcc-like vertical dot sequence.

Thus, remarkable homogeneous trigonal 3D quantum dot crystals are formed, and their lattice constants can be tuned continuously just by changing the spacer layer thickness. Samples were grown by MBE on several-micrometer thick PbTe buffer layers predeposited on (111) oriented $BaF_2$ substrates. All superlattices consisted of five monolayers (ML) PbSe, alternating with $Pb_{1-x}Eu_xTe$ spacer layers with a thickness ranging from 300 to 600 Å and $x_{Eu} = 0.05 - 0.1$. The number of periods in the superlattices ranged from $N = 1$ to 100. The growth rates were 0.08 and 0.96 ML/s for PbSe and $Pb_{1-x}Eu_xTe$, respectively, and a substrate temperature of 360°C was used throughout all growth experiments. [15.150]

During superlattice growth, the evolution of the surface structure was monitored *in situ* by reflection high-energy electron diffraction (RHEED). During the growth of the PbSe wetting layer, the RHEED pattern remains equal to that of the 2D $Pb_{1-x}Eu_xTe$ spacer layer. At the critical coverage of 1.4 ML, weak chevron-shaped 3D transmission spots start to appear in the diffraction image. At 2.3 ML coverage, an abrupt transition to a 3D transmission pattern takes place, and the specular spot completely disappears. During the dot overgrowth with the $Pb_{1-x}Eu_xTe$ spacer layers, these changes are completely reversed, and a 2D reflection diffraction image is recovered after about 200 Å spacer thickness. Thus, a smooth 2D surface is restored once the PbSe islands have been buried by the spacer layer. This characteristic behavior is reproduced during the growth of each superlattice period, and even after 100 periods, the same absolute intensity changes take place. In particular, the critical coverage at which PbSe islanding occurs remains constant throughout superlattice growth. This indicates that the distribution of material between the 2D wetting layer and the 3D islands does not change within the superlattice stack. The lattice mismatch of PbSe with respect to PbTe is $\approx 5.5\,\%$. As a result, for PbSe deposition on PbTe (111), strain-induced coherent islands are formed on the epitaxial surface once a critical coverage of 1.4 ML is exceeded [15.147]. These islands have a pyramidal shape with a triangular base and steep (100) side facets [15.147].

After superlattice growth, the samples were rapidly cooled to room temperature to freeze-in the epitaxial surface morphology. The surface was imaged by atomic force microscopy (AFM) directly after removal of the samples from the MBE system. AFM measurement were carried out using sharpened Micro- and Ultralevers of Park Scientific Instruments. Special image processing software was used for real space statistical analysis of the dot size distributions on the one hand, and for frequency space analysis of the degree of lateral ordering on the other hand. The vertical and lateral correlation of the dots in the superlattices were studied using high-resolution X-ray reciprocal space mapping.

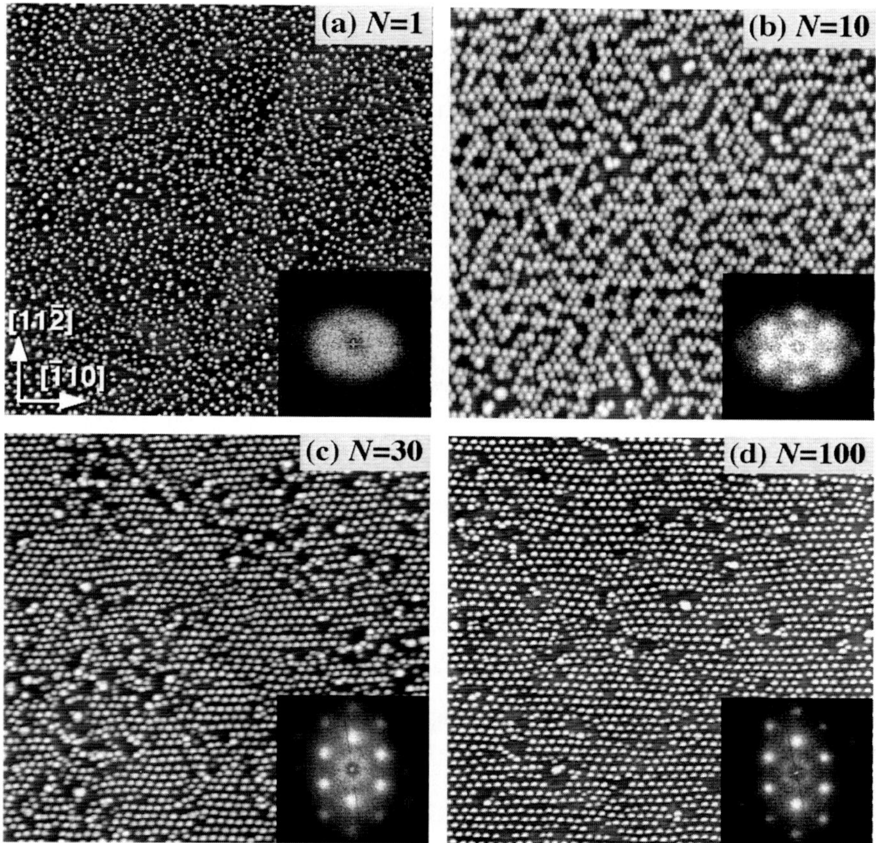


**Fig. 15.10.** AFM images (3×3 μm$^2$) of (**a**) a single layer of self-assembled PbSe quantum dots and the last PbSe layer of a PbSe/$Pb_{1-x}Eu_xTe$ superlattice with (**b**) $N = 10$, (**c**) $N = 30$, (**d**) $N = 100$ periods. The inserts show the Fourier transformation of the AFM images (taken from [15.148])

Figure 15.10 shows a series of AFM images of the last uncapped PbSe dot layers of samples consisting of $N = 1, 10, 30$, and 100 superlattice periods and a constant $Pb_{1-x}Eu_xTe$ spacer thickness of 470 Å.

For the single layer (Fig. 15.10a), the PbSe islands are distributed randomly on the surface without any preferred lateral correlation. With increasing number of superlattice periods, a rapidly progressing ordering of the dots occurs. Already after 10 periods, the dots are preferentially aligned in single and double rows along the <110> directions (Fig. 15.10b). Measurements on samples with less than 10 bilayers show that this ordering commences first with the formation of small patches of hexagonally ordered regions, which subsequently enlarge and join to form these row-type structures. With a further increasing number of bilayers, increasingly larger ordered regions are formed (Fig. 15.10c). For samples with $N \geq 30$, the perfect hexagonal arrangement is disturbed only by single point defects, such as missing dots, dots at interstitial positions or, occasionally, by additionally inserted dot rows ("dislocations").

The development of the lateral ordering was determined by Fourier transformation (FFT) (inserts of Fig. 15.10), as well as autocorrelation (AC) analysis of the AFM images. The FFT power spectrum of the $N = 1$ single dot layer AFM image exhibits a broad and diffuse ring around the frequency origin. By fitting cuts through the ring in several directions with Gaussians, one obtains a mean peak position of about 12.5 $\mu m^{-1}$, which corresponds to an average dot–dot distance of 800 Å. In addition, the AC spectrum of the AFM image does not exhibit any structure outside the central maximum, indicating a lack of any preferred direction of the nearest neighbor dot position.

In contrast, the FFT power spectrum of the 10 bilayer sample (Fig. 15.10b) clearly shows six well separated side maxima at a frequency of 14.5 $\mu m^{-1}$. This corresponds to a mean spacing between the dot rows of 590 Å. Six side maxima appear also in the AC spectrum, indicating that the next-nearest neighbors of the dots are along the <110> directions, with a preferred dot–dot distance of 680 Å within the rows. Apart from the six side maxima, the FFT power spectrum also exhibits a well-defined ring at a spatial frequency of one third of the side peaks, and this ring also exhibits a hexagonal symmetry. From a closer inspection, it is found to be due to the missing rows in the dot arrangement. In analogy with the notation of superstructures on crystalline surfaces, this dot arrangement can be referred to as a $(3 \times 1)$ missing row superstructure indicating that, on average, every third dot row is missing (see Fig. 15.10b). For the 30 and 100 period superlattices, the peaks in the FFT spectra sharpen drastically, and many higher-order satellite peaks are observed (Fig. 15.10c and d).

The vertical correlation of the PbSe dots in the superlattices were characterized by coplaner high-resolution X-ray diffraction (Cu $K_{\alpha_1}$), using a primary Bartels monochromator and a secondary analyzer crystal to record the diffracted intensity as a function of the various diffraction angles. The intensity distribution in reciprocal space (reciprocal space maps, RSMs) around various bulk reflections was recorded. Figure 15.11 shows reciprocal space maps around the (222) bulk reflection recorded for two different azimuth directions within the surface plane. In the maps, the ordinate axis, $q_z$, is parallel to the [111] surface normal, and the abscissa axes, $q_x$, parallel to $[\bar{2}11]$ and $[\bar{1}\bar{1}2]$ for Fig. 15.11a and b, respectively. In these measurements, a large number of satellite peaks not only normal, but also parallel, to the surface is observed. This clearly proves that the dot positions are highly correlated both laterally and vertically, creating a periodic lattice in all three directions.

For the explanation of the X-ray data, a structural model was developed where the dots in each superlattice layer are laterally displaced with respect to

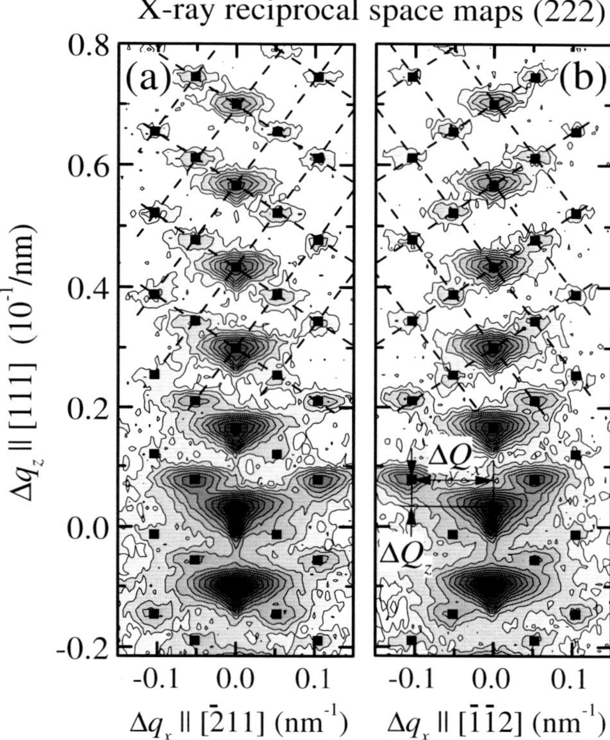


**Fig. 15.11.** Reciprocal space maps of diffracted intensity of a 60 period $PbSe/Pb_{1-x}Eu_xTe$ quantum dot superlattice with a superlattice period of 470 Å recorded around the bulk (222) reflection. The squares indicate the expected intensity maxima for a fcc-like $ABCABC\ldots$ vertical stacking sequence (taken from [15.148])

the dots in the previous layer. Because of the three-fold rotational symmetry and the identical lateral distance $L$ between the dots in each superlattice layer, this in-plane lateral displacement must be equal to $\frac{1}{2}L\cos 30°$ along the $[\bar{1}\bar{1}2]$ directions. This means that nucleation of the dots always occurs in the middle of three adjacent dots in the previous layer, as shown schematically in Fig. 15.12.

This lattice exhibits an fcc-like $ABCABC\ldots$ vertical dot stacking sequence with a periodicity in growth direction that is three times larger than the superlattice period $D$. A detailed analysis of the reciprocal space maps shown in Fig. 15.11 allows us to calculate the distances and angles in real space for the quantum dot crystal. The lateral dot separation within the hexagonal planes in the trigonal lattice is equal to $L = 580 \pm 30$ Å, in good agreement with the AFM results, and the vertical separation of the dot planes

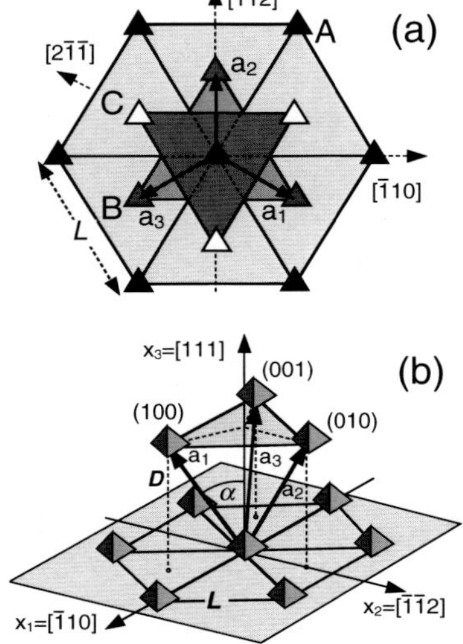


**Fig. 15.12.** Schematic illustration of the trigonal 3D lattice formed by the PbSe dots. (**a**) top view, (**b**) 3D representation. $L$ is the dot distance within the hexagonal dot planes, $D$ is the vertical separation of these planes, and $\alpha$ is the trigonal dot correlation angle (taken from [15.148])

is $D = 470 \pm 10\,\text{Å}$. Thus, a trigonal lattice constant of $a_0 = 610 \pm 20\,\text{Å}$ and a trigonal angle of $\alpha = 39.5°$ is obtained. This non-vertical dot correlation and the $ABCABC\ldots$ vertical stacking was also verified by cross-sectional transmission electron microscopy [15.157].

The excellent optical properties of Bragg mirrors (see Sect. 15.2.2) could be combined with the unique quantum dot crystals to fabricate vertical-cavity-surface-emitting lasers for the mid-infrared spectral region [15.158]. Narrow laser emission at 4.2 – 3.9 μm induced by optical pumping was achieved up to temperatures of 90 K. At a wavelength of 3.1 μm, laser operation was obtained up to a temperature of 65°C [15.159].

### 15.2.3 Organic Semiconductors

Research on conjugated organic systems is a rapidly expanding field at the interface between chemistry, condensed matter physics, materials science and device physics due to the promising opportunities for applications of these $\pi$-electron semiconductors in electronics and photonics. Due to their interdisciplinarity, this class of materials attracted the attention of a large number of researchers and originated in the beginning of a revolution in "Organic Electronics". With an initial focus on the p- and n-doping of conjugated oligomers and polymers, the unique electrochemical behavior of these technological important materials enabled the development of cheap sensors. Because of the progress toward better developed materials with higher order and purity, these organic materials are now also available for "organic electronic" devices. More generally, organic electronics includes now diodes, photodiodes, photovoltaic cells, light emitting diodes, lasers, field effect transistors, electro-optical couplers and all organic integrated circuits and claims thereupon for key technology of the 21th century [15.160–162].

The area of conjugated organic semiconductors can be divided conditionally into two large parts: conjugated polymers and small conjugated organic molecules. Conjugated polymers combine properties of classical semiconductors with the inherent processing advantages of plastics and therefore play a major role in low cost, large area optoelectronic applications [15.163–165]. Unfortunately, polymer films are commonly highly disordered in the solid state and, consequently, show low charge carrier mobilities because of strong Anderson localization [15.166]. Therefore, an inherent part of research in the field of organic electronics focuses on small molecule systems, in which highly ordered crystalline structures can be achieved – in contrast to the disordered, often amorphous phases of the polymers. These molecules are additionally thermally stable up to $300 - 400°C$, can be obtained as pure materials and processed in high-vacuum or ultra high-vacuum conditions. There is a number of papers about significant influence of structural order on the performance of thin film devices based on small molecules [15.167]. In particular, the recent emphasis in research of conjugated oligomer films is founded in the correlation between the electronic structure [15.168] and charge transport through the active layer [15.169]. These studies are motivated by their implications on charge transport in organic field effect transistors (OFETs). It has been demonstrated (mostly for $\alpha$-sexithiophene films) that the carrier mobility in OFETs can be significantly improved if the degree of order in all the film increases [15.170]. The correlation between degree of order and carrier mobility was also found for thin film structures based on other molecules [15.171]. High order, implying enhanced charge carrier transport are also of major importance in other organic electronic devices: photodetectors and photovoltaic devices [15.172, 173] and LEDs [15.174].

In addition to chances in device performance, well-ordered or single-crystalline molecular films allow the investigation of anisotropic optical and electronical properties of $\pi$- conjugated systems [15.175, 176]. It should be mentioned here that such phenomena are commonly not observed in conjugated polymer thin films, which are usually disordered. Therefore, such investigations are also of considerable fundamental interest. The challenging task for the future is to grow epitaxial layers of high crystalline quality. The main difference between organic and inorganic materials with respect to epitaxial growth is the different nature of the bonds. As explained in detail in Sect. 1.1 the inorganic materials are first physisorbed and then chemisorbed on the growing surface. In the case of organic materials only physisorption occurs because no chemical bonds are formed between the molecules. As a result, the growth process is governed by very weak bonds of Van der Waals' type reflected by a very small sticking coefficient. That means that epitaxial growth of organic materials is performed usually at comparable low temperatures. Many attempts have been made so far to grow organic materials on inert surfaces mainly by MBE. An extended overview can be found in [15.177]. MBE growth occurs in an open system far away from thermodynamic equilibrium. Especially in the case of Van der Waals epitaxy it would be a distinct advantage to use a growth method which works as close as possible to thermodynamic equilibrium, which would allow one to grow at relatively high vapor pressures of the organic material in the region of the substrate where the deposition occurs.

A growth method which satisfies these conditions is a modification of the MBE, the so-called hot-wall-beam epitaxy (HWBE) [15.178]. In contrast to the MBE system, HWBE uses the near field distribution of effusing molecules at the orifice of an effusion cell [15.179]. The substrate can also be used to close the tube of the source like a lid forming a semiclosed growth system, which is then known as hot-wall epitaxy (HWE) [15.88]. HWE has proved a very successful growth method for organic materials like $C_{60}$ [15.179–181] and its Ba-containing compounds [15.182]. We will describe here these experiments in more detail. A quartz tube, with the source material at the bottom and the substrate on the top closing it tight with respect to the mean free path of the evaporated source molecules, is placed with three separated heaters into a high-vacuum chamber. The region of the growth reactor between source and substrate, called the hot wall, guarantees a nearly uniform and isotropic flux of the molecules onto the substrate surface. The advantage of such a closed system is the minimization of the loss of source material, which can be very important in the case of new organic materials which are not commercially available. To perform the doping experiments with Ba a slight modification of the HWE system was necessary to add a second evaporation source for a doping material to the usual growth reactor.

A schematic cross-section of the improved HWE system can be seen in Fig. 15.13 together with a typical temperature profile along the quartz tube. The Ba doping source is contained in a concentric quartz ampoule and heated separately by oven 2. In that way the partial pressure of $C_{60}$ and Ba could be controlled independently. The $C_{60}$ source material was at first 99.4 % pure. It had to be cleaned from solvents and impurities by subliming the material three times at 550°C under dynamical vacuum of $1\times10^{-6}$ mbar and by protecting it from visible light in order to minimize photo-induced polymerization. About 200 mg of the cleaned material was loaded into the HWBE system, which was enough to fabricate more than 50 epilayers with an area of 1 $cm^2$ and an average thickness of 200 nm. The Ba-doping material was loaded in a glove box under nitrogen atmosphere into the quartz ampul to avoid any oxidation.

As substrate material, sheets of mica were used because of the inert character of freshly cleaved surfaces, free of unsaturated bonds which is in favor of Van der Waals epitaxy. The topmost layer of mica consists of potassium atoms forming a hexagonal grid with a periodicity which is close to the diam-

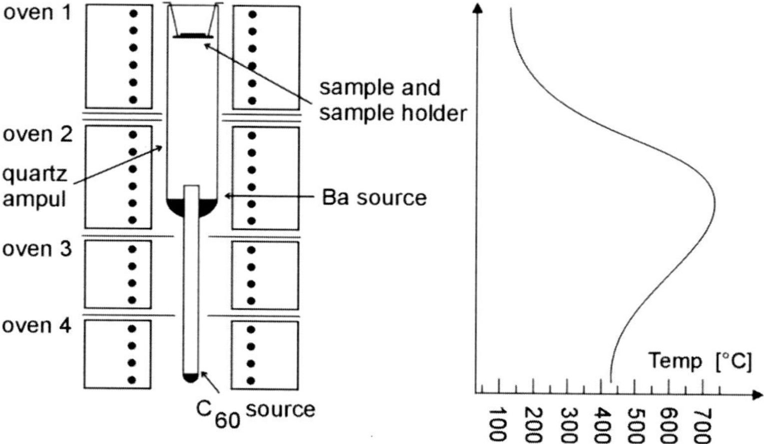


**Fig. 15.13.** Schematic cross-section of the HWE reactor to grow Ba-doped $C_{60}$ films together with a typical temperature profile (taken from [15.182])

eter of $C_{60}$ molecules. Consequently epitaxial growth of $C_{60}$ on mica should be initiated in the (111) direction. The mica sheets were cut into pieces ($15 \times 15\,\mathrm{mm}^2$), cleaved in air with an adhesive tape, and immediately transferred into the vacuum chamber of the HWE system. The substrates, before being transfered into the growth reactor, were preheated in the HWE chamber for 1 h at 400°C in a separate oven to remove adhesives from the substrate surface.

To grow pristine $C_{60}$ epilayers, the substrate temperature was varied from 100°C to 200°C, and the wall temperature from 340°C to 440°C. Out of this wide range of temperatures a set was selected which gave the best crystalline quality. The influence of the substrate temperature on the growth and crystalline quality of $C_{60}$ films was studied by varying the substrate temperature at a fixed source and wall temperature of 400°C.

In Fig. 15.14, the rocking curve FWHM of the (111) reflex of $C_{60}$ films, which are about 120 nm thick, is plotted as a function of $T_{\mathrm{sub}}$. Regarding temperatures between 100°C and 200°C, the $C_{60}$ film grown at 140°C has a minimal FWHM of $200 \pm 20$ arcs, indicating a nearly perfect monocrystalline growth. For lower temperatures, there is a gentle increase of the FWHM, whereas for higher temperatures, a significant jump can be observed between 180°C and 200°C. The 120 nm thick films grown at $T_{\mathrm{sub}}$ between 100°C and 180°C exhibit a narrow Gaussian shape as shown in the inset of Fig. 15.14.

The additional installed annealing oven allowed one to perform post-growth annealing processes without breaking the vacuum. The improvement in crystalline quality of the $C_{60}$ epilayers by annealing was investigated by HRXD. Typical rocking curves of an as-grown and an annealed $C_{60}$ layer of the same thickness of 100 nm are compared in Fig. 15.15. The FWHM of the

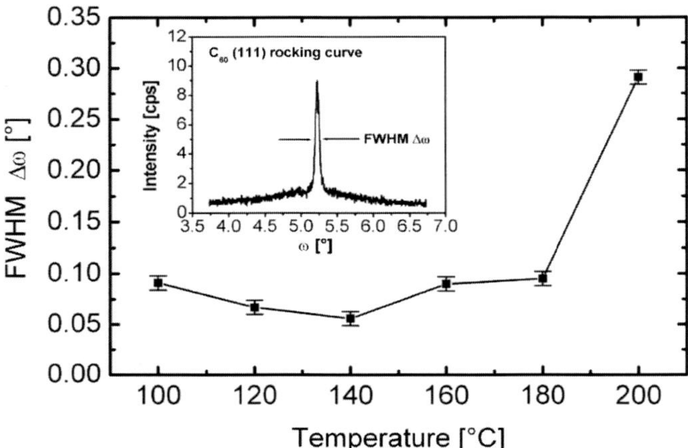


**Fig. 15.14.** FWHM of the $C_{60}$ (111) rocking curve versus the applied substrate temperature (taken from [15.180])

as-grown layer is 240 arcs, comparable with the results reported above. The improvement of the annealed layer, which was baked for 20 min at a temperature of 130°C, is documented by the decrease of the FWHM to 140 arcs and the increase of the peak intensity. The crystalline quality of the annealed $C_{60}$ was also tested by pole-figure measurements. Six sharp maxima indicated twin formation; however, the sharpness of the peaks, and the very low background signal were a clear indication of the high crystalline quality.

For all doping experiments with Ba, the $C_{60}$ source temperature and the substrate temperature were kept constant at 400°C and 140°C respectively. To study the incorporation of Ba the Ba source temperature was varied between 470°C and 750°C, resulting in $Ba_xC_{60}$ compound layers.

It is very important to notice that as soon as Ba was used in the growth reactor, the growth rate of the $C_{60}$ layers was mainly controlled by the Ba source temperature and only slightly dependent on the $C_{60}$ vapor pressure. Figure 15.16 shows the average growth rate of Ba-doped $C_{60}$ layers which were grown for 7 h. For Ba source temperatures smaller than 600°C, the results follow an exponential function: for higher Ba temperatures saturation can be observed. This behavior can be interpreted by the assumption that the incorporated Ba causes a charge transfer to $C_{60}$ [15.183] and controls the sticking coefficient for the $C_{60}$ molecules that are always present as a surplus in the vapor. However, when the surface coverage of Ba is saturated, the growth rate also cannot be increased further. In that way the growth mechanism changes from Van der Waals epitaxy, which is typical of pure fullerenes, to a different mechanism resulting from a stronger bonding type evoked by the charge transfer.

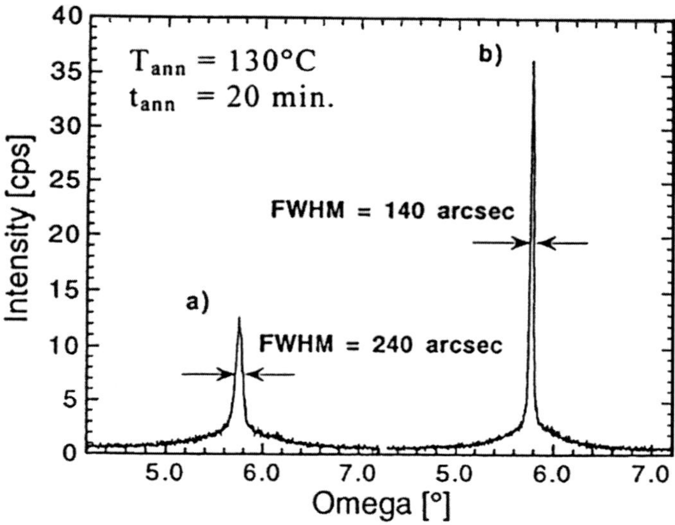


**Fig. 15.15.** X-ray rocking curve of the (111)-reflex for (**a**) as-grown and (**b**) annealed $C_{60}$ layers. The annealing time and temperature are inserted (taken from [15.182])

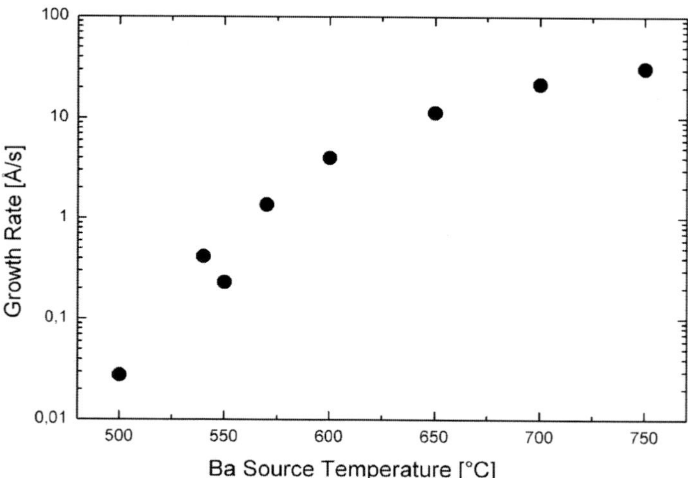


**Fig. 15.16.** Growth rate of Ba-doped $C_{60}$ layers as a function of the Ba source temperature (taken from [15.182])

As a first example of organic molecules $C_{60}$ was discussed as a representative of very symmetric molecules. In comparison an elongated small molecule with a pronounced axis like parasexiphenyl (PSP) should be described. This molecule consists of six carbon rings aligned in a straight direction and interconnected at the corners of the carbon rings forming a so-called para linkage. Each dangling bond of the carbon atoms is saturated by hydrogen. There exist bulk crystals of this material with a monoclinic structure, in which the molecules are aligned parallel to each other with their long axis, the molecular planes are tilted from layer to layer forming a so-called herring-bone structure. Layers of PSP emit highly intense blue light making them very interesting for optoelectronic devices like LEDs.

First attempts to grow thin films of PSP were made by physical vapor deposition on isotropic inorganic substrates like glass and GaAs. Even in those layers a high degree of order was found by high resolution X-ray diffraction [15.184]. Further improvement in structural order of PSP films was achieved by evaporation of PSP on rubbed layers of previously deposited PSP, serving as an orientation-inducing layer. The results reveal that the orientation of the PSP molecules in the top layer is effectively influenced by the rubbed layer. The long axes of the PSP molecules are oriented parallel to the rubbing direction. However, the pole figure measurements document that the rubbing procedure does not determine the direction of all PSP molecules; a certain amount of crystallites possesses a different orientation [15.185].

The next step in increasing the structural order was made by MBE [15.186]. PSP molecules were evaporated on GaAs substrates with a mis-

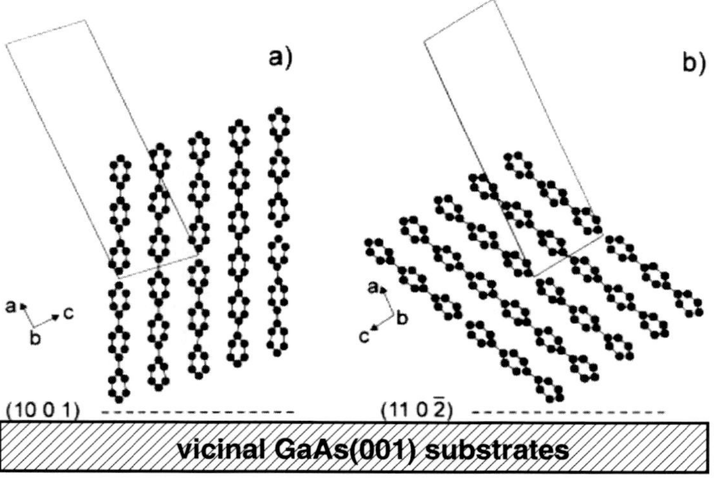


**Fig. 15.17.** Orientation of hexaphenyl molecules with respect to high index vicinal GaAs substrate surfaces for (**a**) the (10 0 1) plane and (**b**) the (11 0 $\bar{2}$) plane parallel to the surface (taken from [15.186])

cut of 2° relative to the (001) surface. After deoxidizing the substrates in As atmosphere, PSP was evaporated from an effusion cell heated to 230°C. Epitaxial growth was found up to a maximum substrate temperature of 150°C. Changing the substrate temperature from 90°C to 170°C the island density increases following an Arrhenius law and the mean thickness of the PSP islands changes from 120 to 400 nm. The structural analysis of these islands was made by electron diffraction in a TEM. The diffraction patterns indicate that the growth of the hexaphenyl islands on the substrate is well defined. Figure 15.17 shows the orientation of the molecules with respect to the GaAs (001) substrate. The vertical orientation of the hexaphenyl molecules, shown in Fig. 15.17a, corresponds to the (10 0 1) contact plane. In Fig. 15.17b the molecular orientation corresponding to the (11 0 $\bar{2}$) contact plane is shown. The molecular chains form an angle of about 40° to the surface of the substrate.

To sum up the experimental evidence, two orientations of the hexaphenyl lattice relative to the substrate were found. They are characterized by contact planes of the type (10 0 1) and (11 0 $\bar{2}$) and correspond to tilts of the hexaphenyl molecules of 0° and 50° relative to the substrate normal. Both orientations should be accounted for in terms of the hexphenyl structure and crystal growth mechanisms and hexaphenyl substrate interactions.

Needle-like structures of PSP were also obtained by the mask-shadowing vapor-deposition technique, as shown in Fig. 15.18 [15.187]. PSP molecules were vapor deposited onto a KCl (001) surface kept at 150°C through a mesh mask having round holes of 0.6 mm in diameter, which faced the KCl substrate with an intervening space of 0.5 mm. A PSP film was formed under the holes of the mask and in addition needle-like crystals were formed in the shadowed region of the KCl surface. The orientation was orthogonal along the KCl [$\bar{1}$10] directions. The length of the needles becomes longer away

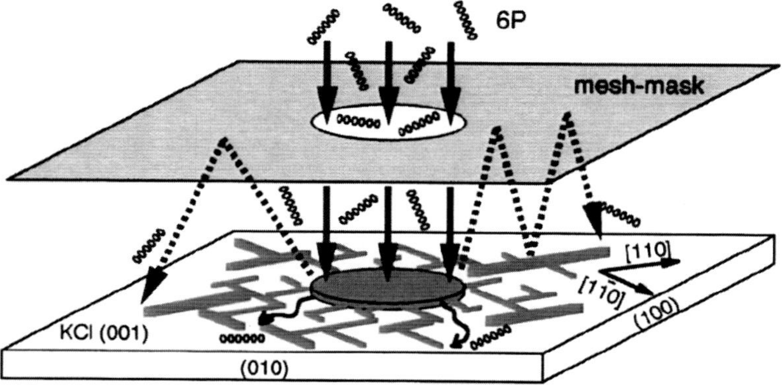


**Fig. 15.18.** PSP needle-like crystals grown on KCl substrate by the mask shadowing vapor deposition technique (taken from [15.187])

from the edge of the hole and reaches more than 100 μm. The growth of the needles is explained by a many-fold reflection of the PSP molecules between the substrate surface and the mask as schematically depicted in Fig. 15.18.

Some portion of PSP molecules further intrudes and migrates into the shadowed region. Those molecules probably desorb from the surface because the temperature of the shadowed region is heated slightly higher than the exposed hole. Due to this temperature gradient, the molecules in the shadowed space would be repeatedly reflected on the KCl surface and the mask wall, intrude into the depth, and then settle at the growing edge of the needles. Under UV light excitation at $\lambda = 365\,\text{nm}$ using a conventional inverse fluorescence microscope, the needles show a blue light emission with bright spots at the tips of the needles. When the excitation was focused on a local region with a round aperture, such a spotty radiation still occurred at the tips of the needles extending outside the excited region. The distance between the emitting tip and the excited edge reaches almost 50 μm. Since it is not believable that the excitons travel so long a distance in organic crystals, this can be attributed to the self-waveguided emission in the needle-like crystals. The light emitted at the excited region is confined inside the crystal and propagated along the needle axis, then radiated from the tip. This self-waveguided effect is based on the uniaxial molecular orientation, and affected by the size and morphology of the needle-like crystal. The electron diffraction pattern taken from a single needle reveals that the molecular axis of PSP lies parallel to the KCl surface and is aligned perpendicular to the needle axis, as schematically shown in Fig. 15.19a.

In order to prove this self-waveguided mode of the blue light emission in the needle-like crystals, the polarization of the emitted light was observed in the cross-section by standing the KCl substrate on a stage of an inverse microscope in the manner that the $[\bar{1}10]$ direction of the KCl substrate was normal to the stage surface. The vertical sample surface was excited by the nonpolarized UV light using a bundle fiber which was set normal to the sample surface. The blue light emission from the crystals was collected by an objective lens of the microscope and introduced through a polarizer to a charge coupled device (CCD) multichannel spectrometer equipped to a side-port of the microscope. In this configuration, the fluorescent light is collected in the directions parallel or perpendicular to the needle axes of the PSP crystals, as schematically shown in Fig. 15.19a. When observed with polarization parallel to the substrate ($E_{||}$), the bright, spotty emissions were seen at the interface. On the other hand, they almost disappear when the polarization is vertically rotated ($E_{\perp}$). Emission spectra taken at both polarizations are shown in Fig. 15.19b. The fluorescence intensity is much higher for $E_{||}$ than $E_{\perp}$. The uniaxial orientation of the elongated PSP molecules in the needle-like morphology is a very suitable configuration to amplify the emitted light along the needle axis. Therefore, such structures are of great interest for a res-

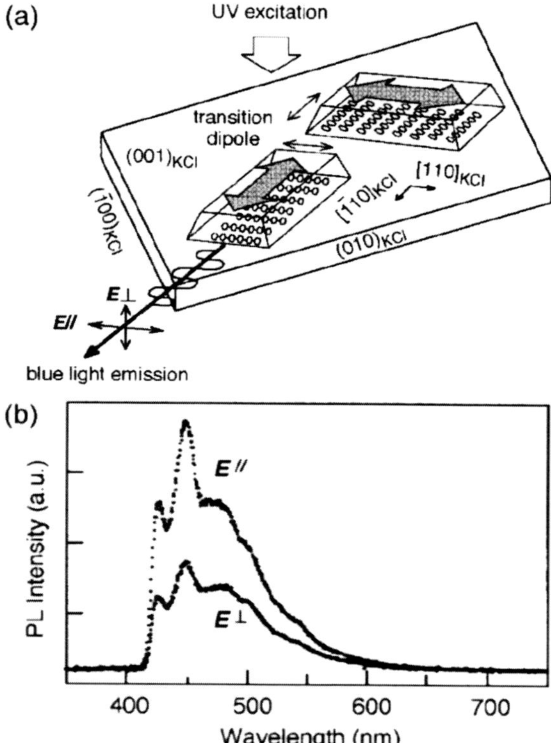


**Fig. 15.19.** (**a**) Schematic representation for the orientation of PSP molecules in the needle-like crystals epitaxially grown on the KCl (001) surface and the configuration of the polarized emission from the crystals. (**b**) Fluorescence spectra taken under the cross-sectional observation of the needle-like crystals with the polarizations of $E_{||}$ and $E_{\perp}$, respectively, shown in (**a**) (taken from [15.187])

onant self-cavity effect and a good candidate to realize blue lasing in organic materials.

The manifold impingement of the PSP molecules caused by the shadowing of the mask described above is automatically given in an HWE system by the local equilibrium of the PSP vapor with the growing surface. Therefore, it is not surprising that similar needle-like structures were obtained for PSP growth on mica substrates by HWE [15.188, 189]. Mica is a layered material with well defined smooth cleaved surfaces and weak surface bonds, which are similar to the Van der Waals bonds of PSP. Under HWE conditions this prerequisite results in highly ordered structures of the deposited layers. The PSP material was evaporated at 240°C and the substrate was held at 90°C. During the deposition process the vacuum was of about $6\times10^{-6}$ mbar. The investigated films were grown 60 min, which results in an average film thickness

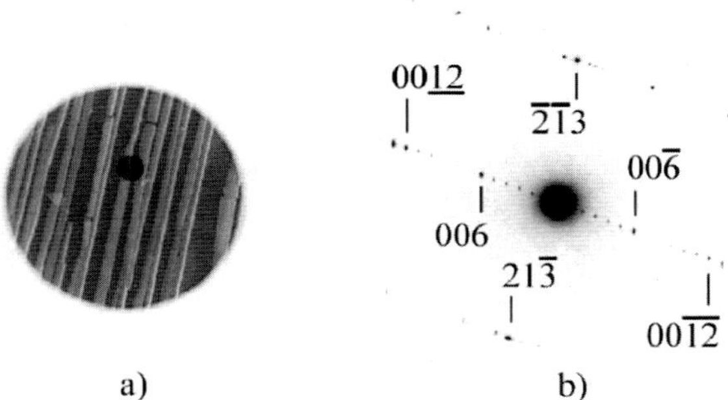


**Fig. 15.20.** (**a**) transmission electron microscopy image of a PSP film in which the needle-like morphology is clearly visible. (**b**) Diffraction pattern taken from the darker spot in (**a**) (taken from [15.188])

of about 120 nm. The film morphology was imaged by AFM, which reveals the typical needle-like morphology. AFM line scans gave typical needle widths, heights and lengths of about 400 nm, 130 nm and more than 200 µm, respectively. Several samples were investigated via transmission electron diffraction (TED) whereby all gave essentially the same results [15.188]. A typical TED image is depicted in Fig. 15.20a. The positions as well as the relative intensities of the diffraction pattern (Fig. 15.20b) are in excellent accordance with the crystal structure of PSP in the $\beta$-phase. Therefore, further analysis by XRD-pole figure measurements (XRD-PF) is based on this crystal structure.

In order to get more detailed information about the epitaxial relationships between the PSP layer and the substrate, XRD-PF measurements were performed at the four strongest reflections of crystalline PSP: $(11\bar{1})$, $(11\bar{2})$, $(20\bar{3})$ and $(21\bar{3})$. The obtained results are of unusual quality for epitaxial grown organic thin films in terms of reflection intensities and sharpness. In particular pole figures are significantly different from those obtained from small molecule films fabricated by physical vapor depositions [15.190–192]: the fiber textures are missing here, which confirms the epitaxial character of HWE films. The detailed analysis reveals only two crystallographic planes of PSP parallel to the substrate surface: $(11\bar{1})$ and $(11\bar{2})$. Within both orientations, there are two antiparallel directions of the PSP crystallites. Moreover, the crystallographic $(11\bar{1})$ orientation is preferred. The simultaneous appearance of mica and PSP reflections enables an easy and accurate determination of the epitaxial relationships between the organic layer and the substrate: PSP $(11\bar{1})$ || mica(001), PSP $[\bar{1}2\bar{1}]$ || mica $[\bar{3}40]$ and PSP $(11\bar{2})$ || mica (001), PSP $[\bar{2}0\bar{1}]$ || mica $[\bar{3}10]$.

Based on the TED and XRD-PF studies one was able to determine the orientation of PSP molecules on the mica surface as schematically depicted in Fig. 15.21. Please note that both crystallographic orientations – $(11\bar{1})$ &

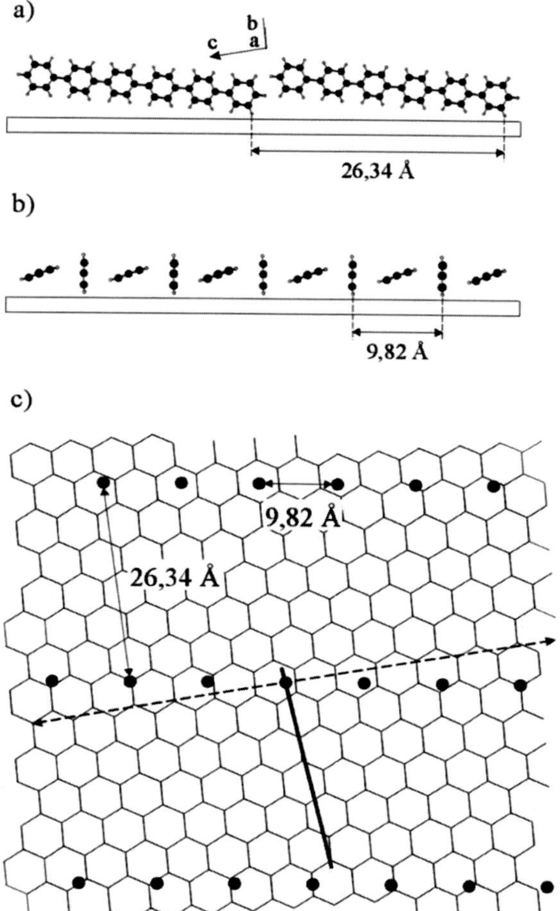


**Fig. 15.21.** Orientation of PSP molecules with respect to the (100) mica surface: (**a**) alignment of a single contact molecule in a side view; (**b**) parallel view; and (**c**) contact point lattice of the PSP molecules for the $(11\overline{1})$ orientation on an idealized hexagonal mica surface (taken from [15.193])

$(11\overline{2})$ || mica (001) – show a very similar alignment of the molecules relative to the substrate surface. Hence, only one of them – the $(11\overline{1})$ orientation – is shown. As demonstrated in Fig. 15.21a the long molecular axes are tilted by 5° with respect to the surface. This finding is also in good agreement with previous optical studies [15.193]. Only the lowest aromatic rings of the molecules are in contact with the substrate surface, as is visible in Fig. 15.21a. This fact can also be clearly seen in Fig. 15.21b, which shows the parallel view of the PSP molecules. It should be noted that the $(11\overline{1})$ plane is the cleavage plane of PSP. Since the contact molecules interact with the mica

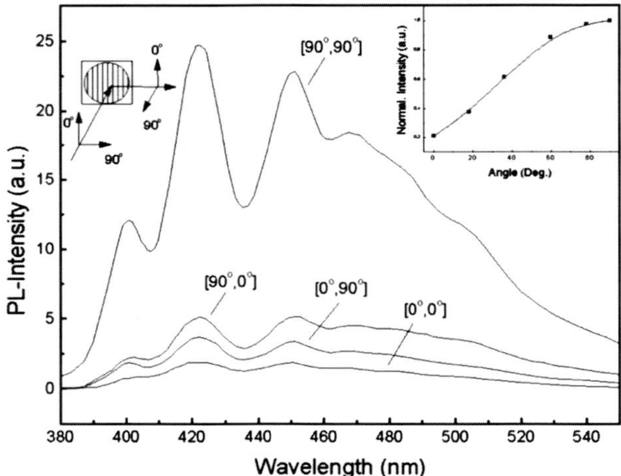


**Fig. 15.22.** Polarized PL spectra of a PSP film at room temperature for excitation at 350 nm. Insert left: a schematic representation of the measuring geometry. Insert right: the angular dependence of the emission for excitation polarized at 90°. (taken from [15.193])

surface only at a single point, a contact point lattice can be generated on top of the mica grid as shown in Fig. 15.21c. The dashed line indicates the needle direction determined by combined TEM and TED investigations with an accuracy of ± 5°. The molecular axes (see the bold line in Fig. 15.21c) are nearly perpendicular to the needle direction, which confirms previous optical results [15.193].

One of the applications of such PSP layers is to develop LEDs with polarized light emission. Obviously, a necessary condition for polarized electrolumienscence is polarized photoluminescence. In order to determine the polarization degree of the light emitted from these highly ordered PSP films, we performed photoluminescence (PL) measurements in a two polarizer geometry for the pump polarization and probe polarization (Fig. 15.22 left, inset). Typical photoluminescence spectra are shown in Fig. 15.22. For all four permutations of polarizations the well-known PL spectrum (upper spectrum) with three pronounced bands [15.194] is observed. The PL emission shows vibronic fine structures. The maximum of emission is observed if the excitation acts perpendicular to the direction of the "needles", consistent with both UV–VIS absorption and IR reflection data. The dichroic ratio for 90°–90° emission compared to 0°–0° is ≈ 14. The strong emission bands have the electric field vector component perpendicular to the film surface, indicating that the PSP molecules are aligned not absolutely flat on the substrate, but are tilted slightly out of plane, which is in excellent agreement with the results obtained by XRD-PF measurements described above.

# 16. Closing Remarks

Present day knowledge on epitaxy, the physical background of this important crystallization technique, as well as its implementation, have been presented in this book in a concise form. The authors idea was to make this presentation clear enough and sufficiently complete to convince the reader that epitaxy, with its modifications related to different materials systems, is still the basis for further development of already known, very sophisticated microelectronic device technology.

Epitaxy has in the past decade made vital contributions in at least three areas of the technology of microelectronic devices. The first is closely related to MBE, in which the UHV environment and the therefore available electron based surface sensitive techniques enabled effective *in situ* control over the growth process at the atomic level in real time. A similar control was later achieved by optical methods for the VPE techniques especially MOVPE. The second is related to, so-called, surface technology, which enabled the preparation of substrate and thin film surfaces with specified orientation, reconstruction, and composition, which are in addition effectively damage free. The third, and by far the most widely exploited area, is related to heteroepitaxy, which is at present the technology of sophisticated heterostructures and quantum well structures with 1D, 2D and 3D quantization.

The experimental and theoretical data presented in this book allow one to conclude that epitaxy is certainly still an open and very broad subject of intensive scientific investigations. This conclusion mainly concerns aspects of epitaxy, which exhibit great application prospects. As examples, let us mention here:

(i) atomic-scale control of growth processes in gas-phase as well as in liquid- or solid-phase epitaxy by submonolayer sensitive, noninvasive, real-time monitoring techniques,

(ii) self-assembling processes during epitaxy as a way to monolithic integration of microelectronic or optoelectronic devices,

(iii) growth of large-scale functional matrix structures for addressed applications, e.g., large lighting panels for billboards or illumination panels, by applying in sequence different epitaxial growth techniques, e.g., LPE, VPE and MBE, for different materials systems, e.g., semiconductors, dielectrics or/and metals.

The epitaxial growth process is basically the very first and decisive in the manufacturing process of a device. Therefore an increasing number of production systems is equipped with surface sensitive characterization techniques which are used for calibrating, monitoring and controlling the growth process, replacing in that way conventional time-consuming post-growth analysis methods. Among the noninvasive techniques, optical methods will become more and more important, because of their high content of information concerning growth rate and temperature, surface morphology and composition and doping efficiency in real time during growth, which allows one to construct a feedback loop to control and correct the growth process. Another advantage of optical *in situ* techniques is their compatibility with growth techniques in UHV like MBE as well as with methods working close to almost atmospheric pressure like MOVPE.

Rapid progress was achieved in the last fiew years in compound semiconductor nanostructure technology as well in the top down (layer growth and structuring) as in the bottom up (self-assembly of quantum structures) techniques. This success provided the realistic hope of constructing novel nanoscale electronics based on quantum devices, in which quantum mechanically controlled motion of electrons within artificially made quantum structures are directly used as their operation principle. Such nanoelectronics is a promising hardware candidate for next-generation information technology.

One of the most recent developments in this field is the replacement of conventional logic gate architectures by hexagonal binary decision diagram quantum circuits. In this approach hexagonal nanowire networks are controlled by nanoscale Schottky gates to implement the binary decision diagram logic architecture. Such structures can be achieved by selective MBE growth on patterned substrates or by self-assembly on specially prepared vicinal surfaces covered with a hexagonal network of surface steps with atomic height. Another new attempt to obtain nanostructures during growth is the geometric deposition with directed molecular beams of compound semiconductors through shallow shadow masks as a means to obtain *in situ* nanostructures of 10 nm in diameter. The formation of such high density nanostructures is definitely a future challenge, because what works now in the lab on a small selected area must work in the production line on the whole diameter of a substrate in a homogeneous, reproducible and reliable way. To develop such techniques is a key factor for future progress towards high density circuits operating at room temperature.

Many more examples of this kind could be given, and new ones can be found in daily news reports, which can never be covered by a standard textbook. Therefore, in conclusion let us emphasize that the whole material presented in this book should give the reader a sufficient basis for further, deeper studies on epitaxy by reading the current international literature on this subject. However, beside the excitement of the latest news, only if we try to understand the nature of the growth processes as deep as possible, can we artificially provide in our growth chambers those environmental conditions that nature does by herself what we want her to do.

# References

## Chapter 1


1.1 D.W. Pashley: *A Historical Review of Epitaxy*, in: *Epitaxial Growth*, part A, ed. by J.W. Matthews (Academic, New York 1975) p. 2

1.2 M.L. Frankenheim: Ann. Phys. **37**, 516 (1836)

1.3 H.J. Scheel: *Historical Introduction*, in: *Handbook of Crystal Growth*, vol. 1a, ed. by D.T.J. Hurle (North-Holland, Amsterdam 1993) p. 36

1.4 F. Wallerant: Bull. Soc. Franc. Min. **25**, 180 (1902); O. Mügge: Neues Jahrb. Mineral. **16**, 335 (1903)

1.5 L. Royer: Bull. Soc. Franc. Min. **51**, 7 (1928)

1.6 F. Bechstedt, R. Enderlein: *Semiconductor Surfaces and Interfaces – Their Atomic and Electronic Structures* (Akademie, Berlin 1988)

1.7 J. Sadowski, M.A. Herman: J. Cryst. Growth **146**, 449 (1995)

1.8 J.M. Howe: *Interfaces in Materials* (Wiley, New York 1997)

1.9 G.B. Stringfellow: Rep. Prog. Phys. **45**, 469 (1982)

1.10 R. Kern, G. Le Lay, J.J. Metois: *Basic Mechanisms in the Early Stages of Epitaxy*, in: *Current Topics in Material Science*, vol. 3, ed. by E. Kaldis (North-Holland, Amsterdam 1979) p. 131

1.11 M. Gebhardt: *Epitaxy*, in: *Crystal Growth: an Introduction*, ed. by P. Hartman (North-Holland, Amsterdam 1973) p. 105

1.12 J.P. van der Eerden: *Crystal Growth Mechanisms*, in: *Handbook of Crystal Growth*, vol. 1a, ed. by D.T.J. Hurle (North-Holland, Amsterdam 1993) p. 307

1.13 A.A. Chernov: *Kinetic Processes in Vapor Growth*, in: *Handbook of Crystal Growth*, vol. 3b, ed. by D.T.J. Hurle (North-Holland, Amsterdam 1999) p. 457

1.14 E. Bauer: Z. Kristallogr. **110**, 372 (1958)

1.15 J.A. Venables, G.D.T. Spiller, M. Hanbücken: Rep. Prog. Phys. **47**, 399 (1984)

1.16 M.A. Herman, H. Sitter: *Molecular Beam Epitaxy – Fundamentals and Current Status*, 2nd ed. (Springer, Berlin 1996)

1.17 H. Lüth: *Surfaces and Interfaces of Solids*, 2nd ed. (Springer, Berlin 1993)

1.18 D. Bimberg, M. Grundmann, N.N. Ledentsov, S.S. Ruvimov, P. Werner, U. Richter, J. Heydenreich, V.M. Ustinov, P.S. Kop'ev, Zh.I. Alferov: Thin Solid Films **267**, 32 (1995)

1.19 G.H. Gilmer: *Atomic-scale Models of Crystal Growth*, in: *Handbook of Crystal Growth*, vol. 1a, ed. by D.T.J. Hurle (North-Holland, Amsterdam 1993) p. 583

1.20 M.A. Sanchez-Garcia, E. Calleja, E. Monroy, F.J. Sanchez, F. Calle, E. Munoz, R. Beresford: J. Cryst. Growth **183**, 23 (1998)

1.21 J.H. Neave, P.J. Dobson, B.A. Joyce, J. Zhang: Appl. Phys. Lett. **47**, 400 (1985)

1.22 T. Shitara, T. Suzuki, D.D. Vvedensky, T. Nishinaga: Appl. Phys. Lett. **62**, 1347 (1993)

1.23 H. Nörenberg, L. Däweritz, P. Schützendübe, H.P. Schönherr, K. Ploog: J. Cryst. Growth **150**, 81 (1995)

1.24 P. Desjardins, J.E. Greene: J. Appl. Phys. **79**, 1423 (1996)

1.25 W. Richter, D. Zahn: *Analysis of Epitaxial Growth*, in: *Optical Characterization of Epitaxial Semiconductor Layers*, ed. by G. Bauer, W. Richter (Springer, Berlin 1996), Sect. 2.4.1
1.26 P.H. Fuoss, D.W. Kisker, F.J. Lamelas, G.B. Stephenson, P. Imperatori and S. Brennan, Phys. Rev. Lett. **69**, 2791 (1992)
1.27 J.J. Harbison, D.E. Aspnes, A.A. Studna, L.T. Florez: Phys. Rev. B **59**, 1687 (1987)

## Chapter 2

2.1 M.J. Stowell: Thin Films **1**, 55 (1968)
2.2 M. Gebhardt: *Epitaxy*, in: *Crystal Growth: An Introduction*, ed. by P. Hartman (North-Holland, Amsterdam 1973) p. 105
2.3 R. Kern, G. Le Lay, J.J. Metois: *Basic Mechanisms in the Early Stages of Epitaxy*, in: *Current Topics in Material Science*, vol. 3, ed. by E. Kaldis (North-Holland, Amsterdam 1979) p. 131
2.4 J.W. Gibbs: in *The Scientific Papers of J.W. Gibbs*, vol. 1 (1902) (Dover Publ. 1961)
2.5 S. Toschev: *Homogeneous Nucleation*, in: *Crystal Growth: An Introduction*, ed. by P. Hartman (North-Holland, Amsterdam 1973) p. 1
2.6 B.K. Chakraverty: *Heterogeneous Nucleation and Condensation on Substrates*, in: *Crystal Growth: An Introduction*, ed. by P. Hartman (North-Holland, Amsterdam 1973) p. 50
2.7 G.B. Stringfellow: Rep. Prog. Phys. **45**, 469 (1982)
2.8 J. Griesche, R. Enderlein, D. Schikora: Phys. Stat. Sol. (a) **109**, 11 (1988)
2.9 J. Sadowski, E. Dynowska, K. Reginski, M.A. Herman: Cryst. Res. Technol. **28**, 909 (1993)
2.10 J.A. Venables, G.L. Price: *Nucleation of Thin Films*, in: *Epitaxial Growth* part B, ed. by J.W. Matthews (Academic, New York 1975) p. 381
2.11 A. Masson, J.J. Metois, R. Kern: Surface Sci. **27**, 463 (1971), and Chap. 2.2 in: *Advances in Epitaxy and Endotaxy*, ed. by V. Ruth, H.G. Schneider, (VEB Deutscher Verlag für Grundstoffindustrie, Leipzig 1971)
2.12 J.J. Metois, M. Gauch, A. Masson, R. Kern: Surface Sci. **30**, 43 (1972), and Thin Solid Films **11**, 205 (1972)
2.13 M. Paunov: Cryst. Res. Technol. **33**, 165 (1998)
2.14 M.J. Stowell: *Defects in Epitaxial Deposits*, in: *Epitaxial Growth*, part B ed. by J.W. Matthews (Academic, New York 1975) p. 437
2.15 S. Sharan, J. Narayan: *Semiconductor Heterostructures: Formation of Defects and their Reduction*, in: *Concise Encyclopedia of Semiconducting Materials and Related Technologies*, ed. by S. Mahajan, L.C. Kimerling, (Pergamon, Oxford 1992)
2.16 U. Gösele: *Point Defect Equilibria in Semiconductors*, in: *Concise Encyclopedia of Semiconducting Materials and Related Technologies*, ed. by S. Mahajan, L.C. Kimerling, (Pergamon, Oxford 1992)
2.17 S. Mahajan: *Defects in Epitaxial Layers*, in: *Concise Encyclopedia of Semiconducting Materials and Related Technologies*, ed. by S. Mahajan, L.C. Kimerling, (Pergamon, Oxford 1992)
2.18 W. Kleber: *An Introduction to Crystallography*, (VEB Verlag Technik, Berlin 1970) p. 292
2.19 D.W. Pashley: Thin Films **1**, 59 (1964)
2.20 D.B. Holt: J. Phys. Chem. Solids **30**, 1297 (1969)
2.21 J.H. van der Merwe: CRC Crit. Rev. Solid State Mater. Sci. **7**, 209 (1978)

2.22 J.W. Matthews: *Coherent Interfaces and Misfit Dislocations*, in: *Epitaxial Growth* part B, ed. by J.W. Matthews (Academic, New York 1975) p. 559, and in: *Dislocations in Solids*, vol. 2, ed. by F.R.N. Nabarro (North-Holland, Amsterdam 1979) chap.7

2.23 B. Lewis, J.C. Anderson: *Nucleation and Growth of Thin Films* (Academic, New York 1978)

2.24 J.M. Howe: *Interfaces in Materials*, (Wiley, New York 1997)

2.25 J. Bloem: J. Cryst. Growth **18**,70 (1973)

2.26 J.L. Regolini, D. Bensahel, J. Mercier: Mater. Sci. Engineering B **4**, 407 (1989), also in *Defects in Silicon*, ed. by C.A.J. Ammerlaan, A. Chantre, P. Wagner (North-Holland, Amsterdam 1989) p. 407

2.27 M.A. Herman, H. Sitter: *Molecular Beam Epitaxy – Fundamentals and Current Status*, 2nd ed. (Springer, Berlin 1996)

2.28 A.J. Pidduck, D.J. Robbins, I.M. Young, A.G. Cullis, A.S.R. Martin: Mater. Sci. Engineering B **4**, 417 (1989), also in *Defects in Silicon*, ed. by C.A.J. Ammerlaan, A. Chantre, P. Wagner (North-Holland, Amsterdam 1989) p. 417

2.29 D.M. Wood, A. Zunger: Phys. Rev. B **40**, 4062 (1989), and D.J. Bottomley, P. Fons, D.J. Tweet: J. Cryst. Growth **154**, 401 (1995)

2.30 J.Y. Tsao: *Materials Fundamentals of Molecular Beam Epitaxy* (Academic, San Diego 1993)

2.31 F.R.N. Nabarro: *Theory of Crystal Dislocations* (Clarendon, Oxford 1967)

2.32 C.B. Duke: CRC Crit. Rev. Solid State Mater. Sci. **8**, 69 (1978)

2.33 F.C. Frank, J.H. van der Merwe: Proc. Roy. Soc., London **198**, 205, 216 (1949)

2.34 A.I. Finch, A.G. Quarrell: Proc. Phys. Soc., London **48**, 148 (1934)

2.35 K.I. Wang, S.G. Thomas, M.O. Tanner: J. Mater. Sci.: Materials in Electronics **6**, 311 (1995)

2.36 S.J. Jain: *Germanium-Silicon Strained Layers and Heterostructures* (Academic, Boston 1994)

2.37 T.E. Whall, E.H.C. Parker: J. Mater. Sci.: Materials in Electronics **6**, 249 (1995)

2.38 K. Ismail, B.S. Meyerson: J. Mater. Sci.: Materials in Electronics **6**, 306 (1995)

## Chapter 3

3.1 S. Mahajan, L.C. Kimerling (eds.): *Concise Encyclopedia of Semiconducting Materials and Related Technologies* (Pergamon, Oxford 1992)

3.2 S.M. Sze: *Semiconductor Devices – Physics and Technology* (Wiley, Chichester 1985)

3.3 R.F.C. Farrow, S.S.P. Parkin, P.J. Dobson, J.H. Neave, A.S. Arrott (eds.): *Thin Film Growth Techniques for Low-Dimensional Structures* (Plenum Press, New York 1987)

3.4 K. Iga, S. Kinoshita: *Process Technology for Semiconductor Lasers – Crystal Growth and Microprocesses* (Springer, Berlin, Heidelberg 1996)

3.5 A.C. Jones, P. O'Brien (eds.): *CVD of Compound Semiconductors* (VCH, Weinheim, New York 1997)

3.6 H. Sakaki, H. Noge (eds.): *Nanostructures and Quantum Effects* (Springer, Berlin 1994)

3.7 H. Koch, H. Lübbig (eds.): *Single-Electron Tunneling and Mesoscopic Devices* (Springer, Berlin 1992)

3.8 U. Meirav, E.B. Foxman: Semicond. Sci. Technol. **10**, 255 (1995)

3.9 M.A. Herman, H. Sitter: *Molecular Beam Epitaxy – Fundamentals and Current Status*, 2nd ed. (Springer, Berlin 1996)
3.10 A.Y. Cho: J. Cryst. Growth **111**, 1 (1991)
3.11 E.M. Stellini, K.Y. Cheng, P.J. Pearah, A.C. Chen, A.M. Moy, K.C. Hsieh: Appl. Phys. Lett. **62**, 458 (1993)
3.12 D. Bimberg, M. Grundmann, D. Ledentsov: *Quantum Dot Heterostructures*, (J. Wiley, New York 1999)
3.13 A.C. Gossard, R.C. Miller, W. Wiegmann: Surf. Sci. **174**, 131 (1986)
3.14 S. Fairweather: III-Vs Rev. **11**, no 3, 18 (1998)
3.15 D.W. Pashley: Adv. Phys. **5**, 173 (1956)
3.16 E. Grünbaum: *List of Epitaxial Systems, in Epitaxial Growth*, part B, ed. by J.W. Matthews (Academic, New York 1975), p. 611
3.17 G.B. Stringfellow: *Organometallic Vapor Phase Epitaxy – Theory and Practice* (Academic, New York 1990)
3.18 M.S. Miller, C.E. Pryor, H. Weman, L.A. Samoska, H. Kroemer, P.M. Petroff: J. Cryst. Growth **111**, 323 (1991)
3.19 S.Y. Hu, M.S. Miller, D.B. Young, J.C. Yi, D. Leonard, A.C. Gossard, P.M. Petroff, L.A. Coldren, N. Dagli: Appl. Phys. Lett. **63**, 2015 (1993)
3.20 K.Y. Cheng, K.C. Hsieh, J.N. Baillaegeon: Appl. Phys. Lett. **60**, 2892 (1992)
3.21 P.M. Petroff, M.S. Miller, Y.T. Lu, S.A. Chalmers, H. Metiu, H. Kroemer, A.C. Gossard: J. Cryst. Growth **111**, 360 (1991)
3.22 J.M. Gaines, P.M. Petroff, H. Kroemer, R.J. Smes, R.S. Geels, J.H. English: J. Vac. Sci. Technol. B **6**, 1378 (1988)
3.23 P.J. Pearah, A.C. Chen, A.M. Moy, K.C. Hsieh, K.Y. Chang: J. Cryst. Growth **127**, 900 (1993)
3.24 D.T.J. Hurle (ed.): *Thin Films and Epitaxy*, in: *Handbook of Crystal Growth* (Elsevier, Amsterdam 1994), vol. 3, Parts A and B

## Chapter 4

4.1 D.T.J. Hurle (ed.): *Thin Films and Epitaxy*, in: *Handbook of Crystal Growth* (Elsevier, Amsterdam 1994) Vol. 3, Parts A and B
4.2 G.L. Olson, J.A. Roth: *Solid Phase Epitaxy*, in Ref. [3.1], Chap. 7
4.3 J.A. Roth, G.L. Olson, D.C. Jacobson, J.M. Poate: Appl. Phys. Lett. **57**, 1340 (1990)
4.4 S. Peterström: Appl. Phys. Lett. 58, 2927 (1991)
4.5 Y. Tsaur, R. McClelland, J. Fan, R. Gale, J. Salerno, B. Vojak, C. Bozler: Appl. Phys. Lett. **41**, 347 (1982)
4.6 Y. Kunii, M. Tabe, K. Kajiyama: Jpn. J. Appl. Phys. **21**, 1431 (1982)
4.7 M. Miyao, M. Moniwa, K. Kusukawa, W. Sinke: J. Appl. Phys. **64**, 3018 (1988)
4.8 J.S. Custer, M.O. Thompson, P.H. Bucksbaum: Appl. Phys. Lett. **53**, 1402 (1988)
4.9 J.S. Williams, J.M. Poate (eds.): *Ion Implantation and Beam Processing* (Academic, New York 1984)
4.10 K.S. Jones, S. Prussin, E.R. Weber: Appl. Phys. A **45**, 1 (1988)
4.11 J.W. Mayer, S.S. Lau: *Electronic Materials Science for Integrated Circuits in Si and GaAs* (Macmillan, New York 1990)
4.12 M.K. El-Ghor, O.W. Holland, C.W. White, S.J. Pennycook: J. Mater. Res. **5**, 352 (1990)
4.13 A.V. Zotov, V.V. Korobtsov: J. Cryst. Growth **98**, 519 (1989)
4.14 G.L. Olson, J.A. Roth: Mater. Sci. Rep. **3**, 1 (1988)
4.15 H. Sitter, G.J. Glanner, M.A. Herman: Vacuum **46**, 69 (1995)

4.16 J.W. Mayer, L. Eriksson, S.T. Picraux, J.A. Davies: Can. J. Phys. **45**, 663 (1968)
4.17 A.Y. Cho: Surf. Sci. **17**, 494 (1969)
4.18 R.A. Smith: *Semiconductors*, 2nd ed. (Cambridge University Press, Cambridge 1976), Chapt. 3.
4.19 I. Suni, G. Göltz, M.A. Nicolet, S.S. Lau: Thin Solid Films **93**, 171 (1982)
4.20 A. Lietoila, A. Wakita, T.W. Sigmon, J.F. Gibbons: J. Appl. Phys.**53**, 4399 (1982)
4.21 W.W. Park, M.F. Becker, R.M. Walser: J. Mater. Res. **3**, 298 (1988)
4.22 Y.J. Jeon, M.F. Becker, R.M. Walser: Mater. Res. Soc. Symp. Proc. **205**, 63 (1992)
4.23 R.M. Walser, Y.J. Jeon: Mater. Res. Soc. Symp. Proc. **205**, 27 (1992)
4.24 E. Nygren, A.P. Pogany, K.T. Short, J.S. Williams, R.G. Elliman, J.M. Poate: Appl. Phys. Lett. **52**, 439 (1988)
4.25 K. Hiramatsu, T. Detchprohm, H. Amano, I. Akasaki: *Effects of Buffer Layers in Heteroepitaxy of GaN*, in: *Advances in the Understanding of Crystal Growth Mechanisms*, ed. by T. Nishinaga, K. Nishioka, J. Harada, A. Sasaki, H. Takei (Elsevier, Amsterdam 1997) p. 399
4.26 Z.Z. Bandič, E.C. Piquette, J.O. McCaldin, T.C. McGill: Appl. Phys. Lett. **72**, 2862 (1998)
4.27 E.C. Piquette, Z.Z. Bandič, J.O. McCaldin, T.C. McGill: J. Vac. Sci. Technol. B **15**, 1148 (1997)
4.28 T. Detchprohm, K. Hiramatsu, H. Amano, I. Akasaki: Appl. Phys. Lett. **61**, 2688 (1992)
4.29 H. Amano, H. Sawaki, I. Akasaki, Y. Toyoda: Appl. Phys. Lett. **48**, 35 (1986)
4.30 S. Nakamura: Jpn. J. Appl. Phys. **30**, L 1705 (1991)
4.31 I. Akasaki, H. Amano, Y. Koide, K. Hiramatsu, N. Sawaki: J. Cryst. Growth **98**, 209 (1989)


## Chapter 5


5.1 M.B. Small, E.A. Giess, R. Ghez:*Liquid Phase Epitaxy*, in Ref. [4.1], Chap. 6
5.2 E. Bauser: *Atomic Mechanisms in Semiconductor Liquid Phase Epitaxy*, in Ref. [4.1], Chap. 20
5.3 K.W. Benz, E. Bauser: *Growth of Binary III-V Semiconductors from Metallic Solutions*, in *Crystals – Growth, Properties and Applications*, ed. by H.C. Freyhardt (Springer, Berlin 1980), Vol. 3, p. 1
5.4 H. Nelson: RCA Rev. **24**, 503 (1963)
5.5 H. Rupprecht, J.M. Woodall, G.D. Pettit: Appl. Phys. Lett. **11**, 81 (1967)
5.6 M.B. Panish, I. Hayashi, S. Sumski: Appl. Phys. Lett. **16**, 326 (1970)
5.7 J.M. Woodal: J. Cryst. Growth. **12**, 32 (1972)
5.8 H.C. Casey Jr., M.B. Panish: *Heterostructure Lasers* (Academic, New York 1978)
5.9 Zh.I. Alferov, V.M. Andreev, V.I. Korolkov, E.L. Portnoi, D.N. Tretyakov: Soviet Phys. Semiconductors. **2**, 1289 (1969)
5.10 G.B. Stringfellow: Rep. Prog. Phys. **45**, 469 (1982)
5.11 E.A. Giess, R. Ghez: *Liquid Phase Epitaxy*, in *Epitaxial Growth*, Part A, ed. by J.W. Matthews (Academic Press, New York 1975),p. 183
5.12 M.B. Small, J.F. Barnes: J. Cryst. Growth **5**, 9 (1969)
5.13 W.W. Mullins, R.F. Sekerka: J. Appl. Phys. **35**, 444 (1964)
5.14 M.B. Small, R. Ghez: J. Cryst. Growth **43**, 512 (1978)
5.15 M.B. Small, R.M. Potemski: J. Cryst. Growth **37**, 163 (1977)

5.16 M.A. Herman, M. Pessa: J. Appl. Phys. **57**, 2671 (1985)
5.17 M.B. Small, J. Blackwell, R.M. Potemski: J. Cryst. Growth **46**, 253 (1979)
5.18 Y. Inatomi, K. Kuribayashi: J. Cryst. Growth **99**, 124 (1990)
5.19 Y. Inatomi, K. Kuribayashi: J. Cryst. Growth **114**, 380 (1991)
5.20 T. Nishinaga,C. Sasaoka, K. Pak: Jpn. J. Appl. Phys. **28**, 836 (1989)
5.21 S. Zhang, T. Nishinaga: Jpn. J. Appl. Phys. **29**, 545 (1990)
5.22 R.A. Smith: Semiconductors, 2nd ed. (Cambridge University Press, Cambridge 1976), Chap. 3
5.23 T. Bryskiewicz: Prog. Cryst. Growth Characterization **12**, 29 (1986)
5.24 L. Jastrzebski, J. Lagowski, H.C. Gatos, A.F. Witt: J. Appl. Phys. **49**, 5909 (1978)
5.25 T. Bryskiewicz: J. Cryst. Growth **43**, 567 (1978)
5.26 Z.R. Zytkiewicz, *Liquid Phase Electroepitaxy*, in *Elementary Crystal Growth*, ed. by K. Sangwal (SAAN, Lublin 1994), p. 521
5.27 T. Bryskiewicz, M. Bugajski, B. Bryskiewicz, J. Lagowski, H.C. Gatos: Inst. Phys. Conf. Ser., No. 91 (Institute of Physics, London 1988), p. 259
5.28 T. Bryskiewicz, P. Edelman, Z. Wasilewski, D. Coulas, J. Noad: J. Appl. Phys. **68**, 3018 (1990)
5.29 G. Bischopink, K.W. Benz: J. Cryst. Growth **128**, 466 (1993)
5.30 Z.R. Zytkiewicz: J. Cryst. Growth **172**, 259 (1997)
5.31 Z.R. Zytkiewicz: J. Cryst. Growth **146**, 283 (1995)
5.32 Z.R. Zytkiewicz: Acta Physica Polonica A **84**, 777 (1993)
5.33 A. Okamoto, S. Isozumi, J. Lagowski, H.C. Gatos: J. Electrochem. Soc. **129**, 2095 (1982)
5.34 T. Bryskiewicz: J. Cryst. Growth. **153**, 19 (1995)

## Chapter 6

6.1 K.F. Jensen: *Transport Phenomena in Vapor Phase Epitaxy Reactors*, in Ref. [4.1], Chap. 13
6.2 G.B. Stringfellow: Rep. Prog. Phys. **45**, 469 (1982)
6.3 L.I. Maissel, R. Glang (eds.): *Handbook of Thin Film Technology* (McGraw-Hill, New York 1970)
6.4 M.A. Herman, H. Sitter: *Molecular Beam Epitaxy – Fundamentals and Current Status*, 2nd ed. (Springer, Berlin 1996)
6.5 H. Hertz: Ann. Phys. **17**, 177 (1882)
6.6 M. Knudsen: Ann. Phys. (Leipzig) **47**, 697 (1915)
6.7 I. Langmuir: Phys. Z. **14**, 1273 (1913)
6.8 M. Knudsen: Ann. Phys. (Leipzig) **29**, 179 (1909)
6.9 E. Rutner: in *Condensation and Evaporation of Solids*, ed. by E. Rutner, P. Goldfinger, J.P. Hirth (Gordon and Breach, New York 1964), p. 149
6.10 R.W. Berry, P.M. Hall, M.T. Harris: *Thin Film Technology* (Van Nostrand, Reinhold, New York 1968)
6.11 G.K. Wehner, G.S. Anderson: *The Nature of Physical Sputtering*, in Ref. [6.3], Chap. 3
6.12 L. Maissel: *Application of Sputtering to the Deposition of Films*, in Ref. [6.3], Chap. 4
6.13 W. Kern, K.K. Schuegraf: *Deposition Technologies and Applications – Introduction and Overview*, in *Handbook of Thin Film Deposition Processes and Techniques*, ed. by K.K. Schuegraf (Noyes Publ., Park Ridge, New Jersey 1988), Chap. 1

6.14 M.H. Francombe: *Growth of Epitaxial Films by Sputtering, in Epitaxial Growth*, ed. by J.W. Matthews (Academic Press, New York 1975), Part A, Sect. 2.5
6.15 M. Konuma: *Film Deposition by Plasma Techniques* (Springer, Berlin, Heidelberg 1992)
6.16 G.S. Anderson, W.N. Mayer, G.K. Wehner: J. Appl. Phys. **33**, 2991 (1962)
6.17 J.A. Thornton: J. Vac. Sci. Technol. A **4**, 3059 (1986)
6.18 D.B. Chrisey, G.K. Hubler (eds.): *Pulsed Laser Deposition of Thin Films* (J. Wiley & Sons, New York 1994)
6.19 R. Kelly, A. Miotello: *Mechanisms of Pulsed Laser Sputtering*, in Ref. [6.18], Chap. 3
6.20 L.C. Chen: *Particulates Generated by Pulsed Laser Ablation*, in Ref. [6.18], Chap. 6
6.21 O. Auciello: *Pulsed Laser Ablation-Deposition of Multicomponent Oxide Thin Films: Basic Laser Ablation and Deposition Processes and Influence on Film Characteristics*, in Ref. [4.1], Chap. 9
6.22 A. Inam, X.D. Wu, T. Venkatesan, S.B. Ogale, C.C. Chang, D. Dijkkamp: Appl. Phys. Lett. **51**, 1112 (1987)
6.23 S.R. Foltyn, R.E. Muenchausen, R.C. Estler, E. Petersen, W.B. Hutchison, K.C. Ott, N.S. Nogar, K.M. Hubbard: in *Laser Ablation for Materials Synthesis*, Mater. Res. Soc. Symp. Proc. No 191, ed. by D.C. Payne, J.C. Bravman (Pittsburg, PA 1990), p. 205
6.24 H. Watanabe: *Halogen Transport Epitaxy*, in Ref. [4.1], Chap. 1
6.25 A.E. Nikolaev, Yu.V. Melnik, N.I. Kuznetsov, A.M. Streldink, A.P. Kovarsky, R.V. Vassilevski, V.A. Dmitriev: *GaN p-n Structures Grown by Hydride Vapor Phase Epitaxy*, in *Nitride Semiconductors*, ed. by F.A. Ponce, S.P. Den Baars, B.K. Meyer, S. Nakamura, S. Strite (Mater. Res. Soc. Symp. Proc., Warrendale, PA 1998), **482**, p. 251
6.26 D.T.J. Hurle (ed.): *Thin Films and Epitaxy*, in: *Handbook of Crystal Growth* (Elsevier, Amsterdam 1994), Vol. 3, Parts A and B
6.27 G.B. Secrest, W.W. Boyd, D.W. Show: J. Cryst. Growth **10**, 251 (1971)
6.28 D.I. Fotiadis, M. Boekholt, K.F. Jensen, W. Richter: J. Cryst. Growth **100**, 577 (1990)
6.29 R.B. Bird, W.E. Stewart, E. Lightfoot: *Transport Phenomena* (Wiley, New York, 1962)
6.30 K.F. Jensen: in Ref. [4.1]
6.31 H. Moffat, K.F. Jensen: J. Cryst. Growth **77**, 108 (1986)
6.32 H. Schlichting: *Boundary Layer Theory* (McGraw-Hill, New York, 1968)
6.33 D.W. Kisker, D.R. McKenna, K.F. Jensen: Mater. Lett. **6**, 123 (1988)
6.34 O.C. Zienkiewicz R.L. Taylor: *The Finite Element Method*, 4th Edition (McGraw-Hill, New York, 1989)
6.35 R. Huyakorn, C. Taylor, R. Lee, P. Gresho: Computers and Fluids **6**, 25 (1978)
6.36 K.F. Jensen, D.I. Fotiadis, D.R. McKenna, H.K. Moffat: Am. Chem. Soc. Symp. Ser. **353**, 353 (1987)
6.37 M. Koppitz, O. Vestavik, W. Pletschen, A. Mircea, M. Heyen, W. Richter: J. Cryst. Growth 68, 6 (1984)
6.38 H. Hardtdegen et al.: in Ref. [8.3], p. 211
6.39 M.R. Leys, C.V. Opdorp, M.P.A. Viegers, H.J.T. van der Mheen: J. Cryst. Growth **68**, 431 (1984)
6.40 N. Puetz, G. Hiller, A.J. Springthorpe: J. Electron. Mater. **17**, 381 (1988)
6.41 L. Stock, W. Richter: J. Cryst. Growth **77**, 144 (1986)
6.42 W. Richter, L. Hünermann: Chemtronics **2**, 175 (1987)

6.43 D.E. Rosner: *Transport Processes in Chemically Reacting Flow Systems* (Butterworth, Stoneham, MA, 1986)
6.44 G.B. Stringfellow: *Organometallic Vapor Phase Epitaxy – Theory and Practice* (Academic Press, New York 1990)
6.45 T.R. Scott, G. King, J.M. Wilson: UK Patent 778.383.8 (1954)
6.46 W. Miederer, G. Ziegler, R. Dotzer: German Patent 1.176.102 (1962)
6.47 W. Miederer, G. Ziegler, R. Dotzer: US Patent 3,226,270 (1963)
6.48 H.M. Manasevit, W.I. Simpson: J. Electrochem. Soc. **116**, 1725 (1969)
6.49 S. Fairweather: III–Vs Rev. **11**, no.3, 18 (1998)
6.50 C.H.L. Goodman, M. Pessa: J. Appl. Phys. **60**, R65 (1986)
6.51 T. Suntola, J. Antson: Finnish Patent 52359 (1974) and US Patent 4058430 (1977) to Lohja Corp. Finland
6.52 T. Suntola: *Atomic Layer Epitaxy*, in Ref. [4.1], Chap. 14
6.53 S. Bedair (ed): *Atomic Layer Epitaxy*, (Elsevier, Amsterdam, 1992)
6.54 T. Suntola: Mater. Sci. Rep. **4**, 261 (1989)

## Chapter 7

7.1 M.A. Herman, H. Sitter: *Molecular Beam Epitaxy – Fundamentals and Current Status*, 2nd ed. (Springer, Berlin 1996)
7.2 K. Ploog: *Molecular Beam Epitaxy of III-V Compounds*, in *Crystals – Growth, Properties and Applications*, ed. by H.C. Freyhardt (Springer, Berlin 1980), Vol. 3, p. 73
7.3 M.A. Herman: Thin Solid Films **267**, 1 (1995)
7.4 B. Lewis, J.C. Anderson: *Nucleation and Growth of Thin Films* (Academic Press, New York 1978)
7.5 D.J. Eaglesham: J. Appl. Phys. **77**, 3547 (1995)
7.6 J.Y. Tsao: *Materials Fundamentals of MBE* (Academic Press, Harcourt Brace Jovanovich, Boston, 1993), Chap. 3
7.7 M.A. Herman, A.V. Kozhukhov, J.T. Sadowski: J. Cryst. Growth **174**, 768 (1997)
7.8 E.H.C. Parker (ed.): *The Technology and Physics of Molecular Beam Epitaxy* (Plenum , New York 1985)
7.9 G.J. Davies, D. Williams: *III-V MBE Growth-Systems*, in Ref. [7.8], p. 15
7.10 V.A. Borodin, V.V. Sidorov, T.A. Steriopolo, V.A. Tatarchenko: J. Cryst. Growth **82**, 89 (1987)
7.11 P.A. Maki, S.C. Palmateer, A.H. Calawa, B.R. Lee: J. Vac. Sci. Technol. B **4**, 564 (1986)
7.12 R. Fernandez, A. Harwit, D. Kinell: J. Vac. Sci. Technol. B **12**, 1023 (1994)
7.13 T.H. Myers, J.F. Schetzina: J. Vac. Sci. Technol. **20**, 134 (1982)
7.14 M.A. Herman, M. Pessa: J. Appl. Phys. **57**, 2671 (1985)
7.15 K.A. Harris, S. Hwang, D.K. Blanks, J.W. Cook,Jr. , J.F. Schetzina, N. Otsuka: J. Vac. Sci. Technol. A **4**, 2061 (1986)
7.16 Y. Ota: Thin Solid Films **106**, 3 (1983)
7.17 J.T. Cheung, J. Madden: J. Vac. Sci. Technol. B **5**, 705 (1987)
7.18 M.B. Panish: J. Electrochem. Soc. **127**, 2729 (1980)
7.19 A.R. Calawa: Appl. Phys. Lett. **38**, 701 (1981)
7.20 M.B. Panish, H. Temkin, S. Sumski: J. Vac. Sci. Technol. B **3**, 687 (1985)
7.21 M.B. Panish, H. Temkin, R.A. Hamm, S.N.G. Chu: Appl. Phys. Lett. **49**, 164 (1986)
7.22 W.T. Tsang: Appl. Phys. Lett. **45**, 1234 (1984)
7.23 W.T. Tsang: J. Cryst. Growth **81**, 261 (1987)

7.24 E. Tokumitsu, Y. Kudou, M. Konagai, K. Takahashi: J. Appl. Phys. **55**, 3163 (1984)
7.25 E. Tokumitsu, T. Katoh, R. Kimura, M. Konagai, K. Takahashi: Jpn. J. Appl. Phys. **25**, 1211 (1986)
7.26 M.B. Panish, H. Temkin: *Gas Source Molecular Beam Epitaxy-Growth and Properties of Phosphorus Containing III-V Heterostructures*, Springer Ser. Mater. Sci., vol. **26** (Springer, Berlin, Heidelberg 1993)
7.27 H. Hirayama, H. Asahi: in *Handbook of Crystal Growth*, ed. by D.T.J. Hurle, Vol. 3, *Thin Films and Epitaxy*, part A, Chap. 5, *MBE with Gaseous Sources* (Elsevier, Amsterdam 1994)
7.28 P. Legay, F. Alexandre, M. Nunez, J. Sapriel, D. Zerguine, J.L. Benchimol: J. Cryst. Growth **148**, 211 (1995)
7.29 H. Ando, A. Taike, R. Kimura, M. Konagai, K. Takahashi: Jpn. J. Appl. Phys. **25**, L 279 (1986)
7.30 F.J. Morris, H. Fukui: J. Vac. Sci. Technol. **11**, 506 (1974)
7.31 M.B. Panish: Prog. Cryst. Growth Charact. **12**, 1 (1986)
7.32 M.B. Panish, R.A. Hamm: J. Cryst. Growth. **78**, 445 (1986)
7.33 M.B. Panish: J. Cryst. Growth **81**, 249 (1987)
7.34 G.B. Stringfellow: *Organometallic Vapor Phase Epitaxy – Theory and Practice* (Academic, New York 1990)
7.35 W.T. Tsang: J. Appl. Phys. **58**, 1415 (1985)
7.36 H. Hirayama, T. Tatsumi, N. Aizaki: Appl. Phys. Lett. **51**, 2213 (1987)
7.37 H. Hirayama, M. Hiroi, K. Koyama, T. Tatsumi: J. Cryst. Growth **105**, 46 (1990)
7.38 S.M. Gates, C.M. Greenlief, D.B. Beach: J. Chem. Phys. **93**, 7493 (1990)
7.39 H. Hirayama, T. Tatsumi, N. Aizaki: Appl. Phys. Lett. **52**, 1484 (1988)
7.40 H. Hirayama, M. Hiroi, K. Koyama: Appl. Phys. Lett. **58**, 1991 (1991)
7.41 M. Zinke-Allmang, L.C. Feldman, M.C. Grabow: Surf. Sci. Rep. **16**, 377 (1992)
7.42 H. Hirayama, M. Hiroi, K. Koyama, T. Tatsumi: Appl. Phys. Lett. **56**, (1990)
7.43 M. Pessa, P. Huttunen, M.A. Herman: J. Appl. Phys. **54**, 6047 (1983)
7.44 T. Sakamoto, H. Funabashi, K. Ohta, T. Nakagawa, N.J. Kawai, T. Kojima: Jpn. J. Appl. Phys. **23**, L 657 (1984)
7.45 J. Nishizawa, H. Abe, T. Kurabayashi: J. Electrochem. Soc. **132**, 1197 (1985)
7.46 Y. Horikoshi, M. Kawashima, H. Yamaguchi: Jpn. J. Appl. Phys. **25**, L 868 (1986)
7.47 M.A. Herman: Vacuum **42**, 61 (1991)
7.48 M. Pessa, O. Jylhä, M.A. Herman: J. Cryst. Growth **67**, 255 (1984)
7.49 M.A. Herman, O. Jylhä, M. Pessa: Cryst. Res. Technol. **21**, 969 (1986)
7.50 C.H.L. Goodman, M. Pessa: J. Appl. Phys. **60**, R 65 (1986)
7.51 M.A. Herman, O. Jylhä, M. Pessa: Cryst. Res. Technol. **21**, 841 (1986)
7.52 P. Juza, H. Sitter, M.A. Herman: Appl. Phys. Lett. **53**, 1396 (1988)
7.53 P.W. Atkins: *Physical Chemistry*, 3rd ed. (Oxford University Press, Oxford 1986)
7.54 M.A. Herman, M. Vulli, M. Pessa: J. Cryst. Growth **73**, 403 (1985)
7.55 T. Yao, T. Takeda: Appl. Phys. Lett. **48**, 160 (1986)
7.56 H. Sitter, W. Faschinger: *Atomic Layer Epitaxy of II-VI Compound Semiconductors*, in *Festkörperprobleme* (Adv. Solid State Phys. ) **30**, 219 (Vieweg, Braunschweig 1990)
7.57 M.A. Herman: Appl. Surf. Sci. **112**, 1 (1997)
7.58 M.A. Herman, J.T. Sadowski: Cryst. Res. Technol. **34**, 153 (1999)

7.59 M.A. Herman, O. Jylhä, M. Pessa: J. Cryst. Growth **66**, 480 (1984)
7.60 M. Pessa, O. Jylhä: Appl. Phys. Lett. **45**, 646 (1984)
7.61 J.T. Sadowski, M.A. Herman: Appl. Surf. Sci. **112**, 148 (1997)
7.62 J.T. Sadowski, M.A. Herman: Thin Solid Films **306**, 266 (1997)
7.63 M. Kawabe, N. Matsuura, H. Inuzuka: Jpn. J. Appl. Phys. **21**, L 447 (1982)
7.64 M. Kawabe, M. Kondo, N. Matsuura, K. Yamamoto: Jpn. J. Appl. Phys. **22**, 64 (1983)
7.65 Y. Horikoshi: Semicond. Sci. Technol. **8**, 1032 (1993)
7.66 A. Salokatve, J. Varrio, J. Lammasuiemi, H. Asonen, M. Pessa: Appl. Phys. Lett. **51**, 1340 (1987)
7.67 J. Varrio, H. Asonen, A. Salokatve, M. Pessa, E. Rauhala, J. Keinonen: Appl. Phys. Lett. **51**, 1801 (1987)
7.68 Y. Horikoshi, H. Yamagouchi, F. Briones, M. Kawashima: J. Cryst. Growth **105**, 326 (1990)
7.69 B.X. Yang, H. Hasegawa: Jpn. J. Appl. Phys. **30**, 3782 (1991)
7.70 M. Lopez, Y. Yamauchi, T. Kawai, Y. Takano, K. Pak, H. Yonezu: J. Vac. Sci. Technol. B **10**, 2157 (1992)
7.71 K. Shiraishi: Appl. Phys. Lett. **60**, 1363 (1992)
7.72 K. Nozawa, Y. Horikoshi: J. Electron. Mat. **21**, 641 (1992)
7.73 T.S. Rao, K. Nozawa, Y. Horikoshi: Appl. Phys. Lett. **62**, 154 (1993)
7.74 A. Fissel, U. Keiser, K. Pfennighaus, B. Schröter, W. Richter: Appl. Phys. Lett. **68**, 1204 (1996)
7.75 J. Nishizawa: Appl. Surf. Sci. **82/83**, 1 (1994)
7.76 J. Nishizawa, T. Kurabayashi: J. Vac. Sci. Technol. B **13**, 1024 (1995)
7.77 J. Nishizawa, T. Kurabayashi: Thin Solid Films **367**, 13 (2000), in Proceedings Issue of the 3rd Inter. Workshop on MBE-Growth Physics and Technology, Warsaw, May 1999, ed. by M.A. Herman
7.78 J. Nishizawa, H. Abe, T. Kurabayashi, N. Sakurai: J. Vac. Sci. Technol. A **4**, 706 (1986)
7.79 K. Fuji, I. Suemune, T. Koui, M. Yamanishi: Appl. Phys. Lett. **60**, 1498 (1992)
7.80 B.Y. Maa, P.D. Dapkus, P. Chen, A. Madhukar: Appl. Phys. Lett. **62**, 2551 (1993)
7.81 J. Nishizawa, T. Kurabayashi, J. Hoshina: J. Electrochem. Soc. **134**, 502 (1987)
7.82 J. Nishizawa, T. Kurabayashi: J. Cryst. Growth **93**, 98 (1988)
7.83 J.C. Bean: J. Cryst. Growth **81**, 411 (1987)
7.84 J.C. Bean, R. Dingle: Appl. Phys. Lett. **35**, 925 (1979)
7.85 N. Matsunaga, T. Suzuki, K. Takahashi: J. Appl. Phys. **49**, 5710 (1978)
7.86 M. Naganuma, K. Takahashi: Appl. Phys. Lett. **27**, 342 (1975)
7.87 Y. Matsushima, S.I. Gonda, Y. Makita, S. Mukai: J. Cryst. Growth **43**, 281 (1978)
7.88 Y. Ota: J. Appl. Phys. **51**, 1102 (1980)
7.89 H. Sugiura: J. Appl. Phys. **51**, 2630 (1980)
7.90 H. Jorke, H.J. Herzog, H. Kibbel: Appl. Phys. Lett. **47**, 511 (1985)
7.91 H. Jorke, H. Kibbel: J. Electrochem. Soc. **133**, 774 (1986)
7.92 E. Kasper: *Silicon Germanium Heterostructures on Silicon Substrates*, in *Festkörperprobleme* (Adv. Solid State Phys.) **27**, 205 (Vieweg, Braunschweig 1987)
7.93 Z. Sitar, M.J. Paisley, B. Yan, J. Ruan, W.J. Choyke, R. F Davis: J. Vac. Sci. Technol. B **8**, 316 (1990)
7.94 Z. Sitar, M.J. Paisley, D.K. Smith, R.F. Davis: Rev. Sci. Instrum. **61**, 2407 (1990)

7.95 J. Amussen, R. Fritz, L. Mahoney: Rev. Sci. Instrum. **61**, 282 (1990)
7.96 Q. Zhu, A. Botchkarev, W. Kim, Ö. Aktas. A. Salvador, B. Sverdlov, H. Morkoc, S.C.Y. Tsen, D.J. Smith: Appl. Phys. Lett. **68**, 1141 (1996)
7.97 L.B. Rowland, R.S. Kern, S. Tanaka, R.F. Davis: Appl. Phys. Lett. **62**, 3333 (1993)
7.98 Z.Q. He, X.M. Ding, X.Y. Hou, X. Wang: Appl. Phys. Lett. **64**, 315 (1994)
7.99 M. Rubin, N. Newman, J.S. Chan, T.C. Fu, J.T. Ross: Appl. Phys. Lett. **64**, 64 (1994)
7.100 H. Morkoc, A. Botchkarev, A. Salvador, B. Sverdlov: J. Cryst. Growth **150**, 887 (1995)
7.101 A. Georgakilas, H.M. Ng, P. Komniuou: *Plasma-Assisted Molecular Beam Epitaxy of III-V Nitrides* in *Nitride Semiconductors*, ed. by P. Ruterana, M. Albrecht, J. Neugebauer (Wiley-VCH Verlag, Weinheim, 2003), Chap. 3

## Chapter 8

8.1 Y. Horikoshi and S. Minagawa (eds.): *Proc. of ICMOVPE VII*, J. Cryst. Growth **145**, (1994)
8.2 J.B. Mullin (ed.): *Proc. of ICMOVPE VIII*, J. Cryst. Growth **170**, (1996)
8.3 R.M. Biefeld and G.B. Stringfellow (eds.): *Proc. of ICMOVPE IX*, J. Cryst. Growth **195**, (1998)
8.4 H. Kawai, K. Onabe (eds.): *Proc. of ICMOVPE X*, J. Cryst. Growth **221**, (2000)
8.5 B. Mullin, A. Krost, M. Weyers (eds.): *Proc. of ICMOVPE XI*, J. Cryst. Growth **248**, (2002)
8.6 G.B. Stringfellow: *Organometallic Vapor Phase Epitaxy – Theory and Practice* (Academic Press, New York 1990), G.B. Stringfellow: *Organometallic Vapor phase Epitaxy*, 2nd ed. (Academic Press, 1999)
8.7 D.W. Kisker, T.F. Kuech: *The Principles and Practice of Organometallic Vapor Phase Epitaxy*, in Ref. [16.1], Chap. 3
8.8 W. Richter, D. Zahn: *Analysis of Epitaxial Growth*, in Ref. [8.18], Sect. 2.4.1
8.9 H. Heinecke, E. Veuhoff, N. Pütz, M. Heyen, P. Balk: J. Electron. Mater. 13, 815 (1984)
8.10 D.I. Fotiadis, S. Kieda, K.F. Jensen: J. Cryst. Growth **102**, 441 (1990)
8.11 P.M. Frijlink: J. Cryst. Growth **93**, 207 (1988)
8.12 S. Fairweather: III-Vs Rev. **11**, no. 3, 18 (1998)
8.13 D.M. Frigo, J.H. Vanberkel, W.A. Maassen, G.P.M. Vammier, J.H. Wilkie: J. Cryst. Growth **124**, 99 (1992)
8.14 D.M. Frigo, G.P.M. Vammier, J.H. Wilkie, A.W. Gal: Appl. Phys. Lett. **61**, 531 (1992)
8.15 P. O'Brian, N.L. Picket, D.J. Otway: Chem. Vap. Deposition **8**, 237 (2002)
8.16 P.A. Lane, P.J. Wright, M.J. Crosby, A.D. Pitt, C.L. Reeves, B. Cockayne, A.C. Jones, T.J. Leedham: J. Cryst. Growth **192**, 424 (1998)
8.17 K. Knorr: PhD Dissertation, Berlin 1998
8.18 G. Bauer, W. Richter (eds.): *Optical Characterization of Epitaxial Semiconductor Layers* (Springer, Berlin, Heidelberg 1996)
8.19 L. Pauling: *The Nature of the Chemical Bond*, 3rd ed. (Cornel University Press, Ithaca 1960)
8.20 R.J. Gillespie: *Molecular Geometry* (Van Nostrand, Reinhold, London 1972)
8.21 W. Richter: Advances in Solid State Physics **26**, 335 (1986)
8.22 W. Richter, P. Kurpas, R. Lückerath, M. Motzkus, M. Waschbüsch: J. Cryst. Growth **107**, 13 (1991)

8.23 K.F. Jensen, D.I. Fotiadis and T.J. Mountziaris: J. Cryst. Growth **107**, 1 (1991)
8.24 M. Zorn: PhD Dissertation, TU Berlin, Mensch und Buch Verlag, 1999
8.25 M. Zorn, T. Trepk, J.-T. Zettler, B. Junno, C. Meyne, K. Knorr, T. Wethkamp, M. Klein, M. Miller, W. Richter: Appl. Phys. A **65**, 333 (1997)
8.26 M. Pristovsek: PhD Dissertation TU Berlin (2001)
8.27 H. Hardtdegen, M. Pristovsek, H. Menhal, J.-T. Zettler, W. Richter, D. Schmitz: J. Cryst. Growth **195**, 211 (1998)
8.28 J.R. Creighton, K. Baucomm: Surf. Sci. **409**, 372 (1998)
8.29 T. Schmidtling, M. Drago, U.W. Pohl, W. Richter: J. Cryst. Growth **248**, 523 (2003)
8.30 S. Peters, T. Schmidtling, T. Trepk, U.W. Pohl, J.-T. Zettler, W. Richter: J. Appl. Phys. **88**, 4085 (2000)
8.31 S.J.C. Irvine: *CRC Critical Reviews in Solid State and Materials Science*, Vol. **13**, 279 (CRC Press LLC, Boca Raton 1987)
8.32 S.J.C. Irvine: *Handbook of Crystal Growth* vol. 3 *Thin Films and Epitaxy*, part B, Chap. 18 (Elsevier, Amsterdam 1994)
8.33 S.J.C. Irvine, H. Hill, G.T. Brown, S.J. Barnett, J.E. Hails, O.D. Dosser, J.B. Mullin: J. Vac. Sci. Technol. B **7**, 1191 (1989)
8.34 S.J.C. Irvine, H. Hill, J.E. Hails, J.B. Mullin, S.J. Barnett, G.W. Blackmore, O.D. Dosser: J. Vac. Sci. Technol. A **8**, 1059 (1990)
8.35 W. Richter, P. Kurpas, M. Waschbüsch: Appl. Surf. Sci. **54**, 1 (1992)
8.36 F. Foulon, M. Stuke: Appl. Phys. A **56**, 267-273, (1993)
8.37 D. Bäuerle: *Laser Processing and Chemistry*, (Springer, Berlin 2000) p. 356ff, p. 405ff
8.38 H. Heinecke: PhD Dissertation RWTH Aachen 1987
8.39 K.L. Hess, R.J. Riccio: J. Cryst. Growth **77**, 95 (1966)

## Chapter 9

9.1 H.E. Duckworth, R.C. Barber, V.S. Venkatasubramanian: *Mass Spectroscopy* (Cambridge University Press, Cambridge 1986)
9.2 J. Sadowski and M.A. Herman: Appl. Surf. Sci. **112**, 148 (1997)
9.3 P.W. Lee, R. Omstead, D.R. McKenna, and K.F. Jensen: J. Cryst. Growth **85**, 165 (1987)
9.4 P. Ho, W.G. Breiland: J. Appl. Phys. **63**, 5184 (1988)
9.5 G. Bauer, W. Richter (eds.): *Optical Characterization of Epitaxial Semiconductor Layers*, (Springer, Berlin, Heidelberg 1996)
9.6 J.E. Butler, N. Bottka, R.S. Sillmon, D.K. Gaskill: J. Cryst. Growth **77**, 163 (1986)
9.7 M.A. Chesters, A.B. Horn, E.J.C. Kellar, S.F. Parker, R. Raval, in: *Mechanisms of Reaction of Organometallic Compounds with Surfaces*, ed. by D.J. Cole-Hamilton, J.O. Williams, NATO ASI Series B, Physics **198**, (Plenum Press, New York 1989)
9.8 D.S. Buhaenko, S.M. Francis, P.A. Goulding, M.E. Pemble: J. Cryst. Growth **97**, 591 (1989)
9.9 H. Patel, M.E. Pemble: J. Phys. IV, Colloq. **1**, 167 (1991)
9.10 G. A Hebner, K.P. Killeen, R.M. Bielfeld: J. Cryst. Growth **98**, 293 (1989)
9.11 K.P. Killeen: Appl. Phys. Lett. **61**, 1864 (1992)
9.12 V.M. Donnelly, R.F. Karlicek: J. Appl. Phys. **53**, 6399 (1982)
9.13 M.L. Fischer, R. Lückerath, P. Balk, W. Richter: Chemtronics **3**, 156 (1988)

9.14 W. Richter: Festkörperprobleme XXVI, Advances in Solid State Physics **26**, 335 (1986)
9.15 W.G. Breiland, M.J. Kushner: Appl. Phys. Lett. **42**, 395 (1983)
9.16 M. Koppitz, W. Richter, R. Bahnen, M. Heyen: in *Springer Series in Chemical Physics* **39**, *Laser Processing and Diagnostics*, ed. by D. Bäuerle, (Springer Verlag, Berlin, 1984), p. 530
9.17 Y. Monteil, R. Favre, P. Raffin, J. Bouix, M. Vaille, P. Gibart: J. Cryst. Growth **93**, 159 (1988)
9.18 P. Ho, W.G. Breiland: Appl. Phys. Lett. **43**, 125 (1983)
9.19 W.G. Breiland, P. Ho, M.E. Coltrin: J. Appl. Phys. **60**, 1505 (1986)
9.20 W.G. Breiland, M.E. Coltrin, P. Ho: J. Appl. Phys. **59**, 3267 (1986)
9.21 P. Ho, W.G. Breiland: Appl. Phys. Lett. **44**, 51 (1984)
9.22 P. Ho, W.G. Breiland: Appl. Phys. Lett. **43**, 125 (1983)
9.23 S. Ishizaka, J. Simpson, J.O. Williams: Chemtronics **1**, 175 (1986)
9.24 K.J. Mackey, D.C. Rodway, P.C. Smith, A.W. Vere: Chemtronics **5**, 85 (1991)
9.25 W. Richter, P. Kurpas, R. Lückerath, M. Motzkus: J. Cryst. Growth **107**, 13 (1991)
9.26 W. Richter: in *Topics in Applied Physics* **70** (Springer Verlag Berlin, 1992)
9.27 S.A.J. Druet, J.P.E. Taran: Prog. Quant. Electr. **7**, 1 (1981)
9.28 R. Brakel, F.W. Schneider: in *Advances in Spectroscopy* vol. **15**, ed. by R.J.H. Clark, R.E. Hester, (Wiley, Chichester 1988)
9.29 M. Alden, A.L. Schawlow, S. Svanberg, W. Wendt, P.L. Zhang: Optics Letters **9**, 211 (1984)
9.30 *Int. Coll. on Optogalvanic Spectroscopy*, J. Physique Colloque C7, 44 (1983)
9.31 D. Fotiadis, M. Boekholt, K.F. Jensen, W. Richter: J. Cryst. Growth **100**, 577 (1990)
9.32 F. Durst, A. Melling, J.H. Whitelaw: *Principle and Practice of Laser Doppler Anemometry*, (Academic Press, London 1976)
9.33 R.J. Schodl: Fluid Eng. **102**, 412 (1980)
9.34 L. Stock, W. Richter: J. Cryst. Growth **77**, 144 (1986)
9.35 D.I. Fotiadis, S. Kieda, K.F. Jensen: J. Cryst. Growth **102**, 441 (1990)
9.36 K.F. Jensen: J. Cryst. Growth **98**, 148 (1989)
9.37 K.C. Chiu, F. Rosenberger: Int. J. Heat and Mass Transfer **30**, 1645 (1987)
9.38 N. Esser, and J. Geurts: in [9.5]
9.39 T.O. Sedgwick, J.E. Smith, R. Ghez, M.E. Cowher: J. Cryst. Growth **31**, 264 (1975)
9.40 J.E. Smith, T.O. Sedgwick: Thin Solid Films **40**, 1 (1977)


## Chapter 10


10.1 H.-J. Günterodt, R. Wiesendanger (eds.): *Scanning Tunneling Microscopy I, II, III* Springer Ser. Surf. Sci. Vols. **20, 28, 29**, (Springer-Verlag, Berlin, Heidelberg 1992)
10.2 S.N. Magonov, M.H. Whangbo: *Surface analysis with STM and AFM*, (VCH, Weinheim, New York, 1996)
10.3 H. Lüth: *Surfaces and Interfaces of Solids*, (Springer-Verlag, Berlin 1993)
10.4 D.W. Kisker, G.B. Stephenson, P.H. Fuoss, F.J. Lamelas, S. Brennan, P. Imperatori: J. Cryst. Growth **124**, 1 (1992)
10.5 G. Bauer, W. Richter (eds): *Optical Characterization of Epitaxial Semiconductor Layers*, (Springer-Verlag Berlin, Heidelberg 1996)
10.6 J.I. Dadap, N.M. Russel, X.F. Hu, J. Ekerdt, M.C. Downer, B. Doris, J.K. Lowell, A.C. Diebold: Proceedings SPIE 2337, 66 (1994)

10.7 J.F. McGilp: Prog. Surf. Sci. **49**, 1 (1995)
10.8 B. Voigtländer: Micron **30**, 33 (1999)
10.9 B. Voigtländer, M. Kästner and P. Smilauer: Phys. Rev. Lett. **81**, 858 (1998)
10.10 U. Köhler, L. Anderson, B. Dahlheimer: Appl. Phys. A **57**, 491 (1993)
10.11 M.A. Herman, H. Sitter: *Molecular Beam Epitaxy – Fundamentals and Current Status*, 2nd ed., (Springer, Berlin 1996)
10.12 D. Pohl: in Ref. [10.1], Vol. II, p. 233
10.13 P. Gruetter, H.J. Mamin, D. Rugar: in Ref. [10.1], Vol. II, p. 151
10.14 K. Akimoto, J. Mizuki, I. Hirosowa, J. Matsui: Rev. Sci. Instrum. **60**, 2362 (1989)
10.15 S. Brennan, P.H. Fuoss, J.L. Kahn, D. Kisker: Ncl. Instrum. Methods A**291**, 86 (1990)
10.16 P.H. Fuoss, D.W. Kisker, F.J. Lamelas, G.B. Stephenson, P. Imperatori, S. Brennan: Phys. Rev. Lett. **69**, 2791 (1992)
10.17 I. Kamiya, L. Mantese, D.E. Aspnes, D.W. Kisker, P.H. Fuoss, G.B. Stephenson, S. Brennan: J. Cryst. Growth **163**, 67 (1996)
10.18 D.W. Kisker, G.B. Stephenson, J. Tersoff, P.H. Fuoss, S. Brennan: J. Cryst. Growth **163**, 54 (1996)
10.19 H. Zabel, I.K. Robinson (eds): *Surface X-Ray and Neutron Scattering*, (Springer Verlag, Berlin, Heidelberg 1992)
10.20 A. Munkholm, G.B. Stephenson, J.A. Eastman, C. Thompson, P. Fini, J.S. Speck, O. Auciello, P.H. Fuoss, S.P. Denbaars: Phys. Rev. Lett **83**, 741 (1999)
10.21 W. Richter: Appl. Phys. A **75**, 129 (2002)
10.22 D.E. Aspnes: Thin Solid Films **233**, 1 (1993)
10.23 H. Patel, M.E. Pemble: J. Phys. IV, Colloq. 1, 167 (1991)
10.24 N. Esser, W. Richter: in *Topics in Applied Physics*, vol. **76**, ed. by M. Cardona, G. Güntherodt, *Raman Scattering from Surface Phonons*, p. 96 (Springer-Verlag, Berlin, Heidelberg 2000)
10.25 D.R.T. Zahn: in B. Kramer (ed.), Advances in Solid State Physics **39**, (Vieweg, Wiesbaden 1999), p. 571
10.26 C.P. Goletti, P. Chirardia, W, Jian, G. Chiarotti: Solid State Commun. **84**, 421 (1992)
10.27 N. Kobayashi, Y. Horikoshi: Jpn. J. Appl. Phys. **30**, L1443 (1991)
10.28 M. Zorn, K. Haberland, A. Oster, A. Bhattacharya, M. Weyers, J.-T. Zettler, W. Richter: J. Cryst. Growth **225**, 25 (2002)
10.29 G.W. Smith, A.J. Pidduck, C.R. Whitehouse, J.L. Glasper, A.M. Keir, C. Pickering: Appl. Phys. Lett. **59**, 3282 (1991)
10.30 C. Lavoie, T. Pinnington, E. Nodwell, T. Tiedje, R.S. Goldman, K.L. Kavanagh, J.L. Hutter: Appl. Phys. Lett. **67**, 3744 (1995)
10.31 K. Haberland, A. Kaluza, M. Zorn, M. Pristovsek, H. Hardtdegen, M. Weyers, J.-T. Zettler, W. Richter: J. Cryst. Growth **240**, 87 (2002)
10.32 M. Pristovsek, B. Han, J.-T. Zettler, W. Richter: J. Cryst. Growth **221**, 149 (2000)
10.33 W. Richter, J.-T. Zettler: Appl. Surf. Sci. **100-101**, 465 (1996)
10.34 K. Hingerl, D.E. Aspnes, I. Kamiya, L.T. Florez: Appl. Phys. Lett. **63**, 885 (1993)
10.35 J.-T. Zettler, T. Wethkamp, M. Zorn, M. Pristovsek, C. Meyne, K. Ploska, W. Richter: Appl. Phys. Lett. **67**, 3783 (1995)
10.36 J.-S. Lee, S. Sugou, Y. Masumoto: J. Cryst. Growth **209**, 614 (2000)
10.37 J.-T. Zettler, M. Pristovsek, T. Trepk, A. Shkrebtii, E. Steimetz, M. Zorn, W. Richter: Thin Solids Films **313-314**, 537 (1998)

10.38 N. Esser, P.V. Santos, M. Kuball, M. Cardona, M. Arens, D. Pahlke, W. Richter, F. Stietz, J.A. Schaefer, B.O. Fimland: J. Vac. Sci. Technol. B **13**, 1666 (1995)
10.39 J.-T. Zettler: Progress in Crystal Growth and Characterization of Materials **35**, 27 (1997)
10.40 D.E. Aspnes: Mat. Sci. Eng. B **30**, 109 (1995)
10.41 K. Haberland, M. Pristovsek, O. Hunderi, J.-T. Zettler, W. Richter: Thin Solid Films **313-314**, 620 (1998)
10.42 N. Habets, T. Schmitt, M. Deufel, M. Lüneburger, M. Heuken, H. Juergensen: Thin Solid Films **409**, 43 (2002)
10.43 T. Bergunde, B. Henninger, M. Lüneburger, M. Heuken, M. Weyers, J.-T. Zettler: J. Crys. Growth **248**, 235 (2003)
10.44 M. Cardona, F.H. Pollak, K.L. Shaklee: J. Phys. Soc. Jap. **21**, 89 (1966)
10.45 R.M.A. Azzam: Optics Communications **19**, 122 (1976)
10.46 H. Wormeester, D.J. Wentink, P.L. deBoeij, C.M.J. Wijers, A. van Silfhout: Phys. Rev. B **47**, 12663 (1993)
10.47 D.E. Aspnes, J.P. Harbison, A.A. Studna, L.T. Folrez: J. Vac. Sci. Technol. **A6**, 1327 (1988)
10.48 D.E. Aspnes: J. Vac. Sci. Technol. **B3**, 1498 (1985)
10.49 D.E. Aspnes, A.A. Studna: Phys. Rev. **B27**, 985 (1983)
10.50 D.E. Aspnes, Y.C. Chang, A.A. Studna, L.T. Florez, H.H. Farrell, J.P. Harbison: Phys. Rev. Lett. **64**, 192 (1990)
10.51 S.M. Koch, O. Acher, F. Omnes, M. Defour, B. Drevillon, M. Razeghi: J. Appl. Phys. **69**, 1389 (1991)
10.52 K. Haberland, PhD dissertation, TU Berlin, (Mensch und Buch Verlag, Berlin, 2002)
10.53 T. Yasuda, D.E. Aspnes, D.R. Lee, C.H. Bjorkman, G. Lucovsky: J. Vac. Sci. Technol. **A12**, 1152 (1994)
10.54 C. Meyne, U.W. Pohl, W. Richter, M. Straßburg, A. Hoffmann, V. Türck, S. Rodt, D. Bimberg, D. Gerthsen: J. Cryst. Growth **214**, 722 (2000)
10.55 C. Meyne, M. Gensch, S. Peters, U.W. Pohl, J.-T. Zettler, W. Richter: Thin Solid Films **364**, 12 (2000)
10.56 C. Meyne, U.W. Pohl, J.-T. Zettler, W. Richter: J. Cryst. Growth **184-185**, 264 (1998)
10.57 V. Emiliani, A.I. Shkrebtii, C. Goletti, A.M. Frisch, B.O. Fimland, N. Esser, W. Richter: Phys. Rev. B **59**, 10657 (1999)
10.58 T. Herrmann, K. Lüdge, W. Richter, N. Esser, P. Poulopoulos, J. Lindner, K. Baberschke: Phys. Rev. B **64**, 184424 (2001)
10.59 I. Kamiya, D.E. Aspnes, H. Tanaka, L.T. Florez, J.P. Harbison, R. Bhat: Phys. Rev. Lett. **68**, 627 (1992)
10.60 D.W. Kisker, G.B. Stevenson, I. Kamiya, P.H. Fuos, D.E. Aspnes, L. Mantese, S. Brennan: phys. stat. sol (a) **152**, 9 (1995)
10.61 B.S. Mendoza, N. Esser, W. Richter: Phys. Rev. B **67**, 165319 (2003)
10.62 W. Richter: Phil. Trans. R. Soc. London A **344**, 453 (1993)
10.63 T. Yasuda, L. Mantese, U. Rossow, D.E. Aspnes: Phys. Rev. Lett. **74** 3431 (1995)
10.64 A.I. Shkrebtii, N. Esser, W. Richter, W.G. Schmidt, F. Bechstedt, B.O. Fimland, A. Kley, R. DelSole: Phys. Rev. Lett. **81**, 721 (1998)
10.65 O. Pulci, G. Onida, R. DelSole, L. Reining: Phys. Rev. Lett. **81**, 5374 (1998)
10.66 F. Bechstedt, R. DelSole, G. Capellini, L. Reining: Solid State Comm. **84**, 765 (1992)

10.67 W.G. Schmidt, N. Esser, A.M. Frisch, P. Vogt, J. Bernholc, F. Bechstedt, M. Zorn, Th. Hannappel, S. Visbeck, F. Willig, W. Richter: Phys. Rev. B **61**, R16335 (2000)
10.68 M. Zorn, K. Haberland, A. Oster, A. Bhattachrya, M. Weyers, J.-T. Zettler, W. Richter: J. Cryst. Growth **235**, 258 (2002)
10.69 K. Haberland, P. Kurpas, M. Pristovsek, J.-T. Zettler, M. Weyers, W. Richter: Appl. Phys. A **68**, 309 (1999)
10.70 K. Haberland, P. Kurpas, M. Pristovsek, J.-T. Zettler, M. Weyers, W. Richter: Appl. Phys. A **68**, 309 (1999)
10.71 U. Rossow: in Ref. [10.5], p. 68
10.72 R.M.A. Azzam, N.M. Bashara: *Ellipsometry and Polarized Light*, (North-Holland Publishing Co, Amsterdam 1977)
10.73 J.C. Maxwell-Garnet: Phil. Trans. Roy. Soc. London **203**, 385 (1904) and **205**A, 237 (1906)
10.74 D.A.G. Bruggeman: Ann. Phys. (Leipzig) **24**, 636 (1935)
10.75 H. Looyenga: Physica **31**, 401 (1965)
10.76 D. Bergman: Phys. Rep. C **43**, 377 (1978)
10.77 K. Knorr: PhD Dissertation, Berlin 1998
10.78 G. Laurence, F. Hottier, J. Hallais: J. Cryst. Growth **55**, 198 (1981)
10.79 J.B. Theeten, F. Hottier, J. Hallais: Appl. Phys. Lett. **32**, 576 (1978)
10.80 D.E. Aspnes, W.E. Quinn, M.C. Tamargo, M.A.A. Pudensi, S.A. Schwarz, M.J.S.P. Brasil, R.E. Nahory, S. Gregory: Appl. Phys. Lett. **60**, 1244 (1992)
10.81 M. Zorn, P. Kurpas, A.I. Shkrebtii, B. Junno, A. Bhattacharya, K. Knorr, M. Weyers, L. Samuelson, J.-T. Zettler, W. Richter: Phys. Rev. B **60**, 8185 (1999)
10.82 A. Bonani, D. Stifter, A. Montaigne-Ramil, K. Schmidegg, K. Hingerl, H. Sitter: J. Cryst. Growth **248**, 211 (2003)
10.83 S. Peters, T. Schmidtling, T. Trepk, U.W. Pohl, J.-T. Zettler, W. Richter: J. Appl. Phys. **88**, 4085 (2000)
10.84 T. Schmidtling, M. Drago, U.W. Pohl, W. Richter: J. Cryst. Growth **248**, 523 (2003)
10.85 E. Steimetz, J.-T. Zettler, F. Schienle, T. Trepk, T. Wethkamp, W. Richter, I. Sieber: Appl. Surf. Science **107**, 203 (1996)
10.86 J.G. Belk, C. McConville, J. Sudijono, T.S. Jones, B.A. Joyce: Surf. Sci. **387**, 213 (1997)
10.87 M. Zorn: PhD dissertation, (Mensch und Buch Verlag, Berlin 1999)
10.88 D.I. Westwood, Z. Sobiesierski, E. Steimetz, J.-T. Zettler, W. Richter: Appl. Surf. Science **123/124**, 347 (1998)
10.89 N. Dietz, K.J. Bachmann: MRS Bull. **20**, 49 (1995)
10.90 S. Selci, F. Ciccacci, G. Chiarotti, P. Chiaradia, A. Cricenti: J. Vac. Sci. Techn. A **5**, 327 (1987)
10.91 N. Kobayashi, Y. Horikoshi: Jpn. J. Appl. Phys. **28**, L1880 (1989); N. Kobayashi, Y. Horikoshi: Jpn. J. Appl. Phys. **30**, L1443 (1991)
10.92 N. Kobayashi, Y. Horikoshi: Jpn. J. Appl. Phys. **29**, L702 (1990)
10.93 N. Dietz, U. Rossow, D.E. Aspnes, K.J. Bachmann: J. Cryst. Growth **164**, 34 (1996)
10.94 S.J.C. Irvine, S. Bjaj, H.O. Sankur: J. Cryst. Growth **124**, 654 (1992)
10.95 H. Sitter, G.J. Glanner, M.A. Herman: Vacuum **46**, 69 (1995)
10.96 L. Mattson, J.M. Bennet: *Introduction to Surface Roughness and Scattering*, (Optical Society of America, Washington D.C 1989)
10.97 J.M. Olson, A. Kibbler: J. Cryst. Growth **77**, 182 (1986)

10.98 D.J. Robbins, A.J. Pidduck, A.G. Cullis, N.G. Chew, R.W. Hardeman, D.B. Gasson, C. Pickering, A.C. Daw, M. Johnson, R. Jones: J. Cryst. Growth **81**, 421 (1987)

10.99 C. Pickering, D.J. Robbins, I.M. Young, J.L. Glasper, M. Johnson, R. Jones: Mater. Res. Soc. Symp. Proc. **94**, 173 (1987)

10.100 A.J. Pidduck, D.J. Robbins, A.G. Cullis, D.B. Glasson, J.L. Glasper: J. Electrochem. Soc. **136**, 3083 (1989)

10.101 A.J. Pidduck, D.J. Robbins, D.B. Glasson, C. Pickering, J.L. Glasper: J. Electrochem. Soc. **136**, 3088 (1989)

10.102 K. Haberland, M. Zorn, A. Klein, A. Bhattacharya, M. Weyers, J.-T. Zettler, W. Richter: J. Cryst. Growth **248**, 194 (2003)

10.103 G.W. Smith, A.J. Pidduck, C.R. Whitehouse, J.L. Glasper, A.M. Keir, C. Pickering: Appl. Phys. Lett. **59**, 3282 (1991)

10.104 K.L. Kavanagh, R.S. Goldman, C. Lavoie, B. Leduc, T. Pinnington, T. Tiedje, D. Klug, J. Tse: J. Cryst. Growth **174**, 550 (1997)

10.105 F. Briones, D. Golmayo, L. Gonzalez, J.L. de Miguel: J. Appl. Phys. Japan **24**, L478 (1985)

10.106 J.E. Epler, T.A. Jung, H.P. Schweizer: Appl. Phys. Lett. **62**, 143 (1993)

10.107 N. Bloembergen, R.K. Chang, S.S. Jha, C.H. Lee: Phys. Rev. **174**, 813 (1968)

10.108 H.W.K. Tom, T.F. Heinz, Y.R. Shen: Phys. Rev. Lett. **51**, 1983 (1983)

10.109 T.F. Heinz, M.M.T. Loy., W.A. Thompson: Phys. Rev. Lett. **54**, 63 (1985)

10.110 H.W.K. Tom, G.D. Aumiller: Phys. Rev. **B33**, 8818 (1986)

10.111 J.F. McGilp , Y. Yeh: Solid State Commun. **59**, 91 (1986)

10.112 J.F. McGilp: Semicond. Sci. Technol. **2**, 102 (1987)

10.113 J.F. McGilp: J. Vac. Sci. Technol. **A5**, 1442 (1987)

10.114 M.C. Downer, Y. Jiang, D. Lim, L. Mantese, P.T. Wilson, B.S. Mendoza, V. Gavrilenko: phys. stat. sol (a) **188**, 1371 (2001)

10.115 D. Guidotti, T.A. Driscoll, H.J. Gerritsen: Solid State Commun. **46**, 337 (1983)

10.116 R.W.J. Hollering, A.J. Hoeven, J.M. Lenssinck: J. Vac. Sci. Technol. **A8**, 3194 (1989)

10.117 M.E. Pemble, D.S. Buhaenko, S.M. Franas, P.A. Goulding, J.T. Allen: J. Cryst. Growth **107**, 37 (1991)

10.118 R. Loudon: Proc. Royal Soc. A **275**, 218 (1963); R. Loudon: Adv. Phys. **13** 423 (1964)

10.119 A. Pinczuk, E. Burstein: *Topics in Applied Physics* Vol. **8**, ed. by M. Cardona, G.üntherodt, (Springer Verlag, Berlin 1975)

10.120 N. Esser, J. Guerts: *Ramanspectroscopy* in Ref. [8.5]

10.121 V. Wagner, W. Richter, J. Geurts, D. Drews and D.R.T. Zahn: J. Raman Spectrosc. **27**, 265 (1996)

10.122 D. Drews, A. Schneider, D.R.T. Zahn: J. Vac. Sci. Technol. B **15**, 1 (1997)

10.123 E. Speiser, T. Schmidtling, K. Fleischer, N. Esser, W. Richter: *Proceedings of the OSI V* phys. stat. sol. (a), (2003), in print

10.124 H. Ibach, D.L. Mills: *Electron Energy Loss Spectroscopy and Surface Vibrations*, (Academic Press, New York 1982)

10.125 A. Förster: PhD Thesis, RWTH Aachen (1988)

10.126 H. Patel, M.E. Pemble: J. Phys. IV, Colloq. **1**, 167 (1991)

10.127 Y.J. Chabal: Surf. Sci. Reports **8**, 211 (1988)

10.128 R.G. Greenler: J. Chem. Phys. **44**, 310 (1966)

10.129 H. Ibach: Surf. Sci. **66**, 56 (1977)

10.130 M.A. Chesters, A.B. Horn, E.J.C. Kellar, S.F. Parker, R. Raval, in: *Mechanisms of Reaction of Organometallic Compounds with Surfaces*, ed. by D.J. Cole-Hamilton, J.O. Williams, NATO ASI Series, Series: B, Physics **198**, (Plenum Press, New York 1989)

## Chapter 11

11.1 R.A. Swalin: *Thermodynamics of Solids*, 2nd ed. (Wiley, New York 1972)
11.2 A.M. Alper: *Phase Diagrams: Materials Science and Technology* (Academic, New York 1970)
11.3 A. Reisman: *Phase Equilibria* (Academic, New York 1970)
11.4 M.L. Hitchman, K.F. Jensen: *Chemical Vapor Deposition – Principles and Applications* (Academic Press, London 1993)
11.5 K.F. Jensen: *Transport Phenomena in Vapor Phase Epitaxy Reactors*, in *Handbook of Crystal Growth*, Vol. 3b, ed. by D.T.J. Hurle (Elsevier, Amsterdam 1994) Chap. 13
11.6 R.B. Bird, W.E. Steward, E N. Lightfoot: *Transport Phenomena* (Wiley, New York 1960)
11.7 Y. Saito: *Statistical Physics of Crystal Growth* (World Scientific, Singapore 1996)
11.8 T. Nishinaga, K. Nishioka, J. Harada, A. Sasaki, H. Takei (eds.): *Advances in the Understanding of Crystal Growth Mechanisms* (Elsevier, Amsterdam 1997)
11.9 W.A. Harrison: *Electronic Structure and the Properties of Solids – The Physics of Chemical Bond* (Freeman Co., San Francisco 1980) Chap. 7
11.10 M.A. Herman: Thin Solid Films **267**, 1 (1995)
11.11 G.B. Stringfellow: Rep. Prog. Phys. **45**, 469 (1982)
11.12 F. Rosenberger: *Fundamentals of Crystal Growth* (Springer, Berlin 1979)
11.13 L. Rosenhead (ed.): *Laminar Boundary Layers* (Clarendon Press, Oxford 1963)
11.14 D.I. Fotiadis, M. Boekholt, K.F. Jensen, W. Richter: J. Cryst. Growth **100**, 577 (1990)
11.15 G.B. Stringfellow: J. Cryst. Growth **68**, 111 (1984)
11.16 P.D. Dapkus: J. Cryst. Growth **68**, 345 (1984)
11.17 H. Schlichting: *Boundary Layer Theory*, 6th ed. (McGraw-Hill, New York 1968)
11.18 M.A. Herman: Vacuum **32**, 555 (1982)
11.19 M.A. Herman, H. Sitter: *Molecular Beam Epitaxy – Fundamentals and Current Status*, 2nd ed. (Springer, Berlin 1996)
11.20 L.I. Maissel, R. Glang (eds): *Handbook of Thin Film Technology* (McGraw-Hill, New York 1970)
11.21 P. Clausing: Z. Physik **66**, 471 (1930)
11.22 P. Clausing: Ann. Phys. (Leipzig) **12**, 961 (1932)
11.23 B.B. Dayton: *Gas Flow Patterns at Entrance and Exit of Cylindrical Tubes*, in *1956 National Symposium in Vacuum Technology Transactions*, ed. by E.S. Perry, J.H. Durant (Pergamon, Oxford 1957) p. 5
11.24 L.Y.L. Shen: J. Vac. Technol. **15**, 10 (1978)
11.25 J. Curless: J. Vac. Sci. Technol. B **3**, 531 (1985)
11.26 T. Yamashita, T. Tomita, T. Sakurai: Jpn. J. Appl. Phys. **26**, 1192 (1987)
11.27 S. Adamson, C.O. Carroll, J.F. McGilp: J. Vac. Sci. Technol. B **7**, 487 (1989)
11.28 Z.R. Wasilewski, G.C. Aers, A.J. Spring-Thorpe, C.J. Miner: J. Vac. Sci. Technol. B **9**, 120 (1991)
11.29 L. Michalak, B. Adamczyk, M.A. Herman: Vacuum **43**, 341 (1992)

11.30 G.C. Aers, Z.R. Wasilewski: J. Vac. Sci. Technol. B **10**, 815 (1992)
11.31 D.A. Porter, K.E. Esterling: *Phase Transformations in Metals and Alloys*, 2nd ed. (Chapman and Hall, London 1992)
11.32 R.T. DeHoff: *Thermodynamics in Materials Science* (McGraw-Hill, New York 1993)
11.33 J.M. Howe: *Interfaces in Materials* (Wiley, New York 1997)
11.34 L.J. Vieland: Acta Metal. **11**, 137 (1963)
11.35 M.B. Panish, M. Ilegems: *Progres in Solid State Chemistry*, Vol. 7, ed. by H. Reiss, J.O. McCaldin (Pergamon, Oxford 1972) p. 39
11.36 G. Scatchard: Chem. Rev. **8**, 321 (1931)
11.37 G.B. Stringfellow: Mater. Res. Bull. **6**, 371 (1971)
11.38 G.B. Stringfellow: Int. Rev. Sci. Inorganic Chem., Series 2, Vol. 10, ed. by L.E.J. Roberts (Butterworth, London 1975) p. 111
11.39 J.C. Phillips, J.A. van Vechten: Phys. Rev. B **2**, 2147 (1970)
11.40 R.J. Young, P.A. Lovell: *Introduction to Polymers*, 2nd ed. (Chapman and Hall, London 1992)
11.41 L.D. Landau, E.M. Lifshitz: *Statistical Physics*, 2nd ed. (Addison-Wesley, Reading, Mass. 1969)
11.42 H. Hertz: Ann. Phys. Leipzig, **17**, 193 (1882)
11.43 M. Knudsen: Ann. Phys. , Leipzig, **47**, 697 (1915)
11.44 R. Kern, O. Le Lay, J.J. Metois: *Basic Mechanisms in the Early Stages of Epitaxy*, in *Current Topics in Materials Science*, Vol. 3, ed. by E. Kaldis (North-Holland, Amsterdam 1979) Chap. 3
11.45 R.C. Cammarata: Prog. Surf. Sci. **46**, 1 (1994)
11.46 J.M. Blakely: *Introduction to the Properties of Crystal Surfaces* (Pergamon, Oxford 1973)
11.47 J.W. Cahn: Acta Metall. **28**, 1333 (1980)
11.48 M. Born, O. Stern: Ber. Berliner Acad. **48**, 901 (1919)
11.49 M. Gebhardt: *Epitaxy*, in *Crystal Growth – An Introduction*, ed. by P. Hartman (North-Holland, Amsterdam 1973), Chap. 3
11.50 M.J. Stowell: Thin Films **1**, 55 (1968)
11.51 B.W. Sloop, C.O. Tiller: J. Appl. Phys. **32**, 1331 (1961)
11.52 M. Grundmann, N.N. Ledentsov, O. Stier, D. Bimberg, V.M. Ustinov, P.S. Kop´ev, Zh. I. Alferov: Appl. Phys. Lett. **68**, 979 (1996)
11.53 R. Nötzel, J. Temmyo, A. Kozen, T. Tamamura, T. Fukui, H. Hasegawa: *Self-Ordered Quantum Dots: A New Growth Mode on High-Index Semiconductor Surfaces*, Adv. Sol. State Phys. Vol. 35, ed. by R. Helbig (Vieweg, Braunschweig, 1996), p. 103
11.54 A.L. Roytburd: J. Appl. Phys. **83**, 228 (1998)
11.55 D.H. Rich, Y. Tang, H.T. Lin: J. Appl. Phys. **81**, 6837 (1997)
11.56 M. Tsuchiya, P.M. Petroff, L.A. Coldren: Appl. Phys. Lett. **54**, 1690 (1989)
11.57 V.A. Shchukin, N.N. Ledentsov, P.S. Kop´ev, D. Bimberg: Phys. Rev. Lett. **75**, 2968 (1995)
11.58 M. Grundmann, O. Stier, D. Bimberg: Phys. Rev. B **52**, 11969 (1995)
11.59 V.A. Shchukin, N.N. Ledentsov, M. Grundmann, P.S. Kop'ev, D. Bimberg: Surf. Sci. **352-354**, 117 (1996)
11.60 M. Grundmann, N.N. Ledentsov, O. Stier, D. Bimberg, V.M. Ustinov, P.S. Kop´ev, Zh.I. Alferov: Appl. Phys. Lett. **68**, 979 (1996)
11.61 V.A. Shchukin, D. Bimberg, V.G. Malyshkin, N.N. Ledentsov: Phys. Rev. B **57**, 12262 (1998)
11.62 D.J. Bottomley: Appl. Phys. Lett. **72**, 783 (1998)
11.63 F. Glas: J. Appl. Phys. **62**, 3201 (1987)
11.64 I.P. Ipatova, V.G. Malyshkin, V.A. Shchukin: J. Appl. Phys. **74**, 7198 (1993)

11.65 K.Y. Cheng, K.C. Hsieh, J.N. Baillargeon: Appl. Phys. Lett. **60**, 2892 (1992)
11.66 E. Carlino, C. Giannini, C. Geradi, L. Tapfer, K.A. Mäder, H. von Känel: J. Appl. Phys. **79**, 1441 (1996)
11.67 K.C. Hsieh, J.N. Baillargeon, K.Y. Cheng: Appl. Phys. Lett. **57**, 2244 (1990)
11.68 J.M. Gerard, Y.M. Marzin, B. Jusserand, F. Glas, J. Primot: Appl. Phys. Lett. **54**, 30 (1989)
11.69 K.Y. Cheng, K.C. Hsieh, J.N. Baillargeon, A. Mascarenhas: in Proc. 18th Int. Sympos. GaAs and Rel. Comp. (Inst. Series, London 1992), p. 589
11.70 V.A. Shchukin, D. Bimberg: Rev. Mod. Phys. **71**, 1125 (1999)
11.71 C. Ratsch, A. Zangwill: Surf. Sci. **293**, 123 (1993)
11.72 D.J. Eaglesham, M. Cerullo: Phys. Rev. Lett. **64**, 1943 (1990)
11.73 C.W. Snyder, B.G. Orr, D. Kessler, L.M. Sander: Phys. Rev. Lett. **66**, 3032 (1991)
11.74 P. Kratzer, E. Penev, M. Scheffler: Appl. Phys. A **75**, 79 (2002)
11.75 L.G. Wang, P. Kratzer, M. Scheffler, N. Moll: Phys. Rev. Lett. **82**, 4042 (1999)
11.76 J.E. Bernard, A. Zunger: Appl. Phys. Lett. **65**, 165 (1994)
11.77 Q. Xie, P. Chen, A. Madhukar: Appl. Phys. Lett. **65**, 2051 (1994)
11.78 Q. Xie, A. Madhukar, P. Chen, N. Kobayashi: Phys. Rev. Lett. **75**, 2542 (1995)
11.79 J. Tersoff, C. Teichert, M.G. Lagally: Phys. Rev. Lett. **76**, 1675 (1996)
11.80 M. Strassburg, V. Kutzer, U.W. Pohl, A. Hoffmann, I. Boser, N.N. Ledentsov, D. Bimberg, A. Rosenauer, U. Fischer, D. Gerthsen, I.L. Krestiukov, M.V. Maximov, P.S. Kop´ev, Zh.I. Alferov: Appl. Phys. Lett. **72**, 942 (1998)
11.81 I. Barin (ed. ): *Thermochemical Data of Pure Substances* (VCH, New York 1989)
11.82 L. Tisza: *Generalized Thermodynamics* (MIT Press, Cambridge, Mass. 1966), Chap. VIII
11.83 C. Delamarre, J.Y. Laval, L.P. Wang, A. Dubon, G. Schiffmacher: J. Cryst. Growth **177**, 6 (1997)
11.84 M.A. Herman: *Semiconductor Superlattices* (Akademie Verlag, Berlin 1986)
11.85 J.S. Speck, A.C. Dyakin, A. Seifert, A.E. Romanov, W. Pompe: J. Appl. Phys. **78**, 1696 (1995)
11.86 N.A. Pertsev, A.G. Zembilgotov: J. Appl. Phys. **78**, 6170 (1995)
11.87 N. Sridhar, J.M. Rickman, D.J. Srolovitz: Acta Mater. **44**, 4085 and 4097 (1996)
11.88 M.A. Herman, T.G. Andersson: Appl. Phys. A **41**, 243 (1986)
11.89 J.S. Langer: Rev. Mod. Phys. **52**, 1 (1980)
11.90 P.G. Shewmon: Trans. Met. Soc. AIME **233**, 736 (1965)
11.91 D.T.J. Hurle, E. Jakeman, A.A. Wheeler: J. Cryst. Growth **58**, 163 (1982)
11.92 D.T.J. Hurle: J. Cryst. Growth **61**, 463 (1983)
11.93 A.A. Wheeler: J. Cryst. Growth **67**, 8 (1984)
11.94 P.W. Voorhees, S.R. Coriell, G.B. McFadden, R.F. Sekerka: J. Cryst. Growth **67**, 425 (1984)
11.95 W.W. Mullins, R.F. Sekerka: J. Appl. Phys. **34**, 323 (1963)
11.96 W.W. Mullins, R.F. Sekerka: J. Appl. Phys. **35**, 444 (1964)
11.97 R.F. Sekerka: *Morphological Stability*, in *Crystal Growth – An Introduction*, ed. by P. Hartman (North-Holland, Amsterdam 1973), Chap. 8
11.98 C.H.J. van den Brekel, A.K. Jansen: J. Cryst. Growth **43**, 364 (1978)
11.99 A.K. Jansen, C.H.J. van den Brekel: J. Cryst. Growth **43**, 373 (1978)
11.100 C.H.J. van den Brekel, A.K. Jansen: J. Cryst. Growth **43**, 488 (1978)
11.101 A.A. Chernov: J. Cryst. Growth **52**, 699 (1981)

11.102 J.S. Langer, H. Müller-Krumbhar: Phys. Rev. A **27**, 499 (1983)
11.103 H.F. Lockwood, M. Ettenberg: J. Cryst. Growth **15**, 81 (1972)
11.104 S.R. Coriell, R.L. Parker: J. Appl. Phys. **37**, 1548 (1966)

## Chapter 12

12.1 P. Bennema, G.H. Gilmer: *Kinetics of Crystal Growth*, in *Crystal Growth – An Introduction*, ed. by P. Hartman (North-Holland, Amsterdam 1973), Chap. 10
12.2 H. Müller-Krumbhar: Phys. Rev. B **10**, 1308 (1974)
12.3 Y. Saito: *Statistical Physics of Crystal Growth* (World Scientific, Singapore 1996)
12.4 W.A. Harrison: *Electronic Structure and the Properties of Solids – The Physics of Chemical Bond* (Freeman and Co, San Francisco 1980) Chap. 7
12.5 R. Kern, O. Le Lay, J.J. Metois: *Basic Mechanisms in the Early Stages of Epitaxy*, in *Current Topics in Materials Science* Vol. 3, ed. by E. Kaldis (North-Holland, Amsterdam 1979) Chap. 3
12.6 A. Madhukar, S.V. Ghaisas: CRC Crit. Rev. Solid State Mater. Sci. **14**, 1 (1988)
12.7 H. Müller-Krumbhar: *Kinetics of Crystal Growth*, in *Current Topics in Materials Science*, Vol. 1, ed. by E. Kaldis (Nort-Holland, Amsterdam 1978), Chap. 1
12.8 T.D. Lee, C.N. Yang: Phys. Rev. **87**, 410 (1952)
12.9 L.K. Runnels: Phys. Rev. Lett. **15**, 581 (1965)
12.10 A. Bellemans, R.K. Nigam: J. Chem. Phys. **46**, 2922 (1967)
12.11 C. Ebner, C. Rottman, M. Wortis: Phys. Rev. B**28**, 4186 (1983)
12.12 J.W. Cahn, R. Kikuchi: Phys. Rev. B **31**, 4300 (1985)
12.13 Y. Saito, T. Ueta: Phys. Rev. A **40**, 3408 (1989)
12.14 J.R. Smith, Jr., A. Zangwill: Surf. Sci. **316**, 359 (1994)
12.15 W.L. Bragg, E.J. Williams: Proc. Roy. Soc. London A **145**, 699 (1934)
12.16 R.A. Swalin: *Thermodynamics of Solids*, 2nd ed. (Wiley, New York 1972)
12.17 Y. Saito: J. Chem. Phys. **74**, 713 (1981)
12.18 R. Venkatasubramanian: J. Mater. Sci. **7**, 1221 and 1235 (1992)
12.19 E. Ising: Z. Phys. **31**, 253 (1925)
12.20 B.M. McCoy, T.T. Wu: *The Two-Dimensional Ising Model* (Harvard University Press, Cambridge, Mass. 1973)
12.21 L. Onsager: Phys. Rev. **65**, 117 (1944)
12.22 M.E. Fischer: Rep. Progr. Phys. **36**, 615 (1967)
12.23 K. Binder, P. Hohenberg: Phys. Rev. B **6**, 3461 (1972)
12.24 H. Leamy, G. Gilmer, K.A. Jackson, P. Bennema: Phys. Rev. Lett **30**, 601 (1973)
12.25 K. Binder, P. Hohenberg: Phys. Rev. B **9**, 2194 (1974)
12.26 J.D. Weeks, G.H. Gilmer, K.A. Jackson: J. Chem Phys. **65**, 712 (1976)
12.27 H. Nakayama, T. Kita, T. Nishino: *Atomic Ordering in Epitaxial Alloy Semiconductors – From the Discoveries to the Physical Understanding*, in Ref. [12.48], p. 163
12.28 K. Binder (ed.): *Monte Carlo Methods in Statistical Physics* (Springer, Berlin 1979)
12.29 M.C. Desjonqueres, D. Spanjaard: *Concepts in Surface Physics*, 2nd ed. (Springer, Berlin 1996), Chap. 6
12.30 A. Zangwill: *Physics at Surfaces*, (Cambridge University Press, Cambridge 1988)

12.31 P.W. Atkins: *Physical Chemistry*, 3rd ed. (Oxford University Press, Oxford 1986)
12.32 S.R. Morrison: *The Chemical Physics of Surfaces* (Plenum Press, New York 1977)
12.33 R. Fowler, E.A. Guggenheim: *Statistical Thermodynamics* (Cambridge University Press, Cambridge 1965)
12.34 A.M. de Jong, J.W. Niemantsverdriet: Surf. Sci. **233**, 355 (1990)
12.35 V.J. Garcia, J.M. Briceno-Valero, L. Martinez: Surf. Sci. **339**, 189 (1995)
12.36 M. Polanyi, E. Wigner: Z. Phys. Chem. A **139**, 439 (1928)
12.37 M.A. Herman: Appl. Surf. Sci. **112**, 1 (1997)
12.38 M.A. Herman, P. Juza, W. Faschinger, H. Sitter: Crystal Res. Technol. **23**, 307 (1988)
12.39 P. Juza, H. Sitter, M.A. Herman: Appl. Phys. Lett. **53**, 1396 (1988)
12.40 W.K. Burton, N. Cabrera, F.C. Frank: Philos. Trans. Roy. Soc., London, Ser. A, **243**, 299 (1951)
12.41 P. Šmilauer: Vacuum **50**, 115 (1998)
12.42 G. Ehrlich, F.G. Hudda: J. Chem. Phys. **44**, 1039 (1966)
12.43 R.L. Schwoebel, E.J. Shipsey: J. Appl. Phys. **37**, 3682 (1966)
12.44 H.J. Gossmann, F.W. Sinden, L.C. Feldman: J. Appl. Phys. **67**, 745 (1990)
12.45 J. Krug: Adv. Phys. **46**, 139 (1997)
12.46 P. Šmilauer, D.D. Vvedensky: Phys. Rev. B **52**, 14263 (1995)
12.47 G.S. Bales, A. Zangwil: Phys. Rev. B **41**, 5500 (1990)
12.48 T. Nishinaga, K. Nishioka, J. Harada, A. Sasaki, H. Takei (eds. ): *Advances in the Understanding of Crystal Growth Mechanisms*, (Elsevier, Amsterdam 1997)
12.49 E.E. Gruber, W.W. Mullins: J. Phys. Chem. Solids **28**, 875 (1967)
12.50 T. Yamamoto, N. Akutsu, Y. Akutsu: *The Terrace–Step–Kink Approach and the Capillary–Wave Approach to Fluctuation Properties of Vicinal Surfaces*, in Ref. [12.48], p. 19
12.51 A. Natori: *Step Structure of Si(111) Vicinal Surfaces*, in Ref. [12.48], p. 61
12.52 A.V. Latyshev, A.L. Aseev, A.B. Krasilnikov, S.I. Stenin: Surf. Sci. **213**, 157 (1989)
12.53 A.V. Latyshev, A.B. Krasilnikov, A.L. Aseev: Thin Solid Films **306**, 205 (1997)
12.54 K. Wada, H. Ohmi: *Crystal Growth Kinetics on Stepped Surface by the Path Probability Method*, in Ref. [12.48], p. 101
12.55 S. Holloway, N.V. Richardson (eds.): *Handbook of Surface Science*, Vol. 1, *Physical Structure*, ed. by W.N. Unertl (Elsevier, Amsterdam, 1996)
12.56 T. Kimoto, A. Itoh, H. Matsunami: Appl. Phys. Lett. **66**, 3645 (1995)
12.57 X.S. Wang, J.L. Goldberg, N.C. Bartelt, T.L. Einstein, E.D. Williams: Phys. Rev. Lett. **65**, 2430 (1990)
12.58 M.C. Tringides: Atomic Scale Defects on Surfaces, in Ref. [12.55], Chap. 12
12.59 M. Uwaha: *Fluctuation and Morphological Instability of Steps in a Surface Diffusion Field*, in Ref. [12.48], p. 31
12.60 E.D. Williams, N.C. Bartelt: *Thermodynamics and Statistical Mechanics of Surfaces*, in Ref. [12.55], Chap 2
12.61 W.W. Mullins, J.P. Hirth: J. Phys. Chem. Solids **24**, 1391 (1963)

## Chapter 13


13.1 W.A. Harrison: *Electronic Structure and the Properties of Solids – The Physics of Chemical Bond* (Freeman and Co, San Francisco 1980) Chap. 7
13.2 A. Messiah: *Quantum Mechanics* (North-Holland, Amsterdam 1967)
13.3 F. Bechstedt, R. Enderlein: *Semiconductor Surfaces and Interfaces*, (Akademie Verlag, Berlin 1988)
13.4 L.I. Schiff: *Quantum Mechanics*, 3rd ed. (McGraw-Hill, New York 1968)
13.5 J.C. Slater, G.F. Koster: Phys. Rev. **94**, 1498 (1954)
13.6 H.H. Farrell, J.P. Harbison, L.D. Peterson: J. Vac. Sci. Technol. B **5**, 1482 (1987)
13.7 A. Kahn: Surf. Sci. Rep. **3**, 193 (1983)
13.8 S. Holloway, N.V. Richardson (eds.): *Handbook of Surface Science*, vol. 1, Physical Structure, ed. by W.N. Unertl (Elsevier, Amsterdam, 1996)
13.9 W.N. Unertl: *Surface Crystallography*, in Ref. [13.8], Chap. 1
13.10 C.T. Chan, K.M. Ho, K.P. Bohnem, *Surface Reconstruction: Metal Surfaces and Metal on Semiconductor Surfaces*, in Ref. [13.8], Chap. 3
13.11 J.P. LaFemina: *Theory of Insulator Surface Structures*, in Ref. [13.8], Chap. 4
13.12 R.L. Lad: *Surface Structure of Crystalline Ceramics*, in Ref. [13.8], Chap. 5
13.13 C.B. Duke: *Surface Structures of Elemental and Compound Semiconductors*, in Ref. [13.8], Chap. 6
13.14 F. Bechstedt: *Principles of Surface Physics* (Springer, Berlin, Heidelberg, New York 2003)
13.15 E. Madelung: Phys. Z. **11**, 898 (1910)
13.16 J.C. Slater: Phys. Rev. **34**, 1293 (1929)
13.17 L.I. Schiff: *Quantum Mechanics*, 3rd ed. (McGraw-Hill, New York 1968)
13.18 R.O. Jones, O. Gunnarsson: Rev. Mod. Phys. **61**, 689 (1989)
13.19 P. Hohenberg, W. Kohn: Phys. Rev. **136**, B 864 (1964)
13.20 W. Kohn, L.J. Sham: Phys. Rev. **140**, A 1133 (1965)
13.21 J.C. Slater, G.F. Koster: Phys. Rev. **94**, 1498 (1954)
13.22 W.A. Harrison: *Electronic Structure and the Properties of Solids – The Physics of Chemical Bond* (Freeman and Co, San Francisco 1980) Chap. 7
13.23 J.P. LaFemina: Surf. Sci. Rep. **16**, 133 (1992)
13.24 C.B. Duke: J. Vac. Sci. Technol. A **10**, 2032 (1992)
13.25 D.J. Chadi: Phys. Rev. B **29**, 785 (1984)
13.26 C.B. Duke: Appl. Surf. Sci. **65/66**, 543 (1993)
13.27 C.B. Duke: *Reconstruction of the Cleavage Faces of Tetrahedrally Coordinated Compound Semiconductors*, ed. by R. Helbig, *Festkörperprobleme, Advances in Solid State Physics*, Vol. 33, p. 1 (Vieweg, Braunschweig, 1994)
13.28 H.B. Gray: *Electrons and Chemical Bonding* (Benjamin, New York 1965)
13.29 D. Haneman: Rep. Prog. Phys. **50**, 1045 (1987)
13.30 C.B. Duke: Chem. Rev. **96**, 1237 (1996)
13.31 E. Pehlke, N. Moll, A. Kley, M. Scheffler: Appl. Phys. A **65**, 525 (1997)
13.32 E. Steimetz, F. Schienle, J.-T. Zettler, W. Richter: J. Cryst. Growth **170**, 208 (1997)
13.33 Y. Temko, T. Suzuki, K. Jacoby: Appl. Phys. Lett. **82**, 2142 (2003) and references therein.
13.34 W.G. Schmidt: Appl. Phys. A **75**, 89 (2002)
13.35 O. Pulci, W.G. Schmidt, F. Bechstedt: Surf. Sci. **464**, 272 (2000)
13.36 W.G. Schmidt, P.H. Hahn F. Bechstedt, N. Esser, P. Vogt, A. Wange,: Phys. Rev. Lett. **90**, 126101 (2003)
13.37 W. Richter: Appl. Phys. A**75**, 129 (2002)

13.38 P. Kratzer, E. Penev, M. Scheffler: Appl. Phys. A**75**, 79 (2002)
13.39 P. Kratzer, C.G. Morgan, M. Scheffler: Phys. Rev. B **59**, 15246 (1999)
13.40 P. Kratzer, M. Scheffler: Phys. Rev. Lett. **88**, 036102 (2002)
13.41 A. Kley, P. Ruggerone, M. Scheffler: Phys. Rev. Lett. **79**, 5278 (1997)
13.42 C.G. Morgan, P. Kratzer, M. Scheffler: Phys. Rev. Lett. **82**, 4886 (1999)
13.43 M. Zorn, P. Kurpas, A.I. Shkrebtii, B. Junno, A. Bhattacharya, K. Knorr, M. Weyers, L. Samuelson, J.-T. Zettler, W. Richter: Phys. Rev. B. **60**, 8185 (1999)

## Chapter 14

14.1 A.I. Finch, A.G. Quarrell: Proc. Phys. Soc. , London **48**, 148 (1934)
14.2 J. Woltersdorf: Thin Solid Films **85**, 241 (1981)
14.3 F.C. Frank, J.H. van der Merwe: Proc. Roy. Soc. , London **198**, **205** and **216** (1946)
14.4 M.A. Herman, H. Sitter: *Molecular Beam Epitaxy – Fundamentals and Current Status*, 2nd ed. (Springer, Berlin 1996)
14.5 M. Gendry, V. Drouot, G. Hollinger, S. Mahajan: Appl. Phys. Lett. **66**, 40 (1995)
14.6 M. Gendry, V. Drouot, C. Santinelli, G. Hollinger: Appl. Phys. Lett. **60**, 2249 (1992)
14.7 W. Li, Z. Wang, J. Liang, Q. Liao, B. Xu, Z. Zhu, B. Yang: Appl. Phys. Lett. **66**, 1080 (1995)
14.8 N. Grandjean, J. Massies, M. Leroux, J. Leymarie, A. Vasson, A.M. Vasson: Appl. Phys. Lett. **64**, 2664 (1994)
14.9 C.A.B. Ball, J.H. van der Merwe: *The Growth of Dislocation-Free Layers*, in *Dislocations in Solids*, ed. by F.R.N. Nabarro (North-Holland, Amsterdam 1983), Vol. 6, Chap. 27
14.10 G.H. Olsen: J. Cryst. Growth **31**, 223 (1975)
14.11 J.H. van der Merwe: CRC Crit. Rev. Solid State Mater. Sci. **7**, 209 (1978)
14.12 J.W. Matthews: *Coherent Interfaces and Misfit Dislocations*, in *Epitaxial Growth* ed. by J.W. Matthews (Academic, New York 1975), Chap. 8
14.13 J.W. Matthews, J.L. Crawford: Thin Solid Films **5**, 187 (1970)
14.14 J.Y. Tsao: *Materials Fundamentals of Molecular Beam Epitaxy* (Academic, San Diego 1993), Chap. 5
14.15 S. Sharan, J. Narayan: *Semiconductor Heterostructures; Formation of Defects and their Reduction*, in *Concise Encyclopedia of Semiconducting Materials and Related Technologies* ed. by S. Mahajan, L.C. Kimerling (Pergamon, Oxford 1992)
14.16 M. Tabuchi, S. Noda, A. Sasaki: J. Cryst. Growth **149**, 12 (1995)
14.17 H.G. Colson, D.J. Dunstan: J. Appl. Phys. **81**, 2898 (1997)
14.18 J.W. Matthews, A.E. Blakeslee: J. Cryst. Growth **27**, 118 (1974); ibid. **29**, 273 (1975); ibid. **32**, 265 (1976)
14.19 G.J. Whaley, P.I. Cohen: Mat. Res. Soc. Proc. **160**, 35 (1990)
14.20 M.J. Ekenstedt, S.M. Wang, T.G. Andersson: Appl. Phys. Lett. **58**, 854 (1991)
14.21 S.M. Wang, T.G. Andersson, M.J. Ekenstedt: Appl. Phys. Lett. **61**, 3139 (1992)
14.22 B. Elman, E.S. Koteles, P. Melman, C. Jaganuath, C.A. Armiento, M. Rothman: J. Appl. Phys. **68**, 1351 (1990)
14.23 Y.H. Lo: Appl. Phys. Lett. **59**, 2311 (1991)
14.24 D. Teng, Y.H. Lo: Appl. Phys. Lett. **62**, 43 (1993)

14.25 C.L. Chua, W.Y. Hsu, C.H. Lin, G. Chriestenson, Y.H. Lo: Appl. Phys. Lett. **64**, 3640 (1994)
14.26 J. Arokiaraj, T. Soga, T. Jimbo, M. Umeno: Appl. Phys. Lett. **75**, 3826 (1999)
14.27 P. Kopperschmidt, S. Senz, R. Scholz, U. Gosele: Appl. Phys. Lett. **74**, 374 (1999)
14.28 Z.H. Zhu, R. Zhou, F.E. Ejeckam, Z. Zhang, J. Zhang, J. Greenberg, Y.H. Lo, H.Q. Hou, B.E. Hammons: Appl. Phys. Lett. **72**, 2598 (1998)
14.29 Z.H. Zhu, F.E. Ejeckam, Y. Qian, J. Zhang, Z. Zhang, G.L. Christenson, Y.H. Lo: IEEE J. Sel. Top. Quantum Electron. **3**, 927 (1997)
14.30 D. Maroudas, L.A. Zepeda-Ruiz, W.H. Weinberg: Appl. Phys. Lett. **73**, 753 (1998)
14.31 J.F. Damlencourt, J.L. Leclercq, M. Gendry, P. Regreny, G. Hollinger: Appl. Phys. Lett. **75**, 3638 (1999)
14.32 M.A. Herman: Cryst. Res. Technol. **34**, 583 (1999)
14.33 F.A. Ponce, S.P. DenBaars, B.K. Meyer, S. Nakamura, S. Strite (eds.): *Nitride Semiconductors*, Mat. Res. Soc. Proc. Vol. **482** (MRS, Warendale 1998)
14.34 H.T. Johnson, L.B. Freund: J. Appl. Phys. **81**, 6081 (1997)
14.35 M. Copel, M.C. Reuter, E. Kaxiras, R.M. Tromp: Phys. Rev. Lett. **63**, 632 (1989)
14.36 N. Grandjean, J. Massies, V.H. Etgens: Phys. Rev. Lett. **69**, 796 (1992)
14.37 M. Copel, M.C. Reuter, M. Horn von Hoegen, R.M. Tromp: Phys. Rev. B **42**, 11682 (1990)
14.38 M. Horn von Heogen: Appl. Phys. A **59**, 503 (1994)
14.39 A. Wakahara, K.K. Vong, T. Hasegawa, A. Fujihara, A. Sasaki: J. Cryst. Growth **151**, 52 (1995)
14.40 R.I.G. Uhrberg, R.D. Gringans, R.Z. Brachrach, J.E. Northrup: Phys. Rev. Lett. **56**, 520 (1986)
14.41 R.D. Maede, D. Vanderbilt: Phys. Rev. Lett. **63**, 1404 (1989)
14.42 A.J. Schell-Sorokin, R.M. Tromp: Phys. Rev. Lett. **64**, 1039 (1990)
14.43 R. Kern, P. Müller: J. Cryst. Growth **146**, 193 (1995)
14.44 J. Falta, M. Copel, F.K. Le Gouses, R.M. Tromp: Appl. Phys. Lett. **62**, 2962 (1993)
14.45 H.J. Osten, J. Klatt, G. Lippert, E. Bugiel: J. Cryst. Growth **127**, 396 (1993)
14.46 J. Massies, N. Grandjean, V.H. Elgens: Appl. Phys. Lett. **61**, 99 (1992)
14.47 H.J. Osten, J. Klatt, G. Lippert, E. Bugiel, S. Higuchi: J. Appl. Phys. **74**, 2507 (1993)
14.48 M. Horn von Hoegen, F.K. Le Gouses, M. Copel, M.C. Reuter, R.M. Tromp: Phys. Rev. Lett. **57**, 1130 (1991)
14.49 Z.R. Zytkiewicz, D. Dobosz: Acta Phys. Polonica A **92**, 1079 (1997), and in *Heterostructure Epitaxy and Devices – HEAD'97*, ed. by P. Kordos, J. Novak (Kluwer Academic Publ., Amsterdam 1998) p. 71
14.50 E. Bauser: *Atomic Mechanisms in Semiconductor Liquid Phase Epitaxy*, in: *Handbook of Crystal Growth* Vol. 3 b, ed. by D.T.J. Hurle (Elsevier, Amsterdam 1994), Chap. 20
14.51 R. Rantamäki, T. Tuomi, Z.R. Zytkiewicz, J. Domagala, P.J. McNally, A.N. Danilewsky: J. Appl. Phys. **86**, 4298 (1999)
14.52 T. Nishinaga, T. Nakano, S. Zhang: Jpn. J. Appl. Phys. **27**, L 964 (1988)
14.53 S. Zhang, T. Nishinaga: Jpn. J. Appl. Phys. **29**, 545 (1990)
14.54 Z.R. Zytkiewicz, J. Domagala, D. Dobosz, J. Bak-Misiuk: J. Appl. Phys. **86**, 1965 (1999)

14.55 Z.R. Zytkiewicz, J. Domagala: Appl. Phys. Lett. **75**, 2749 (1999)
14.56 J.A. Smart, E.M. Chumbes, A.T. Schremer, J.R. Shealy: Appl. Phys. Lett. **75**, 3820 (1999)
14.57 J.A. Freitas,Jr. , O.H. Nam, R.F. Davis, G.V. Saparin, S.K. Obyden: Appl. Phys. Lett. **72**, 2990 (1998)
14.58 S. Nakamura, M. Senoh, S.I. Nagahama, N. Iwasa, T. Yamada, T. Matsushita, H. Kiyoku, Y. Sugimoto, T. Kozaki, H. Umemoto, M. Sano, K. Chocho: Appl. Phys. Lett. **72**, 211 (1998)
14.59 P. Gibart, B. Beaumont, P. Vennégués: *Epitaxial Lateral Overgrowth of GaN* in *Nitride Semiconductors*, ed. by P. Ruterana, M. Albrecht, J. Neugebauer (Wiley-VCH Verlag, Weinheim, 2003), Chap. 2
14.60 K. Hiramatsu, H. Matsushima, T. Shibata, N. Sawaki, K. Tadatomo, H. Okagawa, Y. Ohuchi, Y. Honda, T. Matsue: *Selective Area Growth of GaN by MO VPE and HVPE*, in Ref. [14.33], p. 257
14.61 S. Nakamura: *InGaN/GaN/AlGaN-Based Laser Diodes with an Estimated Lifetime of Longer than 10,000 Hours*, in Ref. [14.33], p. 1145
14.62 Y. Kato, S. Kitamura, K. Hiramatsu, N. Sawaki: J. Cryst. Growth **144**, 133 (1994)
14.63 J.A. Smart, A.T. Schremer, N.G. Weimann, O. Ambacher, L.F. Eastman, J.R. Shealy: Appl. Phys. Lett. **75**, 388 (1999)
14.64 J.A. Smart, E.M. Chumbes, A.T. Schremer, J.R. Shealy: Appl. Phys. Lett. **75**, 3820 (1999)
14.65 J. Griesche, R. Enderlein, D. Schikora: Phys. Status Solidi (a) **109**, 11 (1988)
14.66 J.P. Faurie, C. Hsu, S. Sivananthan, X. Chu: Surf. Sci. **168**, 473 (1986)
14.67 J. Sadowski, E. Dynowska, K. Reginski, M.A. Herman: Cryst. Res. Technol. **28**, 909 (1993)
14.68 J. Sadowski, M.A. Herman: J. Cryst. Growth **146**, 449 (1995)
14.69 A. Zur, T.G. McGill: J. Appl. Phys. **55**, 378 (1984)
14.70 T.H. Myers, Y. Cheng, R.N. Bicknell, J.F. Schetzina: Appl. Phys. Lett. **42**, 247 (1983)
14.71 E.I. Givargizov: *Artificial Epitaxy (Graphoepitaxy)*, in D.T.J. Hurle (ed. ): *Handbook of Crystal Growth*, Vol. 3 b (Elsevier, Amsterdam 1994), Chap. 21
14.72 E.I. Givargizov: *Oriented Crystallization on Amorphous Substrates* (Plenum Press, New York 1991)
14.73 N.N. Sheftal, N.A. Bouzynin: Vestnik Moskovskovo Universiteta **27**(3), 102 (1972)
14.74 V.I. Klykov, N.N. Sheftal: J. Cryst. Growth **52**, 687 (1981)
14.75 M.W. Geis, D.C. Flanders, H.I. Smith: Appl. Phys. Lett. **35**, 71 (1979)
14.76 E.I. Givargizov, N.N. Sheftal, V.I. Klykov: *Diataxy (Graphoepitaxy) and other Approaches to Oriented Crystallization on Amorphous Substrates*, in *Current Topics in Materials Science*, Vol. 10, ed. by E. Kaldis (North-Holland, Amsterdam 1982) p. 1
14.77 J.C.C. Fan, B.Y. Tsaur, M.W. Geis: J. Cryst. Growth **63**, 453 (1983)

## Chapter 15

15.1 D.A. King, D.P. Woodruff (eds.): *The Chemical Physics of Solid Surfaces*, Vol. 8 *Growth and Properties of Ultrathin Epitaxial Layers* (Elsevier Science B.V., Amsterdam 1997)
15.2 J.A. Venables: *Surface Processes in Epitaxial Growth*, in Ref. [15.1] Chapt. 1
15.3 J.A. Venables, G.D.T. Spiller, M. Hanbucken: Rep. Progr. Phys **47**, 399 (1984)

15.4 J.A. Venables, G.I. Price: in *Epitaxial Growth*, part A, ed. by J.W. Matthews (Academic Press, New York 1975) Chapt.4
15.5 A. Schmidt, V. Schunemann, R. Anton: Phys. Rev B **41**, 11875 (1990)
15.6 .R. Anton, A. Schmidt, V. Schunemann: Vacuum **41**, 1099 (1990)
15.7 C.R. Henry, C. Chapon, C. Duriez, S. Gorgio: Surf. Sci. **253**, 177 and 190 (1991)
15.8 C.R. Henry, C. Chapon, C. Goyhenex, R. Monot: Surf. Sci. **272**, 283 (1992)
15.9 E. Bauer: *The Many Facets of Metal Epitaxy*, in Ref. [15.1] Chapt. 2
15.10 G. Le Lay: *Monolayer Films of Unreactive Metals on Semiconductors*, in Ref. [15.1]. Chap. 8
15.11 M. Jalochowski: Progr. Surf. Sci. **48**, 287 (1995)
15.12 M. Jalochowski, H. Knoppe, G. Lilienkamp, E. Bauer: Phys. Rev. B **46**, 4693 (1992)
15.13 M. Jalochowski, M. Hoffmann, E. Bauer: Phys. Rev. Lett. **76**, 4227 (1996)
15.14 G.W. Jones, J.M. Marcano, J.K. Norskov, J.A. Venables: Phys. Rev. Lett. **65**, 3317 (1990)
15.15 H. Noro, R. Persaud, J.A. Venables: Vacuum **46**, 1173 (1995)
15.16 H. Noro, R. Persaud, J.A. Venables: Surf. Sci. **357/358**, 879 (1996)
15.17 H. Brune, H. Roder, C. Boragno, K. Kern: Phys. Rev. Lett. **73**, 1955 (1994)
15.18 H. Brune, H. Roder, K. Bromann, K. Kern, J. Jacobsen, P. Stolze, K.W. Jacobsen, J. Noeskov: Surf. Sci. **349**, L 115 (1996)
15.19 K. Sumitomo, T. Kobayashi, F. Shoji, K, Oura: Phys. Rev. Lett. **66**, 1193 (1991)
15.20 S.J: Pearton, *GaN and Related Materials*, in: *Optoelectronic Properties of Semiconductors and Superlattices*, series ed. M.O. Manasreh (Gordon and Breach, Amsterdam 1997)
15.21 S.J. Pearton, *GaN and Related Materials II*, in: *Optoelectronic Properties of Semiconductors and Superlattices*, series ed. M.O. Manasreh (Gordon and Breach, Amsterdam 1999)
15.22 B. Gil, *Group III Nitride Semiconductor Compounds - Physics and Applications* (Oxford University Press, Oxford 1998)
15.23 J.H. Edgar, *Properties of Group III Nitrides*, emis Datareviews Series No. 11, INSPEC (1994)
15.24 J.I. Pankove, T.D. Moustakas, *Gallium Nitride (GaN) I*, in: *Semiconductors and Semimetals*, Vol. 50 (Academic Press, San Diego 1998)
15.25 J.I. Pankove, T.D. Moustakas, *Gallium Nitride (GaN) II*, in: *Semiconductors and Semimetals*, Vol. 57 (Academic Press, San Diego 1999)
15.26 A.G. Bhuiyan, A. Hashimoto, A. Yamamoto: J. Appl. Phys. **94**, 2779 (2003)
15.27 V. Davydov, A. Klochikhin, S. Nanov, J. Aderhold: *Growth and Properties of InN* in [15.29], Chap. 5
15.28 MRS Internet Journal of Nitride Semiconductor Research, `http://nsr.mij.mrs.org`
15.29 P. Ruterana, M. Albrecht, J. Neugebauer (eds.): *Nitride Semiconductors* (Wiley-VCH Verlag, Weinheim, 2003)
15.30 J.W. Orton, C.T. Foxon, Rep. Prog. Phys. **61**, 1 (1998)
15.31 S. Strite, H. Morkoc, J. Vac. Sci. Technol. B **10**, 1237 (1992)
15.32 H. Yang, L.X. Zheng, J.B. Li, X.J. Wang, D.P. Xu, Y.T. Wang, X.W. Hu, P.D. Han, Appl. Phys. Lett. **74** (17), 2498 (1999)
15.33 D.J. As, A. Richter, J. Busch, M. Lübbers, J. Mimkes, K. Lischka: Appl. Phys. Lett. **76** (1), 13 (2000)

15.34 D.J. As, in: *III-V Nitride Semiconductors: Growth and Substrate Issues*, ed. by M.O. Manasreh and J. Ferguson, *Optoelectronic Properties of Semiconductors and Superlattices*, Vol. 20 (Gordon and Breach Publishers, Amsterdam 2002)
15.35 H. Yang, O. Brandt, K. Ploog: phys. stat. sol. (b) **194**, 109 (1996)
15.36 D. Schikora, M. Hankeln, D.J. As, K. Lischka, T. Litz, A. Waag, T. Buhrow, F. Henneberger, Phys. Rev. B **54** (12), R8381 (1996)
15.37 S.E. Hooper, C.T. Foxon, T.X. Cheng, L.C. Jenkins, D.E. Lacklison, J.W. Orton, T. Bestwick, A. Kean, M. Dawson, G. Duggan, J. Cryst. Growth **155**, 157 (1995)
15.38 H. Okumura, S. Misawa, S. Yoshida: Appl. Phys. Lett. **59** (9), 1059 (1991)
15.39 H. Okumura, K. Ohta, G. Feuillet, K. Balakrishnan, S. Chichibu, H. Hamaguchi, P. Hacke, S. Yoshida: J. Cryst. Growth **178**, 113 (1997)
15.40 S. Strite, J. Ruan, Z. Li, A. Salvador, H. Chen, D.J. Smith, W.J. Choyke, H. Morkoc: J. Vac. Sci. Technol. B **9**, 1924 (1991)
15.41 M.E. Lin, G. Xue, G.L. Zhou, H. Morkoc: Appl. Phys. Lett. **62** (26), 3479 (1993)
15.42 C.H. Hong, K. Wang, D. Pavlides: Inst. Phys. Conf. Ser. No. **141**, 107 (1994)
15.43 T. Lei, T.D. Moustakas, R.J. Graham, Y. He, S.J. Berkowitz: J. Appl. Phys. **71** (10), 4933 (1992)
15.44 Z. Sitar, M.J. Paiseley, B. Yan, R.F. Davis: MRS Symp. Proc. **162**, 2089 (1990)
15.45 K. Ploog, O. Brandt, H. Yang, A. Trampert: J. Vac. Sci. Technol. B **16** (4), 2229 (1998)
15.46 H. Okumura, H. Hamaguchi, T. Koizumi, K. Balakrishnan, Y. Ishida, M. Arita, S. Chichibu, H. Nakanishi, T. Nagatomo, S. Yoshida: J. Cryst. Growth **189/190**, 390 (1998)
15.47 B. Daudin, G. Feuillet, J. Hübner, Y. Samson, F. Widmann, A. Philippe, C. Bru-Chevallier, G. Guillot, E. Bustarret, G. Bentoumi, A. Deneuville: J. Appl. Phys. **84** (4), 2295 (1998)
15.48 J. Wu, H. Yaguchi, H. Nagasawa, Y. Yamaguchi, K. Onabe, Y. Shiriki, R. Ito: J. Cryst. Growth **189/190**, 420 (1998)
15.49 A. Barski, U. Rössner, J.L. Rouviere, M. Arlery: MRS Internet J. Nitride Semicond. Res. **1**, 21 (1996)
15.50 Y. Hiroyama, M. Tamura: Jpn. J. Appl. Phys. **37** part 2, No. 6A, L630 (1998)
15.51 T.S. Cheng, L.C. Jenkins, S.E. Hooper, C.T. Foxon, J.W. Orton, D.E. Lacklison: Appl. Phys. Lett. **66** (12), 1509 (1995)
15.52 J.W. Orton, D.E. Lacklison, N. Bab-ali, C.T. Foxon, T.S. Cheng, S.V. Novikov, D.F.C. Johnson, S.E. Hooper, L.C. Jenkins, L.J. Challis, T.L. Tansley: J. Electron. Mat. **24**, 263 (1995)
15.53 R.C. Powell, N.E. Lee, Y.W. Kim, J.E. Greene: J. Appl. Phys. **73** (1), 189 (1993)
15.54 F.A. Ponce, in: *Group III Nitride Semiconductor Compounds*, ed. by B. Gil (Oxford Science Publications, Oxford 1998), p. 122
15.55 O. Briot, in: *Group III Nitride Semiconductor Compounds*, ed. by B. Gil (Oxford Science Publications, Oxford 1998), p. 70
15.56 G. Popovici, H. Morkoc, S.N. Mohammad, in: *Group III Nitride Semiconductor Compounds*, ed. by B. Gil, (Oxford Science Publications, 1998), p. 19
15.57 H.P. Maruska, J.J. Tietjen: Appl. Phys. Lett. **15**, 367 (1969)
15.58 R.J. Molnar, in: Semiconductors and Semimetals, Vol. **57** (Academic Press, San Diego 1998)

15.59 H. Amano, N. Sawaki, I. Akasaki, Y. Toyoda: Appl. Phys. Lett. **48** (5), 353 (1986)
15.60 A. Nakadeira, H. Tanaka: Appl. Phys. Lett. **70** (20), 2720 (1997)
15.61 J. Wu, H. Yaguchi, K. Onabe, R. Ito, Y. Shiraki: Appl. Phys. Lett. **71** (15), 2067 (1997)
15.62 J. Wu, H. Yaguchi, K. Onabe, Y. Shiraki: Appl. Phys. Lett. **73** (14), 1931 (1998)
15.63 J. Wu, H. Yaguchi, H. Nagasawa, Y. Yamaguchi, K. Onabe, Y. Shiraki, R. Ito: Jpn. J. Appl. Phys. **36**, Part 1 (7A), 4241 (1997)
15.64 A. Nakdaira, H. Tanaka: phys. stat. sol. (a) **176**, 529 (1999)
15.65 D.P. Xu, H. Yang, L.X. Zheng, X.J. Wang, L.H. Duan, R.H. Wu: J. Cryst. Growth **191**, 646 (1998)
15.66 H. Amano, M. Kito, K. Hiramatsu, I. Akasaki: Jpn. J. Appl. Phys. **28**, L2112 (1989)
15.67 S. Nakamura, M. Senoh, T. Mukai: Jpn. J. Appl. Phys. **30** (10A), L1708 (1991)
15.68 H. Tanaka, A. Nakadaira: Proc. 2nd Int. Symp. on Blue Laser and Light Emititing Diodes (Ohmusha, Lts., Tokyo), 669 (1998)
15.69 Z. Yang, L.K. Li, W.I. Wang: J. Vac. Sci. Technol. B **14**, 2354 (1996)
15.70 W. Kim, A. Salvator, A.E. Bothkarev, O. Aktas, S.N. Mohammed, H. Morkoc: Appl. Phys. Lett. **69** (14), 559 (1996)
15.71 C.R. Abernathy, in: *GaN and Related Materials*, ed. by S.J. Pearton, *Optoelectronic Properties of Semiconductors and Superlattices*, series ed. M.O. Manasreh, (Gordon and Breach, Amsterdam 1997) p. 11
15.72 O. Brandt, H. Yang, B. Jenichen, Y. Suzuki, L. Däweritz, K.H. Ploog: Phys. Rev. B **52** (4), R2253 (1995)
15.73 T.D. Moustakas, in: *Semiconductors and Semimetals*, Vol. **57**, (Academic Press, San Diego 1999)
15.74 T.H. Myers, M.R. Millecchia, A.J. Ptak, K.S. Ziemer, C.D. Stinespring: J. Vac. Sci. Technol. B **17** (4), 1654 (1999)
15.75 N. Newman, in: *Semiconductors and Semimetals*, Vol. 50 (Academic Press, New York, 1998), p. 55
15.76 D.J. As, D. Schikora, A. Greiner, M. Lübbers, J. Mimkes, K. Lischka: Phys. Rev. B **54** (16), R11118 (1996)
15.77 H. Siegle, L. Eckey, A. Hoffmann, C. Thomsen, B.K. Meyer, D. Schikora, K. Lischka: Solid State Commun. **96** (12), 943 (1995)
15.78 J. Menninger, U. Jahn, O. Brandt, H. Yang, K.H. Ploog: Phys. Rev. B **53**, 1881 (1996)
15.79 A.P. Lima, T. Frey, U. Köhler, C. Wang, D.J. As, B. Schöttker, K. Lischka, D. Schikora: J. Cryst. Growth **197**, 31 (1999)
15.80 B. Schöttker, J. Kühler, D.J. As, D. Schikora, K. Lischka: Mat. Sci. Forum Vols. **264-268**, 1173 (1998)
15.81 H. Yang, O. Brandt, M. Wassermaier, J. Behrend, H.P. Schönherr, K.H. Ploog: Appl. Phys. Lett. **68** (2). 244 (1996)
15.82 T. Lei, K.F. Ludwig, T.D. Moustakas: J. Appl. Phys. **74** (7), 4430 (1993)
15.83 Y.I. Ravich, B.A. Efimova, I.A. Smirnow: *Semiconducting Lead Chalcogenides* (Plenum Press, New York, 1970)
15.84 G. Bauer, H. Burkhard, H. Heinrich, A. Lopez-Otero: J. Appl. Phys. **47**, 1721 (1976)
15.85 J.N. Zemel: Solid State Surf. Sci. **1**, 291 (1969)
15.86 H. Holloway, in: Physics of Thin Films, Vol. 11, ed. by G. Hass and M.H. Frankombe (Academic Press, New York, 1980), p. 116

15.87 E.M. Logothetis, H. Holloway, A.J. Varga, E. Wilkes: Appl. Phys. Lett. **21**, 318 (1971)
15.88 A. Lopez-Otero: Thin Solid Films **49**, 3 (1978)
15.89 A. Ishida, M. Aoki, H. Fujiyasu: J. Appl. Phys. **58**, 797 (1985)
15.90 P. Pichler, E.J. Fantner, G. Bauer, H. Clemens, H. Pascher, M. v. Ortenberg, M. Kriechbaum: Superlatt. Microstr. **1**, 1 (1985)
15.91 K. Shinohara, Y. Nishijima, H. Ebe: Appl. Phys. Lett. **47**, 1184 (1985)
15.92 H. Clemens, P. Ofner, H. Krenn, G. Bauer: J. Cryst. Growth **84**, 571 (1987)
15.93 A. Ishida, H. Fujiyasu, H. Ebe, K. Shinohara: J. Appl. Phys. **59**, 3023 (1986); and Superlatt. Microstr. **2**, 575 (1986)
15.94 A. Ishida, S. Matsuura, M. Mizuno, H. Fujiyasu: Appl. Phys. Lett. **51**, 478 (1987)
15.95 H. Clemens, H. Krenn, B. Tranta, P. Ofner, G. Bauer: Superlatt. Microstr. **4**, 591 (1988)
15.96 J. Fuchs, Z. Feit, H. Preier: Appl. Phys. Lett. **53**, 894 (1988)
15.97 G. Springholz, G. Bauer: Appl. Phys. Lett. **60**, 1600 (1992)
15.98 D.L. Partin: J. Vac. Sci. Technol. **21**, 1 (1982)
15.99 D.L. Partin: J. Electron. Mater. **13** 493 (1984)
15.100 N. Frank, G. Springholz, G. Bauer: Phys. Rev. Lett. **73**, 2236 (1994)
15.101 G. Springholz, Z. Shi, H. Zogg in: *Thin Films: Heteroepitaxial Systems*, ed. by W.K. Liu and M.B. Santos, *Series on Directions in Condensed Matter Physics*, Vol. 15, (World Scientific, 1999) p. 621
15.102 N. Frank, A. Voiticek, H. Clemens, A. Holzinger, G. Bauer: J. Cryst. Growth **126**, 293 (1993)
15.103 A. Katzir, R. Rosman, Y. Shani, K.H. Bachem, H. Böttner, H. Preier in: *Handbook of Solid State Lasers*, ed. by P.K. Cheo, (Marcel Dekker Inc., New York, 1989) p. 228
15.104 J. Yoshino, H. Munetaka, L.L. Chang: J. Vac. Sci. Technol. B **5**, 683 (1987)
15.105 H. Clemens, P. Ofner, G. Bauer, J.M. Hong, L.L. Chang: Mater. Lett. **7**, 127 (1988)
15.106 P. Müller, A. Fach, J. John, A. Tiwari, H. Zogg, G. Kostorz: J. Appl. Phys. **79**, 1911 (1996)
15.107 P. Müller, H. Zogg, A. Fach, J. John, C. Paglino, A.N. Tiwari, M. Krejci, G. Kostorz: Phys. Rev. Lett. **78**, 3007 (1997)
15.108 A.I. Fedorenko, A. Fedorov, A.Y. Sipatov, O.A. Mironov: Thin Solid Films **267**, 134 (1995)
15.109 K. Iga, F. Koyama, S. Konoshita: IEEE J. of Quantum Electronics **24**, 1845 (1988)
15.110 C.F. Schaus, H.E. Schaus, S. Sun, M.Y.A. Raja, S.R.J. Brueck: Electr. Lett. **25**, 538 (1989)
15.111 J.L. Jewell, K.F. Huang, K. Tai, Y.H. Lee, R.J. Fischer, S.L. McCall, A.Y. Cho: Appl. Phys. Lett. **55**, 424 (1989)
15.112 N.E.J. Hunt, E.F. Schubert, R.A. Logan, G.J. Zydzik: Appl. Phys. Lett. **61**, 2287 (1992)
15.113 E.F. Schubert, Y.H. Wang, A.Y. Cho, L.W. Tu, G.J. Zydzik: Appl. Phys. Lett. **60**, 921 (1992)
15.114 E. Hadji, J. Bleuse, N. Magnea, J.L. Pautrat: Appl. Phys. Lett. **67**, 2591 (1995)
15.115 V.M. Andreev: Semiconductors **33**, 942 (1999)
15.116 S.S. Murtaza, J.C. Campbell, J.C. Bean, L.J. Peticolas: Appl. Phys. Lett. **65**, 795 (1994)
15.117 A. Srinivasan, S. Murtaza, K. Anselm, Y.C. Shi, J.C. Campbell, B.G. Streetman: J. Vac. Sci. Technol. B **13**, 765 (1995)

15.118 M. Seto, W.B. de Boer, V.S. Sinnis, A.P. Morrison, W. Hoekstra, S. de Jager: Appl. Phys. Lett. **72**, 1550 (1998)
15.119 C.R. Pidgeon, S.D. Smith: J. Opt. Soc. of America **54**, 1459 (1964)
15.120 W. Heiss, T. Schwarzl, J. Roither, G. Springholz, M. Aigle, H. Pascher, K. Biermann, K. Reimann: Progress in Quantum Electronics **25**, 193 (2001)
15.121 T. Schwarzl, G. Springholz, H. Seyringer, H. Krenn, S. Lanzerstorfer, W. Heiss: IEEE J. Quantum Electronics **35**, 1753 (1999)
15.122 G. Springholz, T. Schwarzl, W. Heiss, H. Seyringer, S. Lanzerstorfer, H. Krenn: J. Cryst. Growth **201/202**, 999 (1999)
15.123 G. Springholz, G. Bauer: Appl. Phys. Lett. **62**, 2399 (1993)
15.124 T. Schwarzl, W. Heiss, G. Kocher-Oberlehner, G. Springholz: Semicond. Sci. and Technol. **14**, L11 (1999)
15.125 D. Leonard, M. Krishnamurty, C.M. Reaves, S.P. Denbaar, P. Petroff: Appl. Phys. Lett. **63**, 3203 (1993)
15.126 J.M. Moison, F. Houzay, F. Barthe, L. Leprince, E. Andre, O. Vatel: Appl. Phys. Lett. **64**, 196 (1994)
15.127 D.J. Eaglesham, M. Cerullo: Phys. Rev. Lett. **64**, 1943 (1990)
15.128 C.W. Snyder, B.G. Orr, D. Kessler, L.M. Sander: Phys. Rev. Lett. **66**, 3032 (1991)
15.129 D.J. Srolovitz: Acta Metall. **37**, 621 (1989)
15.130 J.Y. Marzin, J.M. Gerard, A. Izrael, D. Barrier, G. Bastard: Phys. Rev. Lett. **73**, 716 (1994)
15.131 V.A. Shchukin, D. Bimberg: Rev. Mod. Phys. **71**, 1125 (1999)
15.132 R.J. Asaro, W.A. Tiller: Metall. Trans. **3**, 1789 (1972)
15.133 M.A. Grinfeld: Sov. Phys. Dokl. **31**, 831 (1986)
15.134 Y.W. Mo, D.E. Savage, B.S. Schwartzentruber, M.G. Lagally: Phys. Rev. Lett. **65**, 1020 (1990)
15.135 A.J. Pidduck, D.J. Robins, A.G. Cullis, W.Y. Geong, A.M. Pitt: Thin Solid Films **222**, 78 (1992)
15.136 M.A. Lutz, R.M. Feenstra, P.M. Mooney, J. Tersoff, O.J. Chu: Surf. Sci. **316**, L1075 (1993)
15.137 L. Goldstein, F. Glas, J.Y. Marzin, M.N. Charasse, G. Le Roux: Appl. Phys. Lett. **47**, 1099 (1985)
15.138 A. Kurtenbach, K. Eberl, T. Shitara: Appl. Phys. Lett. **66**, 361 (1995)
15.139 B.R. Bennett, R. Magno, B.V. Shanabrook: Appl. Phys. Lett. **68**, 958 (1996)
15.140 J.L. Rouviere, J. Simon, N. Pelekanos, B. Daudin, G. Feullet: Appl. Phys. Lett. **75**, 2632, (1999)
15.141 S. Tanaka, S. Iwai, Y. Aoyagi: Appl. Phys. Lett. **69**, 4096 (1996)
15.142 K. Tachibana, T. Someya, Y. Arakawa: Appl. Phys. Lett. **74**, 383 (1999)
15.143 S.H. Xin, P.D. Wang, A. Yin, C. Kim, M. Dombrowolsk, J.L. Merz, J.K. Furdyna: Appl. Phys. Lett. **69**, 3884 (1996)
15.144 F. Flack, N. Samarth, V. Nikitin, P.A. Cromwell, J. Shin, D.D. Awschalom: Phys. Rev. B **54**, R17312 (1996)
15.145 M. Strassburg, V. Kutzer, U.W. Pohl, A. Hoffmann, I. Broser, N.N. Ledentsov, D. Bimberg, A. Rosenauer, U. Fischer, D. Gerthsen, I.L. Krestnikov, M.V. Maximov, P.S. Kopév, Zh.I. Alverov: Appl. Phys. Lett. **72**, 942 (1998)
15.146 G. Springholz, V. Holy, M. Pinczolits, G. Baier: Science **282**, 734 (1998)
15.147 M. Pinczolits, G. Springholz, G. Bauer: Appl. Phys. Lett. **73**, 250 (1998)
15.148 G. Springholz, M. Pinczolits, V. Holy, P. Mayer, K. Wiesauer, T. Roch, G. Bauer: Surf. Sci. **454-456**, 657 (2000)
15.149 G. Springholz, M. Pinczolits, V. Holy, S. Zerlauth, I. Vavra, G. Bauer: Physica E **9**, 149 (2001)

15.150 G. Springholz, K. Wiesauer: Phys. Rev. Lett. **88** (1), 15507 (2002)
15.151 M. Grundman, J. Christen, N.N. Ledentsov, J. Böhrer, D. Bimberg, S.S. Ruvimov, P. Werner, U. Richer, U. Gösele, J. Heydenreich, V.M. Ustinov,.A. Yu, Egorov, A.E. Zhukov, P.S. Kopév, Zh.I. Alverov: Phys. Rev. Lett. **74**, 4030 (1995)
15.152 R. Leon, P.M. Petroff, D. Leonhard, S. Fafard: Science **267**, 1966 (1995)
15.153 J. Tersoff, C. Teichert, M.G. Lagally: Phys. Rev. Lett. **76**, 1675 (1996)
15.154 F. Liu, S.E. Davenport, H.M. Evans, M.G. Lagally: Phys. Rev. Lett. **82**, 2528 (1999)
15.155 V.A. Shchukin, N.N. Ledentsov, P.P. Kopév, D. Bimberg: Phys. Rev. B **57**, 12262 (1998)
15.156 C. Teichert, L.J. Peticolas, J.C. Bean, J. Tersoff, M.G. Lagally: Phys. Rev. B **53**, 16334 (1996)
15.157 G. Springholz, V. Holy, P. Mayer, M. Pinczolits, A. Raab, R.T. Lechner, G. Bauer, H. Kang, L. Salamanca-Riba: Mat. Sci. and Engin. B **88**, 143 (2002)
15.158 G. Springholz, T. Schwarzl, W. Heiss, G. Bauer: Appl. Phys. Lett. **79**, 1225 (2001)
15.159 W. Heiss, T. Schwarzl, G. Springholz, K. Biermann, K. Reimann: Appl. Phys. Lett. **78**, 862 (2001)
15.160 A.J. Heeger: Current Appl. Phys. **1**, 247-267 (2001)
15.161 M.D. McGehee, E.K. Miller, D. Moses, A.J. Heeger in: *Advances in Synthetic Metals: Twenty Years of Progress in Science and Technology*, ed. by P. Bernier, S. Lefraut, G. Bidan (Elsevier, Lausanne 1999).
15.162 M.D. McGehee, A.J. Heeger: Adv. Mat. **12**, 1655 (2000)
15.163 J.H. Burroughes, D.D.C. Bradley, A.R. Brown, R.N. Marhus, K. Mackay, R.H. Friend, P.L. Burn, A. Kraft, A.B. Holmes: Nature **347**, 539 (1990)
15.164 M. Granström, M.G. Harrison, R.H. Friend, in: *Handbook of Oligo- and Polythiophenes*, ed. by D. Fichou (Wiley-VCH, Weinheim, 1999) p. 405 ff
15.165 S.E. Shaheen, C.J. Brabec, N.S. Sariciftci, F. Padinger, T. Fromherz, J.C. Hummelen: Appl. Phys. Lett. **78** (6), 841 (2001)
15.166 H.E. Katz, A. Dodabalapur, Z. Bav, in: *Handbook of Oligo- and Polythiophenes*, ed. by D. Fichou (Wiley-VCH, Weinheim, 1999), p. 459 ff
15.167 D. Fichou, C. Ziegler, in: *Handbook of Oligo- and Polythiophenes*, ed. by D. Fichou (Wiley-VCH, Weinheim, 1999) p. 183 ff
15.168 C. Taliani, W. Gebauer, in: *Handbook of Oligo- and Polythiophenes*, ed. by D. Fichou (Wiley-VCH, Weinheim, 1999) p. 361 ff
15.169 D.J. Gundlach, Y.-Y. Liu, T.N. Jackson, D.G. Schlom: Appl. Phys. Lett. **71**, 3853 (1997)
15.170 C.D. Dimitrakopoulos, P.R.L. Malenfant: Adv. Mat. **14** (2), 99 (2002)
15.171 H.E. Katz, A.J. Lovingre, J. Johnson, C. Kloc, T. Slegrist, W. Li, Y.-Y. Lin, A. Dodalapur: Nature **404**, 478 (2000)
15.172 C. Videlot, D. Fichou, F. Garnier: Synth. Met. **101**, 618 (1999)
15.173 C. Videlot, D. Fichou: Synth. Met. **102**, 885 (1999)
15.174 H. Yanagi, S. Okamoto: Appl. Phys. Lett. **71**, 2563 (1997)
15.175 A. Andreev, G. Matt, C.J. Brabec, H. Sitter, D. Badt, H. Seyringer, N.S. Sariciftci: Adv. Mat. **12** (9), 629 (2000)
15.176 H. Yanagi, T. Morikawa, S. Hotta, K. Yase: Adv. Mat. **13** (5), 313 (2001)
15.177 A. Koma: Prog. Cryst. Growth and Charact. **30**, 129 (1995)
15.178 J. Humenberger, K.H. Gresslehner, W. Schirz, H. Sitter, K. Lischka: Mat. Res. Soc. Symp. Proc., Vol. **216**, 53 (1991)
15.179 J. Humenberge, H. Sitter: Thin Solid Films **163**, 241 (1989)
15.180 D. Stifter, H. Sitter: Appl. Phys. Lett. **66**, 679 (1995)
15.181 D. Stifter, H. Sitter: Fullerene Sci. and Technol. **4** (2), 277 (1996)

15.182 H. Sitter, D. Stifter, T. Nguyen Manh: Thin Solid Films **306**, 313 (1997)

15.183 A. Oshiyama, S. Saito, N. Hamada, Y. Miyamoto: J. Phys. Chem. Solids **53**, 1457 (1992)

15.184 R. Resel, G. Leising: Surf. Sci. **409**, 302 (1998)

15.185 K. Erlacher, R. Resel, J. Keckes, F. Meghdadi, G. Leising: J. Cryst. Growth **206**, 135 (1999)

15.186 K. Erlacher, R. Resel, S. Hampel, T. Kohlmann, K. Lischka, B. Müller, A. Thierry, B. Lotz, G. Leising: Surf. Sci. **437**, 191 (1999)

15.187 H. Yanagi, T. Morikawa: Appl. Phys. Lett. **75** (2), 187 (1999)

15.188 H. Plank, R. Resel, S. Purger, J. Keckes, A. Thierry, B. Lotz, A. Andreev, N.S. Sariciftci, H. Sitter: Phys. Rev. B **64**, 235423 (2001)

15.189 A. Andreev, G. Matt, H. Sitter, C.J. Brabec, D. Badt, H. Neugebauer, N.S. Sariciftci: Synth. Met. **116**, 235 (2001)

15.190 R. Resel, N. Koch, F. Meghdadi, W. Unzog, K. Reichmann: Thin Solid Films **305**, 232 (1997)

15.191 B. Servet, S. Ries, M. Trotel, P. Almot, G. Horowitz, F. Garnier: Adv. Mater. **5**, 461 (1993)

15.192 R. Resel, M. Ottmar, M. Hanack, J. Keckes, G. Leising: J. Mater. Res. **15**, 934 (2000)

15.193 A. Andreev, G. Matt, C.J. Brabec, H. Sitter, D. Badt, H. Seyringer, N.S. Sariciftci: Adv. Mater.**12**, 629 (2001)

15.194 F. Meghadadi, S. Tasch, B. Winkler, W. Fischer, F. Stelzer, G. Leising: Synth. Met. **85**, 1441 (1997)

# List of Abbreviations

| | |
|---|---|
| $a$ | lattice constant |
| a-$\sim$ | amorphous $\sim$ |
| AFM | atomic force microscopy |
| ALE | atomic layer epitaxy |
| APB | antiphase boundary |
| APB | antiphase domain boundary |
| $a_v$ | evaporation coefficient |
| $\boldsymbol{b}$ | Burgers vector |
| BCF model | Burton – Cabrera – Frank |
| BEP | beam equivalent pressure |
| BET | Brunauer – Emmett – Teller |
| $C$ | concentration, constant |
| c-$\sim$ | cubic $\sim$ |
| CARS | coherent anti-Stokes Raman scattering |
| CBE | chemical beam epitaxy |
| CCD | charge coupled device |
| CG-mode | columnar growth mode |
| $C_p$ | specific heat |
| $D$ | diffusion coefficient |
| $d$ | thickness |
| DRS | differential reflectivity spectroscopy |
| $E$ | energy, magnitude of electric field |
| $\boldsymbol{E}$ | electric field |
| ECR | electron cyclotron resonance |
| ELO | epitaxial lateral overgrowth |
| $f$ | misfit, $f(\ldots)$ sometimes short for function of ... |
| FET | field effect transistor |
| FFT | fast Fourier transformation |
| FME | flow-rate modulation epitaxy |
| FM-mode | Frank–van der Merve growth mode |
| FTIR | Fourier transform infrared spectroscopy |
| FWHM | full width half maximum |
| $G$ | Gibbs free energy |
| GIXS | grazing incidence X-ray scattering |
| GSMBE | gas source molecular beam epitaxy |
| GW | derived from: Greens function $G$, Coulomb Potential $W$ |
| $H$ | enthalpy |
| HBT | hetero bipolar transistor |

| | |
|---|---|
| HELO | heteroepitaxial lateral overgrowth |
| HPGS | high pressure gas sources |
| HREELS | high resolution electron energy loss spectroscopy |
| HSMBE | hydride source molecular beam epitaxy |
| HTSC | high temperature superconductor |
| HVPE | hydride vapor phase epitaxy, halide vapor phase epitaxy |
| HWE | hot-wall epitaxy |
| $I$ | ion current |
| IRRAS | infrared reflectance absorption spectroscopy |
| ITDS | isothermal desorption spectroscopy |
| $\boldsymbol{J}$ | current density |
| $K$ | elastic constant |
| $k_B$ | Boltzmann constant |
| $L$,$l$ | length |
| LADA | laser assisted deposition and annealing |
| LCAO | linear combination of atomic orbitals |
| LDA | laser Doppler anemometry |
| LDA | local density approximation |
| LDH | low dimensional heterostructure |
| LED | light emitting diode |
| LEEBI | low energy electron beam irradiation |
| LEED | low energy electron diffraction |
| LG model | lattice gas model |
| LIF | laser-induced fluorescence |
| LIFE | laser-induced flash evaporation |
| LLS | laser light scattering |
| LMBE | laser molecular beam epitaxy |
| LPE | liquid phase epitaxy |
| LPEE | liquid phase electroepitaxy |
| LPGS | low pressure gas sources |
| LRO | long range order |
| LS | light scattering |
| $m$ | mass |
| $M$ | molecular weight |
| $\mu$ | chemical potential |
| MB model | Matthews–Blakeslee model |
| MBE | molecular beam epitaxy |
| MC | Monte Carlo |
| MD | misfit dislocation |
| MEE | migration-enhanced epitaxy |
| MFC | mass flow controller |
| MLE | molecular layer epitaxy |
| MOCVD | metalorganic chemical vapor deposition |
| MOMBE | metalorganic molecular beam epitaxy |

| | |
|---|---|
| MOS | metal on semiconductor, metal-oxide-semiconductor |
| MOVPE | metalorganic vapor phase epitaxy |
| $N, n$ | number |
| $N_A$ | Avogadro constant |
| NSTL | near surface transition layers |
| OFET | organic field effect transistor |
| OMVPE | organometallic vapour phase epitaxy |
| $p$ | pressure |
| PAMBE | plasma assisted molecular beam epitaxy |
| PBN | pyrolytic boron nitride |
| PED | potential enhanced doping |
| PEM | photo elastic modulator |
| PL | photoluminescence |
| PLD | pulsed laser deposition |
| PLE | pulsed laser evaporation |
| $\pi_p$ | Peltier coefficient |
| PRS | p-polarized reflectance spectroscopy |
| PSD | power spectral density |
| PSP | parasexiphenyl |
| PVD | physical vapor deposition |
| $Q$ | heat, deposition rate |
| QD | quantum dot |
| QMS | quadrupole mass spectrometer |
| QWR | quantum wire |
| RAS | reflectance anisotropy spectroscopy |
| RBS | Rutherford back scattering |
| REMS | reflection mass spectrometry |
| RF | radio frequency |
| RHEED | reflection high energy electron diffraction |
| RMBE | reactive ion molecular beam epitaxy |
| RRS | resonance Raman scattering |
| RS | Raman scattering |
| $S$ | entropy |
| SAG | selective area growth |
| SCF-LCAO | self-consistent field – linear combination of atomic orbitals |
| SDA | surface dielectric anisotropy |
| SDR | surface differential reflectivity |
| SE | spectroscopic ellipsometry |
| SEM | scanning electron microscopy |
| SF-mode | step flow growth mode |
| SHG | second harmonic generation |
| SILO | strain induced lateral overgrowth |
| SIMS | secondary ion mass spectrometry |
| SK-mode | Stranski–Krastanov growth mode |

| | |
|---|---|
| SLM | standard liters per minute |
| SOI | silicon on insulator |
| SOM | scanning optical microscopy |
| SOS | silicon on sapphire |
| SOS | solid on solid |
| SPA | surface photo absorption |
| SPE | solid phase epitaxy |
| SPM | scanning probe microspcopy |
| SPS | short-period superlattice |
| SRO | short range order |
| SRPL | spatial resolved photoluminescence |
| SSL | serpentine superlattice |
| SSMBE | solid source molecular beam epitaxy |
| STM | scanning tunneling microscopy |
| $T$ | temperature |
| TDS | thermal desorption spectroscopy |
| TE mode | transversal electric mode |
| TED | transmission electron diffraction |
| TEM | transmission electron microscopy |
| TLV | threshold limit value |
| TM mode | transversal magnetic mode |
| TMAl | trimethylaluminum |
| TMGa | trimethylgallium |
| TMIn | trimethylindium |
| TPD | temperature programmed desorption |
| TRR | time resolved reflectivity |
| TSK model | terrace – step – kink model |
| TSL | tilted superlattice |
| UHV | ultra high vacuum |
| $v$ | velocity, growth velocity |
| VCSEL | vertical cavity surface emitting laser |
| VPE | vapour phase epitaxy |
| VSEPR | valence-shell, electron-pair repulsion |
| VUV | vacuum ultraviolet |
| VW-mode | Volmer–Weber growth mode |
| XRD | X–ray diffraction |
| XRD-PF | X-ray diffraction pole figure measurements |
| YBCO | yttrium barium copper oxide |
| ZMR | zone-melting recrystallization |

# List of Metalorganic Precursors

| Element | Chemical | Name | Short Name |
|---|---|---|---|
| Al | $Al(CH_3)_3$ | trimethylaluminum | TMAl |
| Al | $Al(C_2H_5)_3$ | triethylaluminum | TEAl |
| Al | $Al(C_4H_9)_3$ | tri-iso-butylaluminum | TIBAl |
| As | $As(CH_3)_3$ | trimethylarsine (trimethylarsenic) | TMAs |
| As | $AsH_3$ | arsine (arsane) | $AsH_3$ |
| As | $As(N(CH_3)_2)_3$ | trisdimethylaminoarsine | TDMAAs |
| As | $C_4H_9AsH_2$ | tertiarybutylarsine | TBAs |
| Bi | $Bi(CH_3)_3$ | trimethylbismuth | TMBi |
| Cd | $Cd(CH_3)_2$ | dimethylcadmium | DMCd |
| Cd | $Cd(C_2H_5)_2$ | diethylcadmium | DECd |
| Ga | $Ga(CH_3)_3$ | trimethylgallium | TMGa |
| Ga | $Ga(C_2H_5)_3$ | triethylgallium | TEGa |
| Ge | $GeH_4$ | germane | $GeH_4$ |
| Hg | $Hg(CH_3)_2$ | dimethylmercury | DMHg |
| Hg | $Hg(C_2H_5)_2$ | diethylmercury | DEHg |
| Hg | $Hg(C_3H_7)_2$ | di-iso-propylmercury | DIPHg |
| Hg | $Hg(C_3H_7)_2$ | di-n-propylmercury | DNPHg |
| Hg | $Hg(C_4H_9)_2$ | di-n-butylmercury | DNBHg |
| In | $In(CH_3)_3$ | trimethylindium | TMIn |
| Mg | $Mg(C_5H_5)_2$ | biscyclopentadienylmagnesium | $Cp_2Mg$ |
| N | $(CH_3)_2NNH_2$ | dimethylhydrazine | DMHy |
| N | $C_4H_9NH_2$ | tertiarybutylamine | TBAm |
| N | $NH_3$ | ammonia | $NH_3$ |
| P | $C_3H_7PH_2$ | iso-butylphosphine | IBP |
| P | $C_4H_9PH_2$ | tertiarybutylphosphine | TBP |
| P | $P(CH_3)_3$ | trimethylphosphine | TMP |
| P | $P(C_2H_5)_3$ | triethylphosphine | TEP |
| P | $PH_3$ | phosphine | $PH_3$ |
| S | $CH_3SH$ | methylthiol | MSH |
| S | $C_4H_4S$ | thiophene | $C_4H_4S$ |
| S | $H_2S$ | hydrogen sulfide | $H_2S$ |
| S | $S(CH_3)_2$ | dimethylsulfide | DMS |
| S | $S(C_2H_5)_2$ | diethylsulfide | DES |
| Sb | $Sb(CH_3)_3$ | trimethylantimony | TMSb |
| Se | $H_2Se$ | hydrogen selenide | $H_2Se$ |
| Se | $Se(C_2H_5)_2$ | diethylselenium | DESe |
| Si | $SiH_4$ | silane | $SiH_4$ |
| Si | $Si_2H_6$ | disilane | $Si_2H_6$ |
| Si | $Si_3H_8$ | trisilane | $Si_3H_8$ |

| | | | |
|---|---|---|---|
| Te | $(C_3H_5)Te(CH_3)$ | methylallyltelluride | MATe |
| Te | $H_2Te$ | hydrogen telluride | $H_2Te$ |
| Te | $Te(CH_3)_2$ | dimethyltellurium | DMTe |
| Te | $Te(C_2H_5)_2$ | diethyltellurium | DETe |
| Te | $Te(C_3H_5)_2$ | di-n-propyltellurium | DNPTe |
| Te | $Te(C_4H_9)_2$ | di-tertiarybutyltellurium | DTBTe |
| Tl | $Tl(CH_3)_3$ | trimethylthallium | TMTl |
| Zn | $Zn(CH_3)_2$ | dimethylzinc | DMZn |
| Zn | $Zn(C_2H_5)_2$ | diethylzinc | DEZn |

# Index

## A

## B



## C

## E

## F



## G

# H

## I

## K



## L

## M

## N



## O

## Q

## R



## S

## T

Springer Series in
# MATERIALS SCIENCE

*Editors:* R. Hull R. M. Osgood, Jr. J. Parisi H. Warlimont

10 **Computer Simulation of Ion-Solid Interactions**
By W. Eckstein

11 **Mechanisms of High Temperature Superconductivity**
Editors: H. Kamimura and A. Oshiyama

12 **Dislocation Dynamics and Plasticity**
By T. Suzuki, S. Takeuchi, and H. Yoshinaga

13 **Semiconductor Silicon**
Materials Science and Technology
Editors: G. Harbeke and M. J. Schulz

14 **Graphite Intercalation Compounds I**
Structure and Dynamics
Editors: H. Zabel and S. A. Solin

15 **Crystal Chemistry of High-$T_C$ Superconducting Copper Oxides**
By B. Raveau, C. Michel, M. Hervieu, and D. Groult

16 **Hydrogen in Semiconductors**
By S. J. Pearton, M. Stavola, and J. W. Corbett

17 **Ordering at Surfaces and Interfaces**
Editors: A. Yoshimori, T. Shinjo, and H. Watanabe

18 **Graphite Intercalation Compounds II**
Editors: S. A. Solin and H. Zabel

19 **Laser-Assisted Microtechnology**
By S. M. Metev and V. P. Veiko
2nd Edition

20 **Microcluster Physics**
By S. Sugano and H. Koizumi
2nd Edition

21 **The Metal-Hydrogen System**
By Y. Fukai

22 **Ion Implantation in Diamond, Graphite and Related Materials**
By M. S. Dresselhaus and R. Kalish

23 **The Real Structure of High-$T_C$ Superconductors**
Editor: V. Sh. Shekhtman

24 **Metal Impurities in Silicon-Device Fabrication**
By K. Graff 2nd Edition

25 **Optical Properties of Metal Clusters**
By U. Kreibig and M. Vollmer

26 **Gas Source Molecular Beam Epitaxy**
Growth and Properties of Phosphorus Containing III–V Heterostructures
By M. B. Panish and H. Temkin

27 **Physics of New Materials**
Editor: F. E. Fujita 2nd Edition

28 **Laser Ablation**
Principles and Applications
Editor: J. C. Miller

29 **Elements of Rapid Solidification**
Fundamentals and Applications
Editor: M. A. Otooni

30 **Process Technology for Semiconductor Lasers**
Crystal Growth and Microprocesses
By K. Iga and S. Kinoshita

31 **Nanostructures and Quantum Effects**
By H. Sakaki and H. Noge

32 **Nitride Semiconductors and Devices**
By H. Morkoç

33 **Supercarbon**
Synthesis, Properties and Applications
Editors: S. Yoshimura and R. P. H. Chang

34 **Computational Materials Design**
Editor: T. Saito

35 **Macromolecular Science and Engineering**
New Aspects
Editor: Y. Tanabe

36 **Ceramics**
Mechanical Properties, Failure Behaviour, Materials Selection
By D. Munz and T. Fett

37 **Technology and Applications of Amorphous Silicon**
Editor: R. A. Street

38 **Fullerene Polymers and Fullerene Polymer Composites**
Editors: P. C. Eklund and A. M. Rao

Springer Series in

# MATERIALS SCIENCE

*Editors:* R. Hull R. M. Osgood, Jr. J. Parisi H. Warlimont

39 **Semiconducting Silicides**
Editor: V. E. Borisenko

40 **Reference Materials in Analytical Chemistry**
A Guide for Selection and Use
Editor: A. Zschunke

41 **Organic Electronic Materials**
Conjugated Polymers and Low Molecular Weight Organic Solids
Editors: R. Farchioni and G. Grosso

42 **Raman Scattering in Materials Science**
Editors: W. H. Weber and R. Merlin

43 **The Atomistic Nature of Crystal Growth**
By B. Mutaftschiev

44 **Thermodynamic Basis of Crystal Growth**
*P–T–X* Phase Equilibrium and Non-Stoichiometry
By J. Greenberg

45 **Thermoelectrics**
Basic Principles and New Materials Developments
By G. S. Nolas, J. Sharp, and H. J. Goldsmid

46 **Fundamental Aspects of Silicon Oxidation**
Editor: Y. J. Chabal

47 **Disorder and Order in Strongly Nonstoichiometric Compounds**
Transition Metal Carbides, Nitrides and Oxides
By A. I. Gusev, A. A. Rempel, and A. J. Magerl

48 **The Glass Transition**
Relaxation Dynamics in Liquids and Disordered Materials
By E. Donth

49 **Alkali Halides**
A Handbook of Physical Properties
By D. B. Sirdeshmukh, L. Sirdeshmukh, and K. G. Subhadra

50 **High-Resolution Imaging and Spectrometry of Materials**
Editors: F. Ernst and M. Rühle

51 **Point Defects in Semiconductors and Insulators**
Determination of Atomic and Electronic Structure from Paramagnetic Hyperfine Interactions
By J.-M. Spaeth and H. Overhof

52 **Polymer Films with Embedded Metal Nanoparticles**
By A. Heilmann

53 **Nanocrystalline Ceramics**
Synthesis and Structure
By M. Winterer

54 **Electronic Structure and Magnetism of Complex Materials**
Editors: D.J. Singh and D. A. Papaconstantopoulos

55 **Quasicrystals**
An Introduction to Structure, Physical Properties and Applications
Editors: J.-B. Suck, M. Schreiber, and P. Häussler

56 **$SiO_2$ in Si Microdevices**
By M. Itsumi

57 **Radiation Effects in Advanced Semiconductor Materials and Devices**
By C. Claeys and E. Simoen

58 **Functional Thin Films and Functional Materials**
New Concepts and Technologies
Editor: D. Shi

59 **Dielectric Properties of Porous Media**
By S.O. Gladkov

60 **Organic Photovoltaics**
Concepts and Realization
Editors: C. Brabec, V. Dyakonov, J. Parisi and N. Sariciftci

Printed by Books on Demand, Germany